A NEW HISTORY OF MODERN COMPUTING

HISTORY OF COMPUTING

William Aspray and Thomas J. Misa, editors

A complete list of the titles in this series appears in the back of this book.

A NEW HISTORY OF MODERN COMPUTING

THOMAS HAIGH AND PAUL E. CERUZZI

The MIT Press
Cambridge, Massachusetts
London, England

The MIT Press would like to thank the anonymous peer reviewers who provided comments on drafts of this book. The generous work of academic experts is essential for establishing the authority and quality of our publications. We acknowledge with gratitude the contributions of these otherwise uncredited readers.

This book was set in Adobe Garamond Pro and Berthold Akzidenz Grotesk by Westchester Publishing Services. Printed and bound in the United States of America.

Library of Congress Cataloging-in-Publication Data

Names: Haigh, Thomas, 1972– author. | Ceruzzi, Paul E., author.
Title: A new history of modern computing / Thomas Haigh
    and Paul E. Ceruzzi.
Description: Cambridge, Massachusetts : The MIT Press, [2021] |
    Series: History of computing | Includes bibliographical references and index.
Identifiers: LCCN 2020048457 | ISBN 9780262542906 (paperback)
Subjects: LCSH: Computer science—History. | Electronic digital computers—History.
Classification: LCC QA76.17 .H34 2021 | DDC 004.09—dc23
LC record available at https://lccn.loc.gov/2020048457

10   9   8   7   6   5   4   3   2   1

To Prof. Dr. Erhard Schüttpelz and to the SIGCIS community,
both vital sources of support for our work

CONTENTS

ACKNOWLEDGMENTS

We are particularly grateful to David Hemmendinger, Gerardo Con Diaz, Marc Weber, Michael J. Halvorson, Alan Staiti, Eugene Miya, and the MIT Press reviewers who all read through the entire manuscript offering innumerable useful suggestions. Paul McJones, Clem Cole, David Brock, Troy Astarte, Tom Lean, Bradley Fidler, Tom Van Vleck, Henry Lowood, Jerome Coonen, Forrest Park, Gordon Bell, and Donald B. Wagner provided important input on specific topics and sections. Series editor Tom Misa had an eagle's eye for typos. Much of the material in this book is adapted from Ceruzzi's earlier *A History of Modern Computing*, and so we also remain in debt to those previously acknowledged there. In particular we appreciate William Aspray's work as longtime series editor, which befitted this book and many of the other books we drew on to create it.

The book's creation was generously supported by Siegen University's Media of Cooperation CRC. This included three periods spent working together in Siegen, Germany, to review the potential for a new overview history, outline the new structure of the book, review existing material in the second edition, and assemble two draft chapters as a test of our new approach. Additional support from Siegen allowed Haigh to focus on writing. We happily acknowledge Erhard Schüttpelz, Sebastian Giessmann, and Tristan Thielman for arranging this support and for their collegial contributions to our project. We talked through the new structure with participants at three Early Digital workshops and benefited from the suggestions of Laine Nooney, David Brock, Stephanie Dick, William Aspray, Martin Campbell-Kelly, Doron Swade, Matthew Kirschenbaum, Len Shustek, Rebecca Slayton, and many others. Parts of the book were later discussed in a Siegen workshop on the history of database management systems with the participation of Moritz Feichtinger and Francis Hunger.

For help with image permissions and scans we are grateful to Debbie Douglas, Janice Hussain, Angela Schad, Katherine Taylor, Brian Daly, Erik Rau, Stephanie

Hueter, Andrea Wescott, Ingrid Crete, Lindsi Wyner, Bryan Roppolo, Amanda Wick, and various contributors to Wikimedia Commons.

Haigh happily acknowledges the support of his family members Maria, Peter, and Paul during the five years of sometimes intense work it took to produce this book. Their love and sacrifice made the book possible. Both authors appreciate the efforts of the MIT Press editorial and production teams. Acquisitions editor Katie Helkie advocated early for the need to update Ceruzzi's book, while Laura Keeler gave prompt answers to all our queries on manuscript preparation. We are particularly grateful for the work of Helen Wheeler and Virginia A. Schaefer at Westchester Publication Services. Stefan Swanson worked heroically on proofreading and reference checking, catching dozens of errors.

Fragments of chapters 6 and 12 were adapted from Haigh's contribution to *The Internet and American Business* (edited by William Aspray and Paul Ceruzzi) and Ceruzzi's contribution to *Social Media Archaeology and Poetics* (edited by Judy Malloy), both published by MIT Press. Portions of chapter 3 are modeled after Haigh's "The Chromium-Plated Tabulator: Institutionalizing an Electronic Revolution, 1954–1958." *IEEE Annals of the History of Computing* 23, no. 4 (October–December 2001): 75–104 and "How Data Got Its Base: Information Storage Software in the 1950s and 60s." *IEEE Annals of the History of Computing* 31, no. 4 (October–December 2009): 6–25.

Funded by the Deutsche Forschungsgemeinschaft (DFG, German Research Foundation)—Project-ID 262513311-SFB 1187 Media of Cooperation.

BECOMING UNIVERSAL: INTRODUCING A NEW HISTORY OF COMPUTING

This book is a comprehensive reimagining of *A History of Modern Computing*, first published in 1998 and expanded with a new chapter in 2003. A lot has changed since 1998 when the Web was a novelty, iPhones didn't exist, and the founders of Google and Facebook were in graduate school and high school, respectively. Doing justice to those changes required more than just adding a few more chapters at the end of the book. For example, as the first edition was being written and conceived, the Internet was still quite an obscure system. Today we view the development of computer communications as a central thread in the history of computing, not just in the 1990s but also in the 1960s and 1970s. The wholesale shift of video and music reproduction to digital technologies likewise challenges us to integrate media history into the long history of computing. Since the original book was written, the computer had become something new, which meant that the book also had to become something new.

The unmistakable importance of the Internet, digital media devices, and video games to modern life has driven public interest in their stories. Yet this discussion is rarely grounded in the longer and deeper history of computer technology. For example, as we finalized our revisions to this book, one of us chanced upon *How the Internet Happened: From Netscape to the iPhone* by Brian McCullough, a tech industry insider.[1] It is readable, admirably tight, and solidly researched—based on two hundred interview podcasts McCullough recorded with company founders. We recommend it to you. Yet we were also struck by how little engagement such approaches to history have with the larger story of computing. As his title suggests, McCullough starts the story of the Internet in 1994 with the first commercial Web browser, giving only occasional flashbacks to the first twenty-five years of the Internet (and its precursor, the ARPANET). He says little about where the core technologies, protocols, or algorithms of the Web came from, or about the evolving technologies personal computing, such as new processors and operating systems, that made the rapid spread of Web browsers possible. He says nothing about Web server technology, or the programming languages and practices that evolved alongside Web browsers. Similar observations can, and have, been made about popular histories of video games and personal computing. Our aim here is to integrate Internet and Web history into the core narrative of the history of computing, along with the history of iPods, video game consoles, home computers, digital cameras, and smartphone apps.

To write the history of a technology only as a series of models, inventors, and refinements is to miss the point. Thomas J. Misa once suggested that the great challenge facing historians of computing is to explain "How Computing Has Changed the World."[2] Doing that might not seem so hard. The computer has a relatively short history, which for our purposes begins in the 1940s. Set against technologies such as agriculture, Arabic numerals, or alphabets its span looks like the blink of an eye. Despite its ever-growing importance, its influence on our lives has so far been less fundamental that that of industrial age technologies such as electric light or power, automobiles, or antibiotics.

Important technologies have complex histories. The automobile, for example, was made possible by the development of earlier technologies such as coaches and bicycles. Its mass adoption with the Ford Model T took place decades after its invention, and was made possible by the development of big business and the invention of mass production as a way to build complex machines cheaply in huge quantities. The Model T's users discovered new uses for it, building new body work or adapting it as a portable power source for agricultural machinery. The automobile facilitated, but did not dictate, a mass exodus from America's cities into sprawling suburbs and exurbs. Because most Americans came to rely on cars to shop, get to work, and socialize, their national culture grew up around the technology. The resulting need for massive quantities of oil reshaped American foreign policy and transformed the fortunes of nations from Norway to Nigeria, usually for the worse.[3]

Doing justice to that story would challenge even the most ambitious historian, but the automotive historian has a crucial advantage over the historian of computing: over the century from 1920 to 2020, the typical car had a roughly stable physical form: a large self-propelled metal box able to move between two and eight people over asphalt at a maximum speed that has roughly doubled, from forty miles an hour to a (legally mandated) seventy or eighty. Cars are still built on assembly lines by large, capital-intensive companies. Ford, General Motors, and Chrysler were the "big three" US automakers of the 1920s and retain that status today. Cars are still distributed by franchised dealers. A basic but functional car costs a skilled worker a few months of salary.

The story of computing offers us no comparable continuities. Few, if any, other technologies have changed their scale, dominant applications, and users so often and so fundamentally. The computer started out as esoteric and specialized as the cyclotron and has finished up only slightly less ubiquitous than clothing or food. In the 1940s, computers were used by a few hundred people worldwide to carry out complex numerical calculations. They were built as one-of-a-kind pieces of custom lab equipment, each costing the equivalent of several million present-day dollars.

Computer scientists have adopted a term from Alan Turing, the *universal machine*, to describe the remarkable flexibility of programmable computers. To prove a mathematical point he described a class of imaginary machines (now called Turing machines) that processed symbols on an unbounded tape according to rules held in a table. By encoding the rules themselves on the tape, Turing's universal machine was able to compute any number computable by a more specialized machine of the same ilk. Computer scientists came to find this useful as a model of ability of all programmable computers to carry out arbitrary sequences of operations, and hence (if unlimited time and storage were available) to mimic each other by using code to replicate missing hardware.[4]

In practice, however, the first modern computers faced severe practical limits on their capabilities. As those restraints were gradually lifted, the scope of what could feasibly or economically be computerized grew dramatically as the computer evolved toward what economists call a "general purpose technology" with highly varied applications. Today about half the world's inhabitants use hand-held computers daily to facilitate almost every imaginable human task. They carry out their work millions of times faster than those early models, fit easily in a pocket, and are cheap enough to be thrown away when a cracked piece of glass needs repair.

Computers will never do everything, be used by everyone, or replace every other technology, but they are more nearly universal than any other technology. In that broader sense the computer began as a highly specialized technology and has moved toward universality and ubiquity. We think of this as a progression toward *practical universality*, in contrast to the theoretical universality often claimed for computers as embodiments of Turing machines.

To the extent that it has become a universal machine, the computer might also be called a *universal solvent*, achieving something of that old dream of alchemy by making an astounding variety of other technologies vanish into itself. Maps, filing cabinets, video tape players, typewriters, paper memos, and slide rules are rarely used now, as their functions have been replaced by software running on personal computers, smartphones, and networks. We conceptualize this convergence of tasks on a single platform as a *dissolving* of those technologies and, in many cases, their business models by a device that comes ever closer to the status of universal technological solvent.

In many cases the computer has dissolved the insides of other technologies while leaving their outward forms intact. Although computer technology is universal, most actual computers are configured and deployed to carry out extremely specialized tasks.

They hide inside cars and consumer appliances to replace the guts of many of the technologies of everyday life, such as telephones, photocopiers, televisions, pianos, and even light bulbs. These computers outnumber humans many times over and cost as little as three cents apiece in bulk. They still have processors and memory and run software, but only computer scientists habitually think of them as computers.

This shape shifting makes the construction of a satisfactory overall history of computing exceptionally difficult. How to tell a story when the scale of the stage and the cast of characters changes so fundamentally? The easiest way to write a book like this would be to devote one or two chapters to each decade. But we want to tell a story with a plot, not just arrange a succession of facts and anecdotes in roughly chronological order. Our answer was to focus on constructing each individual chapter to tell the story of a transformation, in which particular communities of users and producers remade the computer into something new. Each chapter tells a coherent story with a stable cast of characters, even though the companies, applications, and communities relevant to that chapter may not appear in others.

For example, the first transformations began in the 1950s, as computers were remade for use as scientific supercomputers capable of feats of number crunching, data processing devices able to automate the work of hundreds of clerks, and real-time control systems used to coordinate air defense. We tell those three parallel stories in three parallel chapters, each reaching into the 1970s. Chapters continue to overlap in time, although as you move through the book you will draw gradually closer to the present. In later chapters the computer becomes a communications medium, a graphical tool, a personal plaything, and so on. The full list is longer, as you can read in the table of contents. We hope that you find the new structure clear and coherent. After sketching out this version, we began by rearranging passages of the existing text within it, filling in the gaps with new material. If you are familiar with Ceruzzi's original book you will be able to find compressed versions of almost all the topics it covered somewhere in our new text.[5]

You will find many examples here of how computerization changed specific parts of the world, but not every part of the world has been changed in the same way. Misa's question, "How did the computer change the world," which was posed to an entire field, admits no single answer. We have instead tried to give a reasonably comprehensive answer to a more tractable question: "How did the world change the computer?" Read together, the smaller stories told in each chapter add up to a larger one. The protagonist of this story is "the computer" itself. To talk about "the computer" might sound a little ridiculous, in a world where some computers are thrown away inside hotel key cards and others cost millions of dollars. Yet at the core of each machine is a package of programming techniques and architectural features reflecting a shared descent. Architectural advances pioneered by Cray supercomputers now help your phone to play Netflix video more effectively. The original *A History of Modern Computing* engaged more deeply than any other overview history of computing with the evolution of computer architecture. Preserving and deepening that focus on the origins and diffusion of new architectural features contributes to knitting the new book together.

Technologies are shaped by societies or, more specifically, by institutions such as governments and corporations, by inventors responding to incentives, and by users who apply and reshape technologies in ways unimagined by their original creators. Another distinctive feature of the original *A History of Modern Computing* was its interest in the stories of computer users, with deeply researched case studies of NASA, the Internal Revenue Service, and other influential organizational users of computers. We have retained these and have also woven shorter examinations of the experiences of computer users into each chapter. This complements our focus on architecture, because new architectural features and software technologies were originally created to serve the specific needs of specific users. This structure builds on the insights of Michael S. Mahoney in his classic paper "Histories of Computing(s)." Mahoney argued that "the histories and continuing experience of the various communities show that they wanted and expected different things from the computer. They encountered different problems and levels of difficulty in fitting their practice to it. As a result, they created different computers or (if we may make the singular plural) computings."[6] Whenever the computer became a new thing it did not stop being everything it had been before. Computers are still used by nuclear weapons labs and banks. These stories intertwined, as new capabilities move from one domain to another.

Our story starts in the 1940s with programmable electronic computers and not, like more traditional overview histories, with mechanical calculators or Charles Babbage. To tell the story of a new technology one would ideally begin by documenting the practices it was applied to and the earlier technologies used and then explore its origin, its spread, and the new practices and institutions that coevolved with it. Decades ago, when the scope of computing was smaller, it made sense to see electronic computing as a continuation of the tradition of scientific computation. The first major history of computing, *The Computer from Pascal to von Neumann* by computing pioneer Herman Goldstine, concluded in the 1940s with the invention of the modern computer. In *A History of Computing Technology*, published in 1985, Michael Williams started with the invention of numbers and reached electronic computers about two thirds of the way through. By the 1990s the importance of computer applications to business administration was being documented by historians, so it was natural for Martin Campbell-Kelly and William Aspray, when writing *Computer: A History of the Information Machine*, to replace discussion of slide rules and astrolabes with mechanical office machines, filing cabinets, and administrative processes.[7]

Giving up their coverage of earlier technologies carries a cost. To understand what changed in the world because of the adoption of a technology, we need to know the "before" as well as the "after." Yet its influence will confront future historians of every kind, whether they are writing about presidential politics or pop music. The breadth of technologies displaced by the computer and practices remade around it makes it seem arbitrary to begin with chapters that tell the stories of index cards but not of televisions; of slide rules but not of pinball machines; or of typewriters but not of the postal system. But to include those stories, each of our chapters would need to become a long book of its own, written by different experts.

A History of Modern Computing was the most widely cited scholarly overview history of computing. For many of the people who picked up a copy in a library, or were assigned it for a class, it gave a first introduction to the topic. We hope our new book is a starting point, and not an end point, for your engagement with this rich history. To help guide you, we have systematically added citations to, and quotations from, some of the many outstanding works of scholarly history on different aspects of the history of computing. Our challenge here is to condense stories big enough to fill entire books into a page or a paragraph. Most histories focus on a specific aspect, occasionally something as broad as the software industry, but more often a single company or computer platform. There are more histories of Google, Microsoft, or Apple than there are of the computer itself. We do not include "further reading" lists for each chapter, but when we mention a book in the text you can safely assume that it is an outstanding and highly relevant source of further reading. These books have inspired and informed us, and we would like to share that gift with you.

Broad as this book is, we must warn you that it is a history of computing technology and practice, not of computer science. Computer science is an academic discipline. It began to come together intellectually in the late 1950s and was institutionalized during the 1960s and 1970s via university departments, corporate research labs, funding agencies, conferences, and journals. When specific work done by computer scientists has a major influence on practice we discuss its contribution, but we cannot try to squeeze into this book the stories of research areas, influential departments, intellectual schools, or the development of subdisciplines such as architecture, theory, graphics, databases, networking, and artificial intelligence. Historians of science have paid remarkably little attention to computer science (and, alas, computer scientists to history), so unfortunately there are no major histories of computer science or of any of its subdisciplines for us to point you toward.[8]

Another question we won't be answering is "What was the first computer?" Arguments about firsts once drove lengthy lawsuits and patent proceedings. They continue to dominate much general discussion of early electronic computing, particularly in

Internet forums. Any answer depends on one's definition of *computer*. In the 1940s the question would not even have made sense, because *computer* usually meant a person employed to carry out complex calculations. The new machines being built at the time were called automatic computers or computing machines. Even those weren't the first calculating machines, which is why we call this a history of *modern* computing.

But we do have to start this book somewhere. We start it in 1945 with the first operation of a machine called ENIAC at the University of Pennsylvania. A truce reached in the 1980s, as professionally trained historians began to engage with this topic, established strings of adjectives to qualifying the "firstness" of the various novel machines constructed during the 1940s. ENIAC is usually called something like the "first electronic, general purpose, programmable computer."[9]

Those qualifying adjectives separate it from two earlier groups of machines. *Electronic* distinguishes it from electromechanical computers whose logic units worked thousands of times more slowly. Often called relay calculators, these computers carried out computations one instruction at a time under the control of paper tapes. They were player pianos that produced numbers rather than music. Among the best known were the Harvard Mark 1, produced by International Business Machines to meet the specification of Harvard's Howard Aiken, and the Z3 designed by German computing pioneer Konrad Zuse. *General purpose* and *programmable* separated ENIAC from special purpose electronic machines whose sequence of operations was built into hardware and so could not be reprogrammed to carry out fundamentally different tasks. The ABC, or Atanasoff-Berry Computer, built at Iowa State, used a fixed program to solve systems of linear equations.[10] The British wartime Colossus machines applied logical tests to inputs from encrypted messages and electronically simulated code wheels. Their basic sequence of operations was likewise fixed.[11]

The ENIAC project introduced the vocabulary of programs and programming and the automation of higher-level control functions such as branches and looping. It was publicized around the world, stimulating interest in electronic computation. Its two main designers founded the first electronic computer company. And even before ENIAC was finished, design work on a planned successor, EDVAC, had defined the key architectural features of the modern computer.

1 INVENTING THE COMPUTER

On February 15, 1946, subscribers to the *New York Times* might have been startled to discover a front-page story titled "Electronic Computer Flashes Answers, May Speed Engineering." It opened with news that "one of the war's top secrets, an amazing machine" was "heralded . . . as a tool with which to begin to rebuild scientific affairs on new foundations."[1] That evening, the new machine, ENIAC (electronic numerical integrator and computer), shown in figure 1.1, was ceremonially switched on and dedicated by Major General Gladeon M. Barnes. He represented the US Army's Ordnance Department, the owner of the new machine and sponsor of the project.

The ceremony was taking place at the University of Pennsylvania where ENIAC had been designed and built, starting in 1943. It would not be reassembled at the Ballistics Research Laboratory in nearby Maryland until 1947, where it enjoyed a long and productive career until being retired in 1955. The machine was demonstrated to reporters and dignitaries by its now-famous team of six female operators, before the visiting men joined the project engineers and managers at a lavish dinner. The 110 guests were entertained in Houston Hall, an elegant stone structure, with lobster bisque, filet mignon, ice cream, and "fancy cakes." The operators were not present, though one woman did make the guest list: Adele Goldstine, who wrote its programming manual and was married to the Army's project lead.

This was not exactly the beginning of the computer age. For one thing, ENIAC had run an elaborate set of hydrogen bomb calculations for Los Alamos starting in December of the previous year. But it was the first time anyone outside a small world of project participants and well-connected specialists with security clearance could begin to think about the possibilities opened up by machines able to carry out complex computations without human intervention while crunching numbers thousands of times faster than mechanical calculators. The press release distributed by the War Department discussed the potential applications of the new technology for nuclear physics, aerodynamics, weather prediction, oil prospecting, and aircraft design, as well as ballistics.[2]

Figure 1.1
ENIAC as installed at the University of Pennsylvania, in a US Army photograph used by the *New York Times* in its 1946 report. This image defined public ideas of what an electronic computer looked like. The machine was configured by setting switches and wiring connections between its many panels, which collectively established a room within a room in which its operators and associated punched card machinery worked. Corporal Irwin Goldstein, an Army maintenance technician, is in the foreground setting data on a "portable function table" later used to hold encoded program instructions. Technician Homer Spence and two operators, Frances Bilas and Betty Jean Jennings (later Jean Bartik), work in the background.

ENIAC

ENIAC's place in computer history rests on more than being the first device to merit check marks for *electronic* and *programmable* on a comparison sheet of early machines. It fixed public impressions of what a computer looked like and what it could do. It even inspired the practice of naming early computers with five- or six-letter acronyms ending with *AC*. During a period of about five years as the only programmable electronic computer available for scientific use, ENIAC lived up to the hype by pioneering applications such as Monte Carlo simulation, numerical weather prediction, and the modeling of supersonic air flow.

Programmability

We should explain what we mean when we say that ENIAC was programmable. The idea of defining and following a program does not originate with automatic computers. Earlier meanings of *program* included a concert program, the program of study for a degree, and the programming of radio stations. In each case the program defined a sequence of actions over time. For example, the concert program might define three pieces of music to be performed, break the largest one into four movements, and specify an intermission. The radio station programmer allocates shows to time slots. Automatic computers similarly followed a program of mathematical operations (like addition or multiplication) rather than courses, radio shows, or musical performances. Discussion of programming a computer first appeared in the ENIAC project. By 1945 it had settled on something like its modern meaning: a computer program was a configuration that carried out the operations needed for a job. The act of creating it was called programming.[3]

ENIAC was not the first programmable computer, but it was the first to automate the job of deciding what to do next after a sequence of operations finished. Consider the context in which it appeared. In the summer of 1944, just as Harvard's Mark I was publicly unveiled, the Navy ordered Grace Murray Hopper to the Computation Lab to assist Howard Aiken with programming it. Hopper had been an assistant professor of mathematics at Vassar College and had taken leave to attend the Navy's Midshipmen's School. According to Hopper, she had just earned her one-and-one-half stripes when she reported to the lab at Harvard. There, Howard Aiken,

> . . . a large object, with three stripes . . . waved his hand and said: "That's a computing machine." I said, "Yes, Sir." What else could I say? He said he would like to have me compute the coefficients of the arc tangent series, for Thursday. Again, what could I say? "Yes, Sir." I did not know what on earth was happening, but that was my meeting with Howard Hathaway Aiken.[4]

Thus began the practice of computer programming in the United States. Hopper wrote out the sequence of codes for that job, and later the codes for more complex mathematical tasks. One of the first was a design problem for James Baker, a Harvard Fellow responsible for the lenses used for top-secret US spy cameras.

Early relay computers relied on human operators to step in when decisions had to be made. The Harvard Mark 1, for example, spent most of its time running loops, created by gluing the ends of the control tape together. It could automatically stop work after finishing that stage in the computation, leaving its operator to figure out which strip of program tape to mount next. As the Mark 1 took six seconds to carry out a multiplication, and a full minute to execute built-in sequences for logarithmic and trigonometric operations, this reliance on human intervention for what we would now call program branching did not initially seem to be a major performance limitation.

ENIAC Design and Construction

During the Great Depression, John Mauchly, the ENIAC project's instigator, was a PhD physicist struggling to carry out meaningful research under a heavy teaching load at a small college. One of his interests was the use of electronics for counting. As war loomed for America, the government sponsored a special summer school in electronics at the University of Pennsylvania. This crash training drew on the Moore School's expertise in the new technology and ties to local industry. Mauchly distinguished himself, which led to a professional step up to a faculty position.

Mauchly learned that hundreds of women were employed at the Moore School to compute firing tables, under contract to the US Army's Ballistics Research Laboratory. When firing shells and mortars, the gun operator would begin with the estimated distance of the target. He would then refer to a book of mathematical tables specific to that combination of gun and munition to find the correct angle. The answer varied depending on change in elevation, temperature, and wind speed.[5] Expressing the motion of a shell as a differential equation was straightforward. Test firing each gun a few times at Aberdeen Proving Ground yielded parameters to plug into this standard equation. The problem was that, like most other real-world differential equations, this one could not be tackled with the methods taught in calculus classes. Because of atmospheric drag the (human) computers had to solve the equation numerically, breaking the shell's flight into several thousand intervals to calculate changes in its speed and position from each instant to the next. The simulated trajectory provided just one row of the firing table, giving the shell's range for one angle of elevation. Producing the entire table by hand took months of work. Hard as the computers worked, their backlog of work grew ever larger. New guns were being shipped to Europe without the tables needed to operate them.

That backlog gave Mauchly the perfect excuse to justify development of an electronic computing device. Since the 1980s, a program so compelling that users acquire an entire computer system to run it has been called a killer application. This term is particularly appropriate to ENIAC, as the aim was literally to kill more efficiently. His chief collaborator was Presper Eckert, a master's degree engineering student who stayed at the Moore School to help run its wartime projects.

Because ENIAC was both electronic and general purpose, its designers faced a unique challenge. In Mauchly's words, "Calculations can be performed at high speed only if instructions are supplied at high speed."[6] That required a control method faster than paper tape. It also meant avoiding frequent stops for human intervention. For example, ENIAC might need to repeat a numerical procedure until a sufficiently precise result was obtained and then automatically move on to the next step of the calculation. Mauchly sketched out several possible mechanisms to select automatically between different preset courses of action depending on the values ENIAC had already calculated. Computer scientists call this conditional branching and view it as a defining feature of the modern computer.[7]

ENIAC's unique control method was based on a careful analysis of the firing table problem. Early in the project, Arthur Burks, one its key designers, worked with Adele

Goldstine, a mathematician who had been overseeing the women calculating firing tables. They produced a draft configuration to tackle the problem and devised methods to diagram the sequence of mathematical operations needed and the physical setup of ENIAC that would generate them. This was implemented by connecting wires to carry "program" pulses from one unit to another. When an addition was followed by a division, a wire ran from an output terminal of an adding and storing unit (accumulator) to an input terminal on the divider. The exact operation triggered by the pulse was determined by switch settings. After the pulse primed the divider to expect input, a separate network of cables fed it the numbers to be divided. Altogether, ENIAC was not so much a single computer as a kit of forty modules from which a different computer was constructed for each problem.

While Eckert and Mauchly are the names usually attached to ENIAC, dozens of people worked to create the machine. A team of about six design engineers was assisted by a secretary, draftswomen, and model builders. Mathematical experts at Penn and the Ballistics Research Laboratory helped to figure out what capabilities the machine would need. University administrators and procurement specialists scrambled to obtain obscure components. By mid-1944, as work shifted from prototyping to full-scale production, around forty people were employed to assemble ENIAC.

ENIAC in Use

It took ENIAC 30 seconds to simulate a trajectory, potentially tracing the shell's journey faster than the shell itself could travel. The war was over by the time ENIAC was finished, so firing table preparation became less urgent. In the end, ENIAC spent something like 15 percent of its production time calculating them and the rest on other, varied jobs, including many to aid the Los Alamos and Argonne laboratories in the development of nuclear weapons and reactors.

ENIAC was also a workplace of around two thousand square feet. Its panels were arranged in a U shape, working like a set of room dividers to enclose an inner space in which its operators worked. Its original home in the Moore School's basement was a little shabby, but the Ballistics Research Laboratory built it a smart and modern home with the only air conditioning unit on the Proving Ground to dissipate heat from its 18,000 or so vacuum tubes. Staff moved their desks inside ENIAC to escape from the summer heat and humidity. Groups of visitors, from cadets to President Truman, were frequently brought to marvel at ENIAC, establishing a tradition of computer rooms as organizational showpieces that lasted through the 1960s.[8]

Today, ENIAC is most often remembered as a machine programmed by women. When in use, like other early computers it needed frequent human attention. ENIAC was typically staffed by two operators, joined frequently by a supervisor or by the scientists for whom the calculation was being carried out. Six women hired in mid-1945 were its operators during an initial period of use in Philadelphia from December 1945 to December 1946. They have become famous as the "women of ENIAC" and are often called the first computer programmers, though remembering them only as programmers doesn't reflect the full range of their work.

Data went in and out of ENIAC on punched cards: small rectangles of cardboard each able to store 80 digits as a pattern of holes. The women spent much of their time punching input data onto cards and running output cards through an IBM tabulating machine to print their contents. Bigger jobs produced thousands of intermediate cards, which they had to process manually using other IBM machines before feeding them back in. They followed configuration diagrams to reconfigure the computer's wires and switches when changing between problems. Because of the expertise they built up, the women often helped scientists and engineers to formulate their problems as sequences of ENIAC operations and convert those sequences to diagrams. On later computers these tasks were usually handled by separate teams of key punch women, computer operators, and programmers. Even ENIAC was eventually operated around the clock by three shifts of operators after it proved itself at Ballistics Research Laboratory.

THE EDVAC APPROACH

By mid-1944, more than a year before ENIAC was working, its creators were already working on ideas for a successor. They recognized that switching ENIAC between jobs would require a lengthy and error-prone period of reconfiguration and that its distributed control method imposed a confusing mass of bottlenecks limiting, for example, the number of branches that could be included in a program.

The proposal to build EDVAC (electronic discrete variable automatic computer) was approved by the government in August. It centered on a new kind of memory proposed by Eckert, the delay line.[9] ENIAC used twenty-eight vacuum tubes to hold each decimal digit. That approach would not scale far. It took years of engineering frustrations to make delay line memory work reliably, but the idea was simple and compelling. Pulses representing several hundred digits moved through a fluid-filled tube. Signals received at one end were immediately retransmitted at the other end, so that the same sequence was cycling constantly. Whenever a number reached the end of the tube it was available to be copied to the computer's processor if needed.

Von Neumann's *First Draft*

How to design a computer around this new storage system? In the summer of 1944, shortly before the formal proposal for EDVAC was written, the ENIAC team welcomed a new collaborator. The famous mathematician John von Neumann was sufficiently intrigued by word of what was going on at the Moore School to have himself brought onto the project, despite his many other urgent collaborations, including the design of the atomic bomb. Engineering work on the new machine was delayed when ENIAC's completion slipped by more than a year, monopolizing the school's resources. During this period, the Moore School team discussed with von Neumann a variety of ideas for how to structure EDVAC. He pulled some of the proposals together into a coherent whole during a trip to Los Alamos, adding a control scheme with a standard instruction format and a simple but flexible set of instructions.

Von Neumann's *First Draft of a Report on the EDVAC* described logical structures rather than the specifics of hardware. One of its most novel features was that, as the team had decided by September 1944, coded instructions were stored in the same storage devices used to hold data. The ideas contained in the *First Draft* are frequently summed up in a single phrase: "the stored program concept." In April 1945, Goldstine circulated von Neumann's text within the Moore School.[10] It is often called the founding document of modern computing. A lightly revised version, dated June 30, went out to about two dozen people interested in electronic computing, most of them already connected to the ENIAC project. As it took more than a year for anyone to get around to filing a patent on ENIAC, this disclosure of the new architecture put the modern computer into the public domain.

The idea of loading a program into main memory was important and set EDVAC apart from existing computer designs. However, following work done by Haigh in collaboration with Mark Priestley and Crispin Rope, we prefer to separate the enormous influence of EDVAC into three clusters of ideas (or *paradigms*).[11]

The first of these, the EDVAC hardware paradigm, specified an all-electronic machine with a large high-speed memory using binary number storage (figure 1.2). Eckert and Mauchly were committed to this before von Neumann arrived. From their viewpoint, everything else in the report was a working through of the implications of this new technology, and they resented the credit that von Neumann received for its contents.

The second was the von Neumann architecture paradigm. Influenced by the emerging field of cybernetics, which asserted the equivalence of brains and mechanical control mechanisms, von Neumann borrowed biological language to describe EDVAC's switching circuits as *neurons*.[12] He broke its structure into *organs*, including a large delay line *memory* for programs and data (the only biological term that stuck). Storage and arithmetic was binary, using what would soon be called bits (a contraction of *binary digits*) to encode information. Each 32-bit chunk of memory (soon to be called a word) was referenced with an address number. The program to be run would be loaded from R, the *recording medium*, into the main memory, M, before being executed. All data being worked on would also be copied into memory. This differed

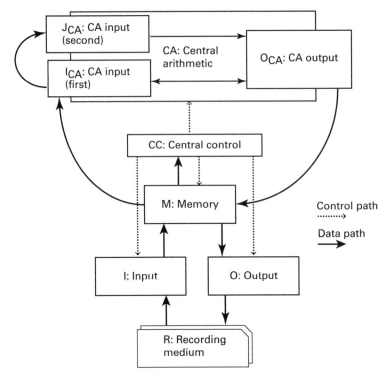

Figure 1.2
The logical architecture of EDVAC as described in the *First Draft*. Reproduced from Thomas Haigh, Mark Priestley, and Crispin Rope, *ENIAC in Action* (Cambridge, MA: MIT Press, 2016, p. 145).

from the relay calculators, which read instructions one at a time from tape, executing each immediately. ENIAC held many data pathways between units, and put adding circuits into every storage unit. EDVAC, in contrast, would have just one connection each way between memory and the *central arithmetic* organ where all arithmetic was centralized. Issues such as the number of adding units and connections between them and memory had been debated during von Neumann's meetings with the Moore School team, but in the report he selected a set of design options and combined them into an elegant and coherent architecture.

The third cluster of ideas was a system of instruction codes: the modern code paradigm. The flow of instructions and data mirrored the way humans performed scientific calculations as a series of mathematical operations, using mechanical calculators, books of tables, and pencil and paper. Even Eckert and Mauchly credited von Neumann with devising the proposed instruction code. It represented each instruction with an operation code, usually followed by parameters or an address. Because the architecture was so simple, even adding two numbers and storing the result took at least four instructions: reading the first number from M into the location marked I_{CA} in the diagram, reading the second number into I_{CA} (which automatically shifted the first into J_{CA}), adding them (which stored the result in O_{CA}), and copying this value back to M (overwriting the second number). In later terminology the special storage circuits, such as I_{CA}, were called *registers*. The one used, like O_{CA}, as the destination for mathematical operations was called the *accumulator*.

Each of these four instructions took several steps to carry out. The basic cycle of an EDVAC-like computer is to fetch the next instruction code from memory (in this case copying it into the *central control* organ), decode that instruction, and execute it. A special circuit, the program counter, holds the number of the address from which the next instruction should be fetched. The program counter is automatically incremented by one every time an instruction is read (hence the term *counter*), so that the next instruction is fetched from the next memory location. Manipulating the number stored in the counter causes a *branch* to a different part of the program code.

Most computers today harness several processor cores running in parallel, but the concept of processing a stream of instructions from an addressable memory remains the most lasting of all the *First Draft*'s contributions. Computer scientist Alan Perlis remarked that "sometimes I think the only universal in the computing field is the fetch-execute cycle."[13]

Early EDVAC-Like Computers

All three paradigms were hugely influential in the early days of computing. Recognizing the eagerness of would-be computer builders to see ENIAC and learn more about the ideas contained in the *First Draft*, in the summer of 1946 the Moore School and the US military co-sponsored a course on the "Theory and Techniques for Design of Electronic Digital Computers."[14] The same year, von Neumann, now overseeing a computer project at the Institute for Advanced Study in Princeton, began work with Arthur Burks and Herman Goldstine on an influential series of reports to develop ideas on computer architecture, instruction sets, and programming.[15] Between them, the Moore School lectures and the IAS reports firmly established the new approach to computer design.

The new computers had simple architectures, but making them work took years of engineering frustration. Early computers had to transmit digital pulses lasting perhaps one hundred thousandth of a second through miles of wire with complete reliability. They relied on digital logic circuits of huge complexity. They included thousands of vacuum tubes, which under normal use would be expected to fail often enough to make any computer useless. Building one meant soldering hundreds of thousands of electrical joints. But the biggest challenge was producing a stable memory able to hold tens of thousands of bits stable long enough to run a program. Delay lines were simple in theory but treacherous in practice. The Moore School tried building a delay line memory for

ENIAC, under contract to the Ballistics Research Laboratory. After two years it was installed, but it never worked well enough to be useful and the laboratory returned it.

As a stopgap, ENIAC was reconfigured in March and April 1948 to run modern code (in the sense previously defined) entered by turning switches on read-only memory panels. It was used in that mode for the rest of its career. Adele Goldstine sketched out the original plan for the conversion and worked to refine it with Jean Bartik, one of the original operators who had been rehired in 1947 to run a contract programming group (the first of its kind) to develop ENIAC applications. Klara von Neumann, wife of John, did the detailed coding for the first modern program executed: a complex Monte Carlo simulation run for Los Alamos to observe virtually and experiment with the chain reaction inside an exploding nuclear weapon. The contributions made by those women, as well as the mostly female team that wired and assembled the machine, make it misleading to talk about just six people as *the* women of ENIAC."[16]

A few months later, a team at the University of Manchester, England, debugged its writable electronic memory with the help of a tiny test computer. This was a stepping stone to a useful EDVAC-like computer, now called the Manchester Mark 1. Even when they could be made to work, delay lines were rather slow, as the computer had to wait for the bits of interest to take their turn cycling through the sensing equipment at the end of the tube. The Manchester group had a faster solution. The device popularly known as the Williams tube, after engineer Freddy Williams, stored bits as charges on a cathode ray tube (CRT) similar to those used in televisions and radar sets of that period. This produced a pattern of visible dots on the tube. By adding a mechanism to read charged spots as well as write them, a single tube could be used to store two thousand bits. The challenge was to read them reliably and to constantly write them back to the screen before they faded away. This technology was licensed by other computer builders, including the Institute for Advanced Study and IBM.[17]

In the spring of 1949, before the full Manchester computer was ready, Maurice Wilkes of Cambridge University managed to get his EDSAC operational. This made it the first full-scale computer to implement all three of the EDVAC paradigms. EDSAC became the basis of a scientific computing service, solving problems in biology, chemistry, and radio astronomy as well as physics and mathematics. One of its beneficiaries, John Kendrew, won a Nobel Prize after EDSAC was programmed to aid in the otherwise infeasible task of deducing the three-dimensional structure of proteins from the scattering of X-rays.[18]

Variations on a Theme

Each computer project launched in the 1940s was an experiment, and designers tried out many variations on the EDVAC theme. Even computers modeled on von Neumann's slightly later IAS design, built at places like Los Alamos, the Bureau of Standards, and the RAND Corporation, diverged by, for example, using different memory technologies.

Successful ideas were widely copied. For example, many programs worked on data held in a matrix structure (essentially a table). The programmer defined a loop to repeat a sequence of operations for each cell in the matrix. Each time it looped, the code had to act on data held in a different memory location. In most computers of the 1940s this was accomplished by having the code modify the instruction held in memory, an idea introduced in the *First Draft*. Beginning with the University of Manchester's Mark 1 computer in 1949, designers added to the processor what would later be called an index register to simplify work with data structures. The instructions stayed the same. Instead, the program updated an offset value in the index register, which was added to the *base address* held in the instruction to determine the actual memory location acted on.[19]

Storing programs and data in addressable memory was a hallmark of the EDVAC approach. A single memory location was a *word* of memory. But computer designers made different choices about how large each word should be and how instructions should be encoded within it. The original EDVAC design called for thirty-two bits in each word. Early computers used word lengths between seventeen (EDSAC) and forty (the IAS computer and the Manchester Mark 1) bits. A larger word size made working with bigger numbers convenient and efficient but demanded more hardware and was wasteful if most numbers and instructions did not fill an entire word. Maurice Wilkes chose a short word length for EDSAC to efficiently store instructions but included hardware support for numbers that filled two words, a practice that later became standard.

One factor influencing word length was the maximum number of addresses specified in a single instruction. Computer architects saw this as a crucial design feature to the extent that they used *one address*, *three address*, and so on as a taxonomy to define classes of computer. The EDVAC instruction set defined in the *First Draft* specified either a source or a destination address when reading or writing main memory, making it a one-address machine. Its arithmetic operations functioned on numbers that had already been loaded into special registers and so were *zero-address* instructions. But EDVAC, as eventually built, had grown into a *four-address* machine. Its addition instruction included memory locations for both numbers to be added and a third for the location in which to store the result. The fourth address specified the location of the next instruction to be read. The example we gave previously, using four instructions to add two numbers and store the result, would take just one instruction in the final machine.

Almost every computer operation involved moving data backward and forward between registers, arithmetic units, and main memory. One crucial engineering decision was whether to move all the bits in a word sequentially on a single wire or send them together along a set of parallel wires. This was the original meaning of *serial* and *parallel* in computer design. EDSAC was serial, a popular choice for machines with delay line memories, which delivered data one bit at a time. Serial transmission of bits was slower but simpler. Most parallel-word computers used Williams tubes, which could retrieve all the bits in a word of memory simultaneously.[20]

You may have noticed that we have not yet mentioned two of the most famous figures in the history of computing: Charles Babbage and Alan Turing. Babbage's efforts a hundred years earlier to build a mechanical computer were remarkable but had no direct influence on work in the 1940s; the ENIAC team didn't know about them and even Howard Aiken, who helped to revive Babbage's reputation as a computer pioneer, designed his computer in ignorance of the details of Babbage's work.[21]

Alan Turing's conceptual work on computability carried out in the 1930s was later foundational to the development of theoretical computer science. While von Neumann was aware of, and intrigued by, Turing's concept of a "universal machine," we see no evidence that it shaped his design for EDVAC.[22] In 1946, however, Turing proposed a radical reinterpretation of the EDVAC approach for the computer project at Britain's National Physical Laboratory. His ACE design coded every operation as a transfer between delay lines. Some transfer destinations carried out special operations, such as adding the number it received. The architecture was commercialized on some early British machines, which were simple and fast but very hard to program because programmers had to track exactly when each value they wanted to work on would be emerging from a delay line and time their code accordingly.[23]

Programming Tools

General purpose computers can do many things. The disadvantage to that flexibility is that getting any particular task done takes minutely detailed programming. It did not take Grace Hopper long to realize that reusing pieces of Mark 1 code for new problems could speed this work. Her group built up a paper tape library of standard sequences, called subroutines, for routine operations such as calculating logarithms or converting numbers between decimal and binary format. In Germany, Konrad Zuse envisioned a "Plan Preparation Machine" (*Planfertigungsgeräte*).[24] It would punch control tapes for his Z4 computer on the basis of commands entered by a user, automatically checking that their syntax was correct.

The arrival of EDVAC-like computers opened up new possibilities for automating program preparation. The computer itself could be programmed to handle the chores involved in reusing code, such as renumbering memory addresses within each subroutine according to its eventual position in memory. These new tools were called assemblers because they assembled subroutines and new code into a single executable program. Assemblers quickly picked up another function. Humans found it easier to refer to instructions by short abbreviations, called mnemonics. For example, LOAD might be the mnemonic for the instruction to copy a number from main memory into a processor register. The list of instruction mnemonics and parameters was called assembly language. The assembler translated each line into the corresponding numerical instruction that the computer could execute.

Of all the 1940s computers, EDSAC had the most convenient programming system. Every time the machine was reset, code wired into read-only memory was automatically triggered to read an instruction tape, translating mnemonics on the fly and loading the results into memory. David Wheeler developed an elegant and influential way of calling subroutines, so that the computer could easily jump back to what it was doing previously when the subroutine finished.[25] EDSAC users built a robust library of subroutine tapes and published their code in the first textbook on computer programming.[26]

The general arrival of *symbolic* and *macro* assemblers in the 1950s made assembly coding less tedious. As instructions were added and removed from a program, the memory locations holding variables or subroutines would change. Symbolic assemblers let programmers use labels rather than numbers to specify addresses, which eliminated the need to edit the code every time locations changed. Unlike a regular mnemonic, which produced a single machine instruction, a macro instruction expanded into an entire block of code. For example, macro instructions made it easier to retrieve information from tape files.

The first symbolic assembler was written by Nathaniel Rochester in 1949 as part of an internal IBM project to experiment with the EDVAC approach. The test assembly lashed the arithmetic from IBM's new 604 Electronic Calculating Punch, its first commercial electronic calculator, to a variety of custom components including a new control unit and larger drum memory. The 604 already had a built-in control mechanism via simple programs wired on its plug board. The team called the EDVAC-style instructions held in the drum memory its stored program, to distinguish it from the plug board controls. As this term was taken up by others, its meaning gradually broadened until "stored program" became short-hand for the whole cluster of new ideas about computer architecture introduced in the *First Draft*.[27]

The Association for Computing Machinery

As interest in electronic computers spread, a new organization, the Association for Computing Machinery, was founded in 1947 for the exchange of information and ideas between those working in the new field. Its first members had come together in a series of special meetings and symposia, but they recognized the benefits that an association, with a regular conference series, could bring.[28] The indefatigable Edmund Berkeley served as its founding secretary. The new association developed slowly at first, beginning publication of a journal in 1954.

The ACM eventually grew into a society of around a hundred thousand members, with a host of special interest groups, publications, and conferences. While our focus here is on computing practice and technology rather than computing research, we will be noting many of the winners of its A.M. Turing Award, first presented in 1966. This is the premier prize in computer science, promoted by ACM as the equivalent of a Nobel Prize. It marks contributions that the computing research community recognizes for exceptional intellectual merit or, in a few cases, undeniable practical influence.

Computer science did not exist as a discipline in the 1950s, so those researching the development and use of computers came from a variety of disciplinary backgrounds. Engineering societies were also important venues, particularly the Institute for Radio Engineers and the American Institute of Electrical Engineers. In 1963 those merged to create the Institute of Electrical and Electronics Engineers (IEEE), whose Computer Society developed as the other main organization in the field.

THE COMMERCIALIZATION OF COMPUTING

"[Y]ou . . . fellows ought to go back and change your program entirely, stop this . . . foolishness with Eckert and Mauchly." That's what Howard Aiken told Edward Cannon of the US National Bureau of Standards in 1948, as part of a committee recommending against supporting their proposal to make and sell electronic computers.[29] Eckert and Mauchly had a vision, not only of how to design and build a computer but also of how a society might benefit from large numbers of them. In Aiken's view, a commercial market would never develop: in the United States there was a need for perhaps five or six such machines, but no more.[30] Aiken was quickly proven wrong. Thousands of computers were sold in the 1950s.

Eckert and Mauchly's drive to build computers was matched by an equal but opposite drive by the University of Pennsylvania to banish commercial interests from the academy. Irwin Travis, an administrator at the Moore School, demanded that members of its staff sign a form relinquishing ownership of inventions made during their work there. Eckert and Mauchly refused to sign. They resigned effective March 31, 1946. Work on the EDVAC was turned over to a succession of other engineers who took five more years to finish it.[31]

The First Computer Start-Up

Eckert and Mauchly could have found work elsewhere, but instead they chose the risky course of founding their own company. They formed a partnership, the Electronic Control Company, in 1946; in December 1948 they incorporated as the Eckert-Mauchly Computer Corporation. Beyond the engineering problems of designing and building a computer and its associated tape drives, memory units, and input-output equipment loomed the bigger problem of raising capital. Their solution was to raise money by preselling the computers. The US Census Bureau was interested, which led to the investigation undertaken by Aiken and his colleagues. The frantic search for funds even took them to American Totalisator Company, who wanted a computer to calculate betting odds at race tracks. On January 12, 1948, John Mauchly wrote a memorandum to his fellow staff at the Eckert-Mauchly Computer Corporation in which he listed a total of twenty-two industries, government agencies, and other institutions that he had contacted.[32] Among them were

> Prudential. [Edmund C.] Berkeley . . . says that considering the number of persons at Prudential who have now expressed themselves in favor of obtaining electronic equipment, he believes there will be no difficulty in getting an order for one Univac.
>
> Oak Ridge . . . it was almost 100% certain that their purchase order would be approved by Army . . .
>
> Army Map Service . . . Army Map Service has taken an interest in Univac equipment.
>
> Bureau of Aeronautics . . . we could possibly obtain a contract . . .
>
> The Metropolitan Insurance Company has a large problem involving a total file of 18,000,000 policies with 2,000,000 changes per week. There are about 20 digits of information for each policy. It appears that this is a natural application for the Univac . . . it would be worthwhile to follow it up . . .
>
> Aircraft Companies. A number of aircraft companies are good prospects . . . there is no doubt that such companies could use Univac equipment. We have had brief contact with Hughes Aircraft, Glen L. Martin, United Aircraft, North American Aviation, and have been told that Grumman goes in for some rather fancy calculations.

The undercapitalized company struggled to turn these prospects into orders. That wasn't enough to save the Eckert-Mauchly Computer Company, even though its main product, the Univac, established the marketplace for computers. It became a division of the business-machines company Remington Rand. But all these organizations, and many more, would soon find compelling reasons to purchase or lease electronic digital computers, if not from Eckert-Mauchly then from someone else.

Eckert and Mauchly, with the help of about a dozen technical employees of their division, designed and built the Univac in a modest factory at 3747 Ridge Avenue in Philadelphia (see figure 1.3). It used four bits to code each decimal digit. In its central processor, four general-purpose accumulators carried out arithmetic. A word was 45 bits long; representing 11 decimal digits plus a sign, or six alphabetic characters. The Univac's clock ran at 2.25 MHz. It could perform about 465 multiplications per second, not a spectacular advance on ENIAC, but its tape system and larger memory made it a much faster machine overall. Delay lines stored one thousand words as acoustic pulses in tubes of mercury, while magnetic tape units stored up to one million characters on each reel of 1/2-inch metal tape. Its central processor contained over

Figure 1.3

The staff of the Eckert-Mauchly Computer Company gather in 1948 at their Philadelphia factory, in front of their first computer, the custom BINAC created for military aircraft developer Northrop. Front row (left to right): J. Presper Eckert (cofounder and ENIAC chief engineer), Frazier Welsh, James Wiener, Bradford Sheppard, and John Mauchly (cofounder and ENIAC project initiator). Back row: Albert Auerbach, Jean Bartik (former ENIAC operator and programming manager), Marvin Jacoby, John Sims, Louis Wilson, Robert Shaw (former ENIAC engineer), and Gerald Smoliar. Courtesy Unisys Corp.

five thousand tubes installed in cabinets arranged in a $10' \times 14'$ rectangle. Inside this rectangle were the mercury delay-line tanks. Many of its novel features later became commonplace, among them alphanumeric as well as numeric processing, magnetic tapes for bulk memory, and circuits called buffers to stop the processor being tied up waiting for data to transfer to tape.[33]

The Univac reflected Eckert's philosophy of loading the vacuum tube circuits conservatively and adding enough redundancy to ensure reliable operation. Statistics gathered by the Metropolitan Life Insurance Company showed that it was available for work 81 percent of the time.[34] The Census Bureau reported, "We never encountered an incorrect solution to a problem which we were sure resulted from an internal computer error."[35]

Univac in Use

On March 31, 1951, Remington Rand turned over the first Univac to the US Census Bureau, beginning commercial sales of large-scale stored program computers in the United States.[36] It wasn't shipped to Washington until late December 1952. After heroic efforts to complete and debug the machine, Eckert and Mauchly were apprehensive about dismantling it, moving it, and setting it up again. They needed it nearby to show to other potential customers. By 1954 about twenty systems were built and sold, at around a million dollars each.[37] Table 1.1 lists Univac installations through 1954.

Because Univac followed the general outline of the EDVAC design, it was flexible enough to be used for business applications as well as scientific computations. This meant broadening the sense of what a computer was for. Aiken could not imagine that "the basic logics of a machine designed for the numerical solution of differential equations [could] coincide with the basic logics of a machine intended to make bills for a department store."[38] Eckert and Mauchly knew otherwise. Univac inaugurated the era of large computers for what were later called "data processing" applications.

The closest technology in widespread administrative use was punched card machines, the core product of IBM. The earliest punched card machines, sold from the

Table 1.1

Univac Installations, 1951–1954

#	Date	Customer
1	Summer 1951	US Census Bureau
2	Late 1952	US Air Force, the Pentagon
3	Late 1952	US Army Map Service
4	Fall 1953	US AEC, New York, NY (at NYU)
5	Fall 1953	US AEC, Livermore, CA
6	Fall 1953	David Taylor Model Basin, Carderock, MD
7	1954	Remington Rand, New York, NY
8	1954	General Electric, Louisville, KY
9	1954	Metropolitan Life, New York, NY
10	1954	Wright-Patterson AFB, OH
11	1954	US Steel, Pittsburgh, PA
12	1954	Du Pont, Wilmington, DE
13	1954	US Steel, Gary, IN
14	1954	Franklin Life Insurance, Springfield, OH
15	1954	Westinghouse, Pittsburgh, PA
16	1954	Pacific Mutual Life Insurance, Los Angeles
17	1954	Sylvania Electric, New York, NY
18	1954	Consolidated Edison, New York, NY
19	1954	Consolidated Edison, New York, NY

Note: This list is compiled from a variety of sources and does not include one or two Univacs that were completed but remained with Remington Rand. In some cases the dates are approximate.

1890s onward to the Census Bureau, were used purely for statistical purposes, tabulating and cross tabulating totals. By the 1930s they were used for many administrative functions and could handle letters as well as numbers and produce printed outputs such as payroll checks. This expansion was reinforced by the deep penetration of IBM's salesforce into the accounting offices of its customers. Punched card machines were often called unit record equipment because a single card encoded information about one thing, such as a sales transaction or employee. A typical small installation consisted of several key punches to get the data onto cards, plus several specialized devices such as tabulators, sorters, and collators. Each machine was configured to carry out the same operation on every card in the deck.

For most customers, what was revolutionary about the Univac was the use of tape in place of punched cards. Univac could scan through a reel of tape, reading selected records, performing some process, and writing the results to another tape.[39] Carrying out all the operations needed for a job meant carrying decks of cards around the room, running them through one machine after another.[40] That made punched card processing labor-intensive. In contrast, the Univac could perform a long sequence of automatic operations before fetching the next record from memory. It replaced not only existing calculating machines but also the people who tended them. Customers regarded the Univac as an information processing system, not a calculator. Published descriptions of the Univac nearly always referred to it as a "tape" machine. For General Electric, for example, "the speed of computing" was "of tertiary importance only."[41]

The Census Bureau, which had been pivotal in getting the fledgling computer company going, hoped to use the Univac for tabulating the 1950 Census. By the time it received its machine in 1951, much of the work had already been put on punched card machines for processing. Then the Bureau had to step aside while the US Air Force and the Atomic Energy Commission commandeered the machine for problems deemed more urgent.[42] Still, Univac #1 was used to cross tabulate census data for four states. Data initially punched onto eleven million cards (one for each person), was transferred to tape for processing by the Univac.[43] The machine was also used for tabulating another subset of the population involving about five million households. Each problem took several months to complete.

Univac #2, installed at the Pentagon for the Air Comptroller in June 1952, was intended for use in Project SCOOP (Scientific Computation of Optimum Problems), which grew out of wartime concerns with getting war materials and men across the Atlantic. The newly created Air Force had mathematically similar challenges in supplying air bases scattered across the globe. Project SCOOP played a key role in the development of linear programming, a widely used technique used to optimize systems involving many variables.[44] Soon after its installation, Univac #2 was put to work on

SCOOP around the clock.[45] Its uniprinter, based on a Remington Rand electric typewriter, could print only about ten characters per second. That proved a poor match for the high speed tape and processor, but in 1954 Remington Rand delivered the Univac High Speed Printer, able to print a full 130-character line at a time. The Univac installed at Air Force's Air Materiel Command at Wright-Patterson AFB in Ohio performed similar tasks. One of its first jobs was calculating "the complete Fiscal 1956 Budget estimate for airborne equipment space parts, involving approximately 500,000 items."[46] The machine did the job in one day, replacing a battery of punched card equipment.

Some Univacs performed classified weapons work, like the one-of-a-kind computers that preceded them. Univac #5, installed at the Lawrence Livermore Labs in April 1953, was one of those. But even that machine did one unorthodox job. In November 1952, before it was shipped to California, Remington Rand used it to predict Eisenhower's victory over Adlai Stevenson in the 1952 presidential election. Narrated on "live" television, the event inaugurated the intrusion of television into national politics, and of computers into the public's consciousness. For a brief period the word *Univac* was synonymous with *computer*, like *Thermos* for *vacuum bottle*.[47]

A final example of the Univac in use comes from the experience at General Electric's Appliance Park, outside Louisville, Kentucky. Under the direction of Roddy F. Osborn in Louisville and with the advice of the accounting firm Arthur Andersen & Co., General Electric purchased a Univac for four specific tasks: payroll, material scheduling and inventory control, order service and billing, and general cost accounting. These were prosaic operations, already being done by punched card equipment. More advanced applications would commence only after such "bread and butter" work reached a "break even point . . . enough to convince management that a computer system can pay for itself in terms of direct dollar savings (people off the payroll) without waiting for the 'jam' of more glamorous applications."[48]

On Friday, October 15, 1954, the GE Univac first produced payroll checks for the Appliance Park employees.[49] Punched card machines had been doing that job for years, but for an electronic digital computer, which recorded data as invisible magnetic spots on reels of tape, it was a significant milestone. Payroll must be done right and on time. GE had rehearsed the changeover thoroughly and they had arranged with Remington Rand that if their machine broke down and threatened to make the checks late, GE people could bring their tapes to another Univac customer and run the job there.[50] Over the course of the next year they had to exercise this option at least once. On several occasions the checks were printed at the last possible minute, and in the early months it was common to spend much more time doing the job with Univac than it had with punched card equipment. No payrolls were late.

An Automated Utopia

GE hoped to do more than replace punched card machines. The Louisville plant was conceived as a showcase for modern methods; that was the motivation to locate it in Kentucky rather than Massachusetts or New York, where traditional methods (and labor unions) held sway. Its Univac heralded an age of *automation*, a term coined at the Ford Motor Company in 1947 and popularized by John Diebold in a 1952 book by that title.[51] Diebold spoke of the 1950s as a time when "the push-button age is already obsolete; the buttons now push themselves."[52] Describing the GE installation, Osborn boasted, "While scientists and engineers have been wide-awake in making progress with these remarkable tools, business, like Rip Van Winkle, has been asleep. GE's installation of a Univac may be Rip Van Business's first 'blink.'" He hoped to apply it to long-range planning, market forecasting based on demographic data, revamping production processes to reduce inventories and shipping delays, and similar jobs requiring a more ambitious use of corporate information. At the end of Osborn's essay the *Harvard Business Review*'s editors appended a quotation from Theodore Caplow's *The Sociology of Work*, published that year. It began,

> The Utopia of automatic production is inherently plausible. Indeed, the situation of the United States today, in which poverty has come to mean the absence of status symbols rather than hunger and physical misery, is awesomely favorable when measured against the budgetary experience of previous generations or the contemporary experience of most of the people living on the other continents.[53]

It would not be the last time that the computer would be promoted as the instigator of a digital utopia. Eckert and Mauchly had taken a scientific instrument, similar to a cyclotron, and turned it into a commercial product used for administrative work as well as technical calculation. This was the first transformation of computer technology, but not the last. The "computer age" is really a series of "computer ages." The history of computing is the story of repeated redefinitions of the nature of the computer itself, as it opened new markets, new applications, and new places in the social order. In the mid-1950s, IBM developed a line of products that met the information-handling needs of American businesses. A decade later, military funding turned the computer into a device with which one interacted: a tool to augment one's intellectual capabilities. In the middle of the 1970s, a group of hobbyists and enthusiasts transformed it into a personal appliance. Around 1980, people transformed it from a piece of specialized hardware to a standardized, consumer product defined by its now-commercialized software. Since the 1990s, it has gone through another transformation: into an agent of a world-wide nexus, a communications medium. This process of reinvention and redefinition is still going on.

2 THE COMPUTER BECOMES A SCIENTIFIC SUPERTOOL

In 1976 the Los Alamos National Laboratory took delivery of the Cray 1, the machine that defined the *supercomputer*. It had more than a million times the computational power of ENIAC, but as well as being the fastest computer of its era it was also the most stylish, an upholstered monolith looking like an abstract sculpture rather than ENIAC's mad scientist tangle of lights and wires.

Yet the Cray had more in common with ENIAC than this comparison suggests. Like the first computers, it was built to compute and nothing else. Even as new markets for computers developed, it was the almost limitless demand of well-funded Cold War laboratories and defense contractors for the biggest and fastest computers that drove the development of the fastest processors and highest performance system architectures. The Monte Carlo simulations run at Los Alamos on the first Cray evolved directly from those first run on ENIAC in 1948. In the 1970s, as in the 1940s, weapons designers and weather forecasters tied up the world's fastest computers for days at a time to these hugely intensive simulations.

Cray sold just 80 copies of its hand-built, five-ton monster but that was enough to make it the clear market leader. The other machines we consider in this chapter, beginning with IBM's multi-million dollar scientific computers of the early 1950s, sold in similar volumes to the same small club of well-funded companies and labs. The needs of this tiny market might seem as relevant to ordinary computing as the performance characteristics of a million-dollar supercar are to your traffic-packed morning commute. But just as now standard features such as anti-lock brakes, turbochargers, and airbags first appeared on luxury or racing cars, the architectural advances developed for supercomputers later made their way into mass market machines. Today's computer graphics and video streaming rely heavily on features such as floating-point arithmetic, input channels, multiple processors, pipelining, and vector processing once found only in the most exotic scientific number crunchers.

THE FIRST SCIENTIFIC COMPUTERS

In May 1952, IBM drew on the experiences of its experimental test assembly to announce its first electronic computer product, the 701. IBM did not sell the machines outright. Initial rental fees were $15,000 per month. Within IBM, the 701 had been known as the Defense Calculator, after its perceived market. True to that perception, of the nineteen computers installed, nearly all went to US Defense Department or military aerospace firms. The first unit was installed at IBM's offices in New York in December, with the first customer shipment in early 1953 to Los Alamos.[1] That set a pattern which endured for decades: more often than not the truck holding the first production model of an exceptionally powerful new computer was headed to Los Alamos.

The 701: IBM's "Defense Calculator"

The 701's capabilities were broadly comparable to the Univac's. Memory reads were faster because it retrieved all the bits of a word at once, rather than the Univac's delay lines that retrieved bits one at a time. Tape operations were less efficient because it funneled all data through a single register in the machine's processor. However, IBM's tape drive was more reliable, using an ingenious vacuum-column mechanism to allow its lightweight plastic tape to start and stop quickly without tearing.

Customers initially used their 701s for problems involving weapons design, spacecraft trajectories, and cryptanalysis. For many of those jobs, processor performance mattered more than tape throughput. Some of that work had previously been done with punched cards, more of it with slide rules and mechanical calculators. Customers also applied the 701 to the same kinds of jobs the Univac was doing: logistics for a military agency, financial reports, actuarial reports, payrolls (for North American Aviation) and even predicting the results of a presidential election for network television (in 1956, the 701 correctly predicted Eisenhower's reelection).[2]

The IBM 704 and Its Successors

By 1956 IBM had installed more large computers than Univac.[3] That owed much to its 704 computer, announced in 1954 as the successor to the 701 but incorporating three key improvements. First, IBM dropped the temperamental Williams tubes in favor of a new technology: core memory. Unfulfilled orders for 701s were immediately switched to the new machine, and those already delivered were retrofitted with the new memory. Second, IBM added hardware support for floating-point arithmetic. Third, the 704 incorporated three index registers (discussed in the previous chapter) to simplify programming. In the sections below we explore these developments in turn.

The *cores* in core memory are small doughnut-shaped pieces of magnetic material, through which several fine wires are threaded, as shown in figure 2.1. Samples of the

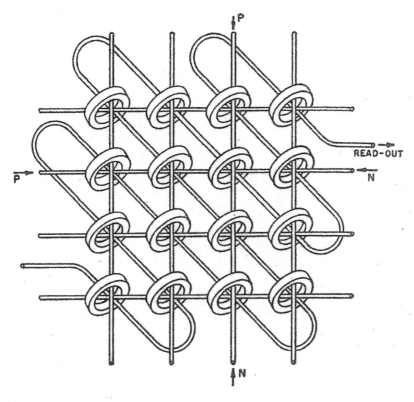

Figure 2.1

Magnetic core memory. ©1953 IEEE. Reprinted, with permission, from Jan A. Rajchman, "A Myriabit Magnetic-Core Matrix Memory," *IRE Proceedings* (October 1953): 1408.

material caught the attention of American researchers, including An Wang, a student of Howard Aiken's at Harvard, who patented the core memory he engineered for the Harvard Mark IV computer.[4] A core memory unit arranges cores in a plane, with wires running vertically and horizontally through the hole in each core. Each core stores one bit of information, by magnetizing it one of two possible ways. The desired bit is selected by passing current along a unique combination of horizontal and vertical wires and is read from a shared output wire.

Core memory had advantages over delay lines, Williams tubes, and drum memory. The cores can be made small. They are *nonvolatile*, retaining information even when power is removed. Connecting multiple planes of core memory let computers read or write all the bits of a word simultaneously. Above all, core memory provides *random access*, taking a small and consistent amount of time to retrieve any word from memory. That differed from delay lines and drums, with which the computer had to wait for the requested data to cycle past the reading device. A large part of the programming effort for some early computers was devoted to positioning instruction and data to minimize this *latency*.

IBM and other manufacturers switched over to core memory during the mid-1950s. Magnetic core memories were retrofitted to ENIAC and MIT's Whirlwind computer in the summer of 1953. Whirlwind's memory, designed by Jay Forrester, used a three-dimensional array that improved on Wang's approach with faster switching speeds, greater storage density, and simpler electronics.[5] Its performance was so much better than the original tube memory that Whirlwind was almost a new machine. The first commercial delivery of a computer designed around core memory was around late 1954, when the ERA division of Remington Rand supplied an 1103 to the National Advisory Committee for Aeronautics (NACA).

Floating Point

Floating-point arithmetic greatly simplifies the planning and programming needed for calculations involving very large and very small numbers. It represents each number as two distinct things: a *significand* in a standard range (for example, between 0 and 10) and an *exponent* that codes the position of the decimal point. This is familiar to scientists and engineers as *scientific notation*, but almost unknown in the commercial world. For example, Avogadro's number, frequently needed in chemistry and physics, is rather large. It is more convenient to work with it as 6.0221×10^{23} than as 602,210,000,000 ,000,000,000,000. The full number is much too long to squeeze into a 32-bit memory location, whereas the five-digit significand and two-digit exponent fit easily. Floating point can also be used directly with binary numbers.

Floating point was not used at all with ENIAC, and its use was initially controversial. When discussing the forthcoming IAS computer, John von Neumann's team suggested that having to manually track the scaling needed for each variable through a computation would take "only a very small percentage of the total time" spent programming, making its claimed advantages "somewhat illusory."[6] They worried that "floating binary point represents an effort to render a thorough mathematical understanding of at least part of the problem unnecessary."[7]

Not everyone had von Neumann's feel for numerical mathematics. Scientific programmers quickly came to depend on floating point. Implementing it meant adding more code to programs or adding more hardware to the computer. Calling floating-point subroutines slows the machine down. Adding circuits to perform floating-point arithmetic allows faster, simpler programs but makes the processor more complicated and expensive. Some early electromechanical computers built by Konrad Zuse in Germany and by Bell Labs in the US had floating-point hardware, but ENIAC, Univac 1, and the IBM 701 did not. As computer engineering advanced, the balance tilted decisively in favor of floating-point hardware. In the words of one computer designer, this was the "biggest and perhaps only factor that separates a small computer from a large computer."[8]

Interrupts and Channels

IBM produced 140 of its model 704 systems. That machine's still more numerous descendants dominated the market for high-end scientific computers for a decade, from the mid-1950s to the mid-1960s. Each new model retained compatibility with the 704 and shared the same basic architecture, including a fixed word length of 36 bits.

Early computers wasted much of their incredibly expensive time waiting for data to arrive from peripherals. Magnetic tapes and drums supplied information much faster than punched cards, but not nearly quickly enough to keep processors busy. Programs that processed data from tape usually spent most of their time running loops of code to repeatedly check if the next chunk of data had arrived. Printers were even slower.

By the mid-1950s, computer manufacturers were adding a new hardware capability, called *interrupts*. Instead of having to repeatedly check whether an input or output operation had finished, the computer could get on with doing something useful. When the tape drive had finished reading new data or the printer was ready for the next line, it would ask the processor to interrupt the program and jump immediately to a subroutine able to handle the situation. The capability is usually traced back to custom modifications made to a Univac 1103 computer for what became NASA. IBM copied the idea for its 704.

One application of interrupts in the late-1950s, on systems such as the IBM 7070, was called SPOOLing: simultaneous peripheral operations on line. SPOOL let programs write their printed output to a tape file, so that the computer could move on to the next job. The name was probably a play on the spools of magnetic tape. Every time the printer was ready for more data it generated an interrupt to briefly switch control to the SPOOL program, which fetched data from the tape.

Other systems required more hardware. However, the 709, introduced in 1958, added an architectural innovation that has defined mainframe computing ever since: input and output *channels* for high data throughput. Its data synchronizer, essentially a small computer in its own right, controlled up to twenty tape drives, a printer, and a set of punched card equipment. Without interrupting the main processor, it could print a file, copy data from one tape to another, or load a file into memory. Users with extraordinary requirements could connect three data synchronizers to a single 709. Long after mainframes lost their edge over cheaper machines in raw processor power, they retained a decisive advantage in throughput on large, data-intensive jobs.

Smaller Drum-Based Computers

Not every engineering company or scientific laboratory could afford a million-dollar computer system like the Univac 1 or IBM's rival 701. Their huge price tags resulted from two unfamiliar technologies that took a great deal of engineering effort: high-speed electronic memory (Williams tubes in the 701 and delay lines in the Univac) and high-speed magnetic tape. Cheaper computers made do without those luxuries. Instead, they used magnetic drums.

In the late 1930s in what may have been the first attempt to build an electronic digital computer, John V. Atanasoff had the idea of using a rotating drum as temporary memory, storing data on 1600 capacitors, arrayed in 32 rows.[9] After World War II, the drum re-emerged as a reliable, rugged, inexpensive but slow memory device. To store data, the new device used magnetic recording techniques similar to those of computer tape, but the magnetic coating was applied to the outer surface of a rotating drum. Each drum storage unit had many read and write heads. Because of this and because drums rotated at high speed, a piece of information could be fetched much more rapidly from drum than from tape.

The leader in this effort was a Twin Cities, Minnesota, firm with its roots in a World War II codebreaking group. Communications Supplementary Activity—Washington (CSAW, pronounced "seesaw") had tackled projects for the Navy from the commandeered campus of a girls' school.[10] In early 1946 two members, Howard Engstrom and William Norris, incorporated Engineering Research Associates (ERA) to keep some of the group working together after the war. The Navy assigned ERA a number of "tasks." Most were highly classified and related to the business of breaking codes. ERA's engineers developed magnetic drum technology for these contracts, starting with a drum on which they had glued oxide-coated paper.[11]

An investor, John E. Parker, had provided space in a St. Paul factory recently used to manufacture wooden gliders for the Normandy invasions. It was cold and drafty, but ERA had little trouble finding and hiring capable engineers freshly minted from the region's engineering schools. Among them was a 1951 graduate of the University of Minnesota who went over to "the glider factory" because he heard there might be a job there. His name was Seymour R. Cray.[12]

Task 13, assigned in August 1947, was to build a programmable electronic computer, code-named Atlas, around the drum memory. In December 1951 it announced a commercial derivative, the Model 1101 (binary notation for the number 13).[13] The strategy required much more capital than its founders could provide, and like the Eckert-Mauchly Computer Corporation, it was purchased by Remington Rand.[14] Remington Rand could then offer two well-designed and capable computer systems: one optimized for science and engineering and the other for commercial use. Installations of its ERA-derived Univac 1103 began in the fall of 1953. That computer was a direct competitor to the IBM 701, equipped with four magnetic tape units and 1024 words of high-speed Williams tube memory in addition to the ERA drum.[15] Around twenty systems were built, most for military agencies or aerospace companies.

Finding a reliable memory was by far the hardest part of putting together a computer in the early 1950s. ERA's big hit was the drum, not the 1101 computer (see figure 2.2).

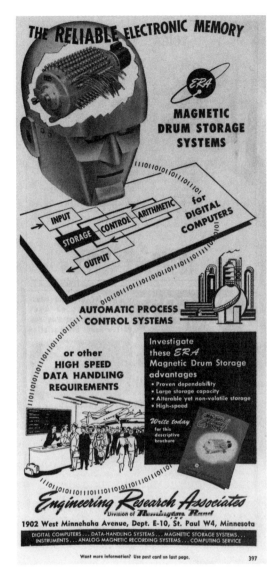

Figure 2.2

Advertisement for ERA Drum memory, 1953. Small electronics firms used ERA drums to enter the computer business. Note the simplified diagram of a von Neumann architecture, which was novel at the time.

Source: *Electronics* magazine, April 1953, p. 397. Courtesy Unisys.

By 1949, ERA was building drums ranging from 4.3 to 34 inches in diameter with capacities of up to 1 million bits, or 65,000 30-bit words. Access time ranged from 8 to 64 milliseconds.[16]

Start-up companies snapped up the drums, kick-starting a new segment of the computer industry by building cheap and simple machines around them. These included the Computer Research Corporation, the Electronic Computer Corporation, and Consolidated Engineering. Many of their computers represented a good value for the money, but there was no getting around the inherent slowness of the drum memory. Providing input/output facilities was another challenge. Attaching a Flexowriter was cheap, but slowed down many jobs. Obtaining punched card equipment improved performance, but meant that a significant portion of the sales cost would go to IBM. The start-ups won a few orders from military agencies and civilian firms, but not usually enough to provide the resources to create advanced follow-on designs. An established company seeking to enter the computer business typically bought out the struggling firm. The three small firms mentioned here were purchased by three office machine companies: National Cash Register, Underwood Corporation (typewriters), and Burroughs (adding machines).

In the mid-1950s a second wave of better-engineered drum computers appeared, and these sold in much larger quantities. The Librascope/General Precision LGP-30, delivered in 1956, had a repertoire of only sixteen instructions and looked like an oversized office desk. Centering the design on a 4096-word drum simplified the rest of the machine hugely. It needed only 113 vacuum tubes and 1,350 diodes, against the Univac's 5,400 tubes and 18,000 diodes. At $30,000 for a complete system, including a Flexowriter for input and output, it was also one of the cheapest early computers. More than 400 were sold.[17] It provided a practical choice for customers unable or unwilling to pay for a large computer.

The G-15, a drum-based computer designed by Harry Huskey and built by Bendix, was perhaps the only computer built in the United States to have been strongly influenced by Alan Turing's ACE design. Huskey had worked on the ACE in 1947 during a stay in England. He designed what became the G-15 while at Wayne State University in Detroit in 1953. The first deliveries were made in 1956 at a basic price of $45,000. Like other computers influenced by the ACE, the G-15 was difficult to program but very fast. Bendix sold more than 400 machines, but it could not come up with a successful follow-up model.[18]

These smaller computers, often overlooked, accounted for a clear majority of those installed during the 1950s. A July 1960 census of computers estimated that a total of 2,704 "medium-scale" computers such as the Bendix model had been installed in the US, greatly exceeding the 466 "large-scale" computers delivered (those costing more than half a million dollars, such as the IBM 700 series and Univac I). Almost half of them were IBM 650 computers, discussed in the next chapter. Small-scale computers (those costing less than $50,000, such as the LGP-30) were almost as common, with 1,583 deliveries.[19]

SCIENTIFIC PROGRAMMING TOOLS

We have already discussed the development of assemblers and other automatic programming tools. Scientific users drove the development and adoption of more ambitious programming tools, called compilers. Whereas assemblers made writing machine instructions faster and more convenient, compilers could translate mathematical equations and relatively easy-to-understand code written in *high-level languages* into code that the computer could execute.

Ease and speed of development were particularly important for scientific applications. Business applications, such as payroll programs, were usually written by full-time programmers working in the data processing department of the user company. Scientific programmers were more likely to be researchers or engineers who needed to apply a computer to one aspect of a larger project. They had less time to learn the quirks of assembly language. In addition, business data processing applications were usually run again and again. Many scientific programs were used for one temporary project, putting a premium on speed of development versus the operational efficiency achievable with carefully crafted assembly code.

Early Compilers

The first programs to be called "compilers" were written by the original Univac software team, led by Grace Hopper, who was recruited from Aiken's lab at Harvard (figure 2.3).[20] Grace Hopper defined *compiler* as "a program-making routine, which produces a specific program for a particular problem."[21] In those days the ideas of assembling, linking, and compiling code were not rigorously separated. Each term referred to the idea of knitting together program code and library subroutines to produce a single executable program. These subroutines packed code for procedures like computing sines, cosines, logs, and above all, floating-point arithmetic. Compilers nonetheless were complex pieces of software. Hopper called the whole activity of using compilers "automatic programming." Beginning in 1952, a compiler named A-0 was in operation on a Univac, followed in 1953 by A-1 and A-2. A version of A-2 was made available to Univac's customers by the end of that year.[22]

The first programming system to operate in the sense of a modern compiler was developed by J. H. Laning and N. Zierler for MIT's Whirlwind computer in the early

Figure 2.3

Grace Hopper, with students Donald Cropper, K. C. Krishnan, and Norman Rothberg, at a Univac I console, c. 1960. Photo courtesy Division of Medicine and Science, National Museum of American History, Smithsonian Institution.

1950s. Their manual, "A Program for Translation of Mathematical Equations for Whirlwind I," was distributed to about one hundred locations in January 1954.[23] It was not a general-purpose programming language, but a way of evaluating algebraic equations. Users entered equations in the familiar algebraic form. The compiler generated Whirlwind machine code to evaluate the equations, automatically keeping track of storage locations. Donald Knuth, who surveyed early programming systems in 1980, observed that the Laning & Zierler system was slower by a factor of ten than other coding systems for Whirlwind.[24] Closing that gap between automatic compilers and hand coding was necessary to win acceptance of compiler systems.

Fortran

The programming language Fortran (from "formula translation"—the early preferred spelling was all capitals) was introduced by IBM for the 704 computer in early 1957. It was a success among IBM customers from the beginning and much modified and extended for many decades after. John Backus, who led the project, was inspired by the work of Laning and Zierler, which he called "an elegant concept elegantly realized." Backus also remarked that their system had been little used, in his opinion because it threatened programmers, who took a perverse pride in their ability to use tricks that few others could fathom.[25]

Backus was determined to produce a system so useful and efficient that nobody could ignore it. He later wrote that Fortran was inspired not by "some brainstorm about the beauty of programming in mathematical notation" but with a "basic problem of economics: programming and debugging costs already exceeded the cost of running a program." His "unusually lazy nature," coupled with "experience with the drudgery of coding," made him look for an easier way.[26] Scientific users proved ready to embrace a system that hid the details of a machine's inner workings, leaving them free to concentrate on solving their own, not the machine's, problems.

Many factors contributed to the success of Fortran. One was that its syntax—the choice of symbols and the rules for using them—was close as possible to that of algebra, given the difficulty of indicating superscripts or subscripts on punched cards. Engineers liked its familiarity; they also liked the clear, concise, and easy-to-read user's manual. Perhaps the most important factor was performance. The Fortran compiler generated machine code that was as efficient and fast as code written by human beings. Backus noted that the 704's floating-point hardware helped this by simplifying the compiler's work.

The word "language" turned out to be a dangerous term. The English word is derived from the French *langue*, meaning tongue, implying that it is spoken. But computer languages are *not* spoken but written, according to a rigidly defined and precise syntax.[27] Hopper once produced a version of the programming language FLOW-MATIC in which all the English terms, such as *Input*, *Write*, and so on, were replaced with their French equivalents. When she showed this to a Univac executive, she was summarily thrown out of his office. Later she realized that the very notion of a computer was threatening to this executive; to have it "speaking" French—a language he did not speak—was too much.[28]

SHARE and the Operating System

In 1955 a group of IBM 701 users in the Los Angeles area, faced with the daunting prospect of upgrading their installations to the new IBM 704, decided that sharing experiences was better than going it alone. Despite working for direct competitors, such as North American Aviation and Lockheed, they had already worked together informally to develop a common assembler for the 701, called PACT. That August, meeting on the neutral territory of the RAND Corporation in Santa Monica, they resolved to create a formal user group for the IBM 704, called SHARE.[29]

Within a year, the membership—all customers of large IBM systems—had grown to sixty-two members. IBM had used its own staff to help tabulator customers apply their equipment to business problems, but supporting scientific computer users was much harder. By letting users take this work on themselves, SHARE hastened the acceptance of IBM's computers. IBM backed up SHARE with administrative support and programming assistance. As SHARE grew, it began to express strong opinions about future directions for IBM computers and software. IBM had to listen, though it did not always obey.

SHARE soon developed an impressive library of routines that each member could use, many of them for mathematical tasks such as inverting matrices. The working practices adopted by SHARE had much in common with later open source projects. There were mechanisms for distributing code and documents, bug reporting procedures, and ad hoc groups of programmers from different companies who pooled their efforts to develop particular systems. Sharing code also meant standardizing configurations and working methods, from the switch settings on the equipment to the conventions used to call subroutines.

SHARE was particularly important in shaping the growth of what was eventually called operating system software. The most innovative early work was done by users, whose systems gradually grew more complicated. Computers of the mid-1950s did not have operating systems in the modern sense. Human operators would load tapes into drives and put input cards into hoppers. They would then set switches to tell the computer where to load its program code from and to start it running. The application code controlled the hardware directly with no intervening levels of software.

The first utility programs to remain in memory while application programs ran were called *monitors*. Memory was scarce, so these programs had to be tiny. Thanks to interrupts, monitors could be used, as their name suggests, to keep tabs on the progress of a program during execution and provide useful debugging information. For example, the FORTRAN Monitor System developed by SHARE in 1959 helped to automate the process of pulling together Fortran and assembly code from different files, compiling programs, and executing them.[30] It was built on an earlier Load and Go system produced by North American Aviation.

When one program finished execution, the monitor triggered code to load the next, without waiting for the human operator to react. One early system designed at the General Motors (GM) Research Laboratories beginning in 1956 was especially significant in establishing what came to be known as *batch* processing.[31] A simple but key element of these systems was the use of special control cards. These codes told the computer that the cards that followed were a Fortran program or were data, or that a new job was starting, or that some other operation followed.

When the IBM 709 arrived in 1956, SHARE launched the SOS (defined variously as standing for Share Operating System and SHARE 709 System) project to develop an ambitious successor to the GM system. SOS aimed to automate much of the work carried out by operators. For that reason, it was the first piece of software to be called an *operating system*. As well as a monitor program and powerful assembler, SOS provided batch control, output buffering, and a large number of input and output subroutines packaged as macro instructions. Using macros meant that changes in computer configuration—for example, to load data from a different tape drive—could be made from one program in a batch to the next without the operator having to pause to adjust console switches.[32]

Getting all the SOS routines to work together was much harder than anticipated. That proved a harbinger of later problems. As operating systems grew more ambitious, they tended to consume ever more memory and processor capacity until there was little left for running the programs the computer was installed to run. Development dragged on from 1956 to mid-1958, with IBM taking on more of the programming work so that it would have something to offer purchasers of its flagship system. When SOS was finally completed, it ran slowly and its tight integration with assembly language limited its usefulness as users shifted to Fortran programming.[33]

When it came time to repeat the process for the 709's successor, IBM assumed primary responsibility for planning and designing the operating system as well as programming it. It built on the SHARE routines to create a suite called IBSYS, which included a batch processing system designed to integrate with Fortran. That system evolved into IBM's Job Control Language (JCL). Many a novice programmer has a vivid memory of JCL cards, with their distinctive double slash (//) or slash-asterisk (/*) punched into special fields. Many also remember the confusion that resulted if a missing card caused the computer to read a program deck as data, or vice versa.

Mathematical Software

High performance scientific computing depended on software as well as hardware. By the 1960s Fortran programming was an increasingly vital skill for scientists and engineers. Computer analysis and simulation underpinned scientific breakthroughs in X-ray crystallography, particle physics, and radio astronomy. Engineers used computers to simulate supersonic airflow, model the stresses on bridges and buildings, and design more efficient engines. To teach programming to large numbers of students, it was essential to find ways to compile their code efficiently, both to get results back to students faster and to get more work done with an expensive computer. Conventional compilers used a lot of the computer's memory and time to generate efficient code that executed quickly. In a teaching environment programs were short and typically did not handle large quantities of data, so it was more important that the program compile quickly than execute quickly. Students usually made syntactic mistakes that prevented their programs from compiling, so it was vital for compilers to give detailed and helpful error reports.

The most innovative work in adapting batch computing to teaching took place at the University of Waterloo in Ontario, Canada. Waterloo's Computer Science Department was among the oldest (founded 1962), and under the leadership of J. Wesley Graham it was teaching computing to more undergraduates than any other school in Canada. The department produced a Fortran compiler called Watfor (Waterloo Fortran), based on a similar compiler developed at the University of Wisconsin in the early 1960s.

> WATFOR was written by four third-year Math students in the summer of 1965. It was a fast in-core compiler with good error diagnostics, which proved especially useful to students for debugging their programs, as well as speeding up execution.[34]

The compiler processed six thousand jobs an hour and by Graham's estimate reduced the cost of running a student program on Waterloo's IBM mainframe from ten dollars to ten cents.[35] Watfor and its successors were widely used at other universities, and Graham's Fortran programming textbook influenced a generation of students.[36]

Another challenge was ensuring the mathematical accuracy of the results obtained. Scientific computing was often called *number crunching*, because real-world problems usually require numerical approximation. Even a single class of equation, such as ordinary differential equations, could be tackled using many different numerical methods.

The spread of computers brought new energy to applied mathematics, as the field of *numerical analysis* developed at the intersection of scientific computing and mathematics. The 1950s and 1960s were a fruitful time in the development of new methods for common tasks, such as calculation of matrix eigenvalues (characteristic values of matrices needed for many scientific computations). But scientists and engineers, for whom applied mathematics was a means to an end, tended to reach for their old textbooks to implement methods designed for the days of hand calculation. These might be inefficient for machine use. In the worst case, being millions of times faster than hand calculation meant that computers could compound errors millions of times faster.

From the late 1960s, this problem was met with concerned efforts to produce reliable and efficient implementations of the modern mathematical methods and package them as software usable by ordinary scientific programmers. One important early project was Eispack for the calculation of matrix eigenvalues, released by Argonne National Laboratory in 1973. It was implemented in Fortran algorithms originally developed in Algol by James Wilkinson, who won a Turing award for his contributions to numerical analysis. Eispack and its sequel Linpack, for linear algebra, were distributed as subroutines ready to be integrated into scientific applications.

One challenge was getting the routines to run effectively on different computer architectures. Fortran was a reasonably consistent language. But being able to compile code on two machines did not guarantee that they would produce the same results. In particular, manufacturers made their own decisions about how to handle the representation, rounding, and truncation of floating-point numbers. Understanding the foibles of particular architectures was also essential to maximizing performance. Linpack dealt with this by segregating machine-specific code in separate libraries, called BLAS (Basic Linear Algebra Subprograms). Jack Dongarra, one of Linpack's creators, still uses the package to benchmark supercomputer performance for the widely referenced TOP500 list of the most powerful supercomputers.[37]

Algol

Algol (for algorithmic language) was a hugely influential programming language. The first version was defined in 1958 following an international gathering of computer experts in Zurich. An improved version was agreed in 1960. Algol 60 failed to displace Fortran for mainstream scientific programming, though it won many loyal adherents. Instead its greatest influence was on computing research. Algol was the seed around which computer science began to crystalize as an academic discipline. A remarkable number of those who worked on Algol went on to become celebrated computer scientists. To name only those who eventually won the ACM's Turing Award, we note that Peter Naur, John Backus, John McCarthy, and Alan Perlis were among the designers of the Algol 60 specifications and that C. A. R. Hoare, Edsger Dijkstra, and Donald Knuth produced compilers based on it. Many other winners carried out research inspired by aspects of Algol.

Fortran had been released first as an IBM package. The language it processed was later copied by other computer manufacturers for their own machines. In contrast, Algol began as an elegant specification, providing a formal definition of its syntax, rather than as a piece of code. Learning from his experience designing Fortran, Backus created the Backus normal form (BNF) notation to use in defining Algol. This is still used to describe computer language grammar.[38]

Algol introduced the block structure, taken up later by languages such as C, Pascal, and Java. Tags of *begin* and *end* marked out chunks of code to be treated as single instructions, which could be looped with the *for* statement or conditionally executed using *if . . . then* structures.

The Algol specification was full of hard-to-implement features. The toughest involved procedures, Algol's block-based refinement of the subroutine. Procedures could be nested inside procedures. Variables declared within procedures, or within other blocks, were inaccessible to the rest of the program and their space was automatically reclaimed. Most radically of all, Algol supported recursion. That meant that procedures could call themselves. Variables could be passed to procedures by copying their current values or by *name* (later termed *call by reference*), which challenged compiler writers to reevaluate complex expressions each time a procedure was called.

Stacks

Dijkstra, then a scientific programmer in Amsterdam with a background in physics and mathematics, was the first to conquer the challenge of Algol implementation. He did it by generalizing the *stack* data structure into a fundamental feature of computer hardware and software. In a stack, the item most recently stored is the first to be retrieved—as in a stack of papers accumulating on a desk. Calling an Algol procedure deposited new information on top of the stack, including space to hold its local variables and other data needed to evaluate its parameters. When the procedure finished, this stored information was removed from the stack, automatically reclaiming memory. The mechanism worked perfectly even if one subroutine called a second, which called a third, and so on. With each call, the stack got higher. A badly designed program, however, in which subroutines called each other endlessly, would produce an ominous pause followed by a "stack overflow" error. Stacks did double duty, useful for compiling programs into machine instructions as well as managing memory when executing them.

The large IBM computers discussed so far evolved from the IBM 704 design, created long before high-level languages were widely used. Like most other computers designed from the 1950s through the 1970s, they were optimized for instructions written in assembly language. Implementing stacks on a conventional computer architecture, like those of the 1950s IBM machines, was clumsy because of the work needed to map the virtual stack onto the actual architecture of addressable memory, registers, and accumulators. This meant that calling subroutines or switching between programs during multiprogramming slowed the computer significantly.

In 1961 the Burroughs Corporation, still best known as a supplier of adding machines despite some success selling small computers, began to work on its first mainframe design. Seeing that high-level languages were the future, it designed what became the Burroughs 5000 computer around Algol, to produce the first architecture created with high-level languages in mind. Uniquely, the machine did not come with an assembler package. Even the operating system was written in a specially extended version of Algol. All variable storage centered on the stack—the top two levels of which were effectively registers and could be acted on by the processor. This meant that subroutine calls could be handled very simply and rapidly. Compiling Algol code was equally efficient, as the actual architecture of the machine matched its internal model.

Despite its novel features, the 5000 struggled in the market place. Its successor, the 5500, was only a modest success. The machines had reliability problems, and like other firms, Burroughs struggled against IBM's market dominance. Burroughs nevertheless stuck with the stack approach into the 1980s for its 6000 and 7000 mainframes. The Burroughs machines won a cult following among programmers who used them. Burroughs was never a major factor in the computer industry, but its mainframes disseminated important ideas in computer architecture and software technology.

TOWARD THE SUPERCOMPUTER

The introduction of the transistor as a replacement for the vacuum tube mirrors the story of core memory. Transistors could serve the same roles as vacuum tubes in digital logic circuits but were smaller, more reliable, and initially more expensive. They used less power and could be driven faster. Bell Laboratories, where the transistor was invented, was not among those considering entering the commercial computer market in the 1950s. As a regulated monopoly, its parent company AT&T weighed every action it took with an eye on the federal courts. In early 1956, after seven years of litigation by the US Justice Department, it signed a consent decree, agreeing not to enter into any business "other than the furnishing of common carrier communications."[39] In part to satisfy federal regulators, Bell Labs made information about transistors available at a nominal cost.

Philco, an electronics firm headquartered in Philadelphia, pioneered the "surface barrier" transistor, which could be made in quantity and performed well. In June 1955 it contracted with the National Security Agency to produce a fast computer based on the architecture of the Univac 1103 (the successor to the 1101). The result, called SOLO, was probably the first general-purpose transistorized computer to operate in the US. Philco marketed a commercial version called the Transac S-1000, followed quickly by an upgraded S-2000 in 1958. The first deliveries of the S-2000 were made in January 1960. These, along with deliveries by Univac of a smaller computer called the Solid State 80, mark the beginning of the transistor era, or Second Generation of computers.[40] IBM was soon to follow, and within a few years the market for bulky, power-hungry, and unreliable vacuum tube computers had essentially vanished.

IBM 7090

According to folklore, IBM discovered this when it submitted a bid to the US Air Force to supply computers for the Ballistic Missile Early Warning System (BMEWS) around the Arctic Circle. At the time IBM had just announced the 709, but the Air Force insisted on transistorized machines. IBM responded by designing a transistorized computer that was architecturally identical to the 709. To speed development it programmed a 709 to simulate the operation of its successor. That let IBM develop and test the software that the new computer would need, while measuring its performance. IBM delivered a computer to a site in Greenland in late 1959, but "IBM watchers" claimed that the company also had to dispatch a large cadre of engineers to finish the machine while it was being installed.

Whether or not that story is true, the company did produce a transistorized computer and market it as the Model 7090, just a year after deliveries of 709 had begun. The 7090 is regarded as the classic mainframe, for its combination of architecture and performance, and for its success: hundreds of machines were installed at an equivalent price of over $3 million each.

The term mainframe for the central processor probably arose because its circuits were mounted on hinged metal frames housed in the cabinets. The frames swung out for maintenance. A typical installation consisted of a cluster of these cabinets standing on a tile false floor raised a few inches to leave room for the numerous thick connecting cables that snaked from one cabinet to another. The room had its own climate control system that, like the ENIAC room at the Ballistics Research Laboratory, made it the most comfortable place to be on a hot summer day.

The operator's console was festooned with an impressive array of blinking lights, dials, gauges, and switches. It looked like what people thought a computer should look like. A nearby cabinet housed the main processor circuits. These were made up of transistors, mounted and soldered along with resistors, diodes, jumper wires, inductors, and capacitors, onto printed circuit boards. The density of these circuits was about ten components per cubic inch. The boards plugged into a *backplane*, where a web of wires carried signals between them. Some mainframes were laboriously wired by hand, but most used a technique called wire wrap; it required no soldering and could be done by a machine, reducing errors. In practice, there would always be occasional pieces of jumper wire, soldered by hand, to fix wiring errors.

Magnetic tape drives dominated a mainframe installation. These tapes were the medium that connected a mainframe computer to the outside world. Programs and data were fed into the computer through tapes; the results of a job were likewise sent to a tape. If a program ran successfully, an operator took the tape and moved it to the drive connected to a 1401 computer, which handled the slower process of printing out results on a chain printer. Results were printed in all capital letters on 15-inch-wide fan-folded paper.

Although marketed as a machine for science and engineering, many customers found the 7094, shown in figure 2.4, well suited for administrative tasks. It could carry out about 50 to 100 thousand floating-point operations per second, making it among the fastest of its day—about as fast as a personal computer in the late 1980s. Its long 36-bit word length was well suited to high-precision scientific calculation and had the further advantage of allowing the processor to address a large memory. It was delivered with up to 32,768 words of core memory. In modern terms, that corresponds to about 150 KB.

The 7094 had seven index registers, extending the architecture inherited from the 704 and 709. However, it lacked a feature that was appearing on some competing machines: general-purpose registers. These arrived in 1956 when the British firm Ferranti, Ltd., announced a machine called Pegasus which had eight registers, seven of which could be used flexibly as accumulators or as index registers. Storing frequently accessed variables in registers provided an appreciable boost to overall performance by eliminating the need to transfer them repeatedly to and from main memory. Other companies slowly copied the idea. By the end of the 1960s it became the standard approach.[41]

Figure 2.4
The transistorized IBM 7094 was the most versatile of IBM's early mainframes. Note the four additional index register readouts, mounted on top of the standard IBM 7090 console.
Source: IBM.

A 7094 installation rented for about $30,000 a month, equivalent to a purchase price of about $1.6 million. With that cost it was imperative that the machine never be left idle. Programs were gathered onto reels of tape and run in batches. Most programmers never touched the machine that ran their programs. They wrote programs in pencil on special coding sheets, which they gave to keypunch operators who typed the code onto cards. The deck of punch cards holding the source code was read by a small IBM 1401 computer and transferred to a reel of tape. The operator took this tape and mounted it on a tape drive connected to the mainframe. The programmer had to wait until a batch was run to get her results. Usually these indicated a mistake or need to further refine the problem. She submitted a new deck and endured another long wait. That method of operation was a defining characteristic of the mainframe era.

Over the course of the 7094's lifespan, many of the operator's jobs were taken over by improvements to its operating system. In normal use the operator mounted and dismounted tapes and occasionally inserted decks of cards into a reader. A few mainframes had a video console, but there was none on the 7094's main control panel. Instead, a row of small lights indicated the status of each bit of each register. The operator could execute a program one step at a time, observing the contents of the registers at each step and, if desired, altering their contents by flipping switches.

Computing at NASA Ames

The government's Ames Research Center, located in Mountain View, California, had been a center for high-speed aerodynamics research since its founding in 1940. In 1955 Ames acquired its first stored-program electronic computer, an IBM 650. In 1958, shortly after the Center became part of the newly founded National Aeronautics and Space Administration (NASA), it acquired an IBM 704, which was replaced in 1961 by an IBM 7090.[42] They were used to calculate things like satellite trajectories, heat transfer, and particle physics.

Throughout the 1960s, the demands on the central IBM installation doubled each year. Meeting that demand was a never-ending headache for the Ames Computation Division. Beginning in 1963, the 7090 was upgraded to a complex of machines called a Direct Couple System. At its heart was an IBM 7094, acquired in July. To maximize its performance, an IBM 7040—itself a large mainframe—was connected to it to handle input and output, putting less strain on the 7094. Reels of tape from the 7040 were transferred in turn to an IBM 1401 connected to a keypunch and printer (figure 2.5). As we show in the next chapter, the 1401 could also operate as a stand-alone computer, and NASA used it to handle the center's administrative work, such as budgeting.[43] An IBM 7740 communications computer handled a connection to several remote terminals.

Figure 2.5
An IBM 7090 Direct Couple System in use at North American Aviation, c. 1962. Note the rows of tape drives—twenty-one visible in this photo. Magnetic tape was the primary mass storage medium for second-generation mainframe computers. Photo by Robert W. Kelley/The LIFE Picture Collection via Getty Images.

This direct couple system served NASA through the dramatic years of the Space Race. By 1968 it was working around the clock (except for weekends), with about 27 hours a month reserved for maintenance. Monthly costs were in the range of around $35,000.

IBM's presence at the Center was mainly in these large computing systems used for general scientific work. Smaller computers came from Honeywell, Digital Equipment Corporation, Scientific Data Systems (SDS), and EAI. They were used for specialized tasks such as controlling experiments, operating a flight simulator, and reducing wind-tunnel data. We discuss more about computers of this kind in subsequent sections.

IBM's Stretch

Even the giant 7090 series mainframes were not the largest scientific computers of the era. Back in 1955, IBM had lost out to Univac on a contract to build a high-performance computer for what became the Lawrence Livermore National Laboratory. Determined not to lose its lead in scientific computing, IBM offered to build for its customers at Los Alamos a machine, nicknamed Stretch, that would provide one hundred times the performance of the lab's current IBM 704 system.

Stretch used transistors rather than vacuum tubes, which delivered a significant increase in performance. The system of small, modular circuit boards that IBM developed for Stretch was later used with great success for other transistorized machines. This made servicing easier and simplified computer design and manufacturing, as machines were assembled from a relatively small number of standard units. Transistorization by itself would not come close to providing the performance boost needed. Stretch's head designer Stephen Dunwell knew that meeting this goal would mean rethinking every aspect of the computer. Some of the innovations his team came up with eventually became standard features of all processors.

One of their tactics is now known as instruction pipelining. Executing a single instruction involves several steps. The Stretch team identified three distinct stages: fetching the bits encoding the instruction from main memory, decoding the control information to set the processor circuits accordingly, and then executing the instruction.

Conventional computers performed these steps in sequence, which was wasteful. For example, the hardware to decode an instruction was sitting idle during the fetch stage. While it executed an instruction, Stretch would use other parts of its processor to fetch and partially decode the next five instructions.[44] That promised a major boost in performance, but at the price of a significant increase in complexity. For example, when the computer processed a jump instruction, work already done to fetch and decode the next instructions from memory would be wasted unless the pipeline had anticipated the upcoming jump. When the jump was conditional (the machine language equivalent of *if . . . then*), rules to figure out which branch was most likely to be executed had to be built into the processor.

Stretch was supplied with the biggest, fastest hard disk drives that IBM could engineer. However, any computer that could calculate so rapidly was still subject to spending most of its time waiting to read or write data. To alleviate this problem, Stretch was the first IBM computer to support multiprogramming. Stretch's enormous core memory of one megabyte was big enough to hold several large programs at once. When a program found itself waiting for a peripheral, Stretch gave a turn to another program. To prevent programs from accidentally overwriting each other, Stretch added memory protection so that each program could work only on its designated block of memory.

To pack data more efficiently into memory, Stretch adopted a flexible word length of anything between 1 and 64 bits. Stretch engineer Werner Buchholz introduced the idea of the *byte*—the smallest unit of information a computer could retrieve and process.[45] Bytes too were originally variable length, up to 8 bits. On later machines, however, IBM standardized the byte at 8 bits. Combining several bytes when needed let more computers manipulate words of 16, 24, 32, or 64 bits. This approach was eventually adopted throughout the computer industry.

In 1960, IBM publicly announced Stretch as the model 7030, priced it at $13.5 million, and promised that it would give sixty times the performance of an IBM 704. But when the first model was tested in 1961, its various new features didn't work together as effectively as anticipated. Stretch was the world's fastest computer by a comfortable margin but provided only about half the promised performance. In response, IBM's leader, Thomas Watson Jr., halved its price. As this would not come close to covering manufacturing costs, he announced that no orders would be taken beyond the eight already placed. Stretch seemed a humiliating failure.[46] Over the next five years, however, innovations from Stretch found their way into the core of IBM's computer line—particularly with the launch of the System/360 line in 1964 (discussed in the next chapter). The cloud of failure hanging over Stretch eventually lifted and today it is remembered as perhaps the most influential computer design since EDVAC.

Most Stretch machines went to the heaviest users of computer simulation: nuclear labs and the US Weather Bureau. One was delivered to the National Security Agency as the heart of a custom system called HARVEST, along with a special data stream coprocessor and an automated system able to load six tapes simultaneously from a library of several hundred. HARVEST remained in operation until 1976, scanning messages for code words and carrying out codebreaking and decryption work.

Virtual Memory and the Atlas

Stretch was not the only source of new architectural thinking. Another influential feature, virtual memory, was pioneered by the Atlas. Designed at Manchester University and built by Ferranti, it entered operation in 1962 as one of the fastest and most influential computers of the era.

Virtual memory is a way to make a computer's fast main memory seem bigger than it is, by swapping data with a slower but larger storage medium such as a disk. A user of the Atlas saw a machine with a virtual memory of one million, 48-bit words. Special hardware translated requests to read or write from these virtual addresses into actions on the machine's much smaller physical memory, a capability known as dynamic address translation.[47]

When there was not enough physical memory to hold all the pages, it looked for pages that were not currently in use to *swap* from core memory out to drum memory. Optimizing this process was crucial. Whenever a program needed to work with a page of memory that had been swapped out, Atlas would have to copy it back from drum memory, which slowed things down enormously. Atlas was designed for multiprogramming, so that it could get on with another program while the page loaded.

The Control Data 6600

When IBM announced Stretch in 1960, it created a new market for what were soon called supercomputers. When it withdrew Stretch the next year, it abandoned that market. Its place was taken by a new company, Control Data (CDC), founded by former members of ERA. Control Data's first model, the 1604, was delivered to the US Navy in 1960. Seymour Cray designed it for speed, as a transistorized sequel to the ERA machines. Until Stretch was finished it was the world's fastest machine. Like ERA before it, CDC's early machines were popular for cryptographic applications.

As Control Data grew, Cray set up an engineering group in Chippewa Falls, Wisconsin, to work without distraction on a new machine that would outperform Stretch. The CDC 6600, announced in 1964, met that goal. Whereas Stretch had pushed performance to new levels by introducing architectural complexity, Cray pared his design down to the minimum. According to Bill Buzbee, former leader of the Computer Research and Applications group at Los Alamos, "For Seymour, throughout his career, simplicity was the essence of his design . . . the [CDC] 6600 only had 64 instructions. The [IBM] 7094 had 300 or so instructions." As a result, Cray's machines were "easy to get your brain around."[48] The CDC 6600 handled the same kinds of work as Stretch, for the same kinds of customer, but ran three times faster. CDC sold more than a hundred systems, and successor machines based on its design were an important part of its product line into the 1980s.

The 6600 did arithmetic very rapidly and most other things not at all. Its ten arithmetic units could work in parallel, for example, by carrying out an addition operation without waiting for its two floating-point multipliers to finish up their work on a previous instruction. Many scientific jobs centered on a small loop of mathematical code run again and again. Cray optimized for this by storing the previous eight instructions in decoded form, to save time if needed again. Tapes, disks, and printers were handled by ten peripheral processors, each a simple computer with its own memory, so that the main processor did not need interrupts or input/output capabilities.

Keeping instructions simple and using high-speed silicon transistors let the main processor run at a 10 MHz clock speed, several times faster than Stretch. This meant that the time it took a signal to traverse a wire was becoming a serious constraint, forcing a design in which components were packed much tighter. That density created serious heat problems, which were solved by threading Freon cooling pipes through the machine.

Under the leadership of William Norris, CDC's supercomputer sales underwrote a program of rapid diversification. Control Data built a healthy business manufacturing tape drives and printers for other computers. In 1968, with its stock riding high, it acquired the much larger finance company Commercial Credit. When IBM used aggressive sales tactics to overhype its own forthcoming supercomputer, CDC launched a long and expensive lawsuit. Unlike most of the companies that brought suit against IBM, Control Data achieved a favorable settlement. That resulted in IBM's transferring its own Service Bureau to CDC in 1973.[49] While other computer companies struggled against IBM, CDC was growing increasingly powerful.

The Cray 1

Seymour Cray didn't care about CDC's plans for diversification. He just wanted to build the world's fastest computers, without worrying about compatibility with older or smaller models. By 1971 CDC management was imposing cuts on his lab, which was struggling to create a machine called the 8600. When Cray decided that continuing to evolve the 6600 architecture wouldn't work, Norris refused to let him start over with a clean sheet design. Cray left Control Data in 1972, but not Chippewa Falls, where he founded a new company: Cray Research. As the world's most celebrated computer designer he had little trouble raising venture capital. Control Data itself invested $250,000 and Cray took six members of his old lab with him.

The first Cray 1 was shipped to Los Alamos, which had also been the biggest single user of ENIAC in the 1940s and the first customer for Stretch and the CDC 6600. Nuclear simulation, like weather forecasting, was a well-funded area that could soak up every available processor cycle on the world's fastest machine and still leave programmers begging for more.

The Cray 1 weighed more than five tons but it was a compact unit, designed as a low ring of cooling equipment around a monolith-like tower six feet tall (figure 2.6). The circular design shortened wire paths, reducing the time needed for signals to travel to increase the computer's maximum speed. Upholstery on top of the ring positioned the Cray 1 for double duty as the "world's most expensive love seat," although only authorized personnel would ever have sat on something that valuable. Unlike the 6600, the Cray 1 could handle its own input and output. It still had to be hooked up to a smaller computer called the maintenance control unit, which loaded the operating system and monitored progress on jobs. Early machines arrived without any operating

system, leaving laboratories to write their own. When it did arrive, the official Cray operating system was relatively simple as the Cray usually ran a single enormous job for days (or longer) at a time.

Cray designed his new machine with 64-bit words. From a computer architecture viewpoint, the biggest change from the 6600 was the incorporation of *vector processing* capabilities, another way of optimizing the performance of inner loops in scientific applications. These loops repeated the same few operations millions of times, each time processing data fetched from a different memory location. A conventional instruction works on a single word of memory. Cray's new vector instructions applied the same operation to each of the 64 words held in one of the computer's eight vector registers. The operation wasn't applied simultaneously to each word—even Cray couldn't squeeze that many arithmetic units into a computer—but the instruction itself had to be fetched and decoded only once. Data could be shuffled between vector registers quickly, which gave further performance boosts when vector operations were batched together. Vector approaches had been tried without huge success on other machines, including a rival CDC model called the Star-100. Cray stripped the idea down to deliver exactly the capabilities needed by scientific users without adding unnecessary complexity.

The Cray 1 was introduced in 1976 and repeated the success of the 6600, selling about eighty copies for approximately eight million dollars each. It was the new standard for nuclear weapons labs, aircraft designers, and weather forecasters. The benefits were undeniable, but exploiting the Cray's power meant going back to writing programs in assembly language, which according to Dongarra, who spent time with the first Cray 1 at Los Alamos, was "rather tedious to write, hard to debug and was only for that one architecture." Adapting to vector processing took years of work for the writers of compilers, mathematical software libraries, and scientific software applications. It took more than a decade to redesign LINPACK with higher level BLAS routines able to deliver full vector performance from applications written in portable Fortran. As Dongarra explained, "Getting the compiler to recognize what you're doing and express it in a way that is efficient on the architecture is a constant struggle."[50]

The Cray 1 remained the world's fastest computer until 1982, when it was succeeded by the Cray X-MP. The MP stood for *multiple processor*. By then, scientific computer users were switching to interactive operating systems and the new modes of computing that we explore in subsequent chapters. So, we halt here with the Cray 1, which had the same single processor architecture and mode of operation as the IBM 704 family, Stretch, and the CDC 6600. All those machines continued the mission of ENIAC: delivering the maximum possible performance when crunching numbers for huge scientific jobs that took hours or days to complete. The Cray was, however, separated from ENIAC by thirty years of rapid improvements in performance, reliability, and ease of use.

Although building a hundred copies of one of those exotic machines would make it a spectacular success, many of the tricks invented by their designers eventually became standard features of ordinary computers. Other key features of modern computers were developed in response to the needs of other kinds of users in business administration, education, and the military. In each of the next three chapters, we rewind the clock from the 1970s back to the early 1950s to look at other strands in the story of the modern computer.

3 THE COMPUTER BECOMES A DATA PROCESSING DEVICE

It wasn't just scientific laboratories that decided to build their own computers rather than wait for the arrival of commercially produced models. One remarkable, and remarkably impatient, company did the same thing. J. Lyons and Company ran a chain of British tea shops and a catering and food distribution business. It had long been a proponent of what in Britain was called *organization and methods* work. That meant empowering experts to study and systematically redesign business procedures, forms, and other administrative systems to optimize the overall efficiency of an organization. Lyons focused on what would later be called *operations research*: the application of statistics and mathematical models to improve the efficiency of its distribution and production networks.[1]

This existing commitment gave its managers an unusual interest in any technology that might automate administrative processes. In 1947 two managers on a trip to the US to look for the latest techniques met Herman Goldstine, then working with von Neumann on the IAS computer. After Goldstine told them about the EDSAC project underway in Cambridge, Lyons commissioned a closely related computer for its own use. LEO ran its first business application in November 1951. It was soon handling payroll, cost accounting, inventory management, and production scheduling.

THE FIRST ADMINISTRATIVE COMPUTERS

We know of no other companies so keen to automate their administrative work that they built their own computers. American businesses were remarkably rapid and aggressive in adopting computer technology, motivated not just by economic rationality but by faith in an automated future in which computers processed every business transaction and provided managers with the data needed for every business decision. Business data processing was a largely separate field from scientific computing, with its own journals, associations, and professional identities. In the computer's march toward universality this was the second major community of computing to develop. From the late 1950s to the 1980s it was unquestionably the largest.

In the United States the administrative use of computers began a few years later in 1954, with an event we have already mentioned: a payroll job run for General Electric's appliance factory in Louisville, Kentucky. General Electric was the first company to order a Univac for administrative use and took delivery of the eighth machine produced. Returning to the United States that year, the LEO team found companies with large budgets and impressive amounts of hardware. Thousands of computers were installed, or scheduled for delivery, by the mid-1950s—long before the economic benefits of computerization were clear or its challenges properly understood. The orders were driven by a wave of enthusiasm for the revolutionary power of computer technology, which began with Edmund Berkeley's 1949 promise in *Giant Brains* that computers would "take a load off men's minds as great as the load that printing took off men's writing."[2] Experts and consultants eager to promote their own careers allied with computer manufacturers to push the vague but compelling idea that computers were about to transform business management. One enthusiast for the new technology warned that any company that decided to take a wait and see approach was likely to be crushed by its competitors. "The ominous rumble you sense" was, he told his readers, "the future coming at us."[3]

When a company placed its order for a piece of the future, what showed up a year or two later was something more tangible: a room full of computer equipment. The LEO report noted that "despite a prodigious expenditure in time and money in the USA the amount of use of automatic computers on office work is negligible . . . Nowhere did we see a computer that was doing jobs to a regular planned schedule." Companies were slow to adapt to the potential of the new technology, instead trying to transfer existing punched card procedures directly to computer.[4]

That did not stop the surge of computer installations: clerical automation soon overtook scientific calculation as the biggest market for computer technology. Computers designed specifically for administrative use ran from relatively affordable machines intended for use alongside existing punched card installations to multimillion dollar computers built to handle the core administrative tasks of large corporations. Data processing spurred developments including the adoption of disk drives, the development of the IBM's dominant System/360 mainframe architecture, and the emergence of software as an industry.

Lyons was an early entrant to this market, establishing LEO Computers in 1954 to market computers to others. It also exploited the strength of its systems analysis team to help customers apply its products effectively. John Aris later summed up the LEO approach: the "system should be thorough and radical . . . Plan it as a whole. Rethink, rather than automate what is there." Making the computer pay for itself meant getting as much work as possible done on each run: "There would always be one or more main files, on tightly packed binary punched cards; the requirement was to pass through these as few times as possible, attempting to carry out all necessary calculations in the time to taken to read/punch the cards and print the results."[5] LEO sold dozens of computers to customers across Britain and its empire. They were powerful machines with some novel features, but in the end, LEO lacked the capital and international reach needed to compete with IBM. It merged with English Electric in 1963, part of a larger process of consolidation taking place across the industry.

IBM's Big Business Computers

IBM had completed the design of a "tape processing machine" for business by March 1950, but by the time a prototype was operating reliably in mid-1952, it had already announced the scientifically oriented 701 as its first computer product.[6] The tape machine eventually followed, as the IBM 702. Like the 701 it was a big vacuum tube computer with Williams tube storage and high-speed magnetic tape drives.

To support the perceived needs of business, the 702's instruction set, addressing system, and arithmetic capabilities were quite different from the 701's. To support accounting work it employed decimal arithmetic, avoiding the frustration and potential rounding errors caused by converting numbers to and from binary. It also had a variable word length. IBM's scientific computers had 36-bit words, equivalent to 9 decimal digits. Its administrative computers were expected to work primarily with quantities smaller than a billion, so keeping variables in fixed length words would waste precious memory space.

By the time of the first 702 installations in 1955, magnetic core memories were beginning to be used in commercial machines. After just fourteen deliveries, IBM shifted to a core-memory successor, the 705.[7] This was the cousin of the scientifically oriented 704, although IBM left out its other big improvement, floating-point arithmetic, on the basis that commercial customers would not pay extra for it. Instead IBM enhanced input and output speeds, so that companies could run tapes holding thousands of records through the computer more rapidly.

By this point IBM management was already beginning to question the efficiency of offering separate large computers for scientific and business applications. The firm designed its next big machine, the 709, to work for both markets. It had excellent input and output capabilities and powerful floating-point arithmetic. As we saw in the last chapter, the 709's transistorized successors were hugely successful. But their capabilities came at a high price and because many customers were committed to the decimal approach of the 705 IBM's product lines continued to proliferate. The firm's first attempt at a transistorized administrative computer, the 7070, flopped because it could not run code written for the 705, forcing it to introduce yet another model, the 7080, to address that need.

IBM's Big Hit: Small Computers

We already discussed the proliferation of cheaper, slower scientific computers based around the drum memory units developed by Engineering Research Associates. IBM used the same drum technology to produce a small business-oriented computer, the IBM 650. Its origins were in the punched card side of the business, coming from IBM's head plant in Endicott, New York rather than the computer group in Poughkeepsie. One proposal, dubbed "Wooden Wheel," was for a plug-programmed machine much like the 604 Multiplier.[8] That evolved into a general-purpose computer storing programs and data on a magnetic drum using ERA technology.[9]

The 650 was launched in 1954. It was the first mass-produced computer, with almost two thousand manufactured. Like ENIAC, the 650 had to be used with a full set of conventional punched card machines to prepare, process, and print cards. For a rental of around $3,500 a month the 650 had to compete with many other inexpensive drum machines, but outsold them all; in part based on IBM's reputation and existing large customer base of punch card users and in part because the 650 was easier to program and more reliable than its competitors.[10]

Thomas Watson Jr. directed that IBM allow universities to acquire a 650 at up to a 60% discount, if the university agreed to offer courses in business data processing or scientific computing.[11] Many universities took up this offer, making the 650 the machine on which thousands of scientists learned to program. Donald Knuth later dedicated his monumental series, *The Art of Computer Programming*, to "the Type 650 computer once installed at Case Institute of Technology, in remembrance of many pleasant evenings."[12]

IBM's 650 demonstrated a large market for computers. Its successor, the model 1401, was delivered to almost every organization using punched card equipment. More than 12,000 were produced. Introduced in 1959, its transistors and core memory outclassed drum machines like the 650. While slower than IBM's big computers, the 1401 was about seven times faster than a 650. Compared with the conventional punched card machines that it typically replaced, the 1401 was faster still, substituting programs for long sequences of laborious manual operations. It was strikingly compact, having replaced vacuum tubes with the new modularized and transitorized circuit boards IBM had devised for its ill-fated Stretch supercomputer.

The 1401's success owed a lot to a piece of peripheral equipment that IBM introduced with it. The type 1403 printer moved a continuous chain of characters laterally across the page. Magnetically driven hammers struck the chain at the precise place where a character was to be printed. In one minute it printed 600 lines, far more than anything else on the market, and could stand up under heavy use.[13] To the layman and to Hollywood, the chattering chain printer joined spinning tapes and flashing lights as symbols of the computer age. Together the 1401 and 1403 constituted an essential upgrade for IBM's core customer base of punched card users.

The 1401 was a flexible workhorse, which IBM supported with software for business and scientific use. It could be used with punched cards for all input and output, like a conventional tabulating machine, but because that trickle of data wasted the potential of the printer most 1401 installations had several tape drives. Magnetic tape offered a high throughput on simple jobs where lack of processing power was not a big handicap. Toward the end of the 1401's life it gained several compatible siblings and disk drive options.

By 1962, just eleven years after IBM leased its first computer, the firm was receiving more income from computers than from punched card machines. By 1966 IBM had installed an estimated 11,300 systems from the 1400 family, accounting for about a third of all US-built computers.[14]

The 1401 was often found in university and scientific computer centers as a supplement to a larger computer, used to process input data onto magnetic tapes, print, and run simple jobs (figure 3.1). At nearly every university computer center, someone figured out a sequence that would sound out the school's fight song when sent to the printer. Printer music, recorded by his father, underpinned Icelandic composer Jóhann Jóhannsson's acclaimed orchestral work *IBM 1401, A User's Manual*.

RAMAC

Data processing systems usually worked through entire files, read sequentially from punched cards or magnetic tape. Update cards were batched and applied to master files weekly or monthly. The arrival of disk drives underpinned a new way of working.[15] An array of spinning disks stored more data at a lower cost than its older cousin, the magnetic drum. Drums had a line of fixed read and write heads, one for each track on the drum. They were rugged and fast but expensive. Disk drives used a single movable head for each side of a disk. To record and read data, each head had to be very close to a disk's surface without touching it, which would cause unacceptable wear. IBM engineers working at a newly established laboratory in San Jose, California, spent 1953 and 1954 experimenting with a bewildering variety of head geometries and positioning mechanisms. They came up with the notion of using a thin film of air as a cushion. Their first product pumped air into the heads. Later IBM disk drives exploited the *boundary layer*—an aerodynamic phenomenon familiar to airplane designers—to float the head on air set in motion by the disk itself.

In 1956 IBM publicly announced the Model 350 disk storage unit. It used a stack of fifty aluminum disks, each 24 inches in diameter, rotating at 1,200 rpm. Total storage capacity was five million characters. The disk was attached to a small drum computer, the 305, better known as the RAMAC (random access method of accounting and control).

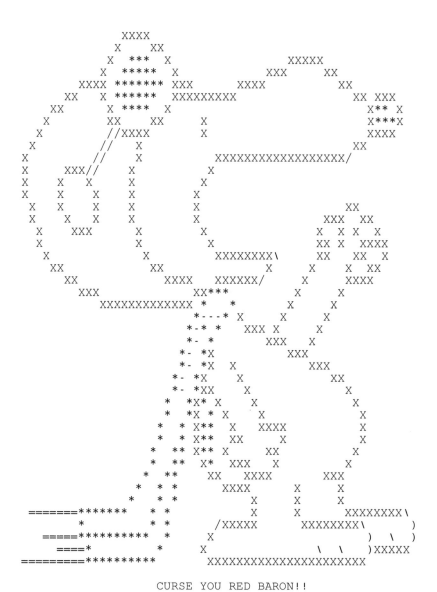

CURSE YOU RED BARON!!

Figure 3.1
Computer installations on college campuses often created images composed of text characters to show off the capabilities of the IBM Chain Printer. Such pictures were later shared online. This widely shared example of what was later called ASCII art was reportedly created by David Wright in 1978.[16]

What IBM meant by "random access" was that data could be fetched rapidly from any part of a disk, in contrast to the sequential operation of a deck of punched cards or a reel of tape. RAMAC made it feasible to work one record at a time. It would have taken a big IBM system six minutes to search a tape and retrieve one value. The relatively inexpensive RAMAC could locate a record and start printing out its contents in less than a second. IBM announced a United Airlines reservations system as the first commercial application. Updates could be entered one by one as they were received, so that RAMAC would always provide current totals. In the spring of 1958, IBM installed "Professor RAMAC" at the US Pavilion at the Brussels World's Fair (see figure 3.2). Visitors could

Figure 3.2
An IBM RAMAC disk unit is loaded onto a Pan American Airways aircraft, for its installation at the Brussels World's Fair, 1957. The drive had a capacity of five million characters and required two additional cabinets holding power supplies and controller electronics to function. Courtesy of International Business Machines Corporation, © International Business Machines Corporation.

enter a query with a keyboard and receive answers in any of ten languages (including Interlingua, an artificial language like Esperanto).[17]

RAMAC presaged a day when direct access to large amounts of data would become not only accepted but essential to the operations of banks, supermarkets, insurance companies, and government agencies. IBM President Tom Watson Jr. heralded its introduction as "the greatest product day in the history of IBM." IBM built only about a thousand RAMACs, but by the early 1960s disk technology was available across its product line.

DATA PROCESSING IN PRACTICE

Companies contemplating computer orders conducted feasibility studies to rationalize their purchases. These were usually based on assumptions about the useful lifespan of computers, the costs of programming, and the potential for clerical cost reduction that would soon seem ludicrously optimistic.[18] After agreeing to order a computer, companies had a waiting period to get ready physically and organizationally for the new equipment. Following a tradition going back to ENIAC, they usually installed it behind plate glass in a modern space with raised floors to hide the cables and suspended ceilings to hide the air conditioning vents. Visitors to the company would file past this spectacle of modernity.

For example, in 1957 Dow Chemical decided to join the computer age. To head its data processing department it chose Charles W. Bachman, an engineer familiar with punched card accounting applications. He recommended the lease of an IBM 709, able to run both administrative and technical jobs. Within a year Bachman had hired more than thirty people and built a computer room. His programmers and analysts were planning applications for the new machine when Dow fell on hard times and canceled its computer order prior to delivery. Bachman, like most others who got involved with data processing projects, built an attachment to computer technology that was stronger than his attachment to his employer. Rather than return to engineering work at Dow he found a new job at General Electric, applying computers to industrial management. We shall return to him later.[19]

Data Processing Work

Many companies, particularly those installing smaller computers, expanded and upgraded their existing punched card departments to data processing departments. The terminology came from IBM, which wanted to tie business computing to its existing strength in punched card machines. Its punched card machines became *data processing equipment*, and its computers *electronic data processing* systems. Thomas Watson Jr., as head of IBM, told punched card supervisors that "the very name that we have applied to our jobs—Data Processing," indicated a professional approach, focused on the "relevant facts on a timely basis, on a basis equal or better than our business competitors." Soon, he promised, executives would "look for data processors to infiltrate into the very tops of their businesses."[20] In 1962 their society, formerly the National Machine Accountants Association, renamed itself the Data Processing Management Association.

There were five main kinds of work in data processing departments. In ascending order of status and pay, these were key punch work, machine operation, programming, systems analysis, and management. Not coincidentally, the proportion of women doing each job fell in exactly the same order. All program code and input data had to be punched onto cards. Because this work had strong parallels with typing it was almost invariably given to women. Keypunch work was the biggest and worst paid job category in data processing. Operating other punched card equipment was treated as men's work by most American companies, although practices varied. Computer operation was seen as an extension of the same labor, so operators were a mix of men and women. Systems analysis was the job of looking at business procedures to redesign them around the new technology. Analysts were not expected to write programs, but they were expected to produce very detailed specifications for computer processes. In big companies this was an extension of the existing work of the overwhelmingly male *systems and procedures* groups, which prior to computerization had taken on jobs such as documenting procedures and redesigning forms. After computerization it remained a job for men.[21]

Programming was the only job without a clear parallel in existing administrative punched card work. It was slotted in conceptually between the existing work of machine operation and systems analysis. Programmers would take the specifications written by the analysts and convert them into computer instructions. Within the ideology of data processing this was seen as a less creative task than analysis   successful programmers would aspire to become analysts and eventually managers, leaving the machine itself further behind as their careers progressed. Early descriptions sometimes mention the lower-status job of *coder*, turning instruction mnemonics into numerical codes, but that task was soon automated by increasingly powerful assembler software.

The openness of programming jobs to women during the 1950s is often exaggerated. As historian Thomas J. Misa has shown, there is no evidence to support the widely repeated assertion that 30 to 50 percent of all programmers were women during the 1950s and 1960s, or that this proportion dropped during the 1970s.[22] However, female programmers were numerous in some computer installations, mostly those focused on technical computing. As we saw in chapter 1, the celebrated "women of ENIAC" reflected a broader reliance on female labor in applied mathematics. This tradition led to opportunities for women in computing centers with similar cultures, such as Bell Labs. Most administrative users of computing, such as banks and insurance companies, likewise approached electronic data as an extension of existing practice. But here that usually meant employing men, as their offices maintained rigid gender separation which barred women from professional and managerial work despite their presence in huge numbers as clerical workers.[23]

Sorting and Report Generation

Data processing jobs worked through one or more input files, processing each record in turn. This followed punched card machine practice. For example, a weekly payroll run started with women punching data from timesheets onto input cards. These cards were sorted into order and combined with a deck of master cards holding the pay rates and names of employees. A series of runs through the tabulator and other specialized machines would print pay checks and a ledger recording the amount paid to each employee. Adding an IBM 650 to the mix was potentially simpler and faster, as a single program could replace several runs through conventional machines. But its processing rate was limited by the speed with which cards could be read.

Bigger computers relied on magnetic tape. The IBM 705 could store five million characters on a reel of magnetic tape. At 15,000 characters per second, it would take about six minutes to process an entire tape. Having six or eight tape drives available let programs access several input and output files simultaneously. But as even the most expensive computers could hold only a few records in memory, it was essential that those files be sorted correctly. When processing payroll, for example, the master employee file and the timesheet data file had to be sorted in the same order. When an employee record was read from one file, the next record read from the other file gave the timecard data for the same employee. The computer could process tens of thousands of payroll records in the six minutes it took to read a tape. Reading through an entire tape to locate a single employee record was not practical—it would be quicker to use a file cabinet.

Sorting records held on punched cards was easy, if slow. Each card generally held a single record, which is why *unit record equipment* was one of IBM's euphemisms for punched card machines. Decks dropped into the sorter were separated into ten output trays based on the value of a selected digit. Running each output deck through the machine again sorted on the next digit. Eventually the records would all be in order and the deck could be reassembled. It wasn't practical to cut a tape into tiny pieces and splice them back together, so finding an effective way to sort tape files was crucial for the efficient application of computers to business.

Compared with some of the other applications of early computers—simulating a nuclear explosion, for example—sorting an employee file might not sound like a complicated problem. But back in 1945, John von Neumann had immediately grasped its importance to the viability of the new kind of computer he had in mind. He took a break from his work for Los Alamos to write a program, trying out his proposed EDVAC instruction set. Von Neumann was pleased with the experience, writing "it is legitimate to conclude already on the basis of the now available evidence, that the EDVAC is very nearly an 'all purpose' machine" and would be faster than a conventional punched card sorter that had been designed for this single task.[24] The *merge sort* algorithm he invented is still widely used.

Going from unsorted input to sorted output required data files to pass through the computer repeatedly. How many times depended on a number of factors: the number of tape drives available to store working files, the number of records that could be squeezed into the computer's memory, how close to being ordered the input file was, and so on. Von Neumann's method was good, but others worked better under certain conditions. We already described the work done by scientific computer installations to share standard code for common tasks. In a similar way, programmers in data processing installations started producing generalized sort routines rather than constantly rewriting the same code for different applications.

Computer manufacturers also developed utilities for common tasks such as sorting. Among the people who followed Eckert and Mauchly from the ENIAC project to the Eckert-Mauchly Computer Corporation was Frances E. (Betty) Holberton, one of the original operations team. By 1952 she had produced a generalized sorting routine for use with the Univac tape drives. Donald Knuth called it "the first major 'software' routine ever developed for automatic programming."[25]

One of the most influential packages was produced by a team working with the IBM 702 at the Hanford Nuclear Reservation, a gigantic plutonium processing complex operated by General Electric. It combined generalized sorting capabilities with a report program generation module. Many administrative programs read through structured data files to compile totals and subtotals. Conventional tabulating machines did this in hardware—operators set switches and wires to specify what to tally and how to format output. Producing a new report in a computerized system meant writing a program. Fred Gruenberger, who worked on the system, noted that managers were "vitriolic" when told it would cost thousands of dollars to tweak a report, because on a tabulator "all you have to do to change your report is move five wires." The new system automatically generated a report program when fed cards describing the desired report format.[26]

In 1957, SHARE set up a Data Processing Committee to support companies applying the IBM 709s to administrative work. Under the direction of Charles Bachman, it adopted the General Electric sort and report generation routines as the basis for a new package, 9PAC, released in May 1959. By 1960, IBM had taken over support and distribution for 9PAC, which became an integral part of the IBSYS operating system. IBM produced a similar package, Report Program Generator (RPG) for its smaller 1401 computers. This made it easy for punched card installations to adopt computer technology without having to write huge amounts of code. RPG took on a life of its own and was implemented across IBM's product range for decades. New versions are still being produced today.

The study of sorting methods also laid the groundwork for what became the cornerstone of theoretical computer science: computational complexity. Because computers spent so much time sorting data, it was essential to understand the properties of different algorithms. *Algorithm* here means a specific step-by-step method to accomplish something. The same algorithm might be coded in many different ways, for different computers, using different languages. Very simple algorithms worked acceptably with smaller input files of a few hundred records, but they slowed down as the input file grew longer. Computer scientists formalized these insights into a science of *computational complexity*, in which they look at the best case, worst case, and average time performance of algorithms, as well as their temporary space needs. The overall performance of an algorithm can be approximated as a mathematical relationship to the number of items being sorted.

Computer scientists categorize algorithms accordingly. For example, von Neumann's merge sort algorithm is said to be have performance $O(n \log n)$ because its performance on a list with n items is proportional to n times the logarithm of n. That makes it a good choice for long, highly disordered inputs. Quicksort was developed by Turing award-winning computer scientist C. A. R. Hoare in 1959.[27] It has better performance under most circumstances, but with the least favorable input, the time it takes will grow proportionally to the square of the input length. The system of time complexity classes was introduced in a 1965 paper by Juris Hartmanis and Richard Stearns.[28] That work earned another Turing award, for extending Turing's own paper on computability to determine what is realistically calculable with a modified Turing machine given plausible constraints on available time. Complexity theory grew into a prominent and, to most programmers, baffling area of computer science. But it began as a formal response to a very practical concern: ranking the performance of algorithms under different circumstances. This illustrates the difference between what we called in this book's introduction the theoretical universality of programmable computers, and their gradual evolution toward practical universality with the development of new capabilities and algorithms.

COBOL

Fortran's success was matched in the commercial world by COBOL (common business oriented language). In May 1959 the US Department of Defense sponsored a group of representatives from different computer manufacturers to develop a shared data processing language. They worked fast, producing preliminary specifications by the end of that year. Jean Sammet, the representative sent by the electronics firm Sylvania, played a key role in shaping COBOL's initial design. In 1961 she joined IBM as a leading expert on programming languages.

As soon as the specification was published, several manufacturers began writing compilers for their computers. The next year, the US government announced that it would not purchase or lease computer equipment unless it could handle COBOL.[29] As a result, COBOL became one of the first languages to be sufficiently standardized that a program could be compiled on computers from different vendors and run to produce the same results. That occurred in December 1960, when almost identical programs ran on a Univac II and an RCA 501.

Part of COBOL's ancestry can be traced to Grace Hopper, who in 1956 had developed a compiler sometimes called FLOW-MATIC, which was geared toward business applications. An IBM project called Commercial Translator also had some influence, as did a Honeywell language called FACT.[30] It was from Hopper that COBOL inherited its most famous attribute, long variable and command names intended to make the code read as English, in contrast with Fortran, which aspired to mimic mathematical notation. COBOL code usually centered on retrieving data from files and manipulating it, for example

```
OPEN INPUT-FILE1.

SORT SORT-FILE 1 ASCENDING FIELD-AA DESCENDING FIELD-BB.

PARAGRAPH 1. READ INPUT-FILE-1 AT END GO TO PARAGRAPH-2.

IF FIELD-A = FIELD-B GO TO PARAGRAPH-1 ELSE

MOVE FIELD-A TO FIELD-AA

RELEASE SORT-RECORD. GO TO PARAGRAPH-1.

PARAGRAPH-2. EXIT.[31]
```

This was intended to reduce the need for comments, as the code itself could be understood by humans as well as by the compiler. Proponents argued that this design made the program understandable especially by managers without programming skills. Neither goal was achieved in practice. It soon became obvious that without careful documentation even the person who wrote the code would struggle to remember what it was supposed to do.

IBM'S SYSTEM/360 SETS THE STANDARD

In early 1965, IBM delivered the first computers in a new range that would ensure the company's domination of the mainframe industry for the next fifty years. That was the System/360, announced in April 1964. It was so named because it was aimed at the full circle of customers, from business to science. IBM initially announced six models, dropping some and adding others by the time deliveries began. It promised that programs written for one model would work on others, preserving its customers' programming investment as their needs grew.

A Compatible Range

The road to System/360 began in the Sheraton New Englander motel in Cos Cob, Connecticut, where an IBM committee known as SPREAD met daily for two months in late 1961. Their deliberations began with a survey of the company's existing products. IBM was fielding a confusing tangle of incompatible machines. Its two most successful ranges have already been described: the small, business-focused 1401 family and the large, scientific 7090 family. When IBM's large business range, the 7070 family, sold disappointingly, it responded by introducing yet another incompatible product, the 7080. Its small scientific machine, the 1620, was doing well, although not as well as the 1401.[32] These product lines had fundamentally different architectures. For example, the 1620 did its arithmetic by looking up results from tables held in core memory. That made it cheaper and slower than conventional machines such as the 650. Customers joked that its informal name CADET stood for "Can't add; doesn't even try!"

These product lines were justified by a notion that business and scientific users had fundamentally different hardware needs, which no longer held up. Business customers were expected to work with large sets of data on which they performed simple arithmetic, whereas scientific customers did advanced calculation on small sets of data. In reality, scientists and engineers were handling large data sets for applications like finite-element analysis, a technique developed for building complex aerospace structures. And business applications were growing increasingly ambitious. In its final report, the SPREAD Committee recommended a single unified product line for science and business.[33]

The SPREAD Committee needed a performance range of 25:1 in the new range. The 7090 and 1401 could be, and eventually were, upgraded to produce compatible successors. But neither architecture seemed likely to stretch to an order-of-magnitude performance increase. The committee was not sure that a single architecture could ever work for both large and small computers. Fred Brooks and Gene Amdahl—later two of the 360's principal architects—initially argued that "it couldn't be done."[34] Few technologies scale over such a range. The engine, transmission, power train, and frame of an 18-wheeler truck are not simply bigger versions of those designed for a subcompact car. Yet by 1970 IBM was offering compatible computers with a 200:1 range.[35]

What changed Brooks's and Amdahl's minds was the rediscovery of a concept almost as old as the modern computer. In 1951 at a lecture given at a ceremony inaugurating the Manchester University computer, Maurice Wilkes argued that "the best way to design an automatic calculating machine" was to build its control section as a tiny computer, so that each machine language instruction was broken down into a series of "micro-operations." The sequence of micro-operations for each instruction was set by a "micro-programme" encoded in a read-only memory.[36] This simplified the control unit, usually the most complex part of the processor.

IBM extended Wilkes's approach to provide comparability across a range of machines rather than simplifying the design of a single machine. Small System 360 computers could use microprogramming to simulate capabilities that larger models ran more efficiently using expensive hardware. This freed engineers to tailor the design of each model to reach an appropriate balance of speed versus cost. Microprogramming gave the 360's designers "the ability to separate the design process ... from the control logic that effectively embodied the instruction-set characteristics of the machine we were trying to develop."[37]

Wilkes mentioned, almost as an afterthought, that a microprogrammed instruction set "need not be decided on finally until a late stage in the construction of the machine."[38] By adopting microprogramming, IBM gained one further advantage: the ability to install a microprogram that would allow the processor to understand instructions written for an earlier IBM computer. In this way IBM could sell new computers to customers without rendering obsolete their investment in application programs. In theory, any modern computer can be programmed to run code originally written for any other computer. In practice, the extra layers of code usually slowed things down unacceptably.

The 360 used a combination of software and the microprogrammed instructions for what Larry Moss of IBM called *emulation* of older machines, implying that it was "as good as" (or even better than) the original, rather than mere "simulation" or worse, "imitation." The 360 Model 65 sold especially well because of its ability to emulate the large business 7070 computer. IBM made sure that the low-end models 30 and 40 effectively emulated the 1401.[39] Their faster circuits could run old programs as much as ten times faster than a real 1401. By 1967, according to some estimates, over half of all 360 jobs were run in emulation mode. Despite its name, software is more permanent than hardware. Decades later, 1401 programs were still running routine payroll and other data processing jobs on computers from a variety of suppliers. The programmers who coded these had no idea how long-lived their work would become.

The System/360 Architecture

Except for the smallest Model 20, each System/360 computer had sixteen general-purpose registers, each usable for almost any purpose. The 360's word length was 32 bits, a slight reduction from earlier large scientific computers that let IBM standardize on 8-bit bytes. Four bytes fit neatly into one word. A byte could encode 256 different combinations: representing uppercase and lowercase letters, the decimal digits one to ten, punctuation, accent marks, and control codes with room to spare. Since 4 bits were adequate to encode a decimal or hexadecimal digit (the 360 supported both) one could pack two into each byte.

IBM represented characters using EBCDIC (extended binary coded decimal interchange code), an extension of a code developed for punched card equipment. It was well designed and offered room for future expansion, but it was not compatible with the ASCII standard adopted by the American National Standards Institute in 1963. ASCII standardized only seven bits, not eight. Punched paper tape was still in common use, and the committee felt that punching eight holes across a standard piece of tape would weaken it too much. The computing world split, with EBCDIC used at IBM and ASCII everywhere else.

The 360 used a 24-bit address, allowing direct access to 16 million bytes of memory. Putting the full address into each instruction seemed extravagant, given the limited memory available on smaller models, so only 12 bits were used in the address field. This was added to the contents of a *base* address register to give the full address. This approach, common on later machines, split the memory into *segments*.

PL/I

When IBM promoted System/360 as a replacement for its data processing and scientific computers, it promised that a single language could handle both application areas. A joint IBM-SHARE committee concluded in early 1963 that Fortran, although popular and heavily used by scientific users, could not be extended to do this job. Instead IBM and SHARE committed to a new language, PL/I (programming language, one), drawing on COBOL and Algol as well as Fortran.

Many IBM installations eventually made PL/I available, but it never was very popular. Its complexity overwhelmed its advantages. Implementing PL/I proved difficult and work progressed slowly. By the time the full language was ready, COBOL and Fortran had reinforced their dominance.

PL/I did introduce one new element that was influential on later language designers: pointers. Pointers are variables that hold memory addresses. As the first general purpose high-level language, PL/I was supposed to be usable for operating systems programming. That required a mechanism to build complex data structures and directly manipulate blocks of memory. Pointers met this need. They added the low-level manipulation of memory, usually possible only with assembly language programming, to the elegance of Algol-style procedures.[40]

IBM's Gamble Pays Off

In an oft-repeated phrase, first used in a *Fortune* magazine article, an IBM employee said "you bet your company" on this line of computers.[41] Had the 360 failed, it would have been a devastating blow. Besides the six computer models, IBM introduced more than 150 other products on the same day. The trade press called them peripherals, but they were central to the System/360 project: a new model keypunch, new disk and tape drives, and teletypes derived from the famous Selectric typewriter with its golf-ball print head (see figure 3.3). The new computers drove them efficiently, standardizing IBM's system of input and output channels across the whole range. These devices defined the 360-era of mainframe computing as much as the beige, slanted control panel.

System/360 did not fail. Instead, orders "exceeded forecasts: over 1100 were received in the first month. After five months the quantity had doubled, making it equal to a fifth of the number of IBM computers installed in the U.S."[42] Most NASA centers, for example, quickly switched over to larger 360 models from their 7090 installations to meet the demands of putting a man on the moon (figure 3.4). Business data processing centers likewise replaced their 7030s.

The success of the 360 threatened IBM's stability almost as much as a failure might have. Manufacturing and delivering the line of computers required enormous resources. The company expanded its production facilities, but delivery schedules slipped and shortages of key components arose. Software production was particularly difficult. The original plan was to stuff all the features of IBM's existing operating

Figure 3.3
An IBM System 360 installation, c. 1965. Well into the third generation, mainframe computers still relied heavily on magnetic tape for mass storage. Note the raised floor, punched card reader, the typewriter-like console printer, and next to it, an IBM 3270 video terminal and the distinctive slope of the main console. At the lower left is a suite of fixed disk drives; at the lower right, a set of chain printers housed in soundproofed cabinets. Courtesy of International Business Machines Corporation, © International Business Machines Corporation.

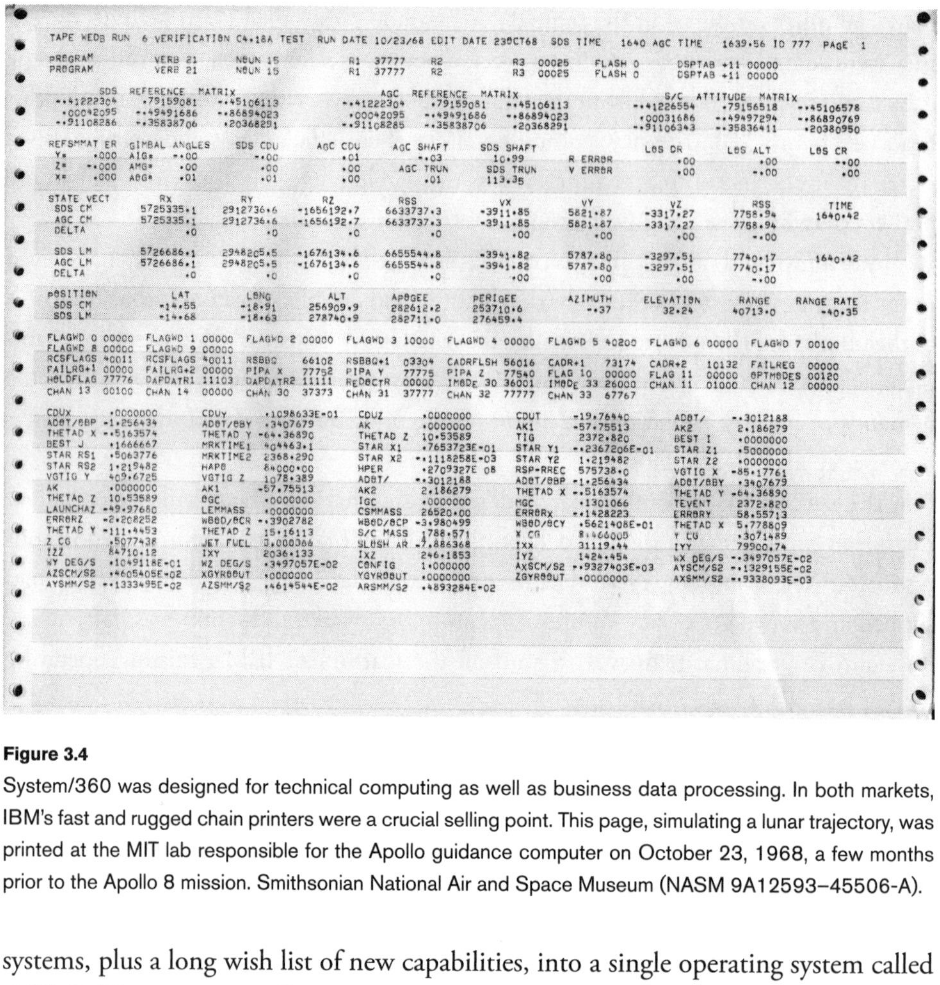

Figure 3.4

System/360 was designed for technical computing as well as business data processing. In both markets, IBM's fast and rugged chain printers were a crucial selling point. This page, simulating a lunar trajectory, was printed at the MIT lab responsible for the Apollo guidance computer on October 23, 1968, a few months prior to the Apollo 8 mission. Smithsonian National Air and Space Museum (NASM 9A12593–45506-A).

systems, plus a long wish list of new capabilities, into a single operating system called OS/360. To run it on smaller computers, some modules would be turned off or loaded only when needed. At the end of 1964, IBM gave up on its plan, announcing that there would be four different operating systems for computers with disk and tape and with larger and smaller memories. The first System/360 machines were delivered without any operating systems. Between 1965 and 1967, four interim systems were released.[43] For employees driven to the breaking point—and there were many—the jump in revenues for IBM may not have been worth the physical and mental stress.

In the end, IBM settled on two operating systems. Most users ran DOS, the disk operating system. Created in just a year by a separate team, it threw out the grand goals of OS/360 in order to provide efficient batch mode processing on the smaller System/360 machines that took over from 1401s as the workhorses of data processing. DOS quickly became the world's most widely used operating system.[44] IBM's promises for OS/360 were finally realized, for its larger computers, with a 1967 release known as MVT for its ability to support multiprogramming with a variable number of tasks. Both systems were extended to support later generations of IBM mainframes, beginning with the System/370 machines of the 1970s, and both are still used and updated today.

Although System/360 was intended to work equally well for scientific and data processing applications, it was much more successful for data processing. The problems began with the System/360 floating point. It used hexadecimal (base 16) rather than binary, which was efficient for smaller, business-oriented machines but created major problems with rounding errors for scientific users. The new general-purpose registers raised more problems with the handling of single and double precision numbers. When IBM described its new architecture, William Kahan, then of Waterloo University, and others "went nuts" as they "recognized something really perverse about the arithmetic." IBM found ways to work around some of the issues in software libraries, but Kahan recalls that after the full scale of the problem was acknowledged in 1966, following lobbying by SHARE, the company spent millions tweaking the hardware of machines already installed.[45] As we saw previously, it was CDC and Cray Research, not IBM, that dominated the supercomputer market from the late 1960s onward. IBM's attempts to counterattack fell victim to infighting or were compromised by the need for compatibility with the System/360 architecture. The Model 91, announced in 1964, was not delivered until 1967, and the most powerful, the Model 195, was first installed in March 1971, a full two years after the equivalent CDC machine.[46]

Fortunately for IBM, data processing was by far the largest sector of computing. From 1965 to 1970, its gross income more than doubled. The computer industry of the mid-1960s was jokingly redefined as IBM and the Seven Dwarfs (Burroughs, Sperry Rand, Control Data, Honeywell, General Electric, RCA, and NCR). Net earnings also doubled, surpassing $1 billion by 1971. By 1970 IBM had an installed base of 35,000 computers, and by the mid-1970s it made sense to describe the US computer industry as having two equal parts: IBM on one side and all the other companies combined on the other.[47] The System/360 architecture served as the anchor for IBM's mainframe product line into the 1990s, growing ever more baroque and cumbersome. That it could remain viable so long is testimony to the strength of the initial effort.

DREAMS OF A MANAGEMENT REVOLUTION

Computers had been sold to American business by people like General Electric's Roddy Osborn as the basis of a managerial revolution. In reality they were usually put to work to speed up jobs already being carried out with punched card machines. As industry analyst Roger Canning put it, even a decade later many computer departments were just "the old tabulating operation with chromium plating."[48]

Consultants and business thinkers tried to revive the idea of a digital utopia for business. In a 1959 *Harvard Business Review* article, two of them indulged in a little science fiction when they imagined "Management in the 1980s." The article introduced the phrase *information technology*, defined as the combination of operations research methods, computer technology, and simulation. That would replace entire levels of middle managers with programmed decision rules. Corporate control would be recentralized as top managers worked with "information engineers" in "staff roles close to the top." In this new world there would be "more impersonal, problem-oriented behavior at the top, with less emphasis on loyalty to the firm and more on relatively rational concern with solving difficult problems."[49] Their vision prophesized elements of today's corporations, such as the hollowing out of middle management and elevation of so-called data science and analytics as the basis of new management cultures.

The Totally Integrated Management Information System

The totally integrated Management Information System (MIS) became the most discussed idea in business computing. The system automated every aspect of a company's operations, hooking together programs so that information flowed effortlessly from one business process to the next. It would be engineered to provide all managers with exactly the reports and statistics they needed. Automating routine transactions would also harvest the data needed for simulation and forecasting systems, to support strategic decision making.

Consultants even pitched Star Trek style decision rooms in which top managers could summon data visualizations by pushing buttons (figure 3.5). According to one, managers were about to trade in desks, papers and tables for comfortable padded chairs facing huge screens in the "information management facility," in which "uneasiness will be replaced by a feeling of confidence in the completeness and timeliness of information." One consultant promised that once "sufficient data has been included in the mathematical expressions placed inside the computer" even "major decisions" would need only a "single, brief meeting."[50]

By the mid-1960s, computer companies were selling their third generation computers, with larger memories and disk drives, as the foundations for these "total management information systems" (see figure 3.6). Univac, for example, ran an advertisement showing on one page a network stretching across a map of the US to symbolize "your decentralized company." On the facing page was a small circular shape, evoking both a reel of magnetic tape and a computer installation. That represented "your decentralized company with a Univac Total Management Information System." Buying a Univac 1108 would ensure that "your total organization is tied together by a total communications system that continuously receives, up-dates and relays management information. Centralized control of decentralized operations becomes a reality. And distance, as a factor in management, becomes irrelevant."[51]

Executive armchair control panel (closeup above) reduces need for paper reporting.

Figure 3.5

Consultant W. Robert Widener promoted the concept of management decision rooms based around computerized screens with armrest controls in his 1968 article "New Management Concepts."

It was a powerful pitch, but by the end of 1968, the impossibility of building such systems had become painfully apparent. A report from the elite consulting firm McKinsey & Company dismissed "the so-called total management information systems that have beguiled some computer theorists in recent years."[52] Even leaving aside the challenges of anticipating the information needs of managers and building forecasting systems, integrating all the operational systems of a large company by tying their input and output files together proved unworkable.

Database Management Systems

Success and failure are hard to judge. Many companies wasted huge sums on doomed projects. But one "total systems" project gave rise to a technological breakthrough that eventually transformed administrative computing practice. As computer management expert Richard L. Nolan noted in 1974, "Writings on MIS have waned recently and have largely been replaced by writings on the Data Base."[53]

When he moved to General Electric, Charles Bachman swapped his data processing management job for a staff position in GE headquarters as a kind of internal consultant, experimenting with new management methods and systems. Managers there were as excited about management information systems as the rest of corporate America.

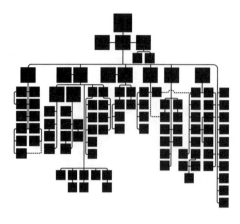

Your business.

Figure 3.6
By the mid-1960s, Univac promoted its computers as "total management information systems" with almost
magical powers to centralize and simplify the control of large diversified corporations. *Fortune*, October 1965,
pages 32–33. Courtesy Unisys.

As General Electric was entering the computer business, a good system would sell computers as well as boosting internal efficiency.

GE's divisions ran a lot of factories, so the effort started at the factory level. Bachman was part of a team dispatched in 1961 to study the operations of a typical plant and produce an integrated manufacturing control system able to handle applications such as production scheduling, order dispatching, inventory management, and parts explosion. To run it they were promised a GE 225 computer, roughly comparable to the IBM 1401, fitted with a prototype disk drive.

Handling data on disk was much more complicated than working with tape. Imagine that a disk holds one hundred thousand customer records. The big benefit of disk storage is random access—a program can request data from any part of the disk. But how does the program know where to find the desired record without having to read its entire contents? Pioneers came up with a variety of methods for structuring and indexing records held on disk, such as hashing, inverted files, and linked lists. Each was hard for a typical programmer to understand and implement.

Thanks to his prior experience with the 9PAC project, Bachman quickly realized that it would be disastrous to expect the programmers of each application to write code to access shared disk files. What if adding a new module to the system forced the addition of new data fields to existing records? This would probably break the existing applications. What if too many records accumulated in one part of the system and had to be restructured or split between two disk drives? That too would require a huge amount of reprogramming.

Bachman's solution to these problems was called the Integrated Data Store (IDS), which he claimed would "deliver the promises we have been making for so many years." IDS took over the entire computer. It handled requests received on punched cards and scheduled the various tasks needed to process data updates. Application programmers did not write code to work directly with data files but instead used IDS commands. As the IDS proposal noted, "the burden of accessing the records is put on the IDS." As a result, "records can only get into the file by being PUT there . . . they are only changed by the MODIFY instruction. They are only deleted by the DELETE instruction. There can be no accidental overwriting." IDS maintained what was later called a "data dictionary," defining all the different record types held by the system and their relationships with each other (for example, that a customer record was related to order records placed by that customer). Bachman's proposal argued that this "level of protection" was "absolutely necessary" for a "multiple usage file" shared between applications.[54]

IDS was operational on a test basis by 1963. General Electric soon tidied it up and released it to users of its computer range, producing updated versions into the 1970s. These computers were not particularly successful, which limited the reach of IDS. Bachman spread his ideas via two groups within CODASYL, the organization best known for developing COBOL. Several influential reports defined the capabilities of the *data base management system* and proposed a standard architecture and data manipulation language closely based on IDS. The reports introduced the idea of a *data definition language* used to specify record formats and linkages, and *sub schemas* so that applications and users could be granted permission to work with only part of a database.[55]

In 1973, Bachman's work was recognized with a Turing award. He was the first winner without a PhD and the first to spend his entire career in industry rather than academia. In his acceptance lecture, "The Programmer as Navigator," Bachman likened the impact of the database management system (DBMS) to the Copernican revolution in astronomy. Previously, business users had seen computers, like punched card machines, as devices through which sequential files flowed. Now the database and its complex structures, not the computer, occupied the center of the data processing universe and everything else revolved around it.[56]

CODASYL-compatible database management systems were common from the mid-1970s onward. They transformed application development for mainframe computer users. Database management systems made it easier for organizations to write integrated applications that shared data with each other. The position of database administrator

(DBA) was originally seen as a managerial role, able to standardize data formats across departments.[57] During the 1970s many companies put a lot of effort into developing their own "enterprise data architectures" to support data sharing between applications.

Bachman's idea of the database management system as the foundation of data processing was amply vindicated, although the specific approach to structuring data he introduced has been superseded. During the 1970s database research became an important area of computer science. The CODASYL approach, later termed a network data model, was challenged by the more elegant but much harder to efficiently implement relational data model devised by Edgar F. Codd and championed by the computer science community. But changes in data processing practice were slow and incremental, as the experience of the Internal Revenue Service illustrates.

Data Processing at the Internal Revenue Service

As far as the eye could see were rows of IBM Model 027 keypunches, boring rectangular holes into 80-column, 3 1/4″ × 7 3/8″ cards. Seated at each station was a woman, her head tilted to the left to scan a piece of paper mounted on a frame, her right hand deftly floating over the keys of the machine. Each press was met with a solid "thunk." When the room was in full swing, said one operator, "there was a certain rhythm, a beat, a sound." A data processing manager said it was "like you had a helmet on and someone was hitting it with a hammer." A film made of the operation suggests that this was, if anything, understated.[58]

The room was in one of several regional centers of the US Internal Revenue Service to process tax returns. In 1964, around the time this film was made, the center was using a set of IBM 7070s. By then, the IRS had one of the most sophisticated and complex computer systems in the world. In contrast to the work in a scientific computing installation, the IRS work involved relatively simple calculations. What really distinguished it was volume.

During the Second World War, the numbers of Americans required to file returns and pay taxes increased from around 8 million to 60 million; the practice of withholding tax from a paycheck also became common. The IRS worked with Friden electromechanical calculators, Burroughs or National accounting machines, and pencils and paper. Punched card equipment arrived in 1948. At each center, around 350 employees keyed basic information from each return onto punched cards.

In 1955, the agency installed an IBM 650 in its regional center in Kansas City, Missouri, where it helped process 1.1 million returns on a test basis. In 1959, the US Treasury Department authorized the IRS to computerize its operations fully. The IRS selected IBM after soliciting bids from 40 manufacturers. At each regional center an IBM 1401 computer with four thousand characters of memory was installed together with a card reader, a punch, a line printer, and two tape drives. A new national center in Martinsburg, West Virginia, received the first of several IBM 7070 mainframes.

Although computers were now processing the returns, the first step in the process was still for keypunchers to enter data from over 100 million returns each year onto more than 400 million cards. The 1401s at each regional center verified that the required information was present, did some data reduction, and transferred the returns to tape. Couriers flew these tapes to Martinsburg, where the 7070 processed the return. The National Center then sent a tape to the Treasury Department to issue refund checks; for less fortunate tax filers, a tape was sent back to a regional center to send out a bill or to ask for more information.

By 1965, the IRS was identifying each taxpayer by a unique Social Security number to avoid confusing persons with the same names. Legislation to require this number on all tax forms led to congressional representatives receiving nasty letters about "Big Brother."

The punching of cards ended in 1967 when machines were installed that allowed direct data entry onto a drum (later a disk), but otherwise this structure remained in place into the 1990s. When the keypunch machines were retired, managers found that productivity did not go up as they had expected. By reintroducing some of the sound that was lost, the operators were able to reestablish the rhythm necessary to maintain high rates of data entry.

Like other data processing users, the IRS processed data sequentially by making bulk updates to record tapes. By the mid-1970s, the master file was updated once a week, producing an updated microfilm that could be held at a regional center to address questions about a specific return. That response time—one week—was not due to any bureaucratic inertia on the part of the IRS; it was built into the structure of the batch processing system.

By 1967, the 1401s had been replaced by compatible Honeywell machines, and IBM 360s replaced the 7070s. With the computerized processing in place and operating smoothly, the agency became more ambitious. Its plan for an all-new Tax Administration System (TAS), projected to cost $650 to $850 million, mirrored the grandiose goals being set out for total management information systems by organizations across the country. The IRS hoped to eliminate data entry work, either by using machines that could read handwritten forms or by having the taxpayer provide returns in a machine-readable form. The agency also intended to put data from returns into an integrated, on-line system. This could immediately apply updates to a shared, but decentralized, database used to answer queries submitted from a network of eight thousand terminals. Other terminals would allow data entry directly into the network without punching cards.[59]

The planners of the Tax Administration System gave much thought to making the system secure from physical damage, malicious intrusion, and simple human errors. But the Watergate hearings revealed that White House operatives were able to obtain the tax records of their political enemies. This time there were more than a few irate letters to members of Congress. Congress directed the General Accounting Office to look at the privacy implication of the system. A copy of its report was leaked to the trade journal *Computerworld*; it ran a lead story under the headline "Proposed IRS System May Pose Threat to Privacy."[60] In the spring of 1977, IRS officials were grilled by Congress with questions such as whether someone could attach an unauthorized terminal to "pull all this information out."[61]

The IRS dropped its plans for the Tax Administration System in January 1978. Instead, it proposed an "equipment replacement and enhancement program." Congress made it drop the word *enhancement*. The old master file on magnetic tape was retained, and online access was abandoned. Patrick Ruttle of the IRS called this "a way of moving into the future in a very safe fashion."[62] Hamstrung by a hostile Congress, the agency limped along. In 1985, the system collapsed: newspapers published lurid stories of returns left in dumpsters and refund checks lost.[63]

Congress then authorized money to develop a new architecture of data handling, but the cycle has repeated since. Even today the chronically underfunded IRS relies on antiquated mainframe systems. In 2018 its systems went down for eleven hours on the day returns were due when its Individual Master File, holding data from a billion tax payer accounts, was overloaded. Its core applications are written in a mix of COBOL and assembly language dating back to the 1960s. Several efforts to replace them with modern technology have failed.

CREATION OF THE SOFTWARE INDUSTRY

In 1967, Larry Welke, a data processing manager and former IBM salesman, launched a new catalog, the *International Computer Programs Directory*, to showcase the products of a nascent industry. The idea that standard computer programs could be sold was not always obvious. Most application programs were written specially for the organization using them. The custom coding work was usually done by programmers working inside the organizations that would use them, and less often by outside companies hired as contractors. System software, such as assemblers, compilers, and database management packages, also wasn't purchased. Manufacturers gave it away to users of their computers at no extra charge. User companies shared it freely with each other, via SHARE and other user groups.

By the end of the 1960s an industry selling packaged software was beginning to develop, and Welke had quit his day job. Most of the first successful software products were utilities that could be used by companies in different industries. The software industry started when companies providing programming services, reusing bits of code from one project to the next, realized that they could produce generalized programs able to handle the needs of multiple companies. Applied Data Research, founded in 1959 to do contract development, had the first hit package: Autoflow. It automatically produced flow charts from program code. Report generation packages gave the nascent industry two other big successes: Pansophic's Easytrieve and Informatics Inc.'s Mark IV.

Early commercial packages targeted niches that IBM's free software either neglected or did not address well. Fortunately for independent software companies, IBM was finding the practice of giving away software increasingly hard to sustain. Development costs were rising rapidly. Now that companies were selling software, giving it away raised antitrust concerns. In 1969, the firm announced its intention to unbundle its software packages and customer education from its hardware by charging for them separately.[64]

In 1970, to promote the new industry, Welke started an awards program for firms achieving more than a million dollars in cumulative sales. That first year, he made twenty-nine awards. As software companies prospered, Welke added new awards for $10 million and later $100 million in sales. Both went first to Mark IV.[65]

By 1982 the software industry was estimated to bring in revenues of more than $5 billion a year. Most large companies still wrote their own application programs even as they relied more heavily on purchased software packages to make these applications easier to write and maintain. Cullinane was one of fastest growing firms, the first packaged software company to be listed on the New York Stock Exchange and, a little later, the first with a billion dollar market valuation.[66] Its main product, IDMS, was a reworking of Bachman's IDS for the dominant IBM System/360 mainframe platform.[67]

Companies started to buy data processing applications as well as systems software. This market was eventually dominated by Systemanalyse und Programmentwicklung (SAP), founded in 1972 by five former employees of IBM Germany. They worked with interested customers to develop application programs to handle their needs. By 1975, it had developed modules for purchasing, inventory management, and financial accounting modules, which were customized for each user company. These were designed so that data entered or updated in one module would flow smoothly to others. SAP realized that it would be more efficient to standardize the core code as a product that could be tweaked for different users by adjusting configuration settings. SAP's modules replaced the custom application programs companies had written for common administrative tasks such as payroll and accounting. Over time, SAP added more modules and more customers, reworking its software to take advantage of the growing power and interactivity of mainframes. By 1982, 250 companies were using SAP's standard R2 software.[68]

By the early 2000s, SAP had displaced most of its competitors in what became known as the enterprise resources planning (ERP) market. Most of the world's largest companies have rebuilt their core administrative processes around SAP. Thousands of companies do nothing but customize SAP or produce add-ins for it, and millions of people earn their living configuring and supporting its software. Together these systems have delivered much of what advocates for "total management information systems" promised back in the early 1960s, although this eventual success rested on fundamental advances in hardware and software during the intervening years. These changes include the gradual remaking of the computer as an interactive tool rather than a batch processing machine—something we discuss in the next two chapters.

4 THE COMPUTER BECOMES A REAL-TIME CONTROL SYSTEM

By the time Stanley Kubrick's *Dr. Strangelove* reached cinemas in 1964, Americans were starting to think of computers as things that directly monitored and controlled things outside climate-controlled data centers. In the darkly comedic masterpiece, a mad general orders his executive officer, played like two other central characters by Peter Sellers, to encrypt and transmit codes ordering B-52 bombers to launch nuclear strikes against the USSR. Much of the action takes place around their base's IBM mainframe, played by a 7090, which appears to be used for secure communication. Special-purpose mechanical computers on board the bombers decrypt the instructions. One persists with the attack even after a recall order is issued because its computer had self-destructed after taking damage. This precipitates the end of the world; the Soviets had secretly built a gigantic computer-controlled "doomsday device" programmed to detonate if it detects a nuclear explosion on Soviet territory.

That computer was fictional, but at the time the movie was released, real-life computers were even more central to the Cold War nuclear standoff. Its circumstances had driven the development of new modes of computing that no commercial project could have justified, propelling computer technology toward universality. Digital computers were reconfigured to respond immediately to changes in the world so that they could track inbound bombers or control the engines of space craft. Minuteman intercontinental missiles were guided by miniaturized programmable computers and disk drives, while the gigantic SAGE network of computer control centers monitored America's borders and coordinated anti-aircraft responses. In contrast, the computers we have discussed previously worked on data that had been punched onto cards or paper tape. The time that elapsed between logging new data and processing it was measured in days or weeks. Bank accounts were credited with interest monthly, when statements were printed. A Cray supercomputer might spend days simulating a minute fraction of a second during a nuclear explosion.

Once developed and proven, Cold War technologies such as computer graphics, networking, and real-time control were quickly transferred to other areas of computing. So were hardware innovations such as reliable electronics, miniaturized circuit boards, and integrated circuits. Many of these technologies were first applied to civilian applications through the *minicomputer*, a new, smaller, and more affordable kind of computer with close technological and institutional connections to the Cold War projects carried out at MIT.

REAL-TIME SYSTEMS

During World War II, defense projects created a series of ingenious analog devices, many of them electronic. ENIAC had been built to calculate the trajectories of shells in order to produce printed firing tables. Other military applications, such as the aiming of anti-aircraft guns, required similar computations to be carried out instantly. The first computers able to compute fast enough for jobs like this were analog machines, to use terminology introduced in the 1940s to differentiate them from the new breed of digital computers such as the Harvard Mark 1 and ENIAC.[1] Digital computers processed data encoded as digits stored in their registers and memory units. They were programmed to execute sequences of mathematical and logical operations needed for a particular task—adding, subtracting, multiplying, and dividing the numbers and shuffling them around in memory.

Analog Computers

Analog computers were so called because they were configured as analogies of the situation being modeled. An increasing voltage in one circuit might, for example, indicate an increase in altitude. The components in an analog computer were interconnected so that changes in one element influenced the others. In an analogy, each part of the story told represents something different in the broader world, and the relationships between those things in the story illuminate their real-world connections. They used a range of media to represent quantities, such as electrical circuits or cams mounted on rotating disks. The influential MONIAC computers used interconnected tanks of water to dynamically depict a Keynesian model economy.[2]

One of America's secret weapons was the Norden bombsight. Its intricate assemblage of gears, gyroscopes, and an integrator performed two functions during a bombing run. The first was to carry out the complex calculations involving altitude, air speed, wind, characteristics of the bomb, and other conditions as the target was imaged by the aircraft's bombardier. The second was to automatically control the aircraft during the bombing run to ensure that the bombs were released at the times specified by those calculations. Sperry

Gyroscope Company, the ancestor of today's Unisys Corporation, built another complex bombsight. A computer to aim the guns of B-29 bombers was designed under contract by General Electric and built by IBM at its Endicott, New York, plant. At the Bell Telephone Laboratories, then located in lower Manhattan, David Parkinson and two collaborators designed an analog device called the M9 gun director (figure 4.1). It coupled gun-aiming machinery with radar that was tracking an enemy aircraft. The combination automatically directed where and when to fire the gun.[3]

The M9 used electronic circuits, including a device called the operational amplifier (op-amp) in place of the mechanical cams, gears, and disk integrators of other analog computers. The op-amp was used in electronic analog computers well into the 1960s and 1970s in the face of increasing digital capabilities. Its last bastion was in custom analog devices for signal processing.[4]

Professor Norbert Wiener of MIT analyzed the fire control problem statistically. Wiener's work, published in a classified report in 1942, was not widely adopted for fire control, but it proved to be influential in post-war development of what we now call information theory.[5] In 1948, Wiener wrote a book that continued this work and discussed its implications for a future world in which humans and intelligent machines

Figure 4.1
This M9 gun was controlled by a director built at Bell Labs: an analog electronic computer.
Source: Bell Labs, AT&T. © Alcatel-Lucent

would exist side by side. *Cybernetics* contained large tracts of dense mathematics interspersed with lucid prose about this future.[6]

Whirlwind and Real-Time Digital Computing

Another wartime analog device, the Link Trainer, became the inspiration for Whirlwind, the first digital computers capable of real-time operation. In 1939, President Franklin D. Roosevelt called for the production of 30,000 warplanes. Those aircraft required trained pilots to fly them, which necessitated orders for thousands of flight simulators from Link Aviation Devices of Binghamton, New York. The simulators look like crude amusement park rides to a modern eye, but they were effective. The pilot trainee placed a hood over the cockpit and had to fly solely on instruments. He received instructions through a headset to fly a certain course. The simulated plane's path was traced over a map by a "crab," to be compared with the course set by the instructor. The Link Trainer used a complex arrangement of pneumatic and electrical devices. It continued to train pilots into the 1950s.[7]

Project Whirlwind's goal was to place the crude, ad hoc design of the Link Trainer on a more theoretical basis. It was led by Jay W. Forrester of the MIT Servomechanisms Laboratory. Whirlwind was authorized in 1946 when the US Navy's Special Devices Division approached the lab with a proposal to build a simulator, not for the purpose of training pilots but for simulating the characteristics of new aircraft while they were still under design. In Whirlwind's initial design, the characteristics of a proposed aircraft would be specified by electronic analog circuits. Reconfiguring the machine to test each new design proved to be a stumbling block. It would require somehow rewiring the circuits of the machine.

A crucial turning point for Whirlwind and for the history of computing came in 1945 when a recent MIT graduate, Perry O. Crawford Jr., joined the project. During the war, Crawford had written a master's thesis, "Automatic Control by Arithmetical Operations," in which he discussed the feasibility of using calculating machine elements to control guns.[8] In the summer of 1945, Forrester & Crawford discussed the feasibility of using digital techniques for fire-control applications.[9] By the end of the war, the Whirlwind team was convinced that the digital approach was superior but realized that fundamental advances in digital circuitry were required to build the proposed simulator.[10]

The Whirlwind computer did not become fully operational until about 1953. By that time it was no longer a machine for simulation, but rather the heart of an air defense system to counter the Soviet Union's threat of atomic attack. Whirlwind influenced the subsequent development of computers in many ways. Whirlwind's original CRT memory system was soon replaced with magnetic cores, which were used throughout the industry for decades after being proven there. Robert Everett, its designer, chose an unusually short 16-bit word length to allow for 2^5 (32) operation codes and 2^{11} (2,048) memory locations. He argued that not all problems required long word lengths and that the computer could be programmed to compute at double or higher precision when necessary.[11] The program needed for higher precision arithmetic was among the reasons Whirlwind stimulated the production of one of the first high-level language compilers.[12]

SAGE

Whirlwind's shift from simulation to air defense was a consequence of the Soviet Union's testing of an atomic bomb in 1949. Earlier in the decade, *Fortune* magazine published a series of polar projection maps by the cartographer Richard Harrison. Looking down from the North Pole showed that the Soviet Union was a lot closer to the United States than generally realized. After 1949, the Harrison maps took on an ominous significance: Soviet bombers flying over the North Pole could threaten US cities with atomic bombs.

The Soviet threat led the US Air Force to commission production versions of the Whirlwind for the SAGE (semi-automatic ground environment) network. SAGE was built to detect, identify, and assist in the interception of enemy aircraft. It relied on computers to process information gathered using radar, aircraft, telephone lines, radio links, and ships and present it in a combination of textual and graphical form.

As historian Paul Edwards has pointed out, the question of whether SAGE worked depends on one's perspective. Its practical military usefulness was minimal. SAGE was designed to detect and guide interceptors to Soviet bombers flying over the North Pole. By the time the first of twenty-four SAGE installations, at McGuire AFB, New Jersey, was declared operational in July 1958, the Soviet Union was already test-launching intercontinental ballistic missiles able to strike the US. SAGE was useless against them and would in any event have been easy to disable with an initial strike, as its control centers were not hardened.[13] Yet, as Edwards noted, "in another, important sense, SAGE did 'work.'" It helped to create the illusion of an active defense against nuclear attack, and it boosted the American leadership in the emerging computer industry.[14]

SAGE introduced more fundamentally new features to computing than any other project of its era, including networking computers and sensors and the development of interactive computer graphics. SAGE's computer was developed from MIT's Whirlwind design. IBM won the contract to build the computers, delivering a prototype in 1955. In another innovation, each system consisted of two identical computers running in tandem to increase reliability, with a switch to transfer control if the primary computer failed. Although the computers used vacuum tubes (55,000 per pair), the reliability of the duplexed system was remarkably good. The last original SAGE computer, operating in North Bay, Ontario, was shut down in 1983.

The important part of *semi-automatic ground environment* was *semi*. The Air Force wanted everyone to know that even with all of the real-time acquisition and processing of radar data, decision making would rest with humans. Within the command centers, operators used video screens to monitor airspace. According to an IBM promotional video, "air defense required split second presentation, as well as split second calculation . . . IBM applied the latest extension of data processing, the display scope: a giant picture tube, on which computer results are instantly and continually translated into visual images." The video described these unfamiliar devices as the "offspring of a marriage between a television tube and a radar screen."[15] Users could switch display options using controls to the side of the screen and select points of interest by shooting them with a light gun (figure 4.2).

The SAGE contract generated half a billion dollars in revenue for IBM during the 1950s, which was more than it was receiving from leasing its regular computers. SAGE's role in getting IBM into the production of core memory had a particularly dramatic effect on its later success with commercial computers. A primary requirement for the system was high reliability, which ruled out mercury delay lines or electrostatic memory. Given the low yields of early cores supplied to IBM, it seemed that millions would have to be made and tested.[16] IBM's own efforts fared much better, producing yields of up to 95 percent by 1954. Among other things, it worked with the Colton Manufacturing Company, which provided pill-making machines to the pharmaceutical industry, to adapt its equipment to press cores with uniform properties.[17]

Figure 4.2
A SAGE console and part of the processor. Note the light gun in the protective box next to the display scope.
Courtesy of the Computer History Museum.

Real-time operation was a challenge for software as much as hardware. IBM subcontracted this to the System Development Corporation, a specially created offshoot of the RAND Corporation. SDC had to develop new techniques and train huge numbers of programmers to apply them. For example, SAGE needed a shared data pool tapped into by different subsystems, a precursor of the later database management systems that became so important in business computing. It has often been claimed that SDC employed or trained most of the programmers in America during the late 1950s.[18] This wasn't literally true. To help its punched card customers to computerize their payroll and accounting systems, IBM trained more than fourteen thousand programmers in 1957 alone, twice the cumulative total SDC reported by 1960. But the IBM training covered basic data processing techniques, described by *Business Week* as courses of three or four weeks aimed at people with "average intelligence, an aptitude for logical thought, and, in most cases, no more than a high school diploma."[19] SAGE provided deeper training in real-time and systems programming, and its veterans played important roles in early interactive computing systems for businesses, such as the SABRE system IBM built to computerize ticket reservations for American Airlines in the late 1950s.[20] Even its name, which ostensibly stood for *semi-automatic business research environment*, was a nod to SAGE.

SAGE was followed by a succession of other military command and control systems such as the SACCS (strategic automated command and control system) used to coordinate America's nuclear weapons. Readers of a certain age may remember this from the 1980s movie *WarGames*. By the early 2000s, real-time control systems had spread throughout the armed forces, tying individual soldiers to battlefield commanders.

NASA Mission Control

In October 1957, the Soviet orbited an earth satellite, Sputnik, ahead of the US Vanguard effort. The successful US Explorer 1, developed at the US Army's Redstone Arsenal in Huntsville, Alabama, raised the stature of that facility and of its team of German engineers led by Wernher von Braun. In 1958 the National Aeronautics and Space Administration (NASA) was established and the Huntsville facility became one of NASA's centers for rocket development. Von Braun and the Center's head of computing, Helmut Hoelzer, established a close relationship with IBM, whose 709 and later 7090 mainframes were in heavy use at the facility.

Most of a space mission consists of unpowered flight, but a lot of computing must be done during the initial minutes of launch, when the engines are burning. If the craft is off course, it must be destroyed. If a launch goes well, the resulting orbit must be analyzed to determine if it is stable. The calculations are formidable and must be carried out in a few seconds. In November 1960 NASA installed two 7090 computers at the newly formed Goddard Space Flight Center in Greenbelt, Maryland. Each could compute trajectories in real time, with one serving as a backup to the other. A backup system, using a single IBM 709, was located in Bermuda, the first piece of land the rocket would pass over after launch.[21] The results went back to NASA's Mission Control in Florida where a flight controller compared planned and actual trajectories to make a series of "go or no go" decisions, beginning ten seconds after engine cutoff and continuing throughout the mission.[22]

Because IBM leased its computers, customers were not permitted to make modifications to the hardware. NASA asked to be exempt from this policy in order to modify the equipment for real-time operations. A 7090 typically processed data from tapes and card readers in the same computer room. In this case the data gathered by radar stations at Cape Canaveral, Florida, were transmitted to Greenbelt, a thousand miles away, at the impressive speed of one thousand bits per second. They also developed system software called Mercury Monitor that used interrupts to immediately handle incoming data to ensure that a life-threatening situation was not ignored. Multiple levels of priority were permitted, as directed by a special trap processor.[23] During the mid-1960s, Mission Control moved to Houston, where a system of three (later five) 7094 computers, each connected to an IBM 1401, was installed.[24] In August 1966, they were replaced by a suite of System/360 Model 75 computers. After the Apollo 8 mission in December 1968, the System/360s in Houston handled the missions' primary navigation functions.

NASA's hardware and software customizations showed for the first time that a commercial mainframe could operate in real time. The Mercury Monitor evolved into a real-time extension of the standard System/360 operating system, taken up by IBM to satisfy the needs of other, more traditional customers. IBM engineers Tom Simpson, Bob Crabtree and three others called this HASP (Houston automatic spooling priority). It allowed the Model 75 to operate as both a batch and a real-time processor. In the 1970s HASP became a fully supported IBM product.

MINIATURIZATION: MISSILES AND MINICOMPUTERS

Miniaturization was one direction of computer technology that SAGE did little to advance. The project's huge budget let IBM push vacuum tube technology far beyond the size limits feasible for civilian applications. Its duplicate computers filled an entire floor of each SAGE blockhouse, coming closer than any other real-world machines to the popular idea, reinforced by science fiction of the period, that 1950s computers filled city blocks.

Guiding Minuteman Missiles

Other Cold War needs pushed real-time computing in the opposite direction: toward the small, light, and portable. Over the course of about twenty minutes, the intercontinental ballistic missiles (ICBMs) that rendered SAGE strategically irrelevant could arc up into space, guide themselves to the other side of the planet, and explode a hydrogen bomb over an enemy target. They showcased breakthroughs in rocket engines, metalworking, and the miniaturization of hydrogen bombs, which fit in the rocket nose cone and weighed around six hundred pounds. An ICBM was also a flying computer. They provided the most dramatic examples of advances in computer electronics driven by aerospace needs for lightweight systems, high computing speeds, low power consumption, and above all, reliability.

The first generation of Minuteman missiles were deployed in 1962. The guidance computer, built by North American Aviation, weighed about sixty-two pounds and was transistorized—one of the reasons that American missiles were much smaller than their Soviet counterparts, which relied on vacuum tube guidance systems. The guidance computer relied on a miniature hard disk drive for its main memory, timing its calculations around the disk's revolutions. More than thirty times a second it processed inputs from the gyroscopes and sent updates to the engine nozzles to keep the missile on track.

The Digital Equipment Corporation

These military projects had another important legacy: inspiring the creation of smaller, cheaper computers intended for real-time control applications. These machines were descendants of Whirlwind, not so much in their design as in the exodus of key people from the MIT Servomechanism Lab to laboratories and companies in the Boston suburbs. This market was created and dominated by the Digital Equipment Corporation (DEC, pronounced "deck"), set up in 1957 by Kenneth H. Olsen (who ran the company for most of its existence), his brother Stan, and Harlan Anderson in a corner of an enormous, largely abandoned 19th-century woolen mill astride the Assabet River in Maynard, Massachusetts. As a graduate student at MIT, Ken Olsen worked on fitting the Whirlwind with core memory. In the mid-1950s he represented MIT's Lincoln Lab in its collaboration with IBM on the computers for SAGE.

In 1955, Olsen took charge of a very early transistorized computer for Lincoln.[25] The TX-0 retained the high speed of Whirlwind, essential for real-time applications, but in a much smaller package. It was built using around 3,600 of Philco's new surface-barrier transistors. When completed in 1957, the TX-0 was one of the most advanced computers in the world. It was moved to the MIT campus in 1958, where it operated until 1975.[26] Reflecting its connections to the interactive SAGE system, the TX-0 had a cathode-ray tube display and a light pen, which allowed an operator to interact directly with a program as it was running. The designer of that display was Ben Gurley, who in 1959 left Lincoln Labs to become one of Digital Equipment Corporation's first employees.

Gurley also designed Digital Equipment Corporation's first computer, the PDP-1, which incorporated many of the TX-0's architectural and circuit innovations. Most transistorized machines, like the IBM mainframes we described previously, retained architectures originally designed for vacuum tubes. The PDP-1 was designed around the possibilities of transistors. It was capable of 100,000 additions per second, fewer than the IBM 7090 but many more than the drum-based computers in its price class. Its core memory held four thousand 18-bit words, later expanded to sixty-four thousand.

Like the TX-0, the PDP-1 allowed efficient data transfers between peripherals and core memory. Its processor defined up to sixteen different priority levels of interrupts, using circuits to handle them in the right order. The processor handled the most urgent needs first, returning to its regular work as soon as the transfer had been dealt with. That solution did not entirely relieve the main processor, unlike IBM's mainframe *channel* systems, but it involved only a small fraction of the cost and complexity. To DEC, this seemed a good tradeoff for a fast but relatively affordable system able to handle real-time operations and batch processing jobs. The PDP-1's initial selling price was $120,000. As DEC engineers later described it, "A single IBM channel was more expensive than a PDP-1."[27]

The PDP-1 ran real-time applications at a fraction of the cost of mainframe systems such as the Mercury Monitor. Its customers were few, with only 55 produced, but they were sophisticated, devising many novel applications. The first customer was the Cambridge consulting firm Bolt Beranek and Newman (BBN) whose Edward Fredkin suggested a number of improvements to the PDP-1. Other users included the Lawrence Livermore Laboratory, Atomic Energy of Canada, and the telecommunications giant ITT. At MIT, it was used for new kinds of interactive real-time applications such as text editing and video games. We discuss these in the next chapter.[28]

DEC not only permitted modification by its customers, it encouraged it. The tiny company could not afford to develop the specialized interfaces, installation hardware, and software that were needed to turn a general-purpose computer into a useful product. Its customers welcomed the opportunity.[29] DEC soon began publishing detailed specifications about the inner workings of its products and it distributed them widely. Stan Olsen said he wanted the equivalent of "a Sears Roebuck catalog" for Digital's products, with plenty of tutorial information on how to hook them up to industrial or laboratory equipment.[30] DEC printed these manuals on cheap newsprint, distributing them liberally to potential customers.

DEC Defines the Minicomputer

Just as CDC was promoting its multimillion dollar supercomputers, DEC created a wave of enthusiasm for high-performance computers at the other end of the market. The product that revealed the size of the market for small, fast, and cheap computers arrived in 1965: the PDP-8 (figure 4.3). It was called a minicomputer. This term was catchy, it fit the times, and it gave the PDP-8 an identity. The miniskirt happened to come along at the time the PDP-8 was beginning to sell. Another source of the name, which fit the PDP-8 perfectly, was also a British export: the Morris Mini-Minor. It was lightweight, responsive, and economical to operate, and it easily outperformed more expensive models. Mini Minors and miniskirts sold around the world. Digital Equipment Corporation did the same with minicomputers.

The PDP-8's circuits followed the pattern set with the PDP-1 by using the latest high-performance transistors. That project was led by C. Gordon Bell, with Edson DeCastro responsible for the logic design. Improvements in logic and core memory technology reduced the memory cycle time to 1.6 microseconds—slightly faster than the IBM 7090 and over one thousand times faster than the Bendix G-15, which had been the fastest inexpensive drum computer of the late 1950s.[31] However, the PDP-8's short word length meant that it could not compete with its mainframe competitors in working with long numbers. DEC had already experimented with a 12-bit design with its PDP-5, which was released in late 1963. The 12-bit word length cut costs, but with five bits of each instruction used for the operation code there were only seven bits left for the address field, coding 128 possible values. A computer with just 128 words of memory would be useless; so, like the System/360 machines discussed in the previous chapter, PDP-8 instructions specified only part of the address. The basic PDP-8 came with four thousand words of memory, divided into thirty-two pages. The short addresses in the instructions applied within the currently selected page. The PDP-8 also got around the limit by allowing *indirect addressing*: the 7-bit address in the instruction could reference a memory location containing the full 12-bit address. Its limited memory steered programmers away from high-level programming languages, toward assembly language or even machine code. Yet the simplicity of the PDP-8's architecture, coupled with DEC's policy of making information freely available, made it an easy computer to understand.

The PDP-8 occupied eight cubic feet of volume and weighed 250 pounds, tiny compared to a mainframe or even a traditional tabulating installation.[32] It earned its designation of *mini* with modularization and miniaturization techniques pioneered by the Minuteman guidance systems. For external memory DEC provided a simple, inexpensive, but capable tape system. A reel of DECtape was light and portable; the drive was compact and could fit into the same equipment rack as the computer itself. Data

Figure 4.3

A Digital Equipment Corporation PDP-8. The original PDP-8 was the size of a refrigerator, earning it the *minicomputer* tag. Its electronics were spread over many small and easily removed circuit boards, visible here without the usual smoked glass cover. Division of Medicine and Science, National Museum of American History, Smithsonian Institution.

could be read or written in either direction in blocks of 128 words. This made it more flexible and convenient than the bulky mainframe tape drives, which were optimized for batch processing applications involving vast amounts of data.

Finally, there was the matter of pricing the PDP-8. A low price would generate sales, but the experiences of many start-up computer companies of the 1950s showed that it might deny DEC the revenue to support continued research and development. DEC decided to take that risk, pricing the PDP-8 at $18,000 including a teletype terminal. Within a few years one could buy a smaller, slower version, the PDP-8/S, for less than $10,000. These prices shocked the computer industry and generated a flood of orders. DEC eventually installed over 50,000 PDP-8 systems, plus uncounted microprocessor-based implementations developed years later (see figure 4.4).[33] It was the world's best-selling computer family, retaining the title until the arrival of personal computers in the late 1970s.

The PDP-8 generated a thriving original equipment manufacturer (OEM) business: a separate company bought minicomputers, added specialized hardware for input and output, wrote specialized software for the resulting systems, and sold them (at a high markup) under its own label. A typical OEM product was the LS-8 from Electronics Diversified, Inc. of Hillsboro, Oregon. It was used to operate theatrical stage lighting by controlling a complex of lights through programmed sequences. The LS-8's abilities were cited as a key element in the success of the long-running Broadway hit *A Chorus Line*.[34] Inside the LS-8 was a PDP-8. Users of the LS-8 did not necessarily know that, because the LS-8 had its own control panel, tailored for theatrical lighting crews. OEM applications ranged across all segments of society, from medical instrumentation to small business record keeping to industrial controllers. One PDP-8–based system was even installed in a potato-picking machine and carried on the back of a tractor.

Established companies, including IBM, eventually entered the minicomputer market to compete, but DEC was able to escape the fate of earlier start-ups. It prospered, in part, by continuing to draw from the knowledge and skills of the MIT research community. As the 1960s drew to a close, the minicomputer was breaking out of its laboratory control and real-time OEM setting and moving into the data processing territory that IBM considered its own. IBM's low-end 360 Model 20 was about as small and cheap as the 360 architecture could go, and even that was only partly compatible with the rest of the line. This was the mirror image of the handicaps that the 360 architecture imposed on IBM's most powerful computers in the supercomputer market. In 1969, IBM responded with System/3, an incompatible computer that could be rented for as little as $1,000 a month. It was "a candid concession that System/360 could not bridge the widening opportunities in the marketplace."[35] System/3 was a successful product because it was easier to adapt for small scale business and

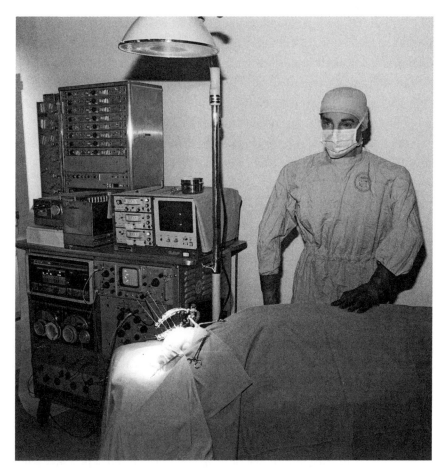

Figure 4.4
The PDP-8e, a miniaturized rack-mount version of the PDP-8, was released in 1970. As one of the smallest and cheapest computers of its day, it helped to establish the minicomputer as an embedded control device. It is the box with the switches on the center left, above two DEC Tape drives. This one was integrated with neurosurgery equipment at Massachusetts General Hospital, but the picture is posed as part of an exhibit at the former Computer Museum in Boston.

accounting jobs than minicomputers, which carried their laboratory workbench heritage with them.

INTEGRATED CIRCUITS

The space consumed by diodes, transistors, and other solid-state electronic components was determined by the need to package them for easy mounting on circuit boards and interconnection with wires. In the early 1960s, the Air Force initiated development of an improved Minuteman, deployed in 1967. Its requirements overwhelmed the existing guidance computer, and so the new missile incorporated the first production computer built with a new technology: the integrated circuit (IC), known popularly as the *silicon chip*. This was the culmination of years of research sponsored by the Air Force, Army, and Navy on ways to shrink electronics. Large Air Force and NASA orders for early chips helped propel the new technology into the commercial marketplace.

Modular Electronics

The first steps in miniaturization involved printed circuit boards, which were used to eliminate wires and pack components more closely together by etching a pattern on a plastic board covered with copper or some other electrical conductor. The board was then dipped into a solvent that removed the conductor from all areas not protected by the etched pattern. This technique was pioneered during the Second World War in several places, including the Centrallab Division of the Globe-Union Company in Milwaukee, Wisconsin, where circuits using miniature vacuum tubes were produced for an artillery fuse. After the war, the printed circuit, as it became known, was adopted by the US Army's Signal Corps for further development. The Army called it Auto-Sembly, emphasizing production benefits rather than miniaturization.[36]

Printed circuits found their first civilian applications in hearing aid production, where miniaturization and portability had long been crucial.[37] They were eventually adopted by the computer industry along with transistorization, initially to create standard modules that each packaged a few components. IBM adopted this approach for Stretch and later rolled it out across its product line. DEC pushed modularity and miniaturization further. The firm's first products were logic modules, not computers. For the PDP-8 it relied on *flip-chip* modules: small printed circuit boards on which transistors, resistors, and other components were mounted. These in turn were plugged into a hinged chassis that opened like a book. The result was a system consisting of a processor, a control panel, and core memory in a package small enough to be embedded into other equipment. To support the huge demand for its systems, DEC developed new mass production techniques including the use of automatic wire-wrap machines for the flip-chip modules.

Printed circuits met the commercial needs of the 1950s and early-1960s well, but military sponsors pushed for more dramatic miniaturization. Jack Kilby was born in Jefferson City, Missouri, and grew up in the farming and oil-well supply town of Great Bend, Kansas. After graduating from the University of Illinois with a degree in Electrical Engineering in 1947, Kilby took a job at Centrallab in Milwaukee—still the industrial leader in printed circuits. He became involved in getting the company involved in making products using germanium transistors. "By 1957," according to Kilby, "it was clear that major expenditures would soon be required. The military market represented a major opportunity, but required silicon devices . . . The advantages of the diffused transistor were becoming apparent, and its development would also have required expenditures beyond the capabilities of Centrallab. . . . I decided to leave the company." The following year he joined Texas Instruments in Dallas, known for having pioneered the shift from germanium to silicon transistors. "My duties," he recalled, "were not precisely defined, but it was understood that I would work in the general area of microminiaturization."[38]

The First Integrated Circuits

Texas Instruments had a project, Micro-Module, to deposit components on a ceramic wafer. Kilby did not find this approach cost effective. In the summer of 1958, he proposed making all the electronic components, not just the transistors, out of germanium or silicon, so that an entire circuit could be fashioned out of a single block of semiconductor material. Increased costs for the individual components would be more than offset by not having to set up separate production, packaging, and wiring processes. Kilby built a circuit with resistors and capacitor made of silicon in August 1958. In September he built another, made of germanium, whose components were laboriously constructed from a wafer and interconnected by fine gold wires. The result, an oscillator, worked. In early 1959 he applied for a patent, which was granted in 1964. Texas Instruments named it the solid circuit.[39]

Robert Noyce had been thinking along the same lines when he heard of Kilby's invention. Noyce also grew up in the Midwest, in Grinnell, Iowa, where his father was a Congregational minister. One of his coworkers at the newly founded Fairchild Semiconductor in Mountain View, California, Swiss-born Jean Hoerni, had developed a process for making silicon transistors. This was well suited for photo-etching production techniques, eliminating the hand construction Kilby had undertaken. It was called the planar process because it produced transistors that were flat. In January 1959, Noyce described in his lab notebook a scheme for using this process to do essentially the same thing as Kilby, but with a piece of silicon (figure 4.5).[40] His planar process included a silicon dioxide protective layer on top of the device. For Noyce the invention of the IC was less the result of a sudden flash of insight than the result of a gradual build-up of engineering knowledge at Fairchild about materials, fabrication, and circuits since the company's founding in 1957.

Noyce applied for a patent in July 1959, a few months after Kilby. Years later the courts would sort out the dispute over who the "real" inventor was, giving each person and their respective companies a share of the credit. However, the idea of incorporating Hoerni's process, thereby allowing one to make the electrical connections in the same

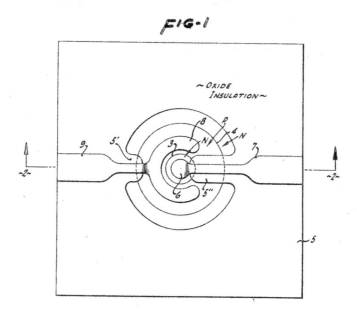

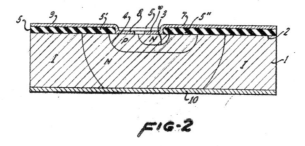

Figure 4.5

A page from Robert Noyce's 1959 patent, which covered a crucial process for the creation of what later became known as integrated circuits.

process as making the devices themselves, was the key to the dramatic progress in integrated electronics that followed. The relationship between printing, photography, and microelectronics has been a close one.

The US aerospace community provided a crucial market for integrated circuits. The improved Minuteman II was really a brand-new missile wrapped around an existing airframe. Autonetics, a division of North American Aviation that was contracted to build its guidance system, chose integrated circuits as the best way to meet project requirements. Its computer used about 2,000 integrated and 4,000 discrete components, as compared with the 15,000 discrete components used in the original Minuteman computer.[41] Thanks in part to this more powerful, smaller, and lighter guidance computer, the new missile was more accurate and more flexible in its targeting, and it could carry a bigger nuclear warhead. Minuteman II first flew in September 1964; a year later, the trade press reported that "Minuteman is Top Semiconductor User," with a production rate of six to seven missiles per week.[42]

According to Kilby, comparisons between the two types of circuits published by Autonetics to explain its decision "seemed very dramatic, and probably did more than anything else to establish the acceptability of integrated circuits to the military."[43] The transistor industry had a history of boom and bust cycles. Were it not for Minuteman II, it would not have quickly established volume production lines for ICs: "Minuteman's schedule called for over 4,000 circuits a week for Texas Instruments, Westinghouse, and RCA."[44] By then Texas Instruments also had adopted planar process technologies. Fairchild was not among the three major suppliers for Minuteman, but before long it too was heavily reliant on the military market: "Military and space applications accounted for essentially the entire integrated circuits market last year [1963], and will use over 95 percent of the circuits produced this year."[45]

Chips for Apollo

Like the Minuteman project, Apollo needed miniaturized computers to control rocket engines. Its guidance system was the product of the MIT Instrumentation Laboratory, which under the leadership of Charles Stark Draper was responsible for the design of guidance computers for the Polaris and Poseidon missiles. Initially, most guidance was to be handled from the ground; as late as 1964 the spacecraft itself was expected to use an analog computer.

As the challenges of the lunar mission became clearer, the computer was redesigned and asked to do a lot more. Eldon Hall, an electronics designer at the Lab already familiar with integrated circuits, ordered the purchase from Fairchild of MicroLogic brand planar silicon chips. Their small size and weight were attractive, although Hall was concerned about the lack of data on reliably manufacturing them in quantity. He persuaded

NASA to accept the use of the new technology. By early 1963 the lab had a model computer running with the new circuits. It had been completely redesigned to exploit the advantages of integrated logic.[46]

Around seventy-five Apollo Guidance Computers were built, of which about twenty-five flew in space. In contrast to the Minuteman computer, for which Texas Instruments was subcontracted to develop more than a dozen types of integrated circuit, each Apollo computer contained about 5,000 identical chips. Thanks to those bulk orders, the price dropped from $1,000 a chip for initial samples to between $20 and $30 a chip for the production run of computers.[47] By August 1964, Fairchild and Philco, an East Coast company given a production contract for the chips, had collectively shipped 110,000 integrated circuits for use in Apollo.[48] This gave semiconductor companies a chance to develop facilities for the bulk production of standard circuits, which in turn they could apply to a civilian market. By the time of the last Apollo flight in 1975 (the Apollo-Soyuz mission), one astronaut carried a pocket calculator (an HP-65) whose capabilities were greater than the on-board computer's. Such was the pace of innovation set in motion by the aerospace community.

Chips Spread

An internal IBM memorandum written in September 1963 stated that ICs "do not constitute a competitive threat either now or in the next five years." For the System/360, announced in April 1964, IBM relied on a less aggressive form of miniaturization. In its solid logic technology, circuits were deposited on a ceramic substrate about half an inch thick, with metallic conducting channels printed on it. In September 1964, another memorandum acknowledged that rapid progress elsewhere with integrated circuits meant that IBM now had a "2–4 year lag in practical experience" with the new technology.[49] By 1966, several competitors were incorporating integrated circuits into their computer designs. IBM eventually followed suit with its System/370 line in the early 1970s.

By then, IC makers had adopted a standard that laid a solid foundation on which the computer industry would grow for the next two decades. That was to supply a set of chips that used transistors to implement all logic: so-called transistor-transistor logic (TTL). TTL chips were inexpensive and easy to design around.[50] Engineers at Fairchild introduced a dual in-line package (DIP) in 1964. This placed chips in an easy-to-handle black plastic or ceramic case, with leads arranged along either side for easy insertion into sockets or directly through the holes of a printed circuit board for soldering. It provided the iconic appearance of chips for the early personal computer age.

By 1970 the printed circuit board had evolved to handle integrated circuits as well as individual components. A minicomputer designer could now lay out a single large printed circuit board, with places for all the ICs necessary for the circuits of a small computer. On an assembly line (usually staffed by women), a person (or machine) would "stuff" chips into holes on one side of the board. She would then place the board in a chamber, where a wave of molten solder would slide across the pins protruding through to the other side, attaching them securely to the printed connections. The process was fast and reliable, and it yielded a rugged product. It was Fairchild engineers who first explored this method of packaging and wiring.

This combination: TTL logic, dual in-line packaging, and large, wave-soldered printed circuit boards remained a standard into the 1990s, before giving way to more compact packages. Minuteman ICs had been packaged in flat configurations, with the leads protruding horizontally. That reduced the overall volume of a device, but flat packs were difficult to install on circuit boards. By the turn of the millennium this problem was overcome, and flat packaging has since become common for consumer products for which compact size is critical.

BUILDING RELIABLE SYSTEMS

Military and control applications demanded consistently high reliability as well as low weight and consistently fast responses. That too goes back to the original Minuteman missiles. In the 1950s, the US Air Force had been repeatedly embarrassed when multi-million dollar rocket launches failed because of components costing just a few dollars. For the Minuteman program it set up new rules that penetrated deep into the production lines of suppliers. At every step of the production, a log was kept that spelled out exactly what was done to each electronic part and by whom. Months later, if a part failed a test, one could go back and find out where it had been. If the failure was due to a faulty production run, then every system that used parts from that run could be removed from service.[51] Suppliers introduced *clean rooms* where workers wore gowns to keep dust away from the materials they were working with. Invented at the Sandia National Laboratories in the early 1960s for atomic weapons assembly, they were washed by a constant flow of ultrafiltered air from the ceiling to the floor. Eventually the industry would build fabrication rooms, or "fabs," that were many times cleaner than a hospital.[52]

The electronics industry developed these procedures to make transistors for Minuteman. Suppliers who could not or would not follow them were dropped. This program came at a steep price. Each Minuteman in a silo cost between $3 and $10 million, of which up to 40 percent was for the electronics. The culture took root. Those who passed the test found an additional benefit: they could charge elsewhere for components meeting the "Minuteman hi-rel" standard. Eventually the 100-fold reduction in failure rates demanded by the Air Force was expected for the commercial world as well.[53]

Reliable Software

The exceptional reliability needed for aerospace control systems challenged software producers as well as hardware designers. Civilian applications and operating systems were full of bugs, many of which were gradually identified and removed during use. An error in the guidance code of a nuclear missile or manned space rocket needed to arise only once during use to make the entire mission a spectacular failure. Programs had to work perfectly even when faced with unanticipated situations or faulty input data. The 1962 explosion of the Mariner I rocket headed for Venus was publicly attributed to a "missing hyphen" in its booster control program.[54] From the 1970s onward the application of formal methods proving that hardware and software met logical specifications was a growth area within computer science.

NASA met challenges by throwing money and people at them. Taking the same approach to software reliability in the Apollo program worked in practice, if not in theory. The staff at the MIT Instrumentation Laboratory, where the computer's hardware was designed, also wrote the software. The software worked without error. In the opinion of Margaret Hamilton (figure 4.6), who led the software team, the success was

Figure 4.6
Margaret Hamilton, working at her desk at MIT's Charles Stark Draper Lab in a photograph taken for its 1973 Annual Report. Hamilton played a crucial role in producing reliable software for the Apollo missions, for which she won the Presidential Medal of Freedom in 2016. After Apollo she founded a business to extend similar methods to other software projects. Courtesy of Draper.

due to a brute force attack, with numerous reviews of the code probing the reliability of every line.[55]

Apollo software was written on mainframes and translated into binary data. The code was wired into read-only *core ropes*: long strings of wires woven through magnetic cores, which stored a binary one if a wire went through it, or a zero if the wire went around it. Hamilton described the technique of scrubbing the code as the *Auge Kugel* method: the term was (incorrect) German for *eyeball*.[56] In other words, one looked at the code and decided if it was correct or not. NASA was not sure whether the academic culture of MIT was disciplined enough for the job. It had the Instrumentation Lab print out the code listings on reams of fan-fold paper and ship them to TRW in Los Angeles, where John Norton, a TRW employee, would scrutinize the code and indicate any anomalies he found. Some at MIT resented Norton's intrusion, but he did manage to find problems with the code.[57] Hamilton was known as the "rope mother." She kept a sense of humor, calling some of the anomalies in the code FLTs or "funny little things," and the women who wired the ropes at the Raytheon plant in suburban Boston, LOLs—"little old ladies." Many of them had worked at nearby Waltham Watch, part of a long tradition of female labor in precision manufacturing.[58]

In 1968 the Apollo 8 Mission carried three astronauts to an orbit around the moon and returned them safely. The on-board Apollo Guidance Computer handled its navigation and insertion into Lunar orbit. The computer gave NASA the confidence to carry out a landing the following year. Some of the work done at MIT for the mission was later formalized into procedures that were adopted elsewhere, although the unique nature of space computing did not always translate well to the commercial sector, in which there were different tradeoffs. The eyeball method evolved into formal code reviews and *walkthroughs*: a process that became known as Validation and Verification.

President Barack Obama awarded Margaret Hamilton the Presidential Medal of Freedom in November 2016 for her work on Apollo software. The award was followed in 2017 by a series of Lego figures honoring her and other women who contributed to space technology. John Norton continued a distinguished career at TRW. Some there claimed that he had the ability to *reverse compile* code: to examine lines of object code and determine which Fortran statements produced it. In 1972 Norton was managing a TRW contract for the Bonneville Power Administration, which used Digital Equipment Corporation PDP-10 computers to manage the power generated by dams on the Columbia River. Bill Gates, then a student at Seattle's Lakeside High School, was assigned to work with Norton. Years later Gates acknowledged Norton's software skills as an influence on his own love of programming.[59]

Redundant Hardware

NASA's follow-on to Apollo, the space shuttle (known officially as the Space Transportation System), was designed to be flown by computer. The shuttle's reusable orbiter flew straight up into space with the aid of booster rockets and a huge disposable fuel tank, but it returned to Earth by gliding down through the atmosphere. It began reentry eighty miles up with a speed of around 17,500 miles an hour. Half an hour later and 5,000 miles away, the orbiter came to a stop at the end of a very long runway. To survive launch and reentry, the orbiter had a very angular shape, which made it inherently unstable as a glider. No human pilot using traditional controls could make the tiny adjustments to flaps and ailerons needed to keep on course and avoid burning up as atmospheric drag converted momentum into heat. The unique shape required extensive wind tunnel testing at the design stage and later drove improvements in computer modeling techniques.[60]

IBM's Federal Systems Division won the contract to supply computers and control software for the shuttle. This had been established in 1959, after the firm's success with SAGE, in order to combine its federal and military work.[61] The computer IBM designed for the shuttle was a variant of a model used in the B-52 bomber and F-15 fighter. It had a long production run, and its reliability and performance were well understood. Its architecture was patterned after the System/360, but much reduced in size, weight, and power consumption. IBM called its avionics computer System/4π, "since there are 4π steradians comprising the surface of a sphere."[62] The Shuttle computers were renamed AP-101. IBM was by then ready to use integrated circuits for processor logic, but for memory it stuck with proven magnetic core technology.

IBM relied on redundancy, a technique it had previously used for a computer that controlled the boosters of Apollo's Saturn V rockets. Its launch vehicle digital computer (LVDC), was installed between the upper stage of the Saturn and the modules that went to the moon. The LVDC controlled and guided the three stages of the Saturn. It had to withstand the forces of a rocket launch and make split-second decisions as the launch proceeded. For reliability, IBM employed what it called triple modular redundancy. Three separate modules handled each logical function. If they did not agree, a "voting" circuit selected the choice of the majority. Tests indicated a mean time between failures (MTBF) of 45,000 hours. Other forms of reliability used for Apollo included a separate abort guidance computer for the lunar module, ground-based mainframes, and manual controls for the command module. These backups became crucial during the Apollo 13 mission in April 1970, when the primary computer had to be shut down to conserve power.

IBM took redundancy further with the shuttle. It was flown by five AP-101 computers, any one of which could bring it back to earth safely. Four ran identical software written by IBM engineers. The fifth computer was programmed by a different team to prevent a common programming error infecting all the shuttle computers. NASA chose five computers, not three, because it hoped the shuttle would become a work-horse "space truck" that could continue a mission after one computer failed. Two failures would end the mission, but the crew could still return to Earth safely.

On its first return from space, the shuttle handled quite unexpectedly, but its computers performed flawlessly in compensating and keeping it on course. Later, in 1983, STS-9 (*Columbia*) landed safely despite two failed computers. The failures were traced to loose pieces of solder or other contaminants in the circuit boards. Over the course of the shuttle program the software worked well, although if there had been an error in *specifications*, not coding, all five would have produced erroneous output. That never happened.

The Legacy of Space

In the early 1960s, aerospace needs for powerful, light, and miniaturized electronics drove dramatic improvements in the state of the art. New technologies such as integrated circuits found their first applications in missiles and space rockets. However, once those technologies were adopted by other industries, space applications became more conservative. Consumer products like smartphones must be endlessly replaced with new models, introducing new bugs in the software, battery problems, or other issues. Once those are straightened out, the cycle repeats itself. Those who build computers for space operations have different priorities, most notably reliability and longevity, but also special requirements such as radiation hardening.

Since the 1970s, space technology has been marked by exceptional reliability rather than raw performance. Thanks to this approach, many systems have achieved lifespans far beyond those originally promised. The space probe Voyager 1 was launched in 1977 on a mission to Jupiter and Saturn. Its main computers have just 4,096 words of memory each, which is limited even by the standards of the 1970s. Many of its systems have now broken and power from its nuclear batteries is dwindling, but thanks to redundant design, it continues to transmit data back from the fringes of the solar system long after the original members of its mission control team retired. In 2017, Voyager I successfully fired thrusters that had sat dormant since the end of its originally planned mission in 1980. To give a more recent example, the Opportunity rover that landed on Mars in 2004 was designed to work for three months but explored for more than fourteen years.

The IBM AP-101 has since become a workhorse of space flight, with a local area network of installations, along with other computers, on the International Space Station. It has been upgraded many times, with microprocessors replacing the medium-scale integrated circuits and semiconductor memory replacing cores. The Shuttle computers were upgraded midway through the Shuttle program, but as with all aerospace applications, the need for rigorous testing and software validation meant that its computers were far behind the state of the art by the time of its last flights in 2011.

To support the Shuttle and Apollo programs NASA pioneered *fly-by-wire* technology. Existing aircraft ran hydraulic lines from controls, such as the throttle, all the way to the engines, rudder, and other controls. NASA flew the first digital fly-by-wire system in 1972. Computer control meant sending digital signals down wires, to circuits controlling electric motors. By the mid-1980s, similar systems were appearing in commercial Airbus planes. Airbus also borrowed NASA's use of redundant controllers, installing a suite of three computers with a voting process.

The complex interactions of hardware and software have contributed to improvements in airline safety but have also made planes harder to test. In the 2010s, Boeing decided to fit gigantic new fuel-efficient engines to update the fifty-year-old 737, still its best-selling model, as the 737-MAX. These changed the aerodynamics of the plane, but to reduce the cost of pilot training and certification Boeing relied on software, the maneuvering characteristics augmentation system, to make the new plane respond in a familiar way when pilots manipulated the controls. Part of the code detected possible stalls and avoided them by driving the nose of the plane down. Two 737-MAX planes crashed after their computers forced them into the ground despite repeated attempts by their pilots to pull up. In 2019 the planes were grounded worldwide, costing Boeing and the airlines flying them tens of billions of dollars. It took almost two years for them to begin to reenter service. Boeing had promised a quick software patch, but investigation showed that cost-cutting moves to make the system activate on the basis of readings from just one sensor had produced a fundamental vulnerability to component failure. Worse, pilots had no easy way to understand what had happened or override the system. Boeing responded by firing its chief executive, conceding the need to retrain pilots for the updated plane, and agreeing to a $2.5 billion settlement to resolve criminal fraud charges for withholding information from government regulators.

Not all the techniques used in space missions to boost reliability have been adopted in other areas of computing. Nobody walks around carrying five cellphones, but more subtle kinds of redundancy underpin the cloud-based systems we now rely on to deliver constant access to our electronic messages and to keep our files safe. If any one computer or drive fails, the systems shift instantly to working copies. The specific techniques are different from and more cost effective than those used by NASA, but the results are similar.

Thanks to massive reductions in the cost of embedded control computers, we are now surrounded by them. Embedded computers are cheap enough to throw away in greeting cards that play snippets of audio. Those shifts relied on the further miniaturization of electronics so that an entire computer could be placed on a single chip. We discuss that process and the idea of Moore's law, with which it is often associated, further on in the book.

5 THE COMPUTER BECOMES AN INTERACTIVE TOOL

The computers we have discussed so far were institutional devices, ordered and controlled by organizations rather than individuals. They were pieces of heavy machinery whose cost could be offset only by running large amounts of work through them. Businesses installing computers for electronic data processing followed templates established for punched card machines, using teams of specialist operators to run computers. The managers on whose behalf jobs were run never interacted directly with the computer. Even programmers wrote their code on paper, receiving printed results hours or days later. Scientific computing installations sometimes took another approach, allowing users to sign up for time on computers as with other kinds of specialist lab equipment. This was called *open shop* operation, versus *closed shop*, in which the computer remained under the control of specialist operators (a reference to unionized workplaces with rules demarcating which workers could do which types of work). For larger computers, the closed shop model eventually won out, even in scientific centers, as newer technologies such as multiprogramming operating systems and hard disk drives made it even less efficient for programmers or engineers to directly control a computer.

Real-time systems worked differently. They were designed to give almost instant responses though application-specific control panels manipulated directly by the people controlling complex systems, such as air defense officers, astronauts, and mission control technicians. Learning how to work the computerized controls was part of being trained for those jobs.

In this chapter we consider two related technologies that spread during the 1960s to put interactive, general-purpose computing in the hands of thousands of users. Minicomputers were small, cheap, and reliable enough to build into industrial control systems or treat as laboratory equipment. Timesharing systems hooked up banks of terminals to larger computers. Users typed commands or entered program code and got back immediate responses from the system. Both provided interactive computing as a commodity, a sharp contrast with the huge expense and custom development involved in systems like SAGE.

THE ORIGINS OF INTERACTIVE COMPUTING

As a product, the *personal computer* did not exist until the 1970s. In another sense, however, what makes a computer "personal" is its relationship to its users. A personal computer is available to individuals to serve their personal needs. Even ENIAC occasionally served as a personal computer. Folklore has it that the mathematician Derek Lehmer was able to commandeer it over a holiday weekend, operating it with the help of his family, to carry out his pet project of sieving for prime numbers.[1]

Hacking Away at MIT

Because of the huge cost of installing and operating a computer, very few people had the opportunity to play around with one, tinkering with programs to satisfy their personal interests or for the simple joy of finding out what the machine could do. In the early 1960s, thanks to its extensive Cold War funding and close connection with DEC, MIT was in the exceptional position of having a spare computer lying around for students to play with freely. Its small TX-0 computer had been built to test core memory and transistor technology. That mission completed, "the atmosphere was loose enough in 1959 to accommodate the strays—science-mad people whose curiosity burned like hunger."[2] Around this machine and a series of successors donated by DEC, beginning with a PDP-1, grew a "hacker" culture that was later celebrated in popular folklore. DEC's Gordon Bell later talked about these machines, and other early models controlled directly by programmers, as part of a tradition of "MIT personal computers."[3]

As Steven Levy explained in his book *Hackers: Heroes of the Computer Revolution*, the "hack," an elaborate and technically inventive prank, was part of MIT's established culture.[4] This practice gave its name to the "computer hackers," students who began to compete obsessively to program the machine to perform new tricks using the tightest possible code. Because they took turns to personally control the computer, they developed interactive programming tools, such as text editors and debuggers, to suit this new way of working. The hackers gave their interactive tools names like "expensive calculator" and "expensive typewriter" to acknowledge the economic absurdity of using precious computer time this way. Hackers, as psychologist Sherry Turkle noted, "loved the machine for itself" rather than as a means to get work done.[5]

Their most celebrated accomplishment was *Spacewar*, the original computer video game (figure 5.1). Two starships fought a torpedo duel around a distant star. The game evolved in late 1961 and 1962 over countless hours of programming effort, as the

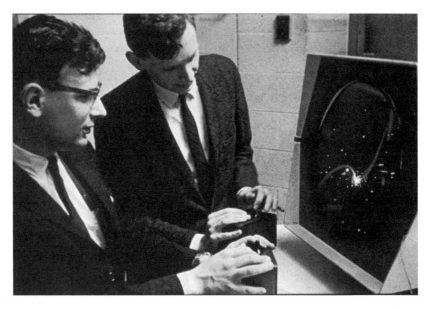

Figure 5.1
Dan Edwards and Peter Samson, two creators of *Spacewar,* play the game on MIT's PDP-1. Courtesy of the
Computer History Museum.

group played around with code to draw and move ships on a screen, draw a star field, and simulate gravity. Steve Russell made the largest contribution. *Spacewar* (or, as Russell likes to call it, "*Spacewar!*") was a compelling showcase for the special features of the PDP-1, including its fast processor, interactive controls, and vector graphics display unit (an evolution of the radar-like displays used with SAGE). The game was soon a favorite at the handful of other sites running PDP-1s with display units and was reportedly used by DEC engineers as a test of new machines.[6]

In 1958, MIT's Lincoln Laboratory finished building TX-2, the follow-up to TX-0. It had 64,000 words of memory, a fast processor, and a video display. A PhD student, Ivan Sutherland, took advantage of its power to produce a computerized drawing system called Sketchpad, shown in figure 5.2. A light pen, similar to the light gun invented for SAGE, was used to draw directly onto the screen. Users selected objects to be manipulated by touching them with the pen, anticipating a crucial element of later graphical user interfaces. As with later "drawing" programs, shapes were stored as vectors, letting users zoom in to see fine detail without loss of image quality. To support computer aided design, Sketchpad let users define master shapes and duplicate them. Changing the master automatically modified the duplicates. The system is remembered as the origin of computer graphics as a research field, and as one of the most influential computer programs ever written. Sutherland eventually won a Turing

Figure 5.2
Ivan Sutherland demonstrates his Sketchpad system running on MIT's TX-2. Shapes could be drawn using either a light pen applied to the screen or a digitizing tablet, as shown here. Courtesy MIT Museum.

Award for it. His company Evans and Sutherland, founded in 1968, pioneered the commercialization of computer graphics for applications such as flight simulation.

Some of the MIT community did try to build cheaper machines intended to be used by researchers as personal tools. Wesley Clark had been captivated by interactive computer use as an early Whirlwind programmer, going on to design the TX computers used at MIT. In early 1961 he came up with a proposal to use Digital Equipment Corporation logic modules to build an interactive single-user computer called the LINC, designing it around a target price of $25,000. A demonstration project, funded by the NIH, made sixteen video screen-based LINCs available to biomedical researchers. DEC produced several commercial variants, which initially sold for around $40,000. DEC had manufactured almost four thousand, primarily for laboratory use, by the time it discontinued the line in 1971.[7] A key feature of the LINC was its compact tape drive, with tapes that users could easily carry with them between computer sessions. As DECtape, this was eventually offered on nearly all DEC's products.[8]

As LINC's price tag shows, *cheap* was relative. Computers that cost millions of dollars were big-ticket purchases even for the largest organizations. Their purchase was approved by senior executives and they were run by centralized computing teams. Minicomputers were indeed cheap in comparison. They could be purchased from departmental budgets and used to meet local needs. Over the course of the 1960s, minicomputer prices dropped further. The $10,000 cost of a PDP-8 was a realistic amount for a well-funded lab to include in a grant request along with other scientific equipment.

The Timesharing Alternative

The first push to make interactive, general-purpose computing widely available did not center on plans to mass produce cheap computers. Nobody during the 1960s was trying to put private minicomputers on (or under) the desks of managers or nontechnical professionals. Instead, the idea was to build very large computers, each able to support many interactive users. The approach was called timesharing because the computer shared its processing cycles, or time, among several users simultaneously. This was proposed in 1959 by Professor John McCarthy of MIT, who argued that "the only way a quick response can be provided" from the university's IBM 709 "at bearable cost is by time-sharing. That is, the computer must attend to other customers while one customer is reacting to some output." McCarthy thought that this would be "the way all computers will be operated in the future" and told his colleagues that "we have a chance to pioneer a big step forward in the way computers are used."[9]

What made that feasible was the ability of a computer to fetch and execute thousands of instructions in the time it took a typist to push one key. While a person paused for a few seconds to ponder the next command to type, the computer could run an entire program for another user. Each user could have the illusion that a complete machine and its software was at his or her disposal. That included whatever programming languages the computer supported and any data sets the user wanted to use, whether supplied by others or by that user.

By 1961, in a widely discussed speech to celebrate the centennial of MIT, McCarthy had become more specific, defining a "time-sharing computer system" as "one that interacts with many simultaneous users through a large number of remote consoles." That, he argued, was the only way to deliver the interactive programming experience currently enjoyed by the small group of hackers using the TX-0 to a larger population. "Such a system" would, according to McCarthy, "look to each user like a large private computer. When the user wants service, he simply starts typing in a message requesting the service. The computer is always ready to pay attention to any key that he may strike." McCarthy recognized that because each program would "do only relatively short pieces of work between human interactions" the computer would need enough memory to hold all the currently used programs simultaneously, as it would be impossible to "shuttle them back and forth continually to and from secondary storage."[10]

In that speech McCarthy introduced not just a new kind of operating system but a new vision of computing as public infrastructure. Timesharing would underpin a new *computer utility* model, in which users purchased time on a timesharing computer to be delivered to terminals in their homes and offices: "computing may someday be organized as a public utility just as the telephone system is a public utility. . . . Each subscriber needs to pay only for the capacity he actually uses, but he has access to all programming languages characteristic of a very large system." This, he promised, would "become the basis of a new and important industry."

Doing all that required some fundamentally new operating system capabilities, supported by significant tweaks to computer architectures. A timesharing system would also need a hard disk drive big enough to store everyone's files, as it would be impractical to expect users to load and unload tapes.

CTSS

McCarthy inspired a team led by Fernando J. Corbató to build a timesharing system. A prototype for what they called the compatible time-sharing system (CTSS) was working by the end of 1961 on MIT's IBM 709. With no disk drive, they had to dedicate a tape drive to each user. That limited the system to four simultaneous users, each operating a modified Flexowriter.

The prototype gave them the confidence to build a practical version of CTSS, released for regular use in the summer of 1963. It ran on MIT's new IBM 7090, using an IBM

Figure 5.3
Corbató poses with an IBM 7090 used to run the CTSS timesharing system. Courtesy MIT Museum.

1301 disk drive able to store 28 million characters (figure 5.3). This was much larger and faster than the disk IBM had introduced with RAMAC back in 1956. Users each received a private area of the disk, known as a *directory*, to store files between sessions. Hardware upgrades followed, as the 7090 was quickly supplanted by two specially tweaked 7094s coupled with even bigger hard drives, but even with IBM's most powerful hardware, CTSS could support only about thirty simultaneous users (though unsuccessful efforts were made to push that up to fifty).[11] Demand for machine access exceeded the available supply.

One of the 7094s running CTSS was operated by a new group within MIT, Project MAC. *MAC* was sometimes said to stand for *machine aided cognition* and sometimes for *man and computer*. Its funding came from the Department of Defense, via the Advanced Research Projects Agency (ARPA). McCarthy's ideas on timesharing as a way of providing broad access to interactive computing had meshed with the personal obsessions of J. C. R. Licklider, a psychologist who led ARPA's first foray into computer research. In 1960, while employed at the Cambridge firm Bolt Beranek and Newman (BBN), Licklider published a paper titled "Man-Computer Symbiosis," forecasting that interactive computing would "involve a very close coupling between the human and electronic members of the partnership." When he joined ARPA in 1962, he began to use its funds to realize this vision of creating a "mechanically extended man."[12] Licklider had been introduced to interactive computing by Wesley Clark but favored timesharing, rather than Clark's aim of building small and cheap computers, as the best way of bringing down the cost.[13]

Existing operating systems, of the kind IBM and SHARE had developed from the late 1950s, provided suites of tools to develop programs and sequence their automatic execution. They did little while the application programs ran. Core memory was so precious that only a stub of the operating system stayed in memory to monitor progress. In contrast, CTSS took over the whole computer and remained in memory throughout. The *compatible* in its name meant that it could run tasks intended for IBM's Fortran monitor system in the background while servicing timesharing users. Almost the whole of the IBM 7094's standard 32,000 word memory was filled by CTSS itself, so a second bank had been custom fitted to hold user programs. Hardware tweaks barred programs in the user memory bank from accessing parts of memory allocated to other programs and prevented them from performing input/output operations directly, forcing them to call operating system routines instead.[14]

CTSS provided the main system for student computing at MIT until 1969. MIT users developed many new applications to exploit the possibilities of timesharing. They built text editors, so that programs could be written and edited interactively on computer terminals instead of punched onto cards, and created a range of programming tools and languages optimized for online use. One of the most popular applications, in place by 1965, was electronic mail. We discuss this and other computer-based communications in the next chapter. CTSS was finally decommissioned in 1973.

Even before CTSS was fully operational, other computer centers launched their own timesharing projects. Programmers who experienced timesharing became instant converts to the convenience of interactive computing, swapping code and ideas back and forth between computer centers as they tried out new ideas. BBN got another simple system up and running on a PDP-1, proving that even a small computer could support the new approach. On the West Coast, the RAND Corporation built a simple but useful system for its aging JOHNNIAC, one of the computers patterned after von Neumann's project at the Institute for Advanced Study.

Universities were the earliest and most aggressive adopters of timesharing systems because of their instructional needs. One computer could handle small programs and data sets from many users. Some, like MIT and Dartmouth, developed their own operating systems and tools rather than waiting for commercial solutions.

Terminals

Timesharing's spread was helped by the arrival in the mid-1960s of cheap and reliable teletype terminals based on the new ASCII standard for character encoding–a contrast with the expensive, special-purpose terminal equipment built for systems like SAGE and SABRE. Timesharing systems and interactive minicomputers were often used with a new device from the Teletype Corporation, the Model 33 ASR (automatic send-receive), shown in figure 5.4.[15]

The Model 33 was cheaper, simpler, and more rugged than the Flexowriter used by earlier small computers. It functioned at up to ten characters a second, working either as a typewriter that sent key codes directly to a computer or in offline mode to punch those codes onto paper tape. It came to symbolize the minicomputer era and the beginning of the personal computer era that followed it. The Control and Escape keys still found on keyboards today owe their ubiquity to the Model 33.

BASIC at Dartmouth

It was not enough simply to have timesharing; one also had to design the system so that students could write programs easily and receive a quick and intelligible diagnosis of their errors. Under the leadership of John G. Kemeny, chairman of its Mathematics Department, Dartmouth College began building such a system in 1963. Kemeny had done calculations using punched card equipment at Los Alamos. That experience led him to believe that "next to the original development of general-purpose high-speed computers the most important event was the coming of man-machine interaction."[16] Kemeny wanted to provide interactive computing to all of Dartmouth's students, not just the quarter of them majoring in science or engineering. Kemeny was aware of the

Figure 5.4
Teletype ASR-33, the principle I/O device for many minicomputers and early personal computers. It allowed upper case only, with a few special characters. Many of the control commands were later adopted by early personal computer operating system commands. The @ sign (Shift-P), was chosen by Ray Tomlinson, an engineer at Bolt Beranek and Newman, to specify the destination host in electronic mail addresses. Charles Babbage Institute, University of Minnesota.

ambitious work being done at MIT, but he and his colleague Thomas E. Kurtz decided to develop a more modest system around a new programming language, BASIC, tailored for the needs of Dartmouth students (figure 5.5).[17]

The language's name was a somewhat tortuous acronym for *beginner's all-purpose symbolic instruction code*. The key word was *beginner*. BASIC was a simple language, designed around the possibilities of interactive computing. Unlike Fortran, which required a significant investment of time to learn enough to write even a trivial program, BASIC had a simple syntax and a more forgiving approach. Its programs were interpreted, not compiled, meaning that they were not converted into machine code until the user entered the RUN command. The interpreter program converted a single line of BASIC into machine instructions, executed them, and then moved on to the next line. Users could type a few lines of code, run them to see what happened, and then type a few more. That approach slowed program execution but eliminated the tedious process of recompiling code each time an edit was made. Because BASIC was intended for timesharing, it was ideal for writing programs that prompted users for input and reacted to it.

Figure 5.5
Teletype terminals used to access the BASIC programming language at Dartmouth College, c. 1964. Professor Kemeny, the codeveloper of BASIC, is standing at left. Courtesy of Dartmouth College Library.

Dartmouth initially used a General Electric 235 computer connected to a smaller GE Datanet computer, which controlled Teletype terminals across the campus. Computer assignments were incorporated into the core curriculum, particularly first-year mathematics courses. Dartmouth built up a large library of shared programs, including popular games such as a football simulator. Unfortunately, most universities were larger and less well funded than Dartmouth and so could not afford to provide computer access to students across campus. But BASIC became one of the most widely used computer programming languages, beginning its spread in 1965 when Dartmouth began to offer access to its system to students elsewhere in partnership with General Electric. According to historian Joy Rankin, by 1971 Dartmouth's computer center was providing access to thirty high schools and twenty colleges. Only three thousand of its thirteen thousand users were on its own campus.[18]

Matlab at the University of New Mexico

Another timesharing application for education was created by Cleve Moler, then on the faculty of the University of New Mexico. Moler's background was in numerical analysis, and he had made important contributions to mathematical software as a contributor to the Eispack project discussed in chapter 2. When teaching an introductory course, Moler discovered that students working on class assignments struggled to write and compile programs to call the Fortran subroutines that made up Eispack. Around 1976 he coded a package called Matlab for interactive use on a timesharing system. Students could work interactively with matrices, typing commands and seeing the results immediately.[19]

Moler distributed several hundred copies of the timesharing version of Matlab, but like BASIC it had its biggest impact once it was reimplemented for personal computers. Several companies commercialized the software. The most successful, The MathWorks, was founded in 1983 and eventually turned Matlab into an essential part of engineering education and a standard tool for practicing scientists, engineers, and data analysts.

SCALING UP TIMESHARING

CTSS made a huge difference to computing at MIT, but to McCarthy and Licklider it was a mere stepping stone toward making interactive computer access an affordable commodity. The cost per user of operating CTSS was far too high for a commercial system. They believed that commercial timesharing services would require very large computers to operate efficiently, just as power companies relied primarily on large generators. Computer experts in the 1960s quoted Grosch's law: the computational capacity of a computer is proportional to the square of its price.[20] Thus a computer ten times more expensive than CTSS's IBM 7094 should support one hundred times more users.

Multics and Project MAC

IBM had a long relationship with MIT, but in 1964 Project MAC informed IBM that it would not be using its forthcoming System/360 hardware for its next timesharing system. The project instead chose a General Electric mainframe.[21] This computer had been designed with multiprocessor capabilities, which was important for the computer utility vision. MIT specified a set of modifications to a GE 635 mainframe to create the prototype for what became the 645, adding features to support security and virtual memory. Timesharing systems had to load many programs at once and so were a natural fit with the system of virtual memory and pages introduced with Atlas. They swapped out the inactive parts of programs while retaining in core memory the parts that were currently being executed.

After losing out to General Electric, IBM launched an urgent program to adapt the System/360 for timesharing. Its architecture had been designed with traditional batch processing in mind and lacked the dynamic address translation needed to support virtual memory. Neither did OS/360 support timesharing, despite its notoriously ambitious list of features. IBM announced in August 1965 that its new Model 67 would provide virtual memory support, support for up to four processors, and an enlarged address space to accommodate up to one megabyte of core memory. In January 1967, GE delivered its first 645 to Project MAC. By then IBM had already fulfilled, with a delivery to the University of Michigan, the first of more than one hundred orders it had received for the Model 67.

Both machines provided the promised hardware features, offering better hardware support for timesharing than any previous computers. The problem was software. GE was developing an operating system called Multics in collaboration with Project MAC and Bell Labs (which was looking for a system to use internally). It hoped that Multics would make its computers the new standard for large-scale timesharing. Their project was unveiled in 1965 with a special session at the Fall Joint Computer Conference, where six papers set out the goals and planned architecture of Multics. By 1968, more than a hundred people spread across the three organizations were working to develop Multics. Looking to trump this, when IBM announced the Model 67 it also announced OS/TSS, the Time Sharing System. IBM reportedly committed a thousand programmers to that project.

TSS was a spectacular disaster. A prototype version was issued in 1967, more than a year after the first Model 67s were delivered, but IBM limited it to "experimental, developmental, or instructional use." Even so, its performance proved so disappointing that IBM decided instead to meet the demand for timesharing with a less ambitious Time Sharing Option for its mainstream OS/360.[22] A few customers such as Carnegie Mellon University worked with IBM to rewrite TSS, producing updates throughout the 1970s, but most Model 67 customers either gave up on exploiting its special features or adopted the Michigan Terminal System (MTS), developed by a consortium of universities.[23] The last MTS site was not shut down until 1999. The hardware features IBM designed for the Model 67 had long since become a standard part of its mainframe architecture thanks to their incorporation into the 370 series from 1972 onward.

Multics fared only marginally better than TSS. A prototype version was running in 1968. By the end of the following year, Multics access was finally being offered to users at MIT. Early versions were less efficient than expected. Multics could support around 25 simultaneous users while one major job ran in the background—around the same number that CTSS had managed on much less powerful hardware. Its complexity undermined its goal of supporting more users. In 1970, Multics was finally delivered outside its development consortium, to a computer that the US Air Force Rome Air Development Center had purchased two years earlier to run it. GE's corporate management had already given up and sold its computer business to Honeywell. The second Multics system installed for an external customer, also for the Air Force, did not go online in the Pentagon until 1973.

The Challenges of Timesharing

Why were systems like Multics and TSS so hard to develop? The most fundamental challenge in the development of any timesharing system was handling the unpredictable demands made by its users. Trying to support more users and capabilities made this dramatically harder.

We already discussed the multiprogramming capabilities added to some large computers in the 1950s, which let a computer load several programs at once to give it something useful to switch to whenever the program it was currently running had to wait for a data transfer. Multiprogramming made batch processing systems significantly more complex, because programs had to coexist in memory. But the aim was to maximize throughput over an entire shift, to which end an operator or job control system could select queued programs with complementary resource requirements to avoid overloading the computer.

A timesharing system, in contrast, had to do its best to execute whatever combination of jobs its users tried to run at any given minute, whatever their combined memory and processor requirements. Even while heavily burdened, it had to give prompt feedback to every key press to maintain McCarthy's illusion of a "large private computer." Accomplishing this required the operating system to give out processing time in small pieces, shifting between users so often that responses seemed instant. Programs had to be suspended and restored at arbitrary points, not just when they were waiting for an input or output—otherwise, one big calculation job could monopolize the whole system indefinitely. This capability was eventually called *pre-emptive multitasking* because the operating system would pre-empt a running program to give another process a turn.

Virtual memory, also challenging to implement, was essential to any hope of supporting large numbers of users. For example, adding virtual memory support to an early version of Michigan's MTS transformed its capacity from five simultaneous users to fifty by keeping only some memory pages from each program in core memory.[24] Writing operating system code to select the most important pages to retain was far from easy. If it made the wrong decisions, the computer would start *thrashing* its disk, frantically loading and unloading pages, and user terminals would freeze.

A lot of the work needed to make large-scale timesharing systems like Multics workable involved refining their process-scheduling and paging algorithms. The shift also required changes to compilers and programmer behavior. For example, experiments showed that a prototype version of Multics started programs slowly because its compiler generated programs that forced the system to load almost every page of their code into memory.[25] Peter Denning worked on Multics while writing his PhD thesis at MIT. He introduced the idea of the *working set* and explained how to optimize page swapping.[26] Problems that seemed so challenging in the 1960s were gradually tamed and written into textbooks for the emerging discipline of computer science, such as Denning's own *Operating Systems Theory* (with E. G. Coffman).[27]

Beyond the inherently difficult problems faced by any large-scale timesharing operating system, the "computer utility" vision saddled Multics with more complexity. A computer utility would hold data and run programs for thousands of users, so keeping programs from snooping or tampering with each other was crucial. Multics coupled a complex system of permissions on files and directory with a *ring* system of privileges within the operating system itself. (This architecture eventually made Multics popular with the Pentagon and the National Security Agency.) As computer utilities were expected to support many users, Multics and TSS tried to exploit economies of scale when several users were running the same code simultaneously, by loading only one copy into physical memory. That set up complex relationships between virtual and physical memory. A computer utility would need to scale up seamlessly, so Multics made multiprocessor support a core requirement. A computer utility would need to work constantly, so Multics had to handle extreme situations without rebooting. Administrators could add and remove processors, disk drives, or memory banks without disrupting the work being carried out by users.

Other issues faced by Multics and TSS had their roots in what Fred Brooks, recalling the problems of OS/360, called the "second system effect."[28] The designers of Multics, having made many pragmatic compromises in creating CTSS, had a backlog of ideas about how to provide more powerful, elegant, or comprehensive implementations of its core capabilities. They threw many ambitious and untested ideas into the mix. For example, most systems use different methods to work with data on disk and data in memory. Multics mapped disk files into the virtual memory spaces of the programs accessing them, so that the same mechanisms were used to manipulate both kinds of data. This made life easier for programmers but harder for the operating system.

Multiprocessor Architectures

We already mentioned that Multics and TSS were both intended to support multiple processors. That was a significant architectural departure from the architectural paradigm for modern computing set out by John von Neumann in the *First Draft*, in which computers executed only one instruction at a time and had just one hardware unit for each operation—one adder, one multiplier, and so on. As discussed in chapter 2, performance considerations pushed processor designers toward more complex approaches such as pipelining to work simultaneously on different steps in the execution of several instructions, and vector processing to apply multiple arithmetic units simultaneously.

Yet the model of a single processor remained dominant. To understand why, one must understand what *processor* meant back in the 1960s. By the 1990s adding another processor meant squeezing an extra chip onto a motherboard. Today a dozen processor cores might be etched onto a single chip. But in the 1960s, the processor of a machine like the GE 645 was not a chip or even a board. It filled perhaps a hundred circuit boards spread over many cabinets, stuffed with thousands of electronic components connected by miles of wire. Even the architectural modifications made for timesharing meant rewiring large sections of the machine and adding cabinets holding circuit boards for things like dynamic address translation. Adding a second processor meant wiring the central parts of an entire second computer to a shared bank of core memory.

The GE 645 and IBM Model 360/67 were not the first computers to support multi-processor configurations. But timesharing was the first compelling application for it. Application programs would not take advantage of the second processor unless they were completely rewritten. Configuring two computers to share memory and disk storage meant coordinating their operation, as problems would result if both tried to use the same memory bank or peripheral at the same time. For most purposes, it made more sense just to run two separate computers. A computer utility, however, might run hundreds of programs at once. The operating system could shift tasks between the processors to keep them all busy, and make sure one processor was available to give quick responses to keystrokes when the others were bogged down running intensive jobs. While even a duplex (two-computer) system was a massive investment in the 1960s, Multics was designed to support even larger configurations and later ran successfully on systems with six mainframe processors each.

The other situation in which multiprocessor capabilities gradually became common was scientific supercomputing. The most popular successor to the single processor Cray 1 was the dual processor Cray X-MP. This was the world's fastest computer from its launch in 1982 to 1985, when the Cray 2, with four processors, was introduced. Supercomputers could really be justified only for organizations like Los Alamos or the National Center for Atmospheric Research with huge individual jobs. Harnessing that power forced programmers to split these application programs into separate parts, called threads, that could run simultaneously on different processors, communicating with each other to coordinate their work. Like the other architectural innovations pioneered by supercomputers, that approach eventually made its way from supercomputers to minicomputers, workstations, personal computers, and eventually smartphones.

Commercial Timesharing Services

The problems afflicting Multics and TSS meant that companies trying to set up commercial timesharing services finished up avoiding mainframes like IBM's 360/67 in favor of cheaper systems designed with timesharing in mind, such as PDP-10s from Digital Equipment Corporation or SDS-940s from Scientific Data Systems (SDS).

SDS was founded by Max Palevsky, who had previously persuaded Packard Bell, a small electronics firm, to enter the computer business. The result, the inexpensive PB 250, was a modest success when it was introduced in 1960.[29] In 1961 Palevsky created SDS, raising around $1 million in venture capital. By 1964, SDS was bringing in more money than DEC. Its main business was the 24-bit scientific computers, not as powerful as Control Data's supercomputers but much more capable than early minicomputers.[30]

In 1964, ARPA funded the University of California-Berkeley's Project Genie, a small-scale West Coast counterpart to Project MAC. After SDS supplied a 930 to Berkeley, a team including Butler Lampson, Peter Deutsch, and Mel Pirtle added support for paged memory and wrote a timesharing operating system for internal use. These modifications underpinned the SDS 940, which sold for up to $250,000 and was used by many of the first commercial timesharing companies.

Offering a commercial service in those days required a great deal of technical expertise and development work. Tymshare, of Cupertino, obtained its first machine and some initial financial support from SDS.[31] So did Comshare in nearby Palo Alto. The firms worked together to rework the Berkeley code to support a new disk drive, supplementing the original storage drum, which increased the number of simultaneous users from around six to twelve.[32] Much of that work was done by Tymshare's Ann Hardy, a veteran of IBM Research and Livermore National Lab where she had worked on its Stretch supercomputer. Hardy recalls that the SDS eventually supported around 38 simultaneous users.[33] The timesharing community was small, and developers and code both moved between installations. For example, all timesharing systems needed an online text editor. Dennis Ritchie later noted that the editor code used at Bell Labs in the late 1970s could be traced back to Lampson and Deutsch's QED for the SDS-940. QED had also been adapted for CTSS at MIT and for Multics.[34]

SDS soon faced new competition as a hardware supplier for this market. Simultaneously with the PDP-8, DEC had introduced a large system, the 36-bit PDP-6 (figure 5.6). Only twenty-three were sold, one to the MIT Artificial Intelligence Lab, where hackers developed the humorously named *incompatible* timesharing system. In its early versions, ITS allowed users to use the system without logging on, to read all files on the system, to chat with each other, and even to spy on each other's screens. Levy called it the "strongest expression yet of the Hacker Ethic."[35]

DEC's follow-up became a favorite of university computer science departments and commercial timesharing services. First delivered in 1966, the PDP-10 was designed from the start to support timesharing. DEC promoted a later version as "the world's lowest cost mainframe computer system."[36] A full-scale PDP-10 installation was indeed more like a small mainframe than a minicomputer, with more than a dozen cabinets holding tape drives, disk drives, memory units and a control panel. Pricing

Figure 5.6
DEC engineers seated in front of a PDP-6, DEC's first 36-bit mainframe scale computer. Gordon Bell, who led the development, is standing third from left, wearing a jacket. Its successor, the PDP-10, was a favorite basis for timesharing systems. Computer History Photographs, Archives Center, National Museum of American History, Smithsonian Institution.

was initially around a quarter of a million dollars for a usable system, and later models were even more expensive and more powerful. CompuServe, founded (as Compu-Serv) in 1969 in Columbus, Ohio, got its start by selling surplus capacity on a PDP-10 to the public. It grew into one of the largest online services of the 1970s and 1980s.[37]

The biggest share of the new market, however, went to General Electric Information Services (GEIS), which began in 1965 as a scheme to sell spare computing time on its demonstration computers. It used a customized version of the timesharing system developed at Dartmouth, including the BASIC language.[38] As the business grew, GEIS produced improved versions of this software to run on newer computers, rather than switching over to GE's own Multics system.

Timesharing expanded rapidly on a wave of hype in the late 1960s, stumbled as that bubble burst during a mild recession in 1970, and settled down in the 1970s as a viable but not particularly large market. GEIS, Tymshare, Comshare, and Compuserve enjoyed considerable success in the 1970s, but many less well-prepared firms had also rushed into the timesharing services market. For a while, when Wall Street was enamored of anything connected with computers, it was easy to raise money to lease a computer and sell its time. Like their counterparts in the electric utility industry, these startups had to install enough capacity to handle peak loads; otherwise, customers would quickly become discouraged. But that meant that during off-peak times the company would be saddled with expensive unused equipment. According to a survey carried out by the trade association ADAPSO in 1967, timesharing firms were losing an average of $275,000 on revenues of $500,000.[39] A presentation by Tymshare, laying out the firm's business model concluded: "Many people wonder how you can make a time-sharing operation profitable with rates that appear to be that low. Well, we wonder about that too!"[40] One of those undercapitalized companies, the Computer Center Corporation, installed PDP-10s in the Seattle area in 1968. While it was getting started, it offered a local teenager, Bill Gates, free time on the computer in exchange for helping rid the system of bugs. C-Cubed folded in 1970, having given Gates a taste of the potential of interactive computing.[41]

As a service industry, timesharing disappointed. As a technology, in contrast, it quickly became ubiquitous as a standard feature of minicomputer operating systems. Indeed, that hurt the growth of commercial timesharing companies because many potential customers found it more economic to install their own minicomputers. The surviving timesharing companies focused on added services such as access to application programs, databases, or communication facilities.

DEC and SDS moved in different directions. In 1967 SDS announced a million-dollar computer, the Sigma 7, intended to carry out business data processing and scientific computing as well as timesharing. In 1969, Palevsky sold the company to Xerox for around $900 million worth of Xerox stock. Xerox spent heavily to underwrite development of a range of Sigma computers to compete directly with IBM's System 360 machines. Meanwhile NASA and timesharing companies, key SDS customers in the late 1960s, cut back heavily on purchases. Xerox closed its computer business in 1975, having entirely failed to expand the 1 percent market share reached by SDS at its peak.[42]

DEC's PDP-10

DEC, in contrast, grew steadily by building on its existing areas of strength and avoiding direct competition against IBM's core markets. DEC supplied a standard time-sharing operating system called TOPS-10 for the PDP-10.[43] It promised to "handle 20, 50, 100 or more users simultaneously. . . . Since response is fast, the user appears to have a dedicated system."[44] That illusion created a mental model of what personal computing could be. TOPS-10 was like a Volkswagen Beetle: basic, simple, and easy to understand and work with. Some of its features and commands were replicated in early personal computing systems.

Using a PDP-10 could not only be fun but addictive: it was no accident that it was the computer on which *Adventure*—perhaps the longest-lived of all computer games—was written by Will Crowther at BBN. Crowther was an experienced cave explorer and keen player of the fantasy roleplaying game Dungeons & Dragons. Players typed simple commands such as "get lamp" and "go west" to explore an underground realm, solving problems and collecting treasure as they went. Crowther's initial version was greatly expanded by Don Woods, a programmer at Stanford's AI lab (another early center of hacker culture).[45] The game became a common feature of timesharing systems and could also be accessed over the fast-expanding ARPANET (discussed in the next chapter). For example, the game was played obsessively by hardware engineers at minicomputer firm Data General, who used their new machine's ability to run the game as proof that it was working reliably.[46]

Ambitious timesharing systems for multiprocessor mainframes turned out to be very complicated answers to the wrong problem. Grosch's law had suggested that bigger computers would always be more cost efficient than small ones, implying that the way to bring down the cost per user of timesharing was to use the biggest feasible computer. The proliferation of timesharing operating systems demonstrated that the cost per user was lower on a PDP-10 or SDS 940 than on a giant mainframe. Mainframes still had the unique ability to move enormous quantities of data, through their channels, back and forth to rows of tape drives and "disk farms." The most demanding individual scientific and commercial jobs would always need the most powerful computers. But a pool of smaller computers could deliver timesharing services more effectively than a single multiprocessor mainframe system. CompuServe, for example, eventually ran more than 200 PDP-10–family machines in its Columbus, Ohio, data center. After DEC stopped producing them in 1983, CompuServe commissioned compatible replacements from other companies.

SOFTWARE ENGINEERING

While commercial timesharing services finished up doing fine with smaller computers, the failure of IBM's TSS effort and the high-profile struggles of Multics combined with other troubled software projects such as IBM's OS/360 debacle to send a message that large-scale industrial teams were unable to develop complex systems software using conventional project management approaches. This opened the door for academically oriented experts to propose a variety of alternative methods under the banner of *software engineering*.

The NATO Conference on Software Engineering

The most famous meeting in software history, the 1968 NATO Conference on Software Engineering, took place in the Grand Hotel Sonnenbichl on the fringes of the alpine resort town of Garmisch, Germany. Its sixty official participants, all of them, as

Janet Abbate has noted, male, were drawn primarily from universities and corporate research labs. They bemoaned the sad state of software construction and shared opinions on how to improve the situation. Computer manufacturers, they noted, were comparatively good at delivering hardware and it usually functioned as promised. Compilers and operating systems turned up months or years late, often missing features, and were usually slower and more memory hogging than expected. The most discussed software projects at the conference were three troubled operating systems (OS/360, TSS, and Multics) and two programming languages (PL/I and Algol).[47]

This almost exclusive focus on what would later be called systems software may surprise readers who are used to thinking of *software* as a synonym for *computer program*. This was not the case in the 1960s. Software first became part of the computing lexicon in the early 1960s, to complement *hardware* by describing the other "wares" sold to computer users. Some definitions suggested that a program was software if purchased from an external company but not if developed in house. As a 1962 Honeywell advertising supplement entitled "A Few Quick Facts on Software" explained

> Software is a new and important addition to the jargon of computer users and builders. It refers to the automatic programming aids that simplify the task of telling the computer 'hardware' how to do its job. . . . Generally there are three basic categories of software: 1) Assembly Systems, 2) Compiler Systems, and 3) Operating Systems.

In 1968, the word was still closely associated with systems programs, mostly created by computer manufacturers, rather than applications programs, which were usually created within the company using them. The share of computer company budgets devoted to software production was rising alarmingly. A few months before the conference, Frank Wagner of fledgling software company Informatics had published a widely copied curve, estimating that software development had accounted for less than 10 percent of overall system development costs in the early 1950s, had reached around half of the total costs, and would finally level off in the mid-1970s at 80 percent of costs.[48]

As Doug McIlroy of Bell Labs remarked during the NATO conference, software creators "undoubtedly get the short end of the stick in confrontations with hardware people because they are the industrialists and we are the crofters." McIlroy was speaking from personal experience. The Multics team had decided to use the new PL/I language to implement their system. Because no PL/I compilers yet existed, a contract development company, Digitek, was hired to produce one. A year later, when Digitek had failed to deliver, McIlroy was part of a crash effort to create a stopgap compiler for the project.

Multics itself wasn't ready for production use by the time of the conference. The other Bell Labs participant, Edward E. David, one of the three overall leaders of the Multics project, wryly observed

Among the many possible strategies for producing large software systems, only one has been widely used. It might be labeled "the human wave" approach, for typically hundreds of people become involved over a several years period. . . . It is expensive, slow, inefficient, and the product is often larger in size and slower in execution than need be. The experience with this technique has led some people to opine that any software system that cannot be completed by some four or five people within a year can never be completed.

Many of the most vocal participants in the conference were veterans of the Algol project, which laid an early foundation for international collaboration in computing research before the establishment of computer science as a discipline. By 1968, however, computer science was a recognized field with PhD graduates, academic departments, and the first standard curriculum from the Association for Computing machinery.[49] In contrast, the Algol effort was falling apart. Its highest profile participants were in the process of quitting over the direction taken by the latest version of the language, soon to be ratified as Algol 68.

The Algol dissenters, such as Peter Naur and Brian Randell (the two editors of the influential book produced from the NATO conference), had distinguished themselves by designing successful compilers (including early Algol implementations) or other systems software—usually in small groups for underfunded computer manufacturers. They identified with scientific and mathematical approaches, and by the late 1960s their careers were shifting from the production of software to research in universities or corporate research laboratories. They expressed particular shock at the huge resources IBM was committing to OS/360 and TSS teams and their conspicuous lack of success. Edsger Dijkstra, a leader among the group, echoed David's criticism of the "human wave" with what he later called the "Chinese army" approach.[50] Both invoked the idea of a large, badly equipped, and poorly trained fighting force whose members were likely to perish in huge numbers when assigned a challenging mission.

What they wanted instead was something like an elite special forces team, the groups of "four or five" mentioned by David. Dijkstra's friends assumed that scientific backgrounds, rigorous thought processes, and exceptional intelligence explained their success in developing software. They suspected that a corresponding lack of those characteristics explained the problems of the IBM teams. Dijkstra, for example, was earning a PhD in physics at Leiden University when he discovered his love of programming. His early breakthroughs were rooted in the practical challenges he overcame developing systems software. First, as we discussed in chapter 2, he introduced the idea of the stack when trying to figure out how to produce an Algol compiler as a staff programmer in Amsterdam. Then, as a professor, he produced the highly influential Technische Hogeschool Eindhoven (THE) Operating System. This introduced semaphore signals used by processes to control access to resources such as peripherals or crucial code blocks, an elegant solution to the tendency of multiprogramming operating systems to freeze unpredictably when processes locked each other out of access to resources.[51]

The *engineering* in *software engineering* was chosen to evoke this productive interplay between theory and practice. THE, for example, grew out of Dijkstra's theoretical work on algorithms for concurrency and informed his later work on process modeling. Naur and Randell wrote in the introduction to its proceedings that the conference evoked the "need for software manufacture to be based on the types of theoretical foundations and practical disciplines" found in engineering.[52] Dijkstra himself favored the Dutch identity of *mathematical engineer* for computing students, but when he shared this at the conference most "began to laugh, because for them it sounded as a contradiction in terms, mathematics being sophisticated and unpractical, engineering being straightforward and practical."[53]

Dijkstra's Crisis

The NATO conference has been remembered as the moment where a "software crisis" was declared, in response to which software engineering, today a popular field at the intersection of computer science and engineering management, was founded. The truth is a little more complex. The phrase *software crisis* appeared only once in the conference proceedings, but was popularized five years later when Dijkstra made it the centerpiece of his Turing Award lecture. According to Dijkstra, at the conference there had "occurred the first open admission of the software crisis. And by now it is generally recognized that the design of any large sophisticated system is going to be a very difficult job." He hoped that new programming languages, the mathematical proving of program correctness, and the systematic use of modularization and abstraction would solve the crisis by making big projects "intellectually manageable." These new methods would "enhance the capacities of the already capable." Most existing programmers would be "edged out of the picture" because of their low "intellectual ceilings."[54] Dijkstra was an unabashed academic elitist, despite spending years in a corporate research laboratory. Decades later, he still believed that "Industry is so incredibly backward and its quality standards are so disgracefully low that, if there is to be a bridge between the industrial world and the academic world, we have to ensure that that bridge will be used for one-way traffic only."[55]

By then Dijkstra and his colleagues such as Wirth and Naur no longer identified with software engineering, having lost interest after a second NATO conference in 1969 failed to reach a consensus on core techniques. According to Randell, "the software engineering bandwagon began to roll as many people started to use the term to describe their work, to my mind often with very little justification." In reaction he "made a particular point for many years of refusing to use the term or to be associated with any event which used it."[56] Instead, they pursued their agenda within a new working group on *programming methodology*. Some of the group pioneered the study of formal methods by which the fidelity of hardware or software to a mathematical specification could be rigorously proved.

An approach known as *structured programming* was probably their most influential contribution of the 1970s. Though often assumed to be primarily about avoiding GOTO statements (an idea captured in the notorious "Go To Statement Considered Harmful" heading Wirth attached to a letter of Dijkstra's) and systematically indenting code, structured programming was originally a much more intellectually ambitious program. The idea was to design programs from the top down, expressing the main program as a series of calls to subroutines, which in turn were made up of lower level subroutines, until the scope of each subroutine was small enough that it could be easily written and understood. These ideas were reflected in program language design, particularly in Wirth's languages Pascal and Modula 2, which were widely adopted in computer science departments.

Software Engineering as a Field

That left the term *software engineering* available for others. During the 1970s it was appropriated by many people who had not been at the 1968 NATO conference, including Fred Brooks and Watts S. Humphrey, who had led work on OS/360; electrical engineer David Parnas; and RAND Corporation manager Barry Boehm—all of whom eventually became celebrated professors of software engineering.

Software engineering's focus shifted from systems software to custom application systems. As institutionalized in textbooks, degree programs, and conferences, the new field was concerned more with channeling the human wave through better performance measurement, improved estimation and control systems, and reproducible development process than with replacing big teams with mathematical commandos. This approach was taken furthest by Humphrey, whose widely influential *capabilities maturity model* treated software development as an organizational accomplishment rather than a technical or intellectual one.[57] Its most enthusiastic adopters were Indian firms such as Infosys, which specialized in the development of custom systems using large teams of programmers earning (by global standards) low wages. Proving that teams could move reliably and consistently from specifications to running code helped to win them customers.[58]

One path from the 1968 conference led to the mathematical rigor of formal methods, and another to software engineering as a management discipline. A third reaction, ultimately more influential, depended on craft methods rather than mathematical proofs or bureaucratic controls. Its adherents stuck with small development teams, looking for practical ways to build powerful systems incrementally from small, reusable pieces. That approach lies behind the development of the Unix operating system at Bell Labs, a project launched the year after the NATO conference and loosely overseen by Douglas McIlroy.

UNIX AND DEC TAKE TIMESHARING MAINSTREAM

The mounting dissatisfaction with Multics apparent at the NATO conference led Bell Labs to withdraw from the project in April 1969. The Bell Labs programmers regretted that giving back their Multics prototype left them without an interactive development system. Two of them, Ken Thompson and Dennis Ritchie, set to work within months on a simple timesharing system for internal use, initially with an already obsolete DEC PDP-7.[59] That system became Unix, the basis for many of today's most widely used operating systems.

Early Work on Unix

The name Unix, pronounced almost identically with "eunuchs," humorously signaled a stripped-down or "castrated" substitute for Multics. Although many Unix features were inspired by Multics, its minimalism was a kind of therapy after the complexity of Multics and PL/I, with their goals of being all things to all people. Before implementation started on Multics, the team "wrote 3000 pages of the Multics System Programmer's Manual . . . while waiting for the PL/I compiler." Then it worked for years to fulfill those promises, during which time many of the originally documented features became "science fiction."[60] It would be unfair, though, to remember Multics only as an overly ambitious system whose spectacular failure gave the creators of Unix a lesson in what not to do. Honeywell, which acquired GE's computer business, eventually sold dozens of Multics systems, and the last one was not shut down until 2000.

For the first Unix prototype, Thompson budgeted a month of his own time: one week to create each of the operating system kernel, a command shell, a text editor, and an assembler. Once those were minimally functional he could use Unix to improve itself. That first system left out a great deal—for example, it held only one program in memory at a time—but it worked well enough to persuade Bell Labs to order a new but still relatively affordable PDP-11/20 from DEC. Because its Multics experience made management reluctant to support software research, the computer and development of the first full version of Unix were justified by a promise to support the work of three clerks in the Bell Labs patent department with processing tools.

The modest funding provided by Bell Labs ensured that Unix remained a small-scale project, aimed at supporting the specific and immediate needs of Bell Labs users (including a suite of programs for text processing) rather than the long-term needs of a diverse population. It became a test of whether the kind of small teams and short development cycles mentioned by Ed David at the NATO conference could deliver a modern operating system.

Unix had short commands, quick to type on a slow teletype, a compact kernel to leave lots of memory free for user programs, and a pervasive stress on efficiency. A lot of ideas from Multics were reimplemented in Unix using much simpler mechanisms. These included its hierarchical file system, the idea of a separate program (called the "shell") to interact with users and interpret their commands, and aspects of its approaches to input and output. In Multics, for example, creating a new process had required the operating system to do so much work that programmers went out of their way to avoid doing it. Unix popularized a mechanism called "forking"—when a process needed to launch another process it simply duplicated the initial process in memory. This kept the kernel fast and simple.

As head of the Computing Techniques Research Group at Bell Labs, McIlroy was the immediate supervisor of the Unix team. His big idea at the NATO conference had been generalized, high-performance software components that could be recombined to build applications. As early as 1964, he had argued for "ways of coupling programs like a garden hose—screw in another segment when it becomes necessary to manage data in another way."[61] In 1972, McIlroy engaged what he called "one of the only places where I very nearly exerted managerial control" over the project by pushing Thompson to implement what became known as the *pipe* mechanism, by which a single Unix command could trigger a series of programs. Each program piped its output to the next as an input. According to McIlroy, Thompson managed this change over one night, and in the morning "we had this orgy of 'one liners.' Everybody had another one liner. Look at this, look at that."[62] Pipes underpinned what was later called the Unix philosophy of software tools, expressed by McIlroy as

1) Make each program do one thing well. To do a new job, build afresh rather than complicate old programs by adding new "features."

2) Expect the output of every program to become the input to another, as yet unknown, program. Don't clutter output with extraneous information. Avoid stringently columnar or binary input formats. Don't insist on interactive input.

3) Design and build software, even operating systems, to be tried early, ideally within weeks. Don't hesitate to throw away the clumsy parts and rebuild them.

4) Use tools in preference to unskilled help to lighten a programming task, even if you have to detour to build the tools and expect to throw some of them out after you've finished using them.[63]

Unix Evolves

The development of Unix shifted gradually from assembler to a new higher-level language. The language Ritchie designed, C, was stripped down to the essentials (like its name, which can be traced, via an intermediate language B, to BCPL). Once Unix was rewritten in C it was easier to *port* it to other computers. Instead of writing a whole operating system, all that was needed was a C compiler able to generate code in the new machine's language and some work to tweak the Unix kernel and standard libraries to accommodate its quirks.

C was optimized for operating systems programming. C code can do almost anything that assembly language can but is easier to write and structure. It gave programmers the convenient features of an Algol-like language, such as block structures, do . . . while loops, and procedure calls without sacrificing direct control over the computer hardware or reducing the efficiency of the final code. Like assembly language programmers, they could access hardware-specific device registers, modify individual bits of memory, and create data structures of arbitrary complexity. Purists complained that C was not a real high-level language at all, but a syntactic wrapper that made assembly language programming more convenient.[64] Its flexibility made it an unforgiving language for novices. C forced programmers to work with memory pointers to manipulate even simple data structures. Any C programmer has probably spent hours staring at code in frustration, eventually discovering that her program is crashing because of a failure to allocate memory to a data structure, or a missing & or misplaced * buried in a cryptic string of variable references.

The shift to C made Unix relatively easy to convert to other computers, but for its first decade and a half it was closely associated with DEC computers, particularly the PDP-11 (see figure 5.7). This 16-bit minicomputer had been launched as a response to increased competition for DEC's PDP-8. It initially had no hardware features to protect memory locations from alteration, either deliberate or accidental, unlike the more expensive PDP-10 and the mainframes that Multics and TSS were supposed to run on. The PDP-11 quickly surpassed its rivals and continued to fuel Digital Equipment Corporation's growth. Sales of the PDP-8 propelled the company from about 900 employees in 1965 to 5,800 in 1970. With the help of the PDP-11, it grew to 36,000 employees by 1977.[65] Over 170,000 PDP-11s were sold in the 1970s.[66] One important new feature was the flexible 56-line Unibus used to connect memory units and peripherals to the processor.

Bell Laboratories was a part of AT&T, a regulated monopoly. In return for steady profits from its telephone business, AT&T had to avoid commercial computing activities. This meant that universities could obtain a Unix license and source code for a nominal fee. As Unix was adapted for more powerful computers and for new applications, it acquired new capabilities such as support for multiprocessor configurations, virtual memory and addressing, and networking. For example, the University of California at

Figure 5.7
Ken Thompson (seated) and Dennis Ritchie (standing), at a PDP-11 installation at Bell Labs, Murray Hill, NJ, working on the Unix operating system. Credit: AT&T.

Berkeley obtained a Unix tape in 1974, following a visit by Ken Thompson. The system was soon running on several PDP-11's on the campus. No one minded if a university customized Unix. That work was what graduate students like Bill Joy were for. When the paper-based Model 33 Teletypes were replaced by video terminals, Joy and his fellow students set out to make Unix more accessible. They didn't stop. By 1978, the university's Industrial Liaison Office was offering tapes of the first Berkeley Software Distribution (BSD) to users around the country.[67]

Despite these improvements, Unix remained best suited to academic or research use rather than commercial timesharing systems. It lacked the capabilities built into mainframe timesharing systems designed for public use, such as billing mechanisms. Its user interface was divisive: McIlroy's modular software tools philosophy meant that different commands were written by different people, each with their own ideas about command formats. Not everybody appreciated that.[68] DEC's own minicomputer operating systems, although less powerful, were designed to be easier to learn and administer.

Students, faculty, and administrators all appreciated the small size, low cost, and interactive capabilities of the minicomputer when it was coupled with the power of the

Unix operating system. It was widely used to teach operating systems courses. *Lions'
Commentary on Unix*, a version of its source code annotated by Australian computer
scientist John Lions, was briefly available from Bell Labs but continued to circulate as
illicit photocopies after suppression by AT&T lawyers.[69] When students graduated,
they took these skills with them, along with an appreciation for the software that
allowed them such freedom. They turned into evangelists for Unix. By the mid-1980s,
Unix had ascended from the humble PDP-11 to become the most widely used operating system for Cray supercomputers.

DEC's VAX

In October 1977, DEC announced the VAX Model 11/780. The name *VAX* (for virtual address extension [of the] PDP-11, signaled that its primary goal was lifting the
64 KB memory cap imposed on the PDP-11 by its 16-bit architecture, but in reality
the VAX introduced a new architecture. It could be made to execute existing PDP-11
software by setting a *mode bit* that called forth the PDP-11 instruction set. The VAX
processor used sixteen 32-bit general registers, like the IBM System/360. It had a rich
set of over 250 instructions, with nine different addressing modes, that allowed a single
instruction to carry out complex operations.[70]

DEC was not the first to offer a so-called *super-mini*. Prime, also located off Route
128 in Framingham, Massachusetts, had shipped a 32-bit minicomputer in 1973. The
impetus for these developments was the growing availability of relatively cheap semiconductor memory, which replaced magnetic core. That made it practical to design machines
with larger main memories, which in turn demanded more address space. The VAX
11/780's performance, roughly calculated at one million instructions per second (MIPS),
became a benchmark against which competitors measured their machines. Even Data
General, whose Nova had been a strong competitor for DEC, had trouble matching it.
It eventually launched the 32-bit Eclipse MV/8000 in 1980, the development of which
was chronicled in Tracy Kidder's classic *The Soul of a New Machine*.[71]

DEC's official operating system was VAX/VMS, a powerful operating system
developed in parallel with the VAX hardware to support its new features. The name,
standing for virtual memory system, highlighted the most important of these features.
Like many other architectural innovations, virtual memory spread from the giant scientific computers. It originated, as we described in chapter 2, with the Manchester
Atlas and reached IBM's core line of mainframes with the System/370 line launched
in 1972. VMS used the new hardware to managing paging efficiently and give each
process its own 32-bit virtual address space of up to two gigabytes. DEC was ambiva-
lent about Unix for the VAX. Ken Olsen allegedly called it "snake oil." But the VAX's
PDP-11 ancestry made it a natural host for Unix, particularly for the BSD variant that
supported its paging hardware (see figure 5.8).

Figure 5.8
Armando Stettner was a leading advocate for Unix within DEC who persuaded the company to distribute thousands of these replica license plates at the USENIX conference. He had the genuine plate on his own car. *UNIX* worked well with the New Hampshire state motto, "Live free or die," a sentiment somewhat undermined by the insistence of AT&T lawyers on the trademark acknowledgment. Photograph: Paul Ceruzzi.

As IBM had done with its System/360, Digital had "bet the company" on a new architecture and product family. The VAX sold around 100,000 over its first decade, outselling all the other 32-bit minis. DEC's plan was to offer the customer a single VAX architecture in solitary or networked configurations that ranged from desktop to mainframe capability. The 9000, which DEC called a mainframe, challenged CDC mainframes for high-end scientific use. Smaller machines like the 11/750, launched in 1980, made VAX technology more attractive for office use. It cost more than a high-end personal computer, but it could drive a room full of terminals. These machines kept DEC profitable and dominant.

Prices started at $120,000 for an 11/730: too expensive for a single engineer, but just cheap enough to serve a team at an aerospace, automotive, or chemical firm. The standard practice had been either to get in line to use the company's mainframe or to sign up for time on a commercial timesharing service. The VAX provided a solid, engineering-oriented operating system and fast input/output facilities for data collection. Finally, DEC sold the VT-100, a fast terminal built around microprocessor technology. Its control codes for text formatting and simple block graphics, based on an emerging standard, were widely copied.

By the start of the 1980s the VAX was the default choice for a company or lab looking for a powerful, expandable computer able to support a reasonably large number of simultaneous users. Many of the applications that needed a computer this powerful also required graphics. For example, particle physics groups used DEC machines to produce high-resolution representations of data from collider experiments.

Many VAX users hooked their machines up to Tektronix graphics terminals. These worked with the same serial cable or modem connections as regular text terminals, but they encoded the instructions needed to draw vector graphics. Tektronix, based in Beaverton, Oregon, had begun as a supplier of oscilloscopes. By the late 1970s, it was selling large numbers of vector graphics terminals. These eliminated expensive video memory by making the contents of the screen itself readable as well as writable. Tektronix called this technology the *direct view storage tube*. The effective resolution of its 4010 series, launched in 1972, was 1,024 by 780 units in the Tektronix coordinate system, which gave a much crisper picture than any early personal computer.[72] By the early 1980s, the cheapest Tektronix terminals sold for a few thousand dollars each and could fit on a desk. Larger models, with even higher resolutions and screens of up to 25 inches, were used for professional work such as chip design, engineering, or architecture. Tektronix's Plot 10 graphics libraries for Fortran programmers also became a de facto industry standard.

By the beginning of the 1980s, the minicomputer had moved far beyond its initial niche as an embedded controller or laboratory tool. Minicomputers took advantage of advances in integrated circuits, packaging, graphics, and processor architecture. Thanks to timesharing, one minicomputer could serve dozens of simultaneous users. Its future seemed bright indeed. What happened next, however, was not what DEC expected. The minicomputer held the seeds of its own destruction, by preparing the way for a new class of machine, the personal computer, that came from an entirely different source.

6 THE COMPUTER BECOMES A COMMUNICATIONS PLATFORM

By 1968, MIT's J. C. R. Licklider had refined his idea of "human computer symbiosis" into a new vision for the power of online systems: "The Computer as a Communications Device." "In a few years," according to a "startling claim" made at the opening of his paper of the same name, "men will be able to communicate more effectively through a machine than face to face."[1] Thanks to Licklider and the community of ARPA-affiliated colleagues he had jokingly referred to a few years earlier as an "intergalactic computer network," many of the applications we now think of as features of the Internet were already running by the 1970s on timesharing systems, including electronic mail, discussion forums, online education, and multiplayer games.[2] The ARPANET itself, from which the Internet evolved, began in the late 1960s as a way to interconnect timesharing systems. Early efforts to get ordinary people online, most successfully with the French Minitel system, rested on providing cheap home terminals able to connect to timesharing computers providing services such as banking, shopping, and news.

COMMUNICATION AND COLLABORATING ON TIMESHARING SYSTEMS

Users of online systems found unanticipated ways to communicate with each other. Even the operators of the SAGE air defense network reportedly gossiped using the AUTODIN network built to interconnect its sites, just as Morse code operators had chatted over telegraph lines and radio links during quiet periods. General purpose timesharing systems were well suited to communication. By definition, a timesharing system allowed multiple simultaneous users to type inputs. "Chat" or "talk" commands, in which messages typed by one user appeared in the output of others, were intended to allow system operators to send warning messages when a shutdown was coming or to help users with problems. A facility of this kind has been documented at

SDC in 1963.[3] These capabilities evolved into later Internet chat systems, and eventually into today's instant messaging applications.

Electronic Mail

The first electronic mail system we know of was created at MIT, as part of CTSS. The phrase *electronic mail* goes back to the 1950s, when the United States Post Office began to plan its response to what was then called the *electronic age*. On November 2, 1959, the Appleton Post-Crescent reported that the Postmaster General was exploring the future possibility of "split second electronic mail." In this new system, "a letter will cost 15 cents. A nickel to send—and a dime to bribe the electronic brain to forget what it read."[4]

The spread of systems like MIT's Mail helped to give the phrase a more specific definition, as a form of text-based online communication. Electronic mail differs from chat because private messages are stored until their recipient logs in, whereas chat displays messages instantly. This was easy to implement because each timesharing user had a personal directory to hold private files, and all timesharing systems included online editing capabilities. All that was needed was a "mailbox" file where messages could be delivered, like paper letters piling up in a real mailbox. MIT's Mail command had been proposed in a staff planning memo at the end of 1964 and was implemented in mid-1965 when Tom Van Vleck and Noel Morris, junior members of the Institute's research staff, took the initiative to write the necessary code. Although Mail, and other systems of the 1960s, could send messages only to other users of the same computer, this was not quite as restrictive as it sounds. Hundreds of people had CTSS accounts and some accessed the system remotely over phone lines from other institutions. The name *mail* was a good one—the system transmitted unstructured text, making it flexible enough to work for official memos, lengthy correspondence, or quick personal notes. As Van Vleck explained, these people used Mail to "coordinate work and share information on all kinds of topics, including personal topics."[5]

By the late 1960s, electronic mail was an almost universal feature of timesharing operating systems. Big systems like Multics had mail facilities, as did timesharing systems for smaller computers, like DEC minicomputers. Unix included an electronic mail mechanism, placing a mail box file in the personal directory of each user.

Like paper mail, electronic mail is well suited to sending a message to a particular person, or to everyone on a distribution list. Other timesharing communication systems were built for group communication and public discussion. These were the ancestors of later bulletin boards, discussion forums, and online communities. One of the most ambitious and influential was EMISARI, created by Murray Turoff in 1971 for the United States Office of Emergency Preparedness. EMISARI combined private electronic messages with a chat system, public postings, voting, and a user directory.

Turoff later became a professor and researcher into what was by then called *computer mediated communication*. His classic 1978 book, *The Network Nation*, written with his wife, the sociologist Starr Roxanne Hiltz, predicted a future where newspapers ceased paper publication, the Post Office abandoned mail delivery, and students earned degrees in online classes.[6] It contributed to a wave of enthusiasm for what was often called the *computer revolution* or *information society*.

Engelbart's Online System

The most famous demonstration of the use of timesharing systems for online collaboration was delivered in December 1968 when Douglas Engelbart of the Stanford Research Institute (SRI) and a crew of over a dozen helpers staged an ambitious presentation of their online system NLS at the Fall Joint Computer Conference in San Francisco. For more than an hour and a half, Engelbart projected interactive computer programs, controlled by a mouse, onto a giant screen using a system of video cameras and a live microwave link to the group's SDS 940 in Palo Alto. The presentation style was as novel as the system being demonstrated. It eventually became legendary as the Mother of All Demos.[7]

Engelbart's system exemplified Licklider's hopes for "human computer symbiosis." He had been one of the first persons to apply for funding from ARPA's Information Processing Techniques Office in late 1962, to support a "conceptual framework" for "augmenting human intellect." The group's earliest experiments with interactive text editing took place before a suitable timesharing system was available. Its most famous invention, first described in 1967, was the *mouse*, which exhaustive tests showed was more efficient and effective than the light pen (used in the SAGE), the joystick, or other input devices. Engelbart recalled that he was inspired by a device called a planimeter, which an engineer slid over a graph to calculate the area under a curve.

The demo was a high point for Engelbart's approach of using timesharing to underpin human collaboration and the structuring of ideas. He showcased not just the mouse but also online text editing, outline processing, the creation of links between different files, and the collaborative editing of a document. NLS took a hierarchical view of documents, as a list of headings which could in turn be expanded into subheadings and eventually into paragraphs of text. Engelbart demonstrated these capabilities with the everyday example of producing a shopping list based on requests received from his wife.

By the 1990s, tools for online collaboration had developed into a major area for research and development, often called computer supported cooperative work (CSCW). Although Engelbart's ideas were powerful, NLS itself was a dead end. To free one hand for the mouse, it used a five-key *chord keyboard* on which each letter was accessed with a combination of fingers; it was productive once mastered but off-putting to casual users. So was the complex and arbitrary command structure of NLS. More fundamentally, the economics were not yet feasible. The cost of hardware to drive video displays was so high that screen images for all eight users flashed in turn on a single output. Carefully timed video cameras grabbed the appropriate frames and displayed them on eight different televisions.

Plato

Plato was an interactive, graphics-based timesharing system intended for education and training from kindergarten to university. It originated at the University of Illinois, a center for research into *computer aided instruction*. Plato III was operational by 1967, using a CDC 1604 computer to support up to twenty users around campus. Its specially built video terminals incorporated microfiche projectors to display lesson slides. Using a new language, TUTOR, instructors coded lessons with the same terminals.[8]

Generous funding from the National Science Foundation underwrote the development of the far more ambitious Plato IV, which entered operation in 1972. The Plato IV terminals still included microfiche projectors, but they also supported exceptionally crisp computer-generated graphics with a resolution of 512×512 pixels. Unlike the vector graphics systems we discussed earlier, they used *bitmap* graphics, treating the screen as a grid of *pixel* dots. That was made possible by a unique display technology developed by the project's leader, Donald Bitzer. Holding the screen image in memory, so that programs could update it by changing the values stored there, would have required at least 32 KB of RAM. Chip memory was not available when Bitzer started his design, and even in 1972 it would have taken 256 of Intel's new memory chips to hold a screen full of data. Some computers in the 1950s had used display tubes as memory. Bitzer took a similar approach, designing a plasma display whose orange pixels could be read as well as written, so that the screen served in effect as its own memory. The screens also incorporated touch sensors, so that students could select options with their fingers, and a computer-controlled slide projector to display full color images.[9]

By 1975, 950 of the new terminals were in use (figure 6.1).[10] That year Bitzer, who had grand plans for Plato IV, predicted that by the 1980s there would be millions of Plato terminals in use around the country.[11] That never happened, but his Computer-based Education Research Laboratory used its CDC Cyber 73 supercomputer to support terminals at remote sites such as schools, colleges, and government agencies as well as those on the University of Illinois campus.

The system's processing power and graphical capabilities were put to good use in a flight simulator developed for the Federal Aviation Administration and in many computer games with space and fantasy themes. Because they shared a computer, players could interact with each other inside the game worlds, a capability that became common on personal computer systems only in the 1990s. Plato was renowned for its communication and collaboration features, which included an online newspaper covering the system

Figure 6.1

A PLATO terminal. Behind the user is a brace, subtly suggesting that the system would allow people with disabilities to be full participants in the Information Age. In the back of the room is a Radio Shack TRS-80 personal computer. Its presence inadvertently suggests why PLATO was unsuccessful: steady improvements to personal computers allowed PLATO-like applications at a much lower cost.
Source: Charles Babbage Institute.

and its users, a conferencing and discussion system, email, and a "personal notes" capability for users to develop shared, searchable repositories of data.[12] These were influential on the development of many later Internet and personal computer based systems for collaborative work, including Lotus Notes (a corporate favorite of the 1990s) which built on ideas that its creator Ray Ozzie had originally explored in Plato.[13]

In 1974, CDC began working to commercialize Plato technology, formalized in a licensing deal signed in 1976.[14] Following the urban turmoil of the late 1960s, its leader Bill Norris was keen to apply computer technology to address social problems. He believed that education and training represented a potentially huge global market that could quickly grow to provide the bulk of CDC's revenues. CDC engineered its own terminals, using a more conventional display, and invested heavily to create computerized lessons and market the Plato system. Unfortunately, the cost of the system remained much higher than that of human teachers. As a result, the expected mass market in schools never developed. The initial cost for a terminal was about $8,000 up front, and CDC leveled high hourly charges to connect it to the Cyber 70 supercomputers in its

Plato data center. By 1981 it employed 500 people to design online teaching materials.[15] CDC sold some mainframes to customers wishing to host their own Plato systems. Users included several universities, American Airlines, and CDC's own training programs.

During the 1980s, CDC ported Plato's educational programs to smaller computers and dropped the cost, but that was not enough to save it. The company began losing large amounts of money in the mid-1980s, and in 1986 Bill Norris resigned. CDC sold the rights to Plato three years later, abandoning its cumulative investment of around $900 million.[16] It's hard to see that Plato could ever have made commercial sense for education without the huge subsidies CDC offered to its test sites, but many of the online applications first explored on the platform proved influential.

LAYING THE INTERNET'S FOUNDATIONS

Advanced as they were, both NLS and Plato were designed around the model of connecting a terminal to a single computer that ran programs and held data for many users. Those users could communicate with each other but not with the outside world. Extending computer communication more generally relied on a new network infrastructure able to connect computers, programs, and users to each other.

Like so much else in computing during the 1960s and 70s, its development was funded and steered by ARPA. Licklider's paper on "The Computer as a Communications Device" centered on a description of his earlier experience using NLS and video conferencing to conduct a grant review meeting. Toward the end of his 1968 demo, Engelbart discussed the work his group planned to support—a new network, the ARPANET, that was about to start operation. He speculated that at the following year's conference in Boston, he would be able to operate NLS over a transcontinental ARPANET link.

Packet Switching

Remote logins of that kind were the original justification for the construction of the ARPANET. All the major centers for computer science research had deployed time-sharing systems for faculty and students. Many were purchased with grant money supplied by ARPA. Each center had developed its own resources, such as language compilers, modeling tools, and text processors. ARPA ran the risk of funding duplicate efforts to replicate at one center capabilities already built at another. Because ARPA-funded centers used many kinds of computers, most with custom or heavily modified operating systems, they could not simply exchange program code. But if the computers were networked, someone logged into a computer at MIT would be able to connect onward to the computer at SRI in California and access its unique resources.

Beginning in 1967, ARPA began a series of meetings to discuss how to link computers across the country in a network (figure 6.2).[17] Previous networking had focused

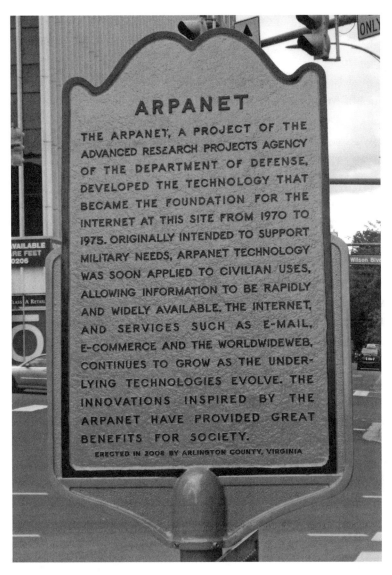

Figure 6.2
Commemorative plaque, 1401 Wilson Blvd. Arlington, Virginia. The ARPANET was initially conceived in the Pentagon, but the Information Processing Techniques Office (IPTO) moved to a nearby office building. Photo by Paul Ceruzzi.

mostly on the connection of terminals to distant computers via modems, and in a few cases (such as SAGE) to the exchange of information between computers as part of a specific application. ARPANET would provide a flexible connection between different computer models, letting them exchange packets holding any kind of data.

The idea of packaging data into packets was itself a novel one. Conventional networking followed the template of the telephone systems: automated switches made a connection between two points, usually a computer and a terminal, for the duration of the session. The ARPANET was the first large-scale trial of the *packet switching* approach to networking. That meant breaking down communication into a series of self-contained packets of data, each including address information specifying its source and destination. It was proposed by Paul Baran of the RAND Corporation in 1960 and independently reinvented in 1965 by Donald Davies of the National Physical Laboratory in England, who came up with the term *packet*.[18] Each packet would be dispatched separately onto the network, to be routed to its destination and reassembled into the correct sequence at the far end. Their full route is not necessarily mapped out in advance. If one link in the path stops working, computers will realize that packets have been delayed and try to resend them by an alternative path. This allowed for reliable, decentralized networks and offered economic benefits by eliminating the need to tie up a connection from one party to the other for the duration of the conversation.

Baran's original proposal stressed the military advantages of a network that could survive the destruction of any individual node and be relied on during a crisis. That is the root of the popular idea that the Internet was built to survive a nuclear war. It wasn't, although packet switching was later put to good use in building robust networks for battlefield use. Davies had proposed a national network spanning the UK, although the lack of available resources in Britain meant that his network was used in practice only to connect one computer to its terminals.

The ARPANET

The early ARPANET was controlled and administered by BBN from its Network Control Center. The actual connections were made between Honeywell DDP-516 minicomputers, known as interface message processors (IMPs). All that the universities joining ARPANET had to do was establish a connection from their timesharing computer to the IMP and tweak its operating system. Four nodes (that is, four IMPS) were operating west of the Rockies in December 1969; a year later there were ten IMPs spanning the country. By 1971 ARPANET consisted of fifteen nodes connecting twenty-three host computers. Of those, nine were PDP-10s, five were IBM System/360s, and the rest were assorted minicomputers and mainframes. ARPANET was unveiled to the public with a demonstration at a hotel in Washington, DC, in October 1972.[19]

ARPA obtained lease lines from AT&T to transmit digital data between computer centers. Even a network with just thirty-three computer centers had to handle more than a thousand possible connections. Instead of renting a line for each possible connection, packet switching allowed a *store and forward* approach to communication. For example, a packet going from UCLA in California to MIT might travel through eight IMPs to get there, including stops at the University of Southern California and Case Western in Ohio.

By today's standards ARPA's computing work seems rife with conflicts of interest: a clubby handful of men, on leave from an even smaller number of universities, took it in turns to shower taxpayers' money on each other and on contractors such as BBN. They had no time for open competition, detailed proposals, or external peer review.[20] Yet it is hard to argue with the results—from the early 1960s to the mid-1970s, ARPA-sponsored projects led to fundamental breakthroughs in interactive computing, operating systems, networking, and computer graphics. No computing research sponsor before or since has enjoyed such a run of success.

Not all the good ideas in computer networking came from ARPA. In France an experimental packet switched network, CYCLADES, was operational by 1974. Louis Pouzin designed it to eliminate the expensive IMPs by making the host computers themselves responsible for routing packets. Although CYCLADES was short lived and never connected more than about twenty computers, Pouzin's approach was later adopted for the Internet.[21]

While the ARPANET was justified by the need for remote logins, its IMPs and interface software provided a general purpose infrastructure for the transmission of packets that users and programmers could rely on without having to worry about the details of how data was being transmitted. Users discovered their own applications—most importantly, the exchange of electronic mail. Electronic mail was already an almost universal feature of timesharing systems. Its extension to transmit messages over the ARPANET to the users of other computers had begun as a quick hack. Discussions to define an elaborate mechanism for *network mail* were moving slowly in 1971 when Ray Tomlinson, a programmer at BBN, realized that an existing ARPANET file transfer capability, to copy files from one computer to another, could also carry electronic mail. Splicing the file transfer code into the popular mail program Sndmsg that BBN had developed for the PDP-10 provided a simple but effective system for network mail. Now that electronic mail could go anywhere on the ARPANET, it was necessary to specify the timesharing system of recipients as well as their usernames. Tomlinson chose the @ sign, a standard but underused part of the teletype keyboard, to separate the two.[22]

As historian Andrew Russell has highlighted, the ARPANET design process was based on "rough consensus and running code."[23] Better to get something working quickly and let it evolve during use than to spend years of committee on a design intended to handle every possible need. Tomlinson's original hack worked on Tenex, which was used by a significant fraction of the early ARPANET sites. That was enough to prove its viability, but the ARPANET worked on common protocols, not common code. Because the network interconnected many computer models, running many operating systems, users could not demand that the recipients of their mail run the same program that they were using. They would still be able to swap messages if both systems stuck to communications *protocols* defining things like how the message would be addressed, the sequence of signals exchanged to begin and end transmission, and how text would be encoded.

These protocols were published as a series of numbered Requests for Comments (RFCs), the centerpiece of an informal approach to ARPANET standard setting. RFCs were archived on the network itself. Some were humorous; others presented new ideas, opened discussion, proposed specifications, or agreed standards. The RFC process was open to all, although in this era the community interested in the network was small, technically minded, and fairly homogeneous. This was a contrast with the approach later taken by vendors such as IBM and DEC, who promised to make networking easy—as long as one standardized on their hardware and software.

Electronic mail quickly and unexpectedly became, according to historian Janet Abbate, the "smash hit" of the ARPANET (see figure 6.3). By 1973, mail constituted a majority of the data transmitted over the network.[24] ARPA administrators embraced it as a way of communicating with their grantees. Soon other Pentagon bureaucrats were having terminals installed just to send and receive email. In 1975 management and oversight of the network shifted from ARPA to the Defense Communications Agency, reflecting its graduation from science project to essential infrastructure.[25] Among ARPANET users, email was beginning to dissolve paper-based memos and displace telephone calls, another step on the path of computers toward practical universality (figure 6.4).

Capabilities evolved rapidly. Sending a message across the network was originally treated as a special instance of transmitting a file, and so a MAIL command was included in RFC 385 on file transfer in 1972. Because it was not always clear when or where a message had come from, RFC 561 in 1973 formalized electronic mail headers including *from*, *date*, and *subject*. In 1975, RFC 680 described fields to help with the transmission of messages to multiple users, including *to*, *cc*, and *bcc*. As well as communication between users of different computers and operating systems, this standards-based approach allowed rapid advances in email client software. Changing to a new and better email program that followed the rules from the RFCs created no problems in continuing to exchange messages with colleagues using older programs. Mail programs evolved rapidly, for example, by integrating the sending and reading of messages within the same program and adding options to reply, forward, or save messages.[26]

Queen Elizabeth II of England became the first head of state to send electronic mail while ceremonially opening a building in the British Royal Signals and Radar Establishment in 1976.[27] Two years later Gary Thuerk, an overly enthusiastic marketer for the Digital Equipment Corporation sent a message remembered as the first spam email. Thuerk tried to invite 600 ARPANET users to come to demonstrations of the latest VAX systems, but he crammed so many recipients into the *to* field that it overflowed, creating a very long message consisting mostly of email addresses. He was condemned by many ARPANET users. Major Raymond Czahor, who oversaw the network for the Defense Communications Agency, called it a "flagrant violation" of the network's restriction to "official US government business only" which was usually enforced loosely and informally.[28]

While many RFCs codified existing practices, some defined new approaches. One of the most important of these, issued in 1982 by Jon Postel, described what remains the

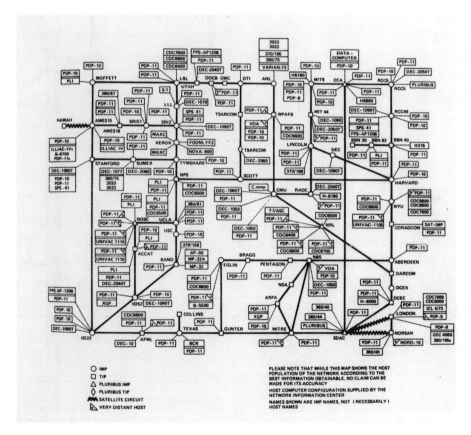

Figure 6.3

Logical map of ARPANET, c. 1978. DEC PDP-11 computers dominated the network at the time, with PDP-10, CDC, and IBM systems also well represented. Note also three satellite links: to Norway, the UK, and Hawaii.

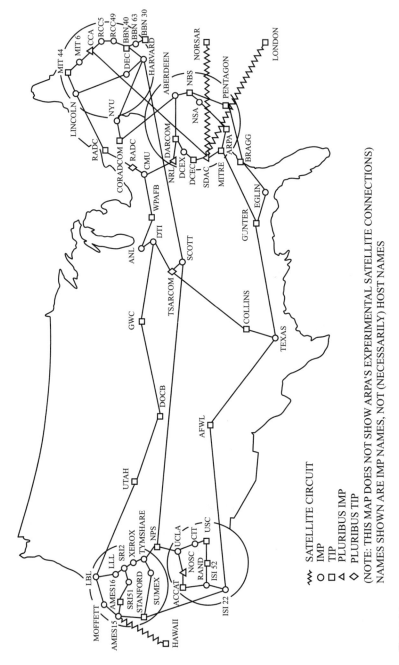

Figure 6.4

Geographical map of ARPANET nodes, c. 1980. This view of the network highlights its concentration in four areas (each highly magnified): Silicon Valley, the Los Angeles basin, Cambridge, Massachusetts, and northern Virginia.

〰〰 SATELLITE CIRCUIT
○ IMP
□ TIP
△ PLURIBUS IMP
◇ PLURIBUS TIP

(NOTE: THIS MAP DOES NOT SHOW ARPA'S EXPERIMENTAL SATELLITE CONNECTIONS)
NAMES SHOWN ARE IMP NAMES, NOT (NECESSARILY) HOST NAMES

standard Internet email transmission method: the simple mail transfer protocol (SMTP). Their adoption was part of the transition from the ARPANET era to the Internet.

Internetworking

The Internet evolved from the ARPANET during the late 1970s and early 1980s. It was conceived not as a single network but rather a collection of different networks, hence the name. During the early 1970s ARPA-funded researchers also investigated new protocols capable of interconnecting networks based on communication media with very different characteristics such as radio links, fast local networks, and long-distance data lines. This required the reliable transmission of messages over unreliable links. One influential system, ALOHAnet, was designed to connect terminals spread across the Hawaiian islands to the University of Hawaii's computer center. Other universities used telephone lines, but this was not practical here. Instead the terminals used shared radio frequencies to send and receive data packets, the first application of *packet radio* communication. Computers received all the packets but ignored those addressed to other computers. Before sending data, a computer first listened to make sure there were no packets being sent. If two computers happened to transmit at the same time, causing a collision, each would back off for a random interval and try again. A follow-up project in the San Francisco area extended the idea of packet radio by using repeater stations to route traffic over the network. Another team experimented with the transmission of data packets via satellite, which could spread the high cost of a satellite link over many simultaneous users.[29]

These new radio networks were incompatible with each other and with the fast-growing ARPANET. That spurred a group led by Vinton Cerf and Robert Kahn to work on *internetworking* to interconnect them. It developed a new *transmission control protocol* (TCP) that was suitable for unreliable radio networks as well as the leased telephone lines used by the ARPANET. Its approach was influenced by the French Cyclades system. TCP would reassemble packets that arrived out of order and ask for the retransmission of data packets that were corrupt or missing. That made it possible to extend ARPANET connectivity over the new networks, via *gateway* connections. After experiments run in the mid-1970s, the group decided to split some of TCP's functions into a second protocol, *internet protocol (IP)*. This allowed for a cleaner separation of functions between those carried out by the gateways (IP) and those carried out by the network nodes (TCP).[30]

As work continued, TCP and IP were being developed not just to collect ARPA's various networks into a single Internet (the origin of the term) but as a replacement for the original network control program running on the ARPANET IMPs. During the 1970s and early 1980s, the Department of Defense increasingly used the ARPANET for its own bureaucratic purposes. It wanted to separate its networks from those of the computing research community but keep the ability to exchange data. Shifting the ARPANET itself to TCP and IP would facilitate that. TCP/IP's separation of network functions into independent layers made it simpler than earlier approaches, as well as scalable over large networks mixing many communications media.[31]

Version 4 of the TCP/IP protocol suite was approved in 1980. ARPANET rapidly shifted over to the new protocol, driven by a 1982 deadline. ARPANET users still struggled initially to implement the new protocols, in part because their host computers now had to deal with routing and transmission reliability issues from which the IMPs had previously shielded them. By the mid-1980s, the Internet consisted of two main networks: the research-oriented ARPANET and the more secure military MILNET. Many other networks would follow.[32]

Many of the most widely used Internet applications of the late 1980s and early 1990s were direct descendants of those pioneered on the ARPANET. These included electronic mail, file transfer (via the file transfer protocol), and remote login (via the Telnet protocol). These established applications were joined by newer ones. For example, internet relay chat (IRC) allowed a small group of users to type messages simultaneously on a shared screen.

Kahn and Cerf shared the 2004 Turing award for their accomplishment, and Cerf remains one of the most prominent representatives of the computing research community thanks to a job as "chief Internet evangelist" at Google. Version 6 of TCP/IP, designed to meet the challenges of the modern Internet, has been available for decades, but as of 2020, version 4 is still used to transmit most Internet traffic.[33]

Other Academic Networks

Because the ARPANET is the direct ancestor of today's Internet, it is easy to assume that it was the only, or at least the most widely used, network of the 1970s and 1980s. It wasn't. Only a tiny fraction of the data transmitted in the U.S. passed over ARPANET. ARPANET was not even the most widely used academic network of the 1980s. Three other services, Usenet, CS Net, and BITNET collectively served many more users.

One of the most popular services of the early Internet, *news*, began as a decentralized system of discussion groups on Usenet, a separate network not based on Internet technologies. Usenet was introduced in 1980 to interconnect minicomputer timesharing systems running Unix. It spread rapidly along with Unix. To keep its sites and their big computers constantly online, ARPANET paid for dedicated local and backbone connections, IMPs, and a network control center. Usenet was a volunteer operation for smaller systems, based around a program called UUCP for Unix to Unix Copy Protocol. At night, when calls were cheaper, Usenet computers made modem connections with their neighbors to exchange messages. This could never work for interactive applications but was good enough for email. Users had to specify their own routing paths when addressing messages, separating each successive stop along the route with a ! (pronounced "bang"). Addressing a message required an up-to-date map of the network, for example, gway!tcol!c anty!uoh!bigsite!foovax!barbox!user. Each name specified one UUCP hop, meaning that a message might take a week to reach its recipient. By 1984, around a thousand timesharing systems were connected to Usenet, approximately the same number that were connected to ARPANET (figure 6.5).[34]

Usenet pioneered a phenomenon that by the early 1990s had become a crucial part of the Internet: newsgroups. The newsgroup system was a hierarchy of thousands of threaded discussion groups, each offering a forum similar to one of the computer conferencing systems of the 1970s but much more widely accessible. Some newsgroups were local, places to post notices of items for sale or discuss university events. Others, like the science fiction discussion forums gathered under rec.arts.sf were subscribed to by many sites across the network. Users read copies of the messages from their local server. At night the computers shared postings made by local users with their neighbors. As with Usenet email, posts eventually propagated across the entire network. As well as messages, newsgroups were used to distribute specially encoded program files and pictures.

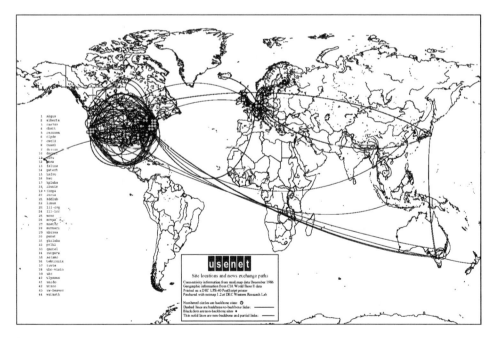

Figure 6.5
Although little remembered today, by 1986 Usenet had built dense connections in North America and Western Europe, with tendrils extending to Australasia and Japan. Reproduced courtesy of Brian Reid.

As more Unix systems were connected to the Internet over the course of the 1980s, the news service continued to work much as before, except that posts were synchronized rapidly between sites over the Internet. Its assimilation into the Internet exemplifies the Internet's growing importance as a way of transmitting information between different kinds of networks via ad hoc gateways. Newsgroup usage peaked in the late 1990s, by which point its content was dominated by spam and pornography.

Another network, CS Net, was founded in 1981 for the many computer science departments that lacked ARPA research contracts and so could not connect directly to ARPANET. It relied on public packet switched networks and dial up connections to run TCP/IP packets to and from universities like Purdue, which was on ARPANET, out to other universities without direct connections. The first sites to be connected were Princeton, Purdue, and Delaware. By 1985, CS Net was providing modem links to more than a hundred host computers, allowing their users to enjoy Internet services such as ftp and telnet. CS Net users could also exchange email with ARPANET users via a gateway service. This was an example of the power of TCP/IP to support *internetworking* in its original sense: the interconnection of different networks. In 1991, CS Net was replaced by NSF Net, which connected universities directly to the Internet.[35]

BITNET (a contraction of "because it's there network"), another important service, was completely separate from the ARPANET. It centered on the City University of New York, which initiated the service in 1981. BITNET connected academic users of IBM's System 370 mainframes using leased telephone lines. To minimize disruption, its creators used an existing operating system feature designed to read decks of punched card input.

Like Usenet, BITNET initiated a service that by the 1990s had become a standard part of the Internet. In this case it was listserv, a program to maintain email discussion lists. Any message sent to the address for a list would automatically be forwarded to all the list subscribers, turning email from a mechanism for one-to-one communication into the underpinnings for online communities with shared interests. In 1989, BITNET merged with CS Net, a step toward the consolidation of academic networks around Internet technologies.

COMMERCIAL NETWORKS

In the previous chapter we described the initial late-1960s boom in commercial timesharing services. This was driven by the idea that it would be most efficient for many companies to share a giant computer. In practice, the availability of timesharing operating systems such as Unix on smaller minicomputers, particularly the PDP-11 family, meant that medium and large organizations found it more cost effective to install their own computers. By the early 1970s, successful timesharing companies marketed access to specialized products and services rather than generic computer time. These included industry-specific application systems, unique programming tools, and access to databases. Tymshare, for example, offered applications for real estate appraisal, railroad rolling-stock record keeping, and personnel administration.[36]

Online Services

Some new services offered only database searches, rather than general-purpose accounts. The two most successful were LEXIS and Dialog. LEXIS gave lawyers access to searchable databases of judgements and other specialized reference sources. Dialog came directly out of the government projects of the 1960s, developed by Lockheed for NASA. By the mid-1970s, Lockheed had launched a public service based on Dialog's search and indexing capabilities, providing access to patents, journals, newspapers, and other electronic text.

Electronic mail was also a popular feature of commercial timesharing services, such as the mainframe operated by the Scientific Time Sharing Corporation. Its main selling point was the cryptic and powerful programming language APL, but STSC also offered an easy-to-use electronic mail facility to its users. In 1976, Jimmy Carter's presidential campaign made heavy use of this service to keep in touch with field operations. The same year, *Business Week* predicted rapid growth in business use of email, in a story titled "When the Interoffice Mail Goes Electronic." By the early 1980s, telecommunications firms such as AT&T and MCI were running their own email services, although the monthly subscription fee and additional charges for messages sent, received, and stored were enough to deter casual use. MCI Mail, which attracted 100,000 subscribers by 1987, was considered a cheap service because it charged only a dollar to send a message within the US. One of its founders, Vinton Cerf, had been a creator of TCP/IP, making this an early attempt to commercialize capabilities originated in the ARPANET world.

Large-scale commercial timesharing businesses began to offer cheap access to home users, initially during off hours to exploit underutilized capacity. One of the most influential was CompuServe, whose origins were discussed in the previous chapter. CompuServe offered consumer services including mail, news, weather, stock market prices, sports scores, and winning lotto ticket numbers. Its popular electronic mail service used the octal (base 8) account numbers of its subscribers as addresses. That reflected its descent from PDP-10 operating systems.

CompuServe also offered an unusual feature called CB Simulator, which allowed users to chat with one another in the manner of users of the Citizens Band radio, which had exploded in popularity, in part as a result of the 55 mile per hour speed limit imposed in 1973. The CompuServe product followed closely the metaphor of the radio, with channels and pseudonyms, called *handles*, to disguise the identities of participants. One 1983 advertisement (figure 6.6) showed a couple sitting in their bathrobes: "Last

Figure 6.6

In 1983, CompuServe was promoting its propriety chat ("CB simulator") and electronic mail ("Email") systems to home computer users across North America.

night, we exchanged letters with mom, then had a party for eleven people in nine different states and only had to wash one glass." The "letters" to mom were sent with the "CompuServe Electronic Mail system (we call it Email™)." CompuServe let the trademark application for *Email* drop, but the popularity of its system popularized this now-standard contraction of *electronic mail*.

While individuals and small companies relied on online services for email access, big companies were increasingly likely to build their own internal data networks.

Email communication was a key motivator for this, alongside other benefits such as access to online applications and shared databases. Computer producers such as IBM and DEC offered their own email systems and network technology, to encourage big companies to standardize on their products.

Videotex and Minitel

People often assume that prior to the opening of the Internet to commercial use in the 1990s nobody foresaw a world in which ordinary people would be sending email, banking and shopping online, reading news online, and so on. Nothing could be further from the truth. If anything, a computer enthusiast of the late 1970s would be surprised that it took almost twenty years for these activities to become common. Books like *The Micro Millennium, Network Nation*, and *The Third Wave* made confident predictions of worlds in which lawyers had been replaced with computers, or workers carried out their jobs from home using terminals.[37]

Plato was a lavishly funded showcase for what was possible with the most advanced technologies of the era. Its mainframes and custom-built graphics terminals were too expensive for widespread use. That made its tens of thousands of users a tiny, and technologically privileged, elite in the rapidly expanding computing world of the mid-1970s. Other projects took the opposite approach: designing online systems around technologies that could, using the technologies of the era, plausibly be mass produced for millions of middle-class households expected to spend a few hundred dollars on a terminal.

Telecommunications companies set to work designing the affordable new chip-based terminals and online systems that were supposed to make this possible. In the US, telephone equipment was traditionally not bought but rather rented from AT&T. Modifications were forbidden. Changes during the 1970s and 1980s opened up telephone networks in the US, and many other countries, for use by data services. AT&T made two far-reaching decisions, the first of which was not to discriminate between voice and data sent over its lines. Both voice and modem signals were in the audio range and in principle no different, even though data might sound funny to a person listening in on the line. The second was to introduce a jack, the RJ11 to connect telephones without the need for a company technician to visit. That allowed a user to connect a modem directly to the phone network, rather than having to dial a number, wait for a high-pitched tone indicating a connection, and place the receiver into a cradle that acoustically coupled the modem to the phone line.

But what kind of terminal would ordinary people be plugging into their phone jacks? In the UK, broadcasting services and the Post Office (which operated the telephone network) agreed on a Videotex standard that offered twenty-five lines of colorful text, each with forty characters. Special characters provided chunky graphics. The low resolution made pages visible on the small and fuzzy television sets of the era, reduced the amount of expensive memory needed to display them, and let pages download quickly. British broadcasters employed unused bandwidth between frames of television to transmit pages over the air to television sets, in a related service called Ceefax.

The Post Office called its online service Prestel. It offered cheap television-connected terminals equipped with modems, intended for browsing use. The modems uploaded at an agonizingly slow 75 bits per second (sending text little faster than a skilled typist) but downloaded at a speedy 1,200. These dialed into Post Office-owned minicomputers scattered around the country. Companies could publish information on Prestel, which users could pay additional charges to access. According to historian Tom Lean, this information was "wide ranging, including news, business information, sports, film reviews, quizzes, gardening advice, and recipes. There were also pages from major retailers . . . , banks and building societies, travel companies . . . and the charity Save the Children." Many other countries purchased the same hardware and software to set up their own national networks. But like Prestel itself, their usage fell far short of predictions. By the end of 1982, more than three years after Prestel's launch, it had only around 18,000 users. The vast majority were businesses rather than the home users it was intended for. Prestel peaked in the late 1980s with about 90,000 users—far short of the millions for which it was designed.[38]

The US followed a different pattern—instead of a single national service run by the government, there were several private viewdata services. The most ambitious was Viewtron, offered by AT&T in partnership with the Knight-Ridder newspaper group. Tests began in 1980, but despite a national rollout in 1985 the service won only about 20,000 subscribers and it was abandoned in 1986.[39] By then the focus of projects intended for home use, Prestel included, was already shifting to systems based on home computers, discussed in the next chapter. British Telecom, the successor to the Post Office, closed down most of Prestel in 1994 and sold off the remains.[40]

Minitel

Only France broke from this pattern of disappointment and failure. France Telecom saturated the country with cheap little Minitel terminals, each combining a small monochrome screen and a simple keyboard. This overcame the problem of low usage that doomed other viewdata systems. Giving away the terminals would boost use of telephone lines and, because Minitel included a searchable telephone directory, eliminate the cost of printing and distributing telephone directories. The project fit a post-War French pattern of building national greatness through high profile, high technology projects such as nuclear power, the Concorde, and high-speed rail. After experimental trials that began in 1980, Minitel was rolled out in Paris at the end of 1983. By 1987 it was available in every part of France. At its peak in mid-1993, almost six and a half million terminals were used for ninety million connection hours.[41]

During the late-1980s, more French people were using online services for banking, shopping, news, and email than in the rest of the world put together. France Telecom was a gateway to other services provided by thousands of other companies, many of which used their own computers to host them. Additional charges were automatically added to the telephone bills of their users, along with standard charges for the connection time. Minitel codes were prominently featured on billboards and in newspaper and magazine advertisements. Its most notorious application was the market for premium priced sex chat services, the textual equivalent of the voice services that became popular in other countries around the same time. France Telecom passed most of the premium charges on to the service providers, meaning that Minitel (unlike the Internet) had a built-in payment mechanism to make electronic publishing viable. Minitel was finally shut down in 2012 after having coexisted successfully with personal computer–oriented and Internet services for most of the 1990s.

Commercial Packet Switched Networks

Thanks in part to the success of the ARPANET, computer companies, governments, and telecommunications firms had by the mid-1970s already recognized the potential of packet switched networks. That did not mean that they intended to adopt Internet-based technologies. Instead, the International Telecommunications Union, which had been negotiating international standards since the days of the telegraph, completed work in 1976 on the first version of a standard known as X.25.

The promise was that data communication could become just another telecommunications service, so that computers plugged into compatible networks could exchange data across national lines. Terminals connected directly to X.25 networks could reach host computers anywhere in the world. As X.25 was intended for commercial use, it included some features missing from Internet technologies, most notably billing mechanisms to charge customers for each packet delivered according to the route taken.[42]

Although X.25 data transmission never became as ubiquitous as telephone calls, it did underpin the most widely used packet switched networks of the 1970s and 1980s. Even BBN, which built the ARPANET, used a protocol closely related to X.25 in creating a spin-off to operate a public data network. Telenet started operation in 1975. Its big competitor was Tymnet, which began as an internal network for Tymshare to bridge its data centers to remote access points, so that customers in smaller markets could connect with local calls. By 1972, Tymnet connected more than forty cities.[43] It was spun off as a separate telecommunications company in 1976 so that it could sell network services to the public, and eventually it surpassed the value of Tymshare's original business of selling computer time. Telenet and Tymnet both offered X.25 interfaces and could interconnect with other networks, including Transpac in France. Minitel users used special telephone numbers to dial into some services, such as the electronic telephone book, but could also dial 3615 to access thousands of approved services via Transpac or 3619 for an international getaway that could even access Internet services.[44]

Building out networks of data centers and access points brought American timesharing businesses up against the heavily regulated telecommunications industry. They had to fight efforts by traditional telecommunications carriers to expand regulation to all firms offering public services over telephone lines. Comshare founder Rick Crandall recalls that to legally share a single long-distance line from the firm's data center between several access modems, it had to build a *store-and-forward* device. Using a standard multiplexer to share a telephone line would be an illegal resale of telecommunications capacity, whereas running each bit through a tiny buffer turned the line-sharing device into a computer.[45] The absurdity of such distinctions led to a series of policy reviews, contributing to a broad deregulation of telecommunications in the early 1980s.

By the mid-1980s, X.25 had become the underpinning for a whole stack of internationally agreed data communications protocols, known as the open systems interconnection (OSI) protocols. These included X.400 for email and X.500 for electronic directory services, as well as more obscure standards for tasks such as the management of network sessions and the encoding of images. Adoption lagged and usable products were slow to appear, but at the time most people involved in telecommunications assumed that the backing of governments, telecom firms, and computer companies meant that the OSI standards would eventually underpin most computer communications. Plans were even drawn up to convert the Internet over to X.25 and the other OSI protocols. That never happened, of course. Instead TCP/IP and other Internet technologies unexpectedly moved beyond the niche markets they dominated in the 1980s to do the job for which OSI was created: standardizing data communications around nonproprietary standards.

INTERNET COMMERCIALIZATION

In 1981, there were about two hundred host computers connected to the Internet. Over the next five years the total rose to around five thousand and by the end of the decade to around 160,000.[46] The rise of the Internet was closely connected to the adoption of the TCP/IP network protocol. In 1980, ARPA threw its support behind the Berkeley standard distribution (BSD) of Unix as a common system the agency could recommend for all its clients, thanks to the portability of Unix to computers from many manufacturers. This protocol, and its bundling with BSD from version 4.2 onward, permanently linked Unix and the Internet.[47] This led to a shift in the dominant community of Internet "hackers" away from PDP-10s running derivatives of Tenex and toward smaller PDP-11 family machines using Unix.

VAX machines running Unix eventually became the most common Internet-connected machines. Unix, born in a collegial environment, made the sharing of files easy; that also made Unix systems vulnerable to viruses and unauthorized intrusions by hackers. The *Internet worm*, written by a student at Cornell and unleashed in 1988, targeted VAX machines and brought the Internet down for several days by exploiting flaws in Unix to infect thousands of hosts.[48] Widespread publicity exposed newspaper readers around the world to two new previously obscure technologies: computer worms and the Internet.

NSFNET

Meanwhile, the Internet itself was rapidly changing. This rapid expansion was made possible by aggressive upgrades to the transcontinental backbone network. The National Science Foundation's NSFNET began operation in 1986 and was rapidly expanded and upgraded. The next year, NSF ran high-capacity T1 lines between the universities where it supported supercomputing centers, increasing capacity from fifty-six thousand bits per second to one and a half million bits per second. By 1989 it was upping that again, to forty-five million bits over T3 lines.[49]

This was the period when other American academic networks rapidly consolidated on Internet technologies. BITNET merged with the NSF's CSNET in 1987, which in turn was subsumed into NSFNET in 1990. By the end of the 1980s, the Internet was well on its way to becoming a true global network through the interconnection of national academic networks such as the British JANET (Joint Academic Network) in 1989. The furthest flung nodes were in Australia and New Zealand, connected via satellite.

With high-speed connections and international reach, NSFNET quickly replaced ARPANET as the hub of the Internet, expanding the three hundred networks connected to its backbone in mid-1988 to more than five thousand by early 1992. The NSF assumed responsibility for the Internet in 1990, and the original ARPANET was decommissioned.

Major upgrades to the capacity of NSFNET took place as part of a political push to create what was then called the information superhighway. Vice President Al Gore Jr. is still mocked for having claimed to invent the Internet during an interview with Wolf Blitzer of CNN in March 1999. His actual words were more careful: "during my service in the United States Congress, I took the initiative in creating the Internet." Gore's vision was reminiscent of the early ARPANET with its focus on remote logins to large computers. He wanted the federal government to sponsor a high-speed network to give researchers to expensive supercomputers. While a senator in 1991, Gore proposed the High Performance Computing Act to create what he called a national information infrastructure so that scientists all across the country could push the frontiers of physics, chemistry, and biomedical research.

The Acceptable Use Policy

These upgrades made the Internet large and fast enough to attract businesses and corporate research labs as well as its traditional academic and government users. The Internet's first big opening to a larger audience came around 1989, not through direct access but with the authorization of "experimental" gateways to transfer email to and from commercial networks such as MCI Mail and CompuServe. As Usenet and the amateur FidoNet collective already had gateways, this turned the Internet into a clearinghouse for different email services. Other online services and corporate networks soon followed. Email gave users access to Internet resources such as listserv discussion groups and file transfer sites, which dispatched files requested via email.

Around the same time, the first efforts were made by entrepreneurial companies to sell Internet access. Two of the first Internet service providers (ISPs), PSINet and UUNET, were established by experienced network users. Their initial customers were universities and state governments. They immediately came into conflict with IBM, which had its own plans to profit from the Internet. By 1991, the ISPs had agreed to interconnect their developing networks with the establishment of the commercial Internet exchange (CIX), which became the hub of a fast-developing private Internet industry.[50]

The NSF had to address the question of how to deal with commercial firms' connecting to and using a network that was conceived, controlled, and paid for by the federal government. It responded with an Acceptable Use Policy for NSFNet, which read in part

> NSF Backbone services are provided to support open research and education in and among U.S. research and instructional institutions, plus research arms of for-profit firms when engaged in open scholarly communication and research. Use for other purposes is not acceptable.[51]

The policy allowed "announcements of new products or activities . . . but not advertising." "Extensive use for private or personal business" was specifically deemed unacceptable. That policy doesn't sound like a great basis for the commercialization of the Internet, but its meaning was altered by a legal change to the charter of the NSF itself, passed in October 1992. This allowed traffic from private networks to flow over its backbone as long as this would "tend to increase the overall capacities of the network to support . . . research and education."[52] NSFNET itself remained a haven for research and scholarship, but the CIX could pass data onto NSFNET and the broader

Internet, as well as exchange it between commercial networks. And those ISPs could sign up businesses and individuals as clients.

Traffic on the Internet, already growing rapidly, grew even faster—from 1 trillion bytes a month in January 1992 to 10 trillion a month in 1994. This traffic was increasingly shifting from NSFNET itself to the fast-growing commercial backbone networks and access points being set up by firms like Sprint, Ameritech, and Pacific Bell. In 1995, the NSFNET was dissolved, as its academic users could buy network access elsewhere. The Internet had been privatized.[53] Meanwhile, Gore left the Senate and became Vice President in January 1993. As Vice President he continued to champion Internet use, insisting that federal agencies provide information online.

The Domain Name System

The Internet Protocol introduced the basic system of numerical addresses still in use in the early twenty-first century. An address such as 129.89.43.3 identifies both the computer to which a data packet should be sent and the network on which it can be found. However, ARPANET users were accustomed to specifying computers and associated email addresses by name, for example, ucbvax or fred@princeton, rather than by numerical addresses. In 1983, a new hierarchical domain name system and associated protocols for name resolution had replaced the increasingly unwieldy list of ARPANET-connected computers maintained by the Network Information Center. Top-level domains had either national codes, such as .uk, or functional codes, such as .mil, .gov, and .edu. Organizations received domain names such as ibm.com or mit.edu to cover their assigned blocks of numerical addresses. Each had a delegated responsibility to allocate those addresses to computers and operate the electronic directory that would translate names within that domain to numbers.[54]

This system required a central authority to create new domains and allocate addresses to them. For many years Jon Postel ran what came to be known as the Internet Assigned Numbers Authority. As the Internet commercialized during the 1990s, the process was formalized, with a greater governmental role in control of the top-level domains and companies competing to offer domain registration services. Commercialization of the Internet sparked many legal disputes over ownership of valuable domain names such as "sex.com" or the registration of company names as domains by opportunistic "cybersquatters." Domain name resolution remains the most centralized aspect of the Internet, and failures of the *root nameservers* (some caused by hackers) have shut down much of the network on several occasions.

Legacy of the Academic Internet

The commercialization of the Internet might seem inevitable, but such a transformation required jumping over many social, political, and technical hurdles. The Internet's robust design put most of the work on the computers and routers that were connected to it, which made the technical hurdle easiest to clear. It grew within a few years from handling a few thousand nodes linked by 56 kbs lines to millions of nodes linked by ever-faster fiber-optic lines.

The speed with which the Internet displaced networks designed for commercial use has more to do with the slow progress of OSI networking efforts by traditional telecommunications firms than with any inherent suitability of Internet technologies to the tasks for which we now use them. The Internet inherited from its academic roots a characteristic set of strengths and weaknesses that together go a long way to explain not only the rapid success of the commercial Internet but also its persistent problems and weaknesses.

1. Unlike networks designed for commercial use, Internet technologies provided no way to charge users according to the network resources they consumed or compensate the providers of network services. Thus, for example, Internet users were not billed by their network provider for use of intercontinental connections or for reading premium content online.

2. The Internet transmitted data packets without regard to their content (often called the *end-to-end principle*). This flexibility was essential in a network designed to serve a research community rather than to perform a specific task.

3. The Internet relied more on social mechanisms than on technical ones to provide security and eliminate troublemakers, having been designed for a homogeneous population of highly educated users given access through their employers or universities. This is why issues such as spam and attacks from hackers proved so hard to deal with after commercialization.

4. The Internet was designed to support many different machine types, achieving compatibility through shared protocol rather than shared code.

5. Any computer connected to the Internet could send data as well as receive it. This *peer-to-peer* operation meant that a computer could turn into a file server, an email server, or (later) a Web server simply by running a new program. Users were expected to publish online information and provide services to each other, rather than to rely on a single central collection of resources.

6. The Internet integrated many different communications media. The abstraction of TCP/IP from media-specific aspects of communication has made it possible to extend the Internet over new technologies such as cellular telephones, Wi-Fi, and fiber-optic cables.

For now, we leave the Internet in 1992, growing rapidly in speed, reach, and number of users and newly opened to commercial use. Around a million computers were connected directly to the network, but those did not yet include many small businesses or home users. It was still not simple to connect to the Internet, nor was it easy to navigate its many different sites, protocols, and services.[55]

One reason for its rapid growth was the greater use of local area networks. Instead of hooking up a single large timesharing computer to the ARPANET via an IMP, universities and research labs would now run a local network to interconnect many minicomputers and personal workstations. Using TCP/IP this entire local network could be interconnected to the Internet. The Domain Name System made it easy to distinguish between different hosts within a particular institutional network. MIT, for example, had two ARPANET IMPs in 1977 to which a total of seven computers were connected. A decade later, the shift to TCP/IP and local networks meant that thousands of MIT workstations and personal computers could send and receive Internet packets. Within another decade the World Wide Web would make the Internet into a core part of the personal computing experience. We tell those stories in subsequent chapters, but first we must return to the 1970s to explain where all those personal computers, local area networks, and workstations came from.

7 THE COMPUTER BECOMES A PERSONAL PLAYTHING

In November 1974, Daniel D. McCracken, author of a series of popular programming textbooks, spent a day in a conference room at the Sheraton-Harbor Island hotel in San Diego with an invited group of computing experts. Though no older than middle aged, the men were, by the standards of their industry, grizzled elders. Their meeting was one of a series known as the RAND Corporation Symposia after their original host back in the 1950s. McCracken dramatically interrupted their musings on "The Future of Programmers" with the announcement that, "he had brought along a computer with him," in the shape of a microprocessor chip. "Ninety percent of all processors ever made," he told his colleagues, "were shipped by Intel in the last two years."[1]

Much of their discussion had centered on a controversial report predicting that in 1985 the world would hold a total 375,000 computers (up from 70,000 in 1970), served by 640,000 programmers. The computers we have discussed so far cost many thousands, sometimes millions, of dollars and were rarely purchased by individuals or for personal use, although hackers, researchers, and students were sometimes able to use computer time for play as well as work. Like the rest of the industry, McCracken's friends were entirely unprepared for the reality that individual computer models would sell more than a million units each by the early 1980s, or that computers would by then greatly outnumber full-time programmers due to a shift to inexpensive, prepackaged software.

NEW ELECTRONIC DEVICES

Owning a computer for personal use became realistic only in the mid-1970s, as improvements in chip technology reduced the size and cost of digital electronics, but as McCracken spoke, the advances in chip density being made by Intel and its competitors had already shaken up the calculator industry. For decades there had been a small market for machines that could perform the four functions of arithmetic.

The most powerful could also calculate square roots. In the 1950s and 1960s, the calculator industry was dominated by firms such as Friden and Marchant in the United States, and Odhner in Europe. Their products were complex, heavy, and expensive. Few mechanical calculator firms survived the transition to electronics.

Electronic Calculators

In 1964, Wang Laboratories, a company founded by An Wang, a Chinese immigrant who had worked with Howard Aiken at Harvard, came out with a LOCI electronic calculator that offered more functions at a lower cost. Its successor, the Wang 300, was even cheaper and easier to use.[2] A few years later Hewlett-Packard, known for its oscilloscopes and electronic test equipment, was selling the HP-9100A calculator for just under $5,000. To display digits, the Wang used Nixie tubes, ingenious tubes invented by Burroughs in 1957. HP used a small cathode-ray tube, as might be expected from a company that made oscilloscopes.

Dramatically cheaper calculators about the size of a paperback book appeared in 1970. They were made possible by a new chip technology, metal-oxide semiconductor (MOS). MOS chips were slower than the bipolar chip technology that then dominated the market, but they were cheaper to produce and could hold more components. A variant, CMOS, required very little power and was ideal for battery-powered devices such as that other 1970s enthusiasm, the digital watch.

Calculators were the first civilian counterpart to the Minuteman and Apollo programs of the 1960s, creating a market that guaranteed chip suppliers a long production run to realize economies of scale. When the Bowmar Brain was advertised for less than $250 for the 1971 Christmas season, the calculator burst into public consciousness. Prices continued to plummet: under $150 in 1972; under $100 by 1973, under $50 by 1976, and finally cheap enough to give away as promotional trinkets.[3] Two Japanese firms with consumer marketing skills, Casio and Sharp, soon dominated. The pioneering calculator companies either stopped making calculators, as did Wang, or went bankrupt, as did Bowmar. Texas Instruments survived by cutting costs to the bone.

Hewlett-Packard stunned the market in early 1972 with the HP-35, a $400 pocket calculator that performed all the logarithmic and trigonometric functions required by engineers and scientists. Within a few years the slide rule joined the mechanical calculator on the shelves of museums. Chuck House, an engineer involved with the early Hewlett-Packard calculators, said, "One could uncharitably say that we invented essentially nothing; we simply took all the ideas that were out there and figured out how to implement them cost-effectively."[4]

The first programmable pocket calculator was Hewlett-Packard's HP-65, introduced in early 1974 for $795. Texas Instruments and others soon followed. These machines could compute logarithms and trigonometric functions and carry out floating-point arithmetic to ten decimal digits of precision. Few mainframes could match that without custom-written software. They could also store and execute short programs, which proved captivating to some users who began to take their calculators home to play with. Most were adult professionals, including civil and electrical engineers, lawyers, financial people, pilots, and so on, with a practical need for calculation (see figure 7.1). They did not fit the profile of hackers as kids with "their rumpled clothes, their unwashed and unshaven faces, and their uncombed hair . . ."[5] But their passion matched the programming enthusiasts' at MIT. Their numbers, increasing as the prices of calculators dropped, were the first indication that personal computing was truly a mass phenomenon. There may never have been more than a few hundred people fortunate enough to be allowed to "hack" on a computer like the PDP-10. By 1975, there were over 25,000 HP-65 programmable calculators in use.[6]

As powerful as the programmable calculators were, the trade press was hesitant to call them *computers*, even though Hewlett-Packard used the term *personal computer* in its advertising as early as 1968.[7] Hewlett-Packard and Texas Instruments sold the machines as commodities; they could ill afford a sales force able to walk a customer

Figure 7.1
Shuttle astronaut Sally Ride aboard the Challenger during mission STS-7 in 1983. Note the three Hewlett-Packard programmable calculators floating next to her. NASA purchased the calculators at a Houston department store and made only minimal modifications. Although the Shuttle carried a suite of IBM 4-pi computers for guidance, navigation, and control, the HP calculators saw heavy use by the crew.
Source: NASA.

through the complex learning process needed to get the most out of one. Calculators were designed to be easy enough to use to make that unnecessary, at least for basic tasks. When customers wanted to do more, they turned to one another. Users' groups, clubs, newsletters, and publications proliferated.

Microprocessors

Back in 1965, Gordon Moore, then running research and development at Fairchild, used a piece of graph paper to plot changes in the economically optimal number of components crammed into a single integrated circuit. Moore noted that it had doubled every year, and he extrapolated that by the mid-1970s one could buy a single chip with the logical complexity of an entire 1950s-era computer. A decade later, Moore's frequently updated charts showed that this exponential growth had indeed been sustained. The phenomenon became famous as *Moore's law*, although as Moore himself wryly observed, that term was applied "to almost any phenomenon related to the semiconductor industry that when plotted on semilog graph paper approximates a straight line."[8]

Moore's law held up for so long that it was viewed by many as a natural feature of chip technology, but Moore himself acknowledged the ever increasing amounts of work, money, and creativity needed to sustain the process.[9] He and Robert Noyce left Fairchild to found Intel, a key moment in the development of what was soon called Silicon Valley. Intel embraced MOS chip technology enthusiastically. It targeted the market for computer memory with its first breakout hit, the 1103 chip launched in 1970. MOS chips were slow in comparison with existing bipolar computer logic, but even slow transistors could flip bits much faster than core memory and, thanks to Intel, were soon cheaper as well.[10]

Each 1103 chip stored only 1,024 bits of data, but the same trajectory that made RAM chips possible in the first place ensured that smaller transistors and higher capacities would follow. Intel enjoyed an early edge in the production of high-density chips due to its development of a *silicon gate* variant on MOS. Demand for ever larger memory chips and other complex devices offered Intel a chance to exploit its strengths to grow both the volume and value of sales over time, unlike the calculator market where, as historian Ross Bassett has noted, revenues collapsed even as sales rose because chips became cheap commodities.[11]

Computer engineer Gordon Bell stated, "The semiconductor density has really been the driving force, and as you reach different density levels, different machines pop out of that in time."[12] MOS brought the concept of a computer on a chip into focus. What was needed was a set of integrated circuits—or even a single integrated circuit— that incorporated the basic architecture of a general-purpose, stored-program computer. By 1971 the idea was realized in silicon. Intel's customer Busicom sought to produce a line of calculators with different capabilities. It wanted a set of custom-designed chips that incorporated the logic for the advanced mathematical functions. Intel's management assigned Marcian E. Hoff, who had joined the company in 1968 (Intel's twelfth employee), to work with Busicom. Gary Boone of Texas Instruments also designed similar circuits around that time.[13]

Intel's focus had always been on general purpose chips with large potential markets, but Hoff realized that by designing fewer, more flexible chips Intel could satisfy Busicom's needs and those of many other potential customers. He proposed a logic chip that implemented the architecture of a general-purpose computer. Hoff was inspired by the PDP-8, which had proven itself useful for control applications despite a very small instruction set. Although the chip computer would be even simpler, Hoff equipped it with a stack so that it could efficiently jump into subroutines to carry out more complex functions, substituting code for specialist mathematical hardware. That was slower, but a person could not press calculator keys that fast anyway. Hoff's idea was an early example of the application of computer technology to "dissolve" the insides of other devices. The new calculators looked like shrunken versions of the old mechanical ones, but they used processors and code to mimic the functioning of special-purpose digital logic, which in turn had recently replaced the rods and gears of traditional calculators.

The detailed design of the 4004 was done by Stan Mazor. Federico Faggin, who left Intel in 1974 to found Zilog, a rival company, was crucial in making the concept practical. Masatoshi Shima from Busicom also contributed. Robert Noyce negotiated a deal with Busicom to provide it with chips at a lower cost, giving Intel in return the right to market the chips to other customers for noncalculator applications.

The result was a set of four chips, first advertised in a trade journal in late 1971, which included "a microprogrammable computer on a chip!"[14] That was the 4004, on which one found all the basic registers and control functions of a tiny, general-purpose, stored-program computer. The other chips were a read-only Memory (ROM), random-access memory (RAM), and a chip to handle output functions. At the time, Intel was not sure what it had invented: Intel's patent attorney resisted Hoff's desire at the time to patent the work as a computer. Intel's model was that an industrial customer bought a microprocessor and wrote specialized software for it. That software was then burned into a read-only memory to give an embedded controller system with the desired functions.

Another Intel team led by Dov Frohman developed a ROM chip that could be easily reprogrammed and erased by exposure to ultraviolet light. Called an erasable programmable read-only memory (EPROM), introduced in 1971, it made it easy for customers to develop microprocessor-based control systems.[15] Until the mid-1980s, EPROMs were Intel's most profitable product line.[16]

The 4004 worked with groups of four bits at a time—enough to code a single decimal digit. At almost the same time as the work with Busicom, Intel entered into a similar agreement to produce a set of chips for a mainframe terminal. To handle the terminal's logic, Mazor and Hoff proposed an 8-bit microprocessor, able to process a full byte at a time and so handle text as well as numbers. The intended customer decided to go with conventional TTL chips, but Intel launched the 8008 microprocessor as a product in April 1972.[17]

Microprocessor suppliers developed packages to support system design. These were essentially complete computers, with the microprocessor, RAM and ROM, support chips mounted on a printed circuit board, bundled with a tutorial manual on how to program the system. The kits were sold for around $200 and given away to engineers who might generate volume sales.[18]

Intel also built fully assembled development systems, on which a customer could test the software for an application. These sold for around $10,000. The company hired Gary Kildall, an instructor at the Naval Postgraduate School in Monterey, California, to develop a language based on IBM's PL/I.[19] He called it PL/M. By 1974, two Intel development systems, Intellec 4 and Intellec 8 included their own resident PL/M compiler. According to a biography of Moore, the team responsible for marketing Intel's microprocessors "had no doubt that this was the future of Intel" and were disappointed when Moore, his face growing red, insisted that "you absolutely are *not* going to talk about these systems as computers." Moore was worried that Intel would be competing with its own customers if it sold the systems to end users.[20]

Small firms did indeed start building computers around Intel's new chip. The first microprocessor-based machine to be sold as a general-purpose computer appeared in 1973, thanks to Thi T. Truong, an immigrant to France from Vietnam. His Micral was a rugged and well-designed computer based on the Intel 8008 microprocessor. It included internal slots for expansion. A base model cost under $2,000, and around 2,000 were sold in the next two years to replace minicomputers for control operations.[21] A compact Canadian machine, the MCM/70, went on sale the next year, with two built-in cassette recorders. It was programmable using a built-in APL language, a cult favorite among users of mainframe timesharing systems, but its one-line display was best suited to numerical calculations. Only a few hundred were produced. Neither firm had hit on a formula to unlock the commercial potential of microprocessor technology.[22]

Early Kit Computers

Electronics hobbyists and enthusiasts had a long history of technical innovation—it was radio amateurs, for example, who opened up the high-frequency radio spectrum for long-distance radio communications after the First World War. After World War II, a cornucopia of war surplus equipment from the US Army Signal Corps found its way into individuals' hands.[23] The hobby expanded beyond amateur radio to include high-fidelity music reproduction, automatic controls, and simple robotics as well. This community supplied a support infrastructure for the personal computer that neither the computer companies nor the chip makers could match. This infrastructure included a variety of electronics magazines, leaving a trace of how hobbyists moved from analog to digital designs. Each issue typically had at least one construction project. The magazine made arrangements with small electronics companies to supply a printed circuit board, already etched and drilled, as well as any hard-to-find components.

One influential hobbyist project was the "TV-Typewriter," designed by Don Lancaster and published in *Radio-Electronics* in September 1973. This device allowed one to display alphanumeric characters, encoded in ASCII, on an ordinary television set. It presaged the advent of video displays and keyboards as the primary input-output devices for personal computers.[24]

The Scelbi-8H, announced in a tiny advertisement in the back of the March 1974 issue of *QST,* a magazine for radio amateurs, used an Intel 8008, and thus may be the first microprocessor-based computer marketed to electronics hobbyists. Prices for a kit started at $440.[25] That July, *Radio-Electronics* featured another kit computer, designed by Jonathan Titus of Virginia Polytechnic Institute in Blacksburg, on its cover with the tag "Build the Mark-8: Your Personal Minicomputer."[26] The thousands of readers who mailed $5.00 for an instruction booklet were told to buy components including custom circuit boards ($47.50), and to contact Intel to buy an 8008 chip (for $120.00). At least one Mark-8 users club sprang up, in Denver.

Other articles described simpler digital devices: timers, games, clocks, keyboards, and measuring instruments, using inexpensive TTL chips. They reflected a concerted effort by the community to bring digital electronics, promising but complex, to amateurs familiar with simpler radio or audio devices.

PERSONAL COMPUTING

1974 was the *annus mirabilis* of personal computing. In January, Hewlett-Packard introduced its HP-65 programmable calculator. That summer Intel announced the 8080, an improved microprocessor. In July *Radio-Electronics* described the Mark-8. In late December subscribers to *Popular Electronics* received their January 1975 issue, with a prototype of the "Altair" minicomputer on the cover. It was designed by Henry Edward Roberts, who had experienced a brief success selling calculator kits from his model-rocket hobby shop in Albuquerque, New Mexico. Roberts hoped that a computer kit might be even more popular.

The MITS Altair
The *Popular Electronics* cover called the Altair the "world's first minicomputer kit." It even looked like the Data General Nova, a fast, compact and affordable minicomputer that had shaken up the industry: a rectangular metal case, a front panel of switches that controlled the contents of internal registers, and small lights indicating binary ones

and zeros (see figure 7.2). Yet the Altair cost one-tenth the price fetched by comparably minimal Nova when it was launched six years earlier. The article, written by Ed Roberts and William Yates pointed out that it was "a full-blown computer that can hold its own against sophisticated minicomputers" and was "not a 'demonstrator' or a souped-up calculator."[27] Not long after the Altair's introduction, journalists began calling these machines "microcomputers," reflecting their small size and reliance on microprocessors.

The magazine offered an Altair for under $400 as a kit, or a few hundred dollars more already assembled. The low cost came mainly from its use of Intel's new 8080 microprocessor, the first whose instruction set and memory addressing capability approached those of the minicomputers of the day. It had more instructions than the 8008, was faster, and had a larger stack. It could address more memory—up to 64,000 bytes—yet required only six instead of twenty supporting chips to make a functional system.[28] Roberts's little company, MITS, bought them for only $75 each, versus a list price of $360.[29]

Inside the Altair's case, one found a machine built mainly of TTL integrated circuits (except for the microprocessor, which was a MOS device), packaged in dual inline packages, soldered onto circuit boards. The Altair used integrated circuits, not magnetic cores, for its primary memory. Assembling the Altair was much more difficult than assembling other electronics kits, such as those sold by the Heath Company. MITS offered to sell "completely assembled and tested" computers for $498, but with such a backlog of orders, one was faced with the choice of ordering the kit and getting

Figure 7.2

The Altair 8800's exterior was modeled after minicomputers, but because it was based on a microprocessor the cost was much lower. Unless additional expansion cards were fitted, these lights were its only means of communication. Photo courtesy Division of Medicine and Science, National Museum of American History, Smithsonian Institution.

something in a couple of months, or ordering the assembled computer and perhaps waiting a year or more. Most ordered the kit.

The audience of electronics hobbyists, toward whom the magazine article was aimed, compared the Altair favorably to building a computer from scratch. At least one organization, the Amateur Computer Society, founded in 1966, had been trying for years to help its members design and build their own minicomputers, but only a handful of amateurs had the drive and skills to pull off this feat.[30] In comparison, the Altair was more like an Ikea chest of drawers: the design work was taken care of, all the pieces were in the box, and anyone with the right skills and tools would eventually be able to put the thing together. Its purchasers looked to one another for support with the inevitable wiring errors and poorly soldered connections. Seen in this light, customers felt, rightly, that they were getting an incredible bargain. It was hard to design a computer, and most semiconductor firms had no distribution channels set up for single-unit or retail sales.

The first thing that customers did with these machines, once they got them running, was play games. Roberts was trying to sell it as a machine for serious work, however. In the *Popular Electronics* article he proposed a list of twenty-three applications, including "Multichannel data acquisition system," "Machine controller," "Automatic controller for heat, air conditioning, dehumidifying," and "Brain for a robot," among others.

When a computer sold for just $400, the extensive support and infrastructure that mini and mainframe companies supplied had to come from users' groups, informal newsletters, commercial magazines, local clubs, specialist book publishers, conventions, and even retail stores. This supporting infrastructure was critical to the success of personal computing. Historians often cite the Homebrew Computer Club, which met near the Stanford campus in the mid-1970s, as especially important.[31] The existing calculator users groups were also important to personal computing. Their newsletters discussed the advantages and drawbacks of each technology: one able to evaluate complex mathematical expressions with ease, the other more primitive but *potentially* capable of doing far more. Many members were bitten by the PC bug.

Ever since timesharing and minicomputers had revealed an alternative to mainframe computing, there were prophets and evangelists who raged against the world of punched cards and computer rooms, promising a digital paradise of truly interactive tools at one's fingertips. The most famous was Ted Nelson, whose self-published book *Computer Lib/ Dream Machines* proclaimed (with a raised fist on the cover): "You can and must understand computers *now*."[32] Nelson glorified computers as "dream machines," but as computer access became a reality it offered more tangible pleasures and challenges to waking users.

Expandability and Modularity

Marketing the computer as a bare-bones kit offered a way for thousands of people to bootstrap their way into the computer age, at a pace that they, not a computer company, could control. The 8080 processor was only one architectural advantage the Altair had over its predecessors. Just as important was its use of a bus. The components of the computer were spread over several small circuit boards, known as cards. These plugged into slots on the bottom of its case, which were interconnected to transmit signals and power between different parts of the machine.

Even when assembled, an Altair was initially mostly empty, with just two cards (one holding the CPU, and the other some memory). MITS had promised many optional cards for additional memory, input/output, and other functions but was too overwhelmed by demand for the Altair itself to produce any for its early shipments. Without them an Altair could do little more than blink a pattern of lights on the front panel. Even that was not easy: one had to flick the toggle switches for each program step, then deposit that number into a memory location, then repeat for the next step, and so on—hopefully, the power did not go off while this was going on—until the whole program (less than 256 bytes long!) was in memory. Bruised fingers from flipping the small toggle switches were the least of the frustrations. When MITS did begin to produce larger memory, they proved extremely unreliable.

Thanks to the sixteen empty slots waiting for additional circuit boards, users could design and build the rest of a useful computer without waiting for MITS to get its act together. Some of those who succeeded went into business themselves, selling cards to other frustrated MITS customers. Plug-in boards offered a way of connecting the machine to a Teletype terminal or paper tape reader, via a serial connection, or to a television set and a keyboard. Another board sent out data in parallel, for connection to a printer. The Altair lost its data when the power was shut off, but before long MITS designed an interface that put out data as audio tones, to store programs on cheap audio cassettes. Cassette storage was slow and cumbersome—one often had to make several tries before successfully loading data into the computer.

Were it not possible for other companies to offer plug-in cards, or even to design improved versions of the Altair itself, as IMSAI and several other companies did, the Altair might have made no greater impact than the Mark-8 before it. IMSAI built more rugged and better equipped computers aimed at small business users. By 1977, MITS, which Roberts had sold to a larger company, was beginning to collapse.

It left behind the first standard for the personal computer industry. Lou Reed famously observed to Brian Eno that although only 30,000 people had purchased the first album by his experimental rock group, the Velvet Underground, each customer had seemingly formed a band after listening to it.[33] The Velvet Underground's final dissolution in 1973 was barely reported, yet its influence eventually outstripped that of more conventionally successful contemporary acts such as Elvis Presley, Led Zeppelin and the Beach Boys.[34] MITS sold a similar number of computers over an equally obscure career. Its revenues were tiny in comparison to IBM or DEC, but it likewise exerts an outsized influence on history thanks to the companies and industries founded by MITS users such as Bill Gates.

Microsoft BASIC

Making the Altair do anything meant writing a program, and as even the most excited users would tire of entering machine instructions by flicking switches, MITS needed to offer a programming language. BASIC was not the only programming language Ed Roberts could have chosen to commission for the Altair. Many computer science professors disparaged BASIC as a toy language that fostered poor programming habits. There were other, more respectable choices. Fortran was still widely used for scientific work. APL, an interactive language invented at IBM by Kenneth Iverson in the early 1960s, had been used for an unsuccessful IBM personal computer in 1973. Gary Kildall had chosen a subset of PL/I for the Intel microprocessor development kit. Roberts considered Fortran and APL but chose BASIC because it was easy to learn and had a track record of running on computers with limited memory.[35]

That choice put Bill Gates on a path to becoming the world's richest man. He was a student at Harvard and had, as we already mentioned, been programming for years when the Altair appeared on the cover of *Popular Electronics*. After his friend Paul Allen saw the magazine, the two set out to write a BASIC interpreter for it.[36] Their BASIC was modeled on a DEC extension of the language that Gates had encountered while programming, rather than directly on the Dartmouth original. DEC's resource sharing time sharing (RSTS-11) operating system was designed for PDP-11 Model 20, which provided just 56 KB of usable memory. The system opened up the PDP-11 to users who would have been daunted by its assembly language. According to Gordon Bell, it "was initially a hard machine to understand and was marketable only to those with extensive computer experience."[37]Following the lead of Dartmouth, its creators used BASIC as the main system interface and programming language. They wrote many of the system capabilities in BASIC, such as the routines to log users in and out. That required extending the language with direct access to hardware and memory. The commands they added, such as SYS, PEEK, and POKE, upset Kemeney and Kurtz by changing the original character of BASIC.

Thanks to Gates, BASIC played a similarly central role as both programming language and user interface for most early personal computers. He and fellow student Monte Davidoff used Harvard's PDP-10 to write the interpreter, based on the written specifications of the Intel 8080. In early 1975, Paul Allen flew to Albuquerque and demonstrated it to Roberts and Yates. It worked. They promised Altair customers that a version of BASIC usable with only 4 KB of memory would be available by June 1975. There were other implementations of BASIC for early microcomputers, but none was as good. It sold for $60 to Altair users. Gates insists that he never became a MITS employee (although Allen was until 1976), and under the name Micro Soft, he and Allen retained the rights to their BASIC.

In a now legendary "Open Letter to Hobbyists," distributed in early 1976, Gates complained about people making illicit copies of his BASIC by duplicating the paper tape (see figure 7.3). They had used more than $40,000 of computer time to develop it. If he and his programmers were not going to be paid, they would have little incentive to develop more software for personal computers. Illicit copying put personal computing at risk: "Nothing would please me more than to hire ten programmers and deluge the hobby market with good software."[38] By 1978 his company, now called Microsoft (figure 7.4), had severed its relationship with MITS and was moving to the Seattle suburb of Bellevue where Gates eventually made good on his promise to hire enough programmers to "deluge" the market with software, beginning with programming languages and system-building tools.

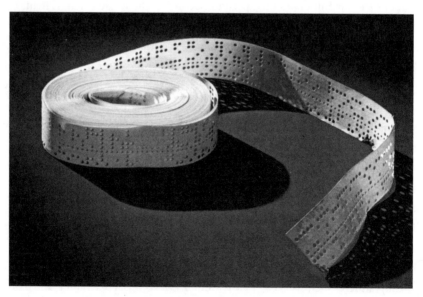

Figure 7.3
Paper tape containing Microsoft BASIC, for the Altair. The ease with which the tape could be copied and distributed without payment to Microsoft led to Bill Gates' famous "Open Letter" to hobbyists. Photo courtesy Division of Medicine and Science, National Museum of American History, Smithsonian Institution.

Figure 7.4

Microsoft employees gather for a group picture in 1978, just before their move from Albuquerque to Seattle. Front row from left: Bill Gates (cofounder), Andrea Lewis (technical writer), Maria Wood (bookkeeper), Paul Allen (cofounder). Middle row: Bob O'Rear (mathematical programmer), Bob Greenberg (programmer), Marc McDonald (first salaried employee), Gordon Letwin (programmer). Back row: Steve Wood, Bob Wallace, Jim Lane (project manager). Used with permission from Microsoft.[39]

CP/M: The Personal Computer Operating System

The floppy disk (or as pedants and IBM called it, *diskette*) symbolized personal computing from the late 1970s well into the 1980s. Yet it was invented, by David L. Noble at IBM, for a completely different purpose. When IBM introduced the System/370, which used semiconductor memory, it needed a way to store the computer's initial control program, as well as to hold the machine's microprogram. From that need came the 8-inch diameter *flexible diskette*, which IBM announced in 1971.[40] Before long, people recognized that it could be used for other purposes. Alan Shugart, who had once worked for IBM, recognized that the floppy's simplicity and low cost made it the ideal storage medium for low-cost computer systems.[41] Nevertheless, floppy drives were rare in the first few years of personal computing. IBM's hardware innovation was not enough; there had to be an equivalent innovation in system software to make the floppy practical.

While working on his PL/M compiler for Intel, Kildall wrote for his own use a small suite of programs that managed the flow of information to and from a floppy disk drive. "It turned out," he explained, "that the operating system, which was called CP/M for Control Program for Micros, was useful, too."[42] Kildall found himself writing variations of it for other customers. He and his wife, Dorothy McEwen, eased themselves into the software business while he worked as an instructor at the Naval Postgraduate School in Monterey, California. Jim Warren, profiling CP/M in the whimsically named *Dr. Dobb's Journal of Computer Calisthenics and Orthodontia*, emphasized its minicomputer heritage. He headlined his piece "Command languages and facilities similar to DECSYSTEM-10" (a derivative of the TOPS system we discussed previously) and concluded that it was "well-designed, based on an easy-to-use operating system that has been around for a DECade."[43]

Suggested prices were well under $100 for the software, or an estimated $800 for a disk system including drive and controller. That wasn't cheap, but disk storage was much faster than tape, and its contents could be modified more easily because the user did not have to run through the entire spool of tape to get at a specific piece of data. CP/M sold well enough to convince Kildall to focus all his efforts on his company Digital Research (initially named Intergalactic Digital Research).[44]

In the following year, 1977, he designed a version with an important difference. IMSAI, the company that had built a "clone" of the Altair, wanted a license to use CP/M for its products. Kildall rewrote CP/M so that only a small portion of it needed to be customized for the specifics of the IMSAI computer. The rest would be common code that did not have to be rewritten each time a new computer or disk drive came along. He called the specialized code the basic input/output system (BIOS). IMSAI's computer system was soon outselling the original Altair, thanks to its rugged power supply, scope for expansion, and support for CP/M.

CP/M was the final piece of the puzzle that, when assembled, made personal computers a practical reality. A personal computer's DOS had little to do with mainframe operating systems such as Multics. There was no need to schedule and coordinate the jobs of many users: an Altair had one user. There was no need to drive a roomful of chain printers, card punches, and tape drives: a personal computer had only a couple of ports to worry about. What *was* needed was rapid and accurate storage and retrieval of files from a floppy disk. A typical file would in fact be stored as a set of fragments, inserted wherever free space was available on the disk. It was the job of the operating system to find those free spaces, store data fragments there, track them, and reassemble them when needed. Doing that gave the user an illusion that the disk was just like a traditional file cabinet filled with paper files.

Mass Market Personal Computers

It took a while for established computer firms to realize the potential of microcomputers. That applied even to DEC, which throughout the 1970s continued to market small computers at successively lower prices and in smaller packages, such as the PDP-8/A, introduced in 1975 for under $3,000.[45] What was DEC working on when Intel announced its 8080, a device with the essentials of a minicomputer on one chip? "We were just in the throes of building the VAX," said C. Gordon Bell, who led its initial design effort.[46] In comparison to VAX, the personal computers of the late-1970s seemed like toys. To build a computer around a standard processor meant surrendering decisions about architecture to a semiconductor firm—how could they allow themselves to do that?

This left the market for personal computers open to firms with no previous experience in the computer industry. In 1977 three companies, Commodore, Apple, and Tandy began to produce relatively affordable and polished personal computers intended to expand the market beyond electronics hobbyists to computer-curious consumers. Each included video circuitry to drive a television or monitor, a cassette interface, and keyboard. This integration of standard hardware and a shift to new, cheaper, and higher capacity *dynamic* RAM chips greatly reduced the cost of a usable computer system. Burning BASIC onto ROM chips made it faster and easier to start using the computer after turning it on.

The most successful of these machines was Tandy's TRS-80, due in part to its wide distribution as a store-brand computer in Tandy's Radio Shack stores, America's leading electronics chain. For a starting price of $400 it offered a Zilog Z-80 chip, more advanced than the Intel 8080. The marketing clout of Radio Shack helped make it an instant hit for the company. Machines like the TRS-80 were appliances that anyone with the money and interest could use to play games or learn the rudiments of computing. Years later one could find TRS-80 computers doing the accounting and inventory of small businesses, for example, using simple BASIC programs loaded from cassettes or a floppy disk. The success of the TRS-80 signaled the end of the experimental phase of personal computing and the beginning of its mature phase.[47]

The TRS-80's main initial rival, the Commodore PET 2001, came complete with monitor, keyboard, and cassette player as a single futuristic and angular assemblage. Chuck Peddle, who designed the machine, built it around the MOS Technology 6502 microprocessor he had co-created a few years earlier. Commodore had purchased MOS in 1976. The original PET's chief drawback was its calculator-style keyboard; its main strength was a powerful built-in version of BASIC. Several generations of improved PETs were introduced over the next five years.

The third member of the trio, the Apple II, was more expensive than either the TRS-80 or the PET but outlasted and eventually outsold both (figure 7.5). Apple's birth in a Silicon Valley garage, by two idealistic young men, Steve Jobs and Steve Wozniak, is part of the folklore of Silicon Valley. Ever since he came across diagrams of a DEC PDP-8 as a high school student, Wozniak's hobby had been redesigning minicomputers to use fewer components. In 1975 he saw an opportunity to put that skill

Figure 7.5

Three pre-assembled microcomputers established the consumer computing market in 1977. Each had a plastic case with built-in keyboard, at least 4 KB of RAM, and BASIC burned into a ROM chip for instant access. From left to right: the Commodore Pet 2001; the Apple II (shown with the Disk II floppy drives introduced the next year); and the original TRS-80 Micro Computer System (later called the Model 1) connected to the optional expansion interface (the box under the monitor, introduced in 1978). The TRS-80 was initially the most successful product. Image by Timothy Colegrove.

into practice when attending the first meeting of a group, the Homebrew Computer Club, founded to give electronics enthusiasts a chance to try out the first Altair to arrive in the area. The club began to meet regularly in a hall on Stanford's campus, where aspiring computer designers shared ideas and prototypes. Eager to impress his peers, Wozniak chose the 6502 processor for his project, largely because of its low cost. Building on his previous experience creating a "TV typewriter," he integrated the processor, memory, video display hardware, and keyboard interface onto a single compact circuit board. Jobs convinced his friend Wozniak to start Apple Computers to manufacture the boards for sale, together with a cassette tape holding a compact BASIC of Wozniak's own devising.[48]

Wozniak's follow-up, the Apple II, combined most of the flexibility of the Altair-style computers with the accessibility of the TRS-80 and PET. It was a tour de force of circuit design. It used fewer chips than its competitors, and yet it outperformed most of them. His ingeniously hacked color graphics capabilities gave it an edge for games, one of the few things that all agreed personal computers were good for. It was attractively housed in a plastic case, unlike the Apple I which had been sold as a bare board without keyboard, power supply, or case. It also had a non-threatening name.

Because Wozniak's BASIC did not support floating-point numbers, the Apple II was also bundled with a tape of Applesoft BASIC, adapted from code licensed from Microsoft. A payment of $10,500 from Apple in August 1977 is said to have rescued Microsoft from insolvency at a critical moment of its history.[49]

Like the Altair, the Apple II had a bus architecture with slots for expansion—a feature Wozniak argued strenuously for, probably because he had seen its advantages on a Data General Nova.[50] The bus architecture allowed Apple and other companies to expand the Apple's capabilities and keep it viable well into the 1980s. For example, Microsoft's popular SoftCard squeezed a Z80 processor into an Apple II expansion slot to run CP/M.

Many personal computer companies, including MITS and IMSAI, offered expensive floppy disk drives. Wozniak set to work on a disk controller to match this capability. Apple purchased the drives (in a new, 5 1/4-inch size) from Shugart Associates. Shugart had its own controller board, but Wozniak had decided that "of the twenty-two or so chips, about twenty of them weren't needed." His design, combining a handful of chips with precisely timed code running on the Apple II's main processor, was a marvel of efficiency.[51] It was faster and cheaper, and it stored more data than standard approaches (figure 7.6).

Apple's 5 1/4" floppy drive could hold 113 KB of data and sold for $495, which included operating system software and a controller that plugged into one of the Apple

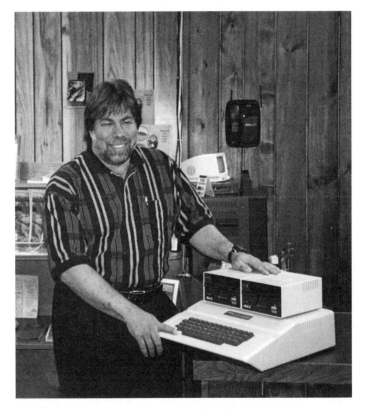

Figure 7.6
The "Woz," Steve Wozniak, holding an Apple II and two floppy disk drives, at the American Computer Museum, 2002. Photo by Paul Ceruzzi.

II's internal slots.[52] An improved controller soon raised that to 140 KB of storage. Apple's sales began to pick up, outstripping the Radio Shack TRS-80 series by 1982. Most purchasers opted for expanded systems with a disk drive and 48 KB of RAM. That made the Apple a favorite platform for the development of complex games and applications. These applications included the VisiCalc spreadsheet, which we will consider in the next chapter along with other business-oriented applications of microprocessor technology.

The Apple II was particularly successful in the education market, as the firm made efforts to supply discounted computers to schools so that middle-class parents would purchase the same computers to use at home. One game developed for the Apple II is still a staple of elementary school classrooms today. *The Oregon Trail* was released in 1978 as an improved reimplementation of a game initially developed for timesharing use by the Minnesota Educational Computing Consortium in the early 1970s. Players guided a wagon packed with supplies and five family members for thousands of miles, through rivers and over mountains. They could trade, hunt for food, and (in the improved 1985 version) admire scenic landmarks. The game gave a powerful impression of the dangers of life in 1847—few wagons arrived without at least one of the family falling victim to rattlesnake bites, broken limbs, drowning, or dysentery.[53]

Bulletin Boards

A personal computer equipped with a modem provided a flexible alternative to dedicated video terminals such as those used with Minitel. The computers ran "terminal emulation" software, usually free or inexpensive packages such as Xmodem, developed by Ward Christensen, to duplicate the functions provided by traditional terminals. These let personal computer users access online services of all kinds without having to buy a separate terminal.

This created a new community of personal computer users looking for online experiences. Hobbyists discovered that they could use their machines to host amateur online systems as well as to connect to them. Bulletin boards systems (BBS) were named after the cork boards found on a college campus, where students posted notes on all sorts of topics. The first, the Computerized Bulletin Board System, went online in Chicago in 1978 "in an effort to generate material for our computer club's newsletter."[54] It used an S-100 system running CP/M. BBSs were typically grassroots activities, often set up by small businesses or electronics hobbyists. Bulletin board systems were usually hosted on personal computer systems, rather than the minicomputers used by commercial online services. The content of these BBSs varied according to the whims of the *sysop* (system operator). A shopping mall might have a list of shops, phone numbers, and show times for movies. A small business would have information about its services, hours, location, and so on. Adult-themed sites and areas for the exchange of pornography were also common. Computer or electronic repair shops set up BBSs with multiple phone lines, and these evolved into network hubs for a community. One could send and receive simple text messages.

The hardware requirements were minimal, typically an extra phone line and a personal computer equipped with a hard disk drive. Most relied on a new product from the Georgia electronics company Hayes Microcomputer Products: a so-called smart modem that one could leave connected to the line so that it automatically connected and disconnected the computer without the need for a person to intervene. Bulletin board software was available for free or at a modest cost.

BBSs provided a way for owners of modest personal computers to enter a world that hitherto had been accessible only to military bases or university users. They created a genuine sense of community among the users. Personal computer users in the US could stay connected as long as they wished, as long as the connection was a local call. As a regulated monopoly, the telephone network was shaped by political as well as technical decisions. A local phone call was not billed by the minute but was part of a flat rate charged to the customer. Long-distance calls were billed by the minute, and they were priced higher to subsidize local calls.

BBS sysops developed a way to circumvent the long-distance telephone charges by gathering messages and sending them to various distant destinations late at night, when phone rates were low. Because the messages were batched, the call could be completed quickly. The developer of the most popular store-and-forward software called it FidoNet, suggesting a dog that dutifully fetched one's newspaper from the front lawn. By the mid-1980s a local FidoNet-connected bulletin board offered users without the connections needed to obtain Internet access the potential for global electronic messaging. Because it needed only intermittent connections, FidoNet was especially influential in developing countries with less reliable telecommunications infrastructures.

Bulletin board culture underpinned the 1983 movie *WarGames*, in which a high school nerd almost triggers nuclear war. The plot is kickstarted when he programs his IMSAI computer to dial every number in a designated area code, hoping to discover the private system of a particular video game company. Instead he stumbles into a lonely Department of Defense mainframe eager to play games with him. The movie helped establish a new stereotype of the hacker as a young man with almost mystical powers to circumvent computer security and a pathological need to tamper with or steal information from computer systems. The new hacker role had roots in the original MIT sense of the word but was altogether more threatening.[55]

One was more likely to see a hacker in a movie than fall victim to one in real life. Media historian Kevin Driscoll has estimated a cumulative bulletin board user population of around 2.5 million in the United States, from their origin in 1977 through

1998. That's a sizable online population but a small fraction of the total number of personal computer users. In 1989, for example, the US Census Bureau found that fewer than 6 percent of households owning personal computers had used them to access bulletin boards.[56]

VIDEO GAMES

Many people had their first sustained interaction with a computer not by typing on a keyboard but by smashing buttons and waggling joysticks in a video game arcade. Arcades had a long history as a place for young people to gather for coin-operated amusements.[57] In the US, politicians were suspicious of them—gambling machines were forbidden outside Las Vegas casinos, and even pinball, which emerged in modern form shortly after the Second World War, was banned in New York, Los Angeles, and Chicago. By the early 1980s, coin-operated video games had replaced pinball machines as the main draw in amusement arcades and were becoming ubiquitous in other public spaces such as bars and fast-food restaurants. Video arcades were an important part of the experience of childhood for many growing up in the US, where more than ten thousand arcades operated, and other developed countries.

Early Arcade Games

The first video arcade games were designed before microprocessors made general-purpose computer technology cheap enough to use. *Computer Space*, introduced in 1971, implemented the basic functionality of the popular computer game *Spacewar*. It was reasonably successful, selling more than a thousand copies, but the game was too complicated for most potential players, who had never seen a video game before. The new industry's first big hit came with *Pong*, a simple two-player video tennis game released in 1972. It was the first product of a new company, Atari, founded by the team behind *Computer Space*: Nolan Bushnell and Ted Dabney. *Pong* was simple to learn and compelling, important characteristics for the bars where many machines were installed. Both machines used digital electronics, rather than microprocessors and code, to implement the game logic.[58] *Pong* had been inspired by a game featured in the Magnavox Odyssey home video game system. It inspired in turn a series of home *Pong* consoles, made easier after a Japanese firm started selling chips with the essentials built in. During the mid-1970s, most home games consoles stuck with the sports theme, tweaking *Pong* to represent doubles tennis, football, squash, and so on.

Another hit Atari game, *Breakout*, moved the bat to the bottom of the screen and added a wall of bricks for the player to demolish. Unlike *Pong* it needed only a single player. Colorful cellophane strips placed over the monochrome video screens enlivened the simple graphics of *Breakout*, and other games of the era. It has also been remembered for the efficiency of its electronic design: early in their partnership Steve Jobs, then working at Atari, submitted an astonishingly efficient hardware design produced by his friend Steve Wozniak.[59] Wozniak called the *Breakout* code he wrote a few years later to try out the color graphics of his Apple II prototype, "the biggest, earthshaking Eureka moment ever," because "software games were going to be incredibly advanced compared to games that were hardwired."[60] The game code was printed in the Apple II manual and, for the first few years, each machine came with game paddles to play it with.

Microprocessor-Based Games

By the late 1970s, the falling price of computer components made it possible to produce much more elaborate video arcade games, defined by computer code burned into ROM chips. The boom began with *Space Invaders*, designed by Tomohiro Nishikado for the Japanese company Tatio. Its components were mostly standard, including an Intel 8080 processor and Texas Instruments sound chip. A wall of alien invaders marches across the screen, inching closer to the player's laser cannon. They drop bombs; the players fire back at them with a laser cannon. The main invasion force starts off slowly. As its members are picked off, the machine's limited processing power moves the survivors faster and the sound chip accelerates its heart-beat rhythm. The invaders always conquered earth in the end, but the highest scoring players got to enter their initials to taunt others. Within a few years of its 1978 release, Tatio had sold hundreds of thousands of *Space Invaders* machines, and its customers had collected several billion dollars in coins from eager players.

Thanks to the success of Space Invaders, alien blasting games dominated arcades for the next few years. Competition drove rapid advances in rival games such as *Galaxians*, *Galaga*, *Astro Blaster*, *Phoenix*, *Scramble*, *Moon Patrol*, and *Space Firebirds*: color graphics, speech, throbbing music, aliens that swirled rather than marched in regiments, planetary surfaces to fly over, and distinct levels rather than endless repetitions of the same task. As players got better, some could tie up machines for long periods with a single coin. Games got harder and more complex in response, culminating in *Defender* from Williams Electronics, a shockingly fast game with a menagerie of lethal aliens and an unforgiving control system involving five action buttons as well as a joystick. Casual players typically died within seconds, putting their next quarter into a less intimidating machine, but crowds would gather to watch experts play.

As the industry boomed, the same basic hardware (joystick, a fire button or two, screen, processor, memory, sound chip) was applied to quirkier scenarios such as *Frogger*, in which a frog has to cross a busy road and dangerous river, or *Q*Bert* where a pink alien with a huge nose hops around a pyramid to change its color. Another Williams hit, *Joust*, featured knights flying on ostriches. Vanquished opponents turned back into eggs.

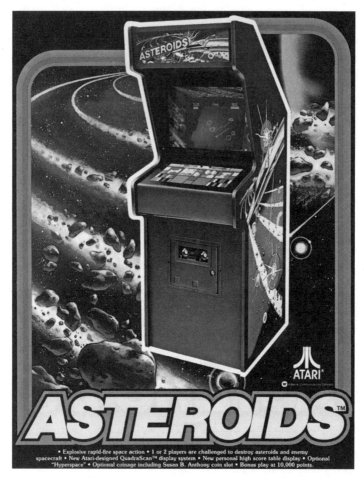

Figure 7.7

This *Asteroids* arcade cabinet sold three lives for a quarter. Early cabinets relied on bold, colorful graphics on the cabinet and around the edge of the screen to supplement their limited graphics. *Asteroids* offered a crisp, but monochrome, vector display inspired by *Spacewar*.

In 1980, *Pac-Man* gave the Japanese firm Namco the industry its biggest ever hit. The game was reportedly designed to attract female players, who were expected to respond better to a game centered on eating than on shooting.[61] The titular mouthy blob runs around a maze, eating dots. Maze games had been a mainstay of videogaming since the 1970s. *Pac-Man*'s twist was the temporary reversal of hunter and hunted: after eating one of four blobs at the corner of the screen its hero could gobble the ghosts he usually fled. *Pac-Man* spawned many sequels and its protagonist was widely licensed for merchandise. He remains an instantly recognizable part of popular culture.

Pac-Man and *Space Invaders* machines used bitmapped graphics technology, just like personal computers such as the Apple II. The arcades of the late 1970s and early

1980s also held machines with a different display technology: vector graphics. As with the displays used by many minicomputers of the era, these drew fine lines by directing a beam of electrons around the tube. The original *Spacewar* game had used such a display. Atari specialized in these games, including classics such as *Lunar Lander* (copied directly from a popular game DEC used to demonstrate its graphics); *Asteroids* (one of the best-selling games of all time), shown in figure 7.7; *Tempest*; and *Star Wars*. The sparse, crisp graphics worked best with spacescapes, although the Atari tank duel game *Battlezone*, released in 1980, was the first use of true-perspective graphics in an arcade video game. The development of vector graphics games ceased around 1985, after bitmap graphics improved.

The Atari VCS Console

In 1977, Atari launched a new home video game system that would establish the model for home videogaming, turning living rooms and basements across America into sites for frantic games of *Space Invaders* and *Missile Command*. Its video computer system, or VCS, was not the first home console but it was the first widely used console designed for plug-in game cartridges. The same model has been used for games consoles ever since: price the machine itself aggressively when first introduced and then recoup the cost by selling each game at a high profit margin. Consoles were sold as peripherals for television sets, turning television from a passive experience into a new, participatory kind of family experience.[62]

The machine launched for $199, cheap by the standards of the era (though equivalent to more than $800 today). To reach that price, the Atari designers had to make some major tradeoffs. The VCS was a real computer with a Mostek 6507 microprocessor (a slightly cheaper version of the 6502 later used in many home computers) and a custom chip for graphics and sound. Atari budgeted for only 128 bytes of RAM, much less than any personal computer. The program code was burned onto relatively spacious 2 KB or 4 KB ROM chips inside the cartridges.

That tiny RAM couldn't come close to holding a full screen of graphics information, even at the VCS's chunky 160 by 192 pixel resolution. The VCS hardware was designed with simple games in mind. It had special facilities to display two relatively detailed *sprites*, one controlled by each player. In the *Combat* game supplied with the system (figure 7.8), sprites were things like tanks or airplanes. The hardware also tracked the position of "ball" and "missile" sprites and drew these on the appropriate points. Everything else on the screen was background graphics drawn at a lower resolution, representing things like clouds or maze walls in the early games. As there wasn't enough memory to store the whole background at once, programmers had to time their code with an awareness of the progress of the television's electron beam across the screen. For example,

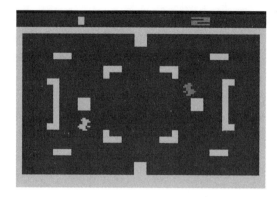

Figure 7.8

The *Combat* game (1977), bundled with many VCS consoles, used its hardware as designed: a symmetrical blocky playfield, two sprites controlled by players, and two missiles to fire at each other. Later games used the same ingredients to draw surprisingly complex screens.

the TV began to draw a line of pixels every 0.00013 seconds. In the interval between finishing one line and starting the next, the program had to load a pattern of bits representing the background graphics of the next line into a 20-bit register used to drive the graphics hardware. The background was 40 pixels wide, but by default the VCS drew a symmetrical pattern—anything else required even more precise timing to update the register as the beam moved across the screen. After the machine had output an entire video frame, the program had a relatively long interval of almost a hundredth of a second to get everything else done before it had to start drawing the screen again.[63]

Most early VCS games used the hardware much as intended—two tanks, racing cars, or bats moved around an otherwise static screen. The VCS really started to take off when its programmers figured out how to replicate popular arcade games with these unpromising ingredients, starting in 1980 with a remarkably faithful rendition of *Space Invaders*, which sold several million copies. The trick was to change the position of the player, ball, and missile sprites even as the screen was being drawn, so that they could be reused within the same frame. Instead of using the laser cannon to shoot at a single invader, the player could battle a whole fleet of aliens.[64] Precisely timed updates to the register used to draw the background yielded the bomb-pocked defensive shields at the bottom of the screen.

As Ian Bogost and Nick Montfort have related, programmers called this process "racing the beam" because they had to manipulate data just as the television's electron beam reached the appropriate part of the screen. Although Atari's *Pac-Man* conversion sold seven million copies, the most of any VCS title, its four ghosts flickered wildly because the sprite used to draw them moved only between video frames, meaning each ghost was present in only a quarter of the frames. Thanks to advances in programming techniques, the VCS version of *Ms. Pac-Man* released the next year was far superior, with smoother animations and a much closer resemblance to the arcade game. VCS programmers continued to get more creative throughout the machine's lifetime, coaxing elaborate landscapes full of moving shapes from the handful of graphical objects at their disposal in games such as *Pitfall II* (1984).[65]

While more technologically advanced, the later VCS titles were enjoyed by much smaller audiences. The public's enthusiasm for new video games had begun to dwindle, even as Atari and its competitors increased their production. Atari expected to follow its previous successes with another smash hit over the holiday period at the end of 1982: adaptation of the movie *E.T.* Unfortunately, the game was even more hurriedly produced and even less fun than the first *Pac Man* adaptation. Most of the time spent playing it involved falling into holes and failing to get out of them. Consequently, most copies remained unsold. This heralded an industry crash the following year, symbolized in popular history by Atari's stealthy burial of hundreds of thousands of unsold cartridges, including many *E.T.* games, in a land fill in Alamogordo, New Mexico. Atari reported a huge loss, shaking the finances of its corporate parent Warner Communications. The next year Warner sold Atari's games console and home computer divisions at a knock-down price to Jack Trameil, the hard-nosed founder of Commodore, who eliminated most of the company and stabilized what was left.

Atari made several efforts to launch newer consoles to replace the VCS, making its last attempt in 1993 with a console called the Jaguar, but none succeeded in replicating its success. The VCS, eventually renamed the 2600, remained in production for a remarkable 15 years, during which around 30 million systems were sold worldwide. More recently it has been repeatedly revived for nostalgic games, through licensed products such as a replica of the classic, and hugely uncomfortable, Atari joystick holding a one-chip recreation of the VCS and a library of emulated games. A small but enthusiastic community of programmers continues to push the limits of the VCS hardware, coding new games to run on emulators or load into the original hardware.

COMPUTERS COME HOME

The first wave of successful personal computers, the S-100 bus machines patterned after the Altair, sold to electronic tinkerers for whom the challenge of putting together a working system was part of the appeal. The next wave, such as the Apple II and TRS-80, cost several thousand dollars for a complete system including memory expansions, a floppy disk drive or two, a printer, and a monitor. A well-equipped system could cost as much as a new car, meaning that even these "mass market" computers sold mostly to upper-middle-class families with an unusual enthusiasm for new technology.

The true mass market emerged a few years later, thanks to a short-lived computer genre: the home computer. Dozens of home computers were marketed, most entirely incompatible with rival machines and even with other models sold by the same company. To purchase a machine was to commit to an ecosystem of unique hardware, software, user groups and magazines. Dedicated users would defend the honor of their chosen machine against rivals, building a part of their identity around it. Most historical discussion has followed the same boundaries, with books devoted to a single platform written by people who retain an emotional bond with it. (The idea of a standard model was tried and failed: a consortium of Japanese companies worked with Microsoft to replicate the success of their VHS videotape standard by introducing a home computer standard called MSX in 1983. At least twenty consumer electronics companies produced compatible machines. Most vanished quickly, although implementations by local producers sold well in Japan, Brazil and Korea.)

The dividing line between a "personal computer" and a "home computer" was initially determined by marketing and customer response. The Atari 800 personal computer, introduced in 1979 as a competitor to the Apple II, was better built, faster, and had more standard features, although it offered less scope for expansion. However, because of Apple's head start with small business and education users, and the Atari's superior chips for sound and animated graphics, the Atari 800 was treated as a gaming machine and sold almost exclusively to home users.

We aim to give a sense of some of the distinctive quirks and charms of the major home computers, but from a historical perspective what they had in common was more important than what divided them. Unlike the Apple machines, which sold to businesses, schools, and wealthy homes, these machines were designed to cost little more than an Atari VCS games console. Home computers came equipped with BASIC burned into ROM and were marketed as an affordable entry point into the computer age. They could be used with peripherals already in the home: a television rather than a monitor, and a tape cassette machine to load and save programs. Most came as sealed units, so that add-on hardware such as disk drives and memory expansion units was plugged in externally.

In practice, however, home computers were used primarily to play games. Commercial games were usually written in machine language, which was faster and more compact than BASIC and allowed access to hardware features hidden from BASIC programmers. Early games were developed primarily by so-called bedroom programmers, hacking away on their home computers and sending the results to publishers, although production was professionalized by the mid-1980s as software companies grew larger and began to use development systems running on more powerful computers to write and assemble game code.

Making the Most of the Micro

Under the banner of promoting "computer literacy," many governments funded efforts to interest their populations in computer technology and spread the skills needed to use and program microcomputers. A generation of children were widely exposed to programming in schools or at home, unlike those born before or after them. BASIC was by far the most widely used language, although Logo, a Lisp-influenced language used to teach programming by drawing objects on screen, was also popular. The computer literacy movement was truly international—even the Soviet Union, where personal computers were very rare, had its own version focused on cybernetic thinking.[66]

The movement was particularly strong in the UK, where the Thatcher government looked forward to the rise of high technology and the eclipse of more traditional, and heavily unionized, industries. Books like *The Micro Millennium* and shows like the BBC's "Making the Most of the Micro" tried to imagine a world remade by these new machines, predicting that homes, schools, and small businesses would use them to carry out their main daily activities.[67]

Homes and computers both had long, almost entirely separate, histories. For the new idea of a "home computer" to make sense, the computer obviously had to change, becoming cheaper, smaller, and less intimidating. Less obviously, the home itself had to be reimagined as a place that needed a computer. Computer enthusiasts and advertisers struggled to do this plausibly. One early idea was to apply the digital control functions of computers, proven in industrial and laboratory settings, to the home. With extra hardware, computers could control heating systems, turn on and off lights, and open garage doors. Such projects appealed only to electronics hobbyists.

Computer producers tried to imagine more universally appealing applications. Homes were seen as the hub of family interaction, so advertisements and advice books often stressed the benefits computers could bring to all family members—helping children with school work, fathers with personal finance, and mothers with programs like "recipe," "pantry inventory," and "shopping."[68] The Texas Instruments advertisement shown in figure 7.9 draws on the sitcom stereotypes of a perfect heterosexual white family living in a spacious suburban home, selling the idea that a home computer would bring family members together. A now famous 1977 advertisement introducing the Apple II showed a man charting stock prices with the machine next to a copy of the *Wall Street Journal* while a woman preparing food on the kitchen countertop paused to stare at him and smile reassuringly.[69] Machines being sold as a way to get ready for a new social order driven by information technology were, ironically, being marketed in a way that doubled down on a conservative vision already being eroded by the social trends of the 1970s such as widespread divorce and more women working outside the home. There was even a coloring book, *The Magic Machine*, to introduce young children to the machines (figure 7.10). It too presumed the continuation of traditional gender roles.[70]

Basic Programming — The TI Home Computer was designed to be used for extensive personal programming. The built-in BASIC language makes the TI-99/4 a valuable desktop tool — at home or at the office. It also gives you a great way to teach your children about computers. And in a society constantly making greater use of computers, familiarity with them will be a tremendous asset.

Personal Finance/Home Business — The TI Home Computer can help pay for itself over and over by helping you with tax and financial planning. It can help you weigh investment alternatives. The TI-99/4 helps take the guesswork out of problem-solving, whether they're personal or business problems. And it can assist with everything from record-keeping and budgeting, to making decisions about major purchases.

Education — Computer-assisted education is now available right in the home. For instance, your TI Home Computer can help students sharpen their grammar, math, spelling and reading skills. The computer guides the student through activities at his or her own pace. Colorful, entertaining pictures — plus sound effects and rewards — make learning fun and encourage achievement.

Entertainment — One of the many capabilities of a powerful computer system like this is its ability to add a new dimension to entertainment — from chess to football. The TI Home Computer is a challenging, "thinking," ever-ready opponent. And, unlike ordinary home TV games, it can also be a patient instructor in a variety of exciting games. Use Solid State Software or program your own.

Figure 7.9

Advertisements like this one, for the Texas Instruments home computer, attempted to showcase the new machine as a new gathering place for the nuclear family. The applications areas highlighted were programming, personal finance, education, and entertainment. Courtesy Texas Instruments, scan by Bryan Roppolo.

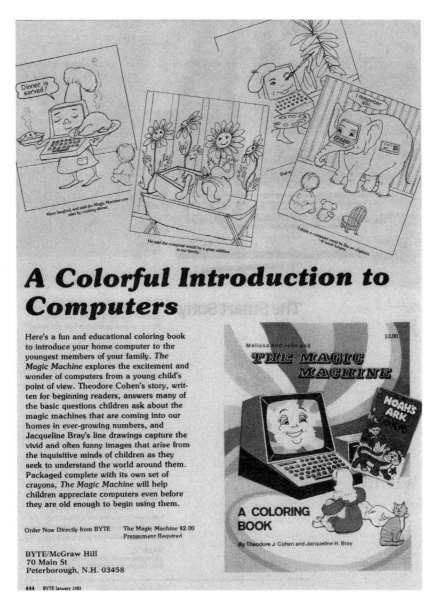

Figure 7.10

The Magic Machine was promoted as a "fun and educational coloring book to introduce your home computer to the youngest members of the family." The image on the top left depicts the text "Mum laughed and said the Magic Machine can start by cooking dinner." From *Byte*, January 1983, p. 444. Used courtesy of Informa.

Early Home Computers

The first really successful home computers were launched in 1981: the Commodore VIC-20 and Texas Instruments TI-99/4A in the United States and the Sinclair ZX81 in the UK. The Sinclair, a tiny plastic wedge designed around lower British disposable incomes, was the purest expression of cost-driven minimalism.[71] It sold for just £70 assembled (even less as a kit), at the time around $140 (figure 7.11).

The ZX81's low price was made possible by advances in chip technology: it used just four chips. Three were the ROM, RAM, and processor. The fourth, a single application-specific integrated circuit (ASIC) chip produced by Ferranti, replaced seventeen off-the-shelf chips used in Sinclair's previous computer. It was an *uncommitted logic array*, which superimposed a layer of custom connections onto a standard grid of available logic components. ASICs had their biggest initial impact with electronic devices and mass market computers, where production runs were large enough to cover the upfront costs of producing custom chips and margins were low.

The ZX81 had an allegedly touch-sensitive membrane keyboard without any actual keys, a crude monochrome display, and no sound output (figure 7.12). Standard memory was just 1 KB, which was barely enough to hold the contents of the display, so filling the screen with text could cause an out of memory error. This could be fixed by filling the computer's only expansion socket with a 16 KB RAM pack, but because the ZX81 had an angled back and the expansion had a straight edge the computer was likely to reset when touched. This was called "RAM pack wobble." Two years later, having cut the price further to £40, Sinclair claimed to have sold more than a million ZX81s. Timex, which manufactured the machine, sold another half million under its own name in the United States.

Sinclair's follow-up, the Spectrum, launched in 1982, was named for its ability to display color graphics. Mitigating the ZX81's most glaring deficiencies, Sinclair also upped the built-in memory on most models to 48 KB, added a beeper, and put little rubbery nubs with a passing resemblance to keys over the membrane sensors.[72] The Spectrum dominated the British home computer market, selling more than five million copies in a country with about a fifth the population of the US and kickstarting a vibrant local video games industry.[73] Its extreme simplicity appealed to lower income countries, including Spain and Portugal, while the socialist countries of Eastern Europe produced unlicensed clones. A Spectrum clone introduced by Microdigital Electrônica in 1985 became the first affordable Brazilian computer, thanks to strong trade barriers erected encourage domestic production.

Middle-class British families and schools were more likely to choose a rival machine, the BBC Micro. The British government, locked in a series of battles with industrial unions, was particularly active in promoting the idea that the arrival of

Figure 7.11

The ZX81 sold very well in the United Kingdom, where its extraordinarily low price more than offset limitations such as a memory capacity of just one kilobyte. W.H. Smith, a nationwide British chain, was advertising not just the computer but also programs, magazines, and blank tapes to use with it. *Computer & Video Games* 1, November 1981, p. 38, used courtesy of W. H. Smith.

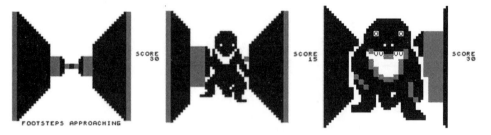

Figure 7.12

The programmers of home computers displayed great creativity. The ZX81 had a text-based display with special characters to represent block shapes and shading. *3D Monster Maze*, programmed by Malcolm Evans in 1981, used these unpromising ingredients to animate a 3D maze, home to a rampaging Tyrannosaurus rex.

microprocessors was accelerating a transition to a postindustrial "information society." The British Broadcasting Corporation, the nation's public service broadcaster, created several educational series to help the population get to grips with the new technology. Rather than adopt any of the existing models for these shows, it sponsored Acorn Computers to develop a machine with all the required features. The BBC Micro had a huge variety of ports and expansion options, with connections for printers, modems, joysticks, additional processors, disk drives, analog instruments, and ROM chips holding applications or additional programming languages. One optional hardware kit connected it to the BBC's teletext broadcasts, which included downloadable software. The BBC Micro cost much more than a Sinclair (although far less than an Apple II), but the machine's quality and flexibility made it the choice of hackers and tinkerers.[74]

American home computers, while less Spartan than Sinclair's offerings, were likewise designed to be sold cheaply. To compete with the Atari VCS for space in front of the television, both the TI-99/4A and the VIC-20 had color, sound, joystick ports, and a cartridge connector. For users who wanted to program, they offered real keyboards and built-in BASIC. Neither was well equipped for applications such as word processing. The VIC displayed just twenty-two characters across its screen (ensuring visibility on a television set) and came with only 5 KB of memory.[75] Nevertheless, Commodore promoted the VIC-20 to parents as something that would be better for their video game–obsessed children than an Atari console. One advertisement featured an intimidating manager in a suit asking a job applicant why his resume mentioned having "shot down two billion aliens from the planet Mondo." "You are good at computer games," asked the man, "but what do you know about computers?" The tagline was "A real computer for the price of a toy."[76]

Texas Instruments originally set a price of $525 for its computer, but aggressive competition from Commodore forced down the price of both machines. By 1982 they were both available for $200, and early the next year retail prices dipped below $100.[77]

Texas Instruments had two major handicaps. First, its solidly built computer had been designed to sell at a much higher price. Commodore, which built its own microprocessors and other chips, had designed its computers for low cost production by using custom chips and saved more money by assembling them in Hong Kong (presaging a general shift to Asian manufacturing later in the 1980s). Second, Texas Instruments had tried to prevent competition for its own, generally mediocre, software titles, whereas Commodore had encouraged other companies to produce games for the VIC. After announcing an annual loss of more than a $100 million as a result of its misadventure in the home computer market, TI surrendered in October 1983.[78]

The Commodore 64

By then Commodore had the American home computer market largely to itself, having vanquished many challengers (figure 7.13). The VIC was still on sale, and like the ZX81 has been claimed as the first computer to sell a million, but its new mainstay was the Commodore 64.

Judged by specifications alone the Commodore 64 might seem a formidable competitor for the Apple II. Commodore reused the VIC's case, but the rapidly falling cost of RAM let Commodore stuff a full 64 KB inside. Custom sound and graphics chips gave it smoothly moving graphics and exceptional musical capabilities. In practice, they sold to different markets. The Commodore 64 was unpolished and unreliable but very cheap—a combination most attractive to home users. To save money and speed development, Commodore stuck with the same minimal BASIC as the VIC-20, giving no easy access to disk storage, sound, or advanced graphics. Prices fell rapidly, from a list price of $595 on introduction in 1982 to a retail price of under $200 in mid-1983. Commodore's single-minded focus on cost led it to rush products to market without careful testing, and then keep redesigning them to reduce manufacturing costs. The *Wall Street Journal* reported that 30 percent of Commodore 64s were being returned immediately as nonfunctioning.[79] Many more failed over time.

Commodore's disk drives, which originally cost $400 each, were even shoddier. Disks tended to stick inside, and drives failed as their read/write heads drifted out of alignment. A bug in the controller chip meant that even when functioning perfectly they transferred data with unrivalled sloth, taking two minutes to load a program.[80] That was about thirty times slower than Apple's disk drives, and barely faster than the tape interfaces of other computers. On the other hand a disk-based Commodore system was barely a third of the price of a comparable Apple II system. European Commodore users mostly stuck with tape, but their American counterparts overwhelmingly purchased disk drives, making disk storage a standard feature of home computing for the first time.

Figure 7.13

Eventually priced at less than $200, the Commodore 64 became the bestselling desktop computer model in history. Commodore stressed the savings its computers offered against better built rivals with similar memory capacities. Notice the family members crowded around a television set.

Sales peaked in 1984, but Commodore was still building the Commodore 64 when it declared bankruptcy a decade later. More than twelve million were produced, making it the bestselling desktop computer model in history. Commodore's first follow-ups, in 1984, confused customers and were rapidly withdrawn. The Commodore 16 and 116 were designed as cheaper, incompatible alternatives but by the time they launched, the low cost of the Commodore 64 left customers in developed countries with little motivation to buy anything worse. The Commodore Plus 4 had built-in business applications targeted to small business users, but that model was not clearly superior to the Commodore 64 and couldn't run the same programs. Commodore eventually offered a series of compatible replacements, most notably the Commodore 128, which had more RAM, an eighty-column display, and thanks to a second processor, a CP/M mode. But few programs took advantage of the new features, and most purchasers stuck with the cheaper original.[81]

Home Computer Games

In practice, the most compelling and widely used applications for home computers were video games. Many of the most popular programs for personal computers were recreations of popular arcade games like *Space Invaders*, *Frogger*, and *Asteroids*. Originally these were unauthorized clones, often with slightly modified names, but arcade companies stepped up legal efforts to control their intellectual property and began to sell licensed conversions.

While some excellent renditions of arcade favorites were produced for home computers, the new platforms liberated game designers to invent many new genres. Arcade games had to be accessible to new players but grow quickly more challenging, delivering an average game length of a few minutes to ensure a steady flow of quarters. Each of the popular platforms inspired several classic games whose designs exploited its particular strengths while minimizing the impact of its weaknesses. *Elite*, launched in 1984, was a hugely influential 3D perspective space simulation and trading game that freed its players to roam around eight galaxies full of planets, pirates, asteroids, and space stations (see figure 7.14). Its smooth high-resolution graphics exploited the BBC Micro's fast processor and flexible graphic modes.[82] The Commodore 64 was good at moving brightly colored sprites over smoothly scrolled but chunky backgrounds, which inspired many classic fighting, shooting, and jumping games set to vibrant soundtracks. The Atari 800 had similar strengths but a faster processor, useful for perspective-based games such as the alien mountainscapes of *Rescue on Fractalus!* The Spectrum's large memory and unusual combination of relatively high-resolution graphics with low-resolution color led to the development of many complex, largely monochrome games with overhead perspectives, such as *Knight Lore* (a werewolf jumps

Figure 7.14

Elite was one of the most complex games of the 8-bit era, beloved for the open universe of trading, exploration, and space combat it squeezed into 32 KB of memory. The original BBC Micro version, shown here, made use of a unique split-resolution feature, combining higher resolution monochrome graphics in the upper part of the screen, used for its smoothly animated 3D wireframe view of space, with colorful but clunky graphics for the radar scope and instruments at the bottom.

around a castle) and the fantasy war game *Lords of Midnight* with its beautifully rendered landscapes.[83] Hit games were inevitably converted for rival platforms, but the results were usually inferior.

Adventure games of the kind popularized on timesharing systems were hugely popular. The most renowned were produced by Infocom, a company with close ties to the MIT hacker community. Infocom authors relied on floppy disk storage to hold relatively large volumes of text to describe their game worlds. Its first product and biggest hit, *Zork*, began on MIT's Incompatible Timesharing System. Reimplemented for microcomputers, beginning with a TRS-80 version in 1980, it eventually sold around four hundred thousand copies. Infocom packaged its games as *Z-code* files, containing instructions for a virtual machine optimized for adventure games. To convert its entire game library for a new machine, all it had to do was to write an interpreter to simulate this machine. As result, its games ran on almost every computer system with a disk drive, from the PDP-11 to the Apple Macintosh.[84]

The Apple II was not a great choice for action games—the graphics that impressed in 1978 made quirky tradeoffs to produce color efficiently, and its speaker beeped and chirped. However, its fast disk drive made it the platform of choice for text adventures in the early 1980s. Apple II games including *Wizardry* (itself inspired by several popular Plato games), the Ultima series, and *Bard's Tale* established another hugely popular genre, the computerized role-playing game. These games, modeled directly on Dungeons & Dragons, saw players assign statistics to parties of adventurers and then guide them on quests through tunnels and across landscapes, battling monsters and collecting weapons and treasure.[85] As games grew larger, they sprawled over several disks, telling users which to insert and remove. *Ultima IV-Quest of the Avatar* of 1985 required Apple users to shuffle between four disks while questing to achieve an almost divine virtue. The Commodore 64's ample memory and slow but affordable disk storage made it a popular choice for role-playing and text adventure fans with smaller budgets.

The proliferation of home computers changed the way people were first exposed to computer technology. In the 1940s and 1950s, most of the people hired as programmers had their first experience of computing on the job. In the late 1960s, as computer science developed as a field within universities, students might program for the first time in a science or engineering class and then decide to major in computing. In the 1970s, more students encountered programming in high school, usually via a timesharing system. In each case, access to computers was limited and took place outside the home. Until the mid-1980s, the proportion of computer science students who were women was rising, following the pattern in other technical and professional subjects. In the 1980s, however, the trend reversed in computing. Some scholars have linked this to the routine exposure of children to home computing, sending an unspoken message to girls that computers were either a mechanism for obsessive hacker-style programming or a device for playing video games. Both activities were practiced disproportionally by boys, giving them the experience and motivation to pursue computer courses in college.[86] Spaces and technologies within the home tend to be strongly gendered, and it has been suggested that parents were more likely to encourage boys than girls to master a newly purchased home computer.[87]

Home Computer Online Services

The videotext systems we discussed in the last chapter, such as Prestel, were originally developed for use with cheap video terminals but soon became popular with home computer users. The most popular part of Prestel, for example, was Micronet 800, a supplemental service for home computer users, which had attracted 18,000 users by 1985. Conceived as an online magazine, it offered online chat, software downloads, multiplayer games, personal computing news, and an area where users could post their own content.[88]

After most videotext systems failed during the mid-1980s, the next generation of online systems shifted away from terminals entirely toward projects aimed at the users of personal computers. Home computer users could do things like stock trading, tax preparation, or household budgeting with their computers. They could send and receive email, book plane tickets, or download software. By the end of the 1980s, the CompuServe Information Service, increasingly popular with hobbyists and small businesses, had around half a million users. Software companies began to offer online support options, while online shopping and banking were aggressively promoted by some of the largest American corporations. Prodigy, developed from a joint venture launched in 1984 by IBM, retailer Sears, Roebuck, and broadcaster CBS, was designed around online news and shopping. Because it was intended for home computers rather than terminals, it could rely on local processing to display a more complex user interface and graphics loaded from disk.[89] These systems saw steady growth, although none came close to matching the popularity of Minitel.

One nominally commercial online service with the ethos of a bulletin board was The WELL ("Whole Earth 'Lectronic Link.") located in Sausalito, California. It was founded in late 1984 as an electronic version of the Whole Earth Catalog. Stewart Brand, who helped start it and gave it its name, played a key role in fusing computer technology with countercultural values. From its beginnings, it was notable not just for its Bay Area countercultural flavor but also for the quality of the discussion and sense of community among its members. It attracted many professional writers, among them Howard Rheingold who used his experiences there to argue for the existence of virtual communities mediated by online interactions.[90]

When personal computing and online services finally became mainstream in the mid-1990s, their appeal rested on several interconnected developments that we discuss in subsequent chapters. Computers became more powerful and affordable, as business-oriented models that evolved from the IBM PC eventually set a new standard for home computing as well as office work. Graphical user interfaces made complex computing tasks easier to control. The Internet and World Wide Web replaced self-contained online services like Prodigy, creating a single online platform for commerce and communication.

Transition to Games Consoles

Sales of home computers peaked in 1983, as prices fell; sales slumped the following year once the market was saturated. Casual users did not find the machines very useful and found little reason to upgrade. Video game enthusiasts were shifting back to game consoles, abandoning both video arcades and home computers. The Commodore 64 had been an attractive alternative to the Atari VCS console, offering better games at lower prices. In practice, games were often free, because tapes and disks (unlike ROM cartridges) could be copied by home users. In 1985, the market for console gaming was revived with the US launch of the Nintendo Entertainment System (following its 1983 introduction in Japan as the Family Computer). The console sold for less than $100. Its games were consistently high quality (thanks to a vetting scheme enforced by Nintendo), introducing many game series and characters that are still among the best-selling today, including the *Mario Kart, The Legend of Zelda, Final Fantasy,* and *Super Mario Brothers* series.[91] Nintendo and its edgier competitor Sega shifted the dominant mode of home videogaming in the US back to consoles. The Super Nintendo Entertainment System (launched in 1991 in the US) reinforced Nintendo's dominance.[92]

As games consoles continued to improve, arcade machines eventually lost their edge in graphics and sound, though a new wave of popular martial arts–oriented games briefly revived the fortunes of video arcades. By the late 1990s, new arcade releases focused on configurations hard to replicate at home: racing games with pedals and steering wheels, hunting or combat games played with big guns, and rhythm games where the player jumps on a tiny dance floor. It was not enough. By 2001 the companies behind the so-called golden age of arcade games, including Sega, Atari, Williams, and Midway, had all stopped manufacturing arcade machines.

The home computer market collapsed further and faster, as the whole idea of the home computer as a separate class of machine dwindled in the mid-1980s. When domestic sales of personal computers really began to take off again in the early 1990s, people were buying cheap versions of computers designed for business use.

8 THE COMPUTER BECOMES OFFICE EQUIPMENT

The IBM Personal Computer, announced in August 1981, was the most consequential computer launch ever: the direct ancestor of most of today's desktop computers, laptops, and servers. IBM grossly underestimated sales: it estimated a total of 250,000 units; "as it turned out, there were some *months* when we built and sold nearly that many systems."[1] The new machine was so successful that in later years, the term *PC* was understood to refer specifically to an IBM or compatible model. (We follow that usage in this book—*personal computer* describes the genre, including non-IBM compatible models by Apple, Atari, and Commodore. The abbreviation *PC* does not.) It was said that if IBM's PC division were a separate company, it would have been ranked number three in the industry in 1984, after the rest of IBM and Digital Equipment Corporation. It gave birth to an entire industry and underpinned the rise of Microsoft and Intel to positions of strategic dominance.

Paradoxically, it was also a conservative design that broke little new ground. Nobody foresaw how successful it would be or that others would copy its architecture to make it the standard for the next decade and beyond. To explain that apparent paradox, we need to step back a few years in time to the mid-1970s, to look at the initial entry of computers into offices. There was nothing new in the administrative use of computer technology, but during the 1950s and 1960s computerizing meant shifting work out of offices and into data processing centers. Most office workers never even saw the computer. Offices sent paper forms to the data processing department. Every week, month, or quarter they received back stacks of fanfold paper printout with information on sales, accounts, and everything else the computer was tracking.

As real-time and interactive timesharing applications became more common, computer terminals gradually began to appear in offices. Sometimes office workers, rather than key punch operators, would enter information into the computer. But mainframe and timesharing systems remained too expensive to replace office tools like desk calculators or typewriters. This started to change in the mid-1970s. The same microprocessors, RAM chips, video interfaces, small printers, and floppy disk drives that made enthusiast and home computing possible also produced computers cheap enough to sit on the desks of office workers. Two application areas were particularly important in laying the groundwork for IBM to introduce a general-purpose personal computer of its own. The first was word processing, which assembled those same technologies to produce buttoned-down office machines rather than hobbyist personal computers. The second was the invention of the spreadsheet, the first compelling business application for regular personal computers. Both helped to change perceptions of personal computers, shifting their main market from enthusiasts and home users to office workers.

PERSONAL COMPUTERS FOR BUSINESS

The arrival of ready-made, consumer-friendly computers like the Apple II broadened the market for personal computing but did not replace the CP/M operating system and the modular, S-100 bus computers patterned after the original Altair. On the contrary, CP/M remained the standard for business-oriented personal computing throughout the 1970s and into the 1980s.

CP/M computers could be fitted with good-quality video terminals and keyboards, able to display a full eighty characters of text. This made them increasingly attractive for word processing applications. Some users of primarily hobbyist computers such as the IMSAI range ran business applications. Other makers of CP/M systems, such as Cromemco, specialized in higher-end computers. They packed floppy drives, and sometimes hard disks, into the main computer box. By the early 1980s, Cromemco computers were in use by the US Air Force and the Chicago Mercantile Exchange.

Word Processing

Word processing is a concept with a complicated history. Before a word processor was a software package, like Microsoft Word, it was a special kind of computer. But before even that, a word processor was an office worker and word processing was an idea about typing pools that ran like factory assembly lines. That idea gained traction after the American Management Association and an obscure publication called *Administrative Management* began to promote it. Companies had invested large sums in specialized equipment to make their manufacturing workers more productive. Office work, in contrast, remained inefficient. According to the American Management Association, word processing could solve this. Personal secretaries would be eliminated and their work transferred to word processors in a central typing pool. This higher volume of work would justify investment in expensive technology, further boosting their productivity.[2]

The word processing idea was tied to machinery, but not originally to computers. In 1971 IBM began to call its dictating machines and automatic typewriters "word processing machines" in its advertisements. At that point its most advanced office device was the MTST—a cousin of its Selectric typewriter that could store and recall sequences of keystrokes on a magnetic tape cartridge. The phrase originated with IBM typewriter salesman and former German fighter pilot Ulrich Steinhilper. Seinhilper claims to have come up with it in the 1950s, when he realized that IBM's new data processing division was stealing the spotlight from his office machines.[3]

The term was not widely used until the early-1970s, taking off just as Cuisinart's *food processors* began to appear in American kitchens. By this point, the falling cost of interactive computing was making it more cost effective to use computers to store, edit, and print various kinds of text. From the MIT hackers, with their "expensive typewriter" program, onward, programmers had loved to edit their programs online. Those technologies also made it possible to edit letters and reports interactively, but computers weren't economically viable for 1960s office work. This is another illustration of the distinction between theoretical and practical universality. Not everyone foresaw that this would ever change: some members of the committee working on the ASCII standard argued that codes for lower-case letters were a waste of space.[4]

Legal documents were the first big market for computer text editing, as they were complicated, went through many drafts, and had lots of money attached to them. Recall that Bell Labs funded the development of Unix to support the work of its patent group. The market for text editing systems designed for office work grew separately but parallel to the enthusiast market for personal computers. The template was set in 1973 by Vydec, a start-up led by former Hewlett-Packard engineers, which offered the first system able to display a full page of text on screen, store it on floppy disk, and print it. Its small and relatively affordable daisywheel printer, a recent invention, was named after a disk that rotated to punch the correct letter. This produced typewriter-quality output, albeit slowly and noisily.

At $18,000 the Vydec, which used minicomputer-style circuits to build its central processor, was expensive. Thanks to the arrival of microprocessors and the rapidly falling cost of RAM chips, many other firms had entered the market for video screen word processors by 1977, including NBI ("Nothing But Initials") in Colorado, Lanier in Atlanta, and CPT in Minneapolis. Lanier was initially the most successful, but the lion's share of the corporate word processing market was eventually taken by Wang Labs.

Dr. An Wang, the founder of Wang Labs, had an astute sense of knowing when to get out of one market and into a new one about to open up. By 1971 the company recognized that its electronic calculators were becoming a commodity, with razor-thin profit margins dependent on packaging more than on technical innovation. Wang

Labs made the crossover to minicomputers with its Model 2200 *computing calculator*.[5] Dr. Wang then directed the company toward word processing. His engineers realized that existing systems confused their users. In 1981, ex-President Jimmy Carter lost a few pages of his memoirs—"I had labored over them for a couple of days"—by pressing the wrong key on his $12,000 Lanier No Problem [sic] word processing system. An anxious phone call to Lanier produced a utilities disk that allowed him to recover the data from the original diskette.[6] Wang came up with a design that would make such disasters unlikely. Commands were accessed by a simple screen of menus.[7]

To keep costs viable for office work, Wang needed to share an expensive hard disk drive and printer among several users. A timesharing minicomputer could handle that, but another requirement was a consistently speedy response. When timesharing systems got busy, their response lagged. The answer was to put processing power into the terminal itself, with the central computer serving primarily for data storage—a radical idea in 1975. The Wang Word Processing System (WPS), shown in figure 8.1, was unveiled at a trade show in New York in June 1976 and, according to some accounts, nearly caused a riot.[8]

A Wang cluster, including hard disk storage, cost $30,000—a price justified by the promise of labor savings and service improvements achieved by centralizing typing work. Its polished packaging and easy-to-use menu system were a world away from the self-assembly personal computing kits of the same era. Wang Labs was ranked 45th in data processing revenues in 1976. It reached eighth place in 1983, just below IBM, DEC, and the remaining mainframe companies. Calling it the "Orient Express of office automation," one analyst predicted that Wang would take the industry's number-three spot by 1990.[9]

WordStar

Word processing programs for personal computers arrived quickly, but initially failed to match the features and polish of the dedicated systems. That changed in June 1979 with the advent of WordStar, written by Rob Barnaby. It included capabilities such as built-in help screens to guide users, on-screen display of page break information, and justification (to produce a neat right margin). Even so, the text displayed on the screen gave little idea of how the printed output would be formatted. To maximize the potential market, WordStar was designed to rely only on standard CP/M capabilities. Instead of cursor keys WordStar used key combinations such as Control-E and Control-D, which were guaranteed to be present on all keyboards. The flipside of CP/M's support for many hardware configurations was endless fiddling to configure WordStar and other application programs.[10]

In his history of literary word processing, *Track Changes*, Matthew Kirschenbaum suggests that science fiction writers were the most active adopters of word processing.

Figure 8.1
A Wang Word Processing System in use. Its screen could display a full eighty-column line of text. Photo: Charles Babbage Institute, University of Minnesota.

They had to be prolific to make a living and had a natural predisposition toward new technology.[11] One of them, Jerry Pournelle, took up a second career writing the "Chaos Manor" column for *Byte* magazine. It grew to be a little magazine inside a big one, with its own letters page, reviews, and gossip. Reading it is one of the most immediate ways to experience the excitement and frustration of personal computing in this period. Trying to make a particular combination of printer, computer, and software package work together could turn into an epic quest taking several pages and many telephone calls and experts to solve.

Pournelle turned his computers into characters in the unfolding narrative, like family pets. His favorite was a CP/M machine he named Zeke, after its Z-80 processor, which was reportedly the first computer used to write a science fiction novel. As late as

May 1983, Pournelle wrote "you can get a good S-100 system "for the price of a comparable IBM," and I for one prefer the more flexible system. . . . You can upgrade for less than with anything else."[12] Pournelle eventually moved on, and Zeke went on display in the Smithsonian, but fellow science fiction writer George R.R. Martin, author of *Game of Thrones*, made headlines in 2014 when he revealed that he uses a vintage version of WordStar to avoid the visual distractions of modern computing. According to Martin, WordStar "does everything I want a word processing program to do and it doesn't do anything else."[13]

CP/M remained popular well into the 1980s for cheaper personal computers, particularly portable systems. The first successful portable was the Osborne 1, released in 1981. It looked a lot like a sewing machine: a bulky box with a handle on one end (figure 8.2). Releasing catches detached a keyboard to reveal two floppy disk drives and a tiny five-inch screen. Its portability was limited. *Byte*'s reviewer wrote "I doubt I would carry an Osborne 1 in the same way I would carry a briefcase . . . [it] weighs about 24 pounds which would tire out all but the most athletic of travelers." He also expressed doubts about marketing claims that it could "fit under an airline seat." The machine wasn't particularly powerful but was an amazing deal because the initial $1,795 purchase price included WordStar and the Supercalc spreadsheet program.

Figure 8.2
Portability is a relative concept. The Osborne 1, an early and affordable portable computer, weighed around 24 pounds and had a tiny five-inch screen. Image created by Wikimedia user Biby, reproduced under Creative Commons Attribution 3.0 Unported license.

As *Byte* concluded, "in a way you are getting a software package with a computer thrown in for (almost) free."[14]

The Osborne Computer Company went bankrupt in 1983. Contemporaries named the *Osborne effect* after the mistake it made by announcing a much superior replacement long before it was ready for sale. The premature announcement was said to have immediately killed sales of its existing computer, although whether that was really the main factor in Osborne's demise has been debated.

The most widely used CP/M product line was released long after the high end of the PC market had moved on. The Amstrad PCW 256, launched in 1985, built a computer, a floppy disk drive, and printer control circuitry into a chunky little green-screen monitor. Amstrad was a British consumer electronics firm. Its combative founder, Alan Sugar, embraced the spirit of a street market trader: "We work on this philosophy—pile 'em high and sell 'em cheap."[15] Amstrad used a television display tube for the monitor and procured what one journalist called "the nastiest keyboard I have ever used."[16]

Many functions of the PCW 256 were combined onto a single custom (ASIC) chip, which greatly reduced manufacturing cost, doing for office computing what Commodore and Sinclair had done for home computing a few years earlier. By this point the semiconductor industry, following Moore's law, had achieved what was termed very large scale integration (VLSI) with tens or hundreds of thousands of transistors on a chip. Designing complex chips was made much easier by methodologies developed from the 1970s by Lynn Conway, a Xerox PARC researcher, and Carver Mead of Caltech. Their textbook, *Introduction to VLSI Systems*, was one of the most influential computer science publications in history.[17] These new production techniques, computerized tools, and design rules were particularly important in the production of special-purpose chips.

Amstrad sold millions after launching the package (including printer) for £399, about a fifth of the usual cost for a system able to handle word processing. It came with a quirky word processing program but could run standard CP/M programs, including WordStar. One Amstrad advertisement showed a trash heap full of abandoned typewriters, and another used the tag line "More than a word processor for less than a typewriter." Both suggested that computer technology was finally cheap enough to dissolve the typewriter rather than supplement it. The long-predicted end of CP/M as a commercial computing platform came only with the withdrawal of the Amstrad PCW line in 1998.

VisiCalc

Hobbyists loved Apple IIs, but they were a poor choice for word processing. Their standard displays showed half a line of text, forty characters, all in block capitals. Users with time, money, and a love of fiddling around could fit expansion boards to overcome these limitations. Even then the only way to capitalize a letter was to push the shift button twice before entering it and twice afterwards to shift back. Printing business-quality documents meant adding a high-quality daisywheel printer and serial card for at least $3,000. As one enthusiast admitted, "an Apple II can be tortured into a decent word processor—at great expense."[18] The total cost expense could exceed that of more polished competitors. Some Apple users added a CP/M expansion card, essentially a whole other computer miniaturized to fit inside the case, just to run WordStar software.[19]

There was exactly one great reason for a business user to get hold of an Apple. Visi-Calc launched in October 1979. Its creators were Daniel Bricklin and Robert Frankston, who had met while working on Project MAC at MIT. Bricklin had worked for Digital Equipment Corporation and in the late 1970s attended the Harvard Business School. There he came across the calculations that generations of business school students had to master: performing arithmetic on *spreadsheets*: rows and columns of numbers, typically documenting a company's performance for a set of months, quarters, or years. He recalled one of his professors posting, changing, and analyzing such tables on the blackboard, using figures that his assistant had calculated by hand the night before.[20]

Bricklin conceived of a program to automate these spreadsheets. Frankston agreed to help write it. In January 1979, Bricklin and Frankston formed Software Arts, based in Frankston's attic in Arlington, Massachusetts. That spring the program took shape, as Frankston and Bricklin rented time on the MIT Multics system. They approached Dan Fylstra, a second-year student, who ran a publishing business called Personal Software. Programs written by personal computer enthusiasts provided Fylstra and his competitors with their first products. This was a new industry, quite separate from mainframe software companies. Some software publishers worked like book publishers, paying royalties to the authors of programs, and others purchased the rights for a flat fee. They began as tiny operations, duplicating disks and packing them into ziplock bags, to be sold with ads in specialist magazines or through the network of dealers that sprang up to handle the new machines.

Personal Software already published a large catalog of programs, the most successful of which was a chess program, but VisiCalc almost immediately eclipsed its other titles. Priced at $200, sales passed the 100,000 mark by mid-1981. Steven Levy recorded its initial impact in the 1984 article "A Spreadsheet Way of Knowledge."

> Don Jackson is a certified public accountant in Cincinnati. He has between 40 and 50 clients, mostly small businesses. Before he bought an Apple three years ago, he painstakingly did his calculations on light green crosshatched ledger sheets. A client would come in to work out a billing procedure, and after Jackson had put the relevant numbers on a sheet—in light pencil, so erasures could be easily made—various questions would come up. For example, if the billing procedure was based on a 15 percent interest rate, what

would happen if the rate went up to 18 percent? To find out, the whole sheet would have to be redone. Each figure would have to be punched into a hand calculator and then checked by one of Jackson's employees. "I would work for twenty hours," Jackson said. "With a spreadsheet, it takes me 15 minutes."[21]

VisiCalc played wonderfully to the Apple's strengths and minimized its weaknesses. Fylstra noted that "the Apple II was essential to VisiCalc."[22] Spreadsheets were small, so its limited memory capacity and disk storage was not a handicap, as it would be for database work. They used text mostly for labels, so the all-caps display was not a problem (figure 8.3). The screen served as a scrollable window onto a larger spreadsheet, so the forty-column display worked much better for spreadsheets than word processing. Because the Apple drove the display directly rather than sending text to a terminal like a CP/M machine or timesharing system, the spreadsheet experience was smoother on it than it would have been on a more expensive platform. This fluidity encouraged users to play around with models and data to answer *what if* questions. Levy noted that the users he observed were becoming obsessed with trying to create "the ultimate model,

Figure 8.3
VisiCalc (1979), the original spreadsheet program, running on an Apple IIe (1983) with monochrome monitor and dual Disc II drives. VisiCalc was ideally suited to the Apple II's combination of a low-resolution, forty-column display with rapid scrolling. The / key triggered the rather cryptic main command menu, BCDEFGIMPRSTUW, seen here in the editing area at the top of the screen. Photograph Thomas Haigh.

the spreadsheet that behaves just like an actual business." One called his model "my pet, in a way. Scratching its ears and brushing its code . . . it's almost an obsession."

The apparently objective computer output and attractive charts helped spreadsheet users to present their ideas forcefully, but as Levy noted, spreadsheets had an important difference from earlier modeling software: they hid the formulas created by users. Printed output showed the numbers produced by the model but not the assumptions used to generate them, making it easy to tweak the formulas to get the desired results. The ability of "junk bond king" Michael Milken (an inspiration for the character Gordon Gecko in the movie *Wall Street*) to value the component parts of underperforming companies and argue for the safety of a portfolio made of risky bonds was often attributed to his team's mastery of spreadsheets. The *New York Times,* reporting on his indictment for insider trading, credited Milken for creating a new Wall Street where "sharp elbows and a working knowledge of computer spreadsheets suddenly counted more than a nose for dry sherry or membership in Skull and Bones."[23]

Fylstra also noted the importance of working with the increasingly professional network of stores, such as the fast growing Computerland chain, that sold both hardware and software. He claimed that Personal Software already had around 500 dealers selling its products before VisiCalc launched.[24] VisiCalc inspired the idea of the *killer app* (short for *killer application*), defined as a software package so compelling that large numbers of users would purchase an entire computer system just to run it.[25] VisiCalc sold Apples the same way that *Space Invaders* and *Pac-Man* sold Atari VCS consoles. People entered computer stores asking for VisiCalc and left with the Apple II needed to run it. The computers were cheap enough that managers and analysts could purchase them from departmental budgets, rather than going through a formal approval process or working with the central data processing department.

VisiCalc symbolized a shift toward packaged application software as the driving force behind personal computing. Unlike mainframe users, companies buying a personal computer were not usually going to hire a team of programmers to write custom software for it. Neither could most users realistically satisfy their needs by writing their own programs in BASIC. Hardware was getting cheaper all the time, but programmers only got more expensive. The future lay in packages like this that could sell hundreds of thousands, and eventually millions, of copies and so spread their development cost over a huge user base.

THE IBM PERSONAL COMPUTER FAMILY

The PC began when an IBM team in Boca Raton, Florida embarked on a project code-named Chess. They saw how fast the personal computer field was developing and recognized that further delay would be fatal. To speed development, the Chess team went outside IBM for nearly every part of its computer, including the software. The floppy disk drives, keyboard, and screen were all variants of components used before.[26]

The First IBM PC

In 1981, when IBM launched its Personal Computer, its main competition at the upper end of the personal computer market came from CP/M computers and from Apple's increasingly popular II Plus. The latter packaged the core components of a personal computer, including graphics hardware, onto a single board that filled the bottom of a compact case, leaving room for a power supply and expansion boards. The keyboard was part of the case, but disk drives themselves were separate boxes. In contrast, CP/M computers such as the IMSAI mostly followed the Altair model. They were sturdy metal boxes at the bottom of which lay a motherboard that was little more than a bus connector. The computer logic was spread over separate boards for the processor, memory chips, terminal interface, printer interface, and so on.

IBM's PC split the difference: like the CP/M machines it was a chunky metal box with an external keyboard (figure 8.4), but like an Apple II its motherboard included

Figure 8.4
Launched in 1981, the IBM PC set the standard for an entire industry. Disk drives and expansion cards fitted inside its beige box. Photo courtesy Division of Medicine and Science, National Museum of American History, Smithsonian Institution.

core computer components such as a processor, a tape interface, and memory chips. Like the Apple II, the IBM included BASIC burned onto a ROM chip for the convenience of budget-minded users with no floppy drive or little RAM. Like the CP/M machines, it relied on an additional card to provide its video interface. Like both CP/M and Apple machines it was highly expandable and needed extra cards for common functions such as the serial and parallel ports needed to hook up modems or printers.

Reviewers saw the IBM PC as an assemblage of the best traits of existing machines rather than a revolutionary advance. *Byte*'s review began "What microcomputer has color graphics like the Apple II, an 80-column display like the TRS-80 Model II, a redefinable character set like the Atari 800, a 16-bit microprocessor like the Texas Instruments TI 99/4, an expanded memory space like the Apple III, a full-function uppercase and lowercase keyboard like the TRS-80 Model III, and BASIC color graphics like the TRS-80 Color Computer? Answer: the IBM Personal Computer, which is a synthesis of the best the microcomputer industry has offered to date."[27]

The IBM PC came with five expansion slots. Aside from the keyboard and tape connectors, every interface was an option, so its slots quickly filled up with cards. Even connecting a display, without which the computer was quite useless, meant fitting an expansion card. The initial choices were a crisp text-only output for business users who chose IBM's green-screen monitor or mediocre color graphics for display on a television or color monitor.

The cheapest configuration, a 16 KB computer with a color graphics adapter for the TV and no disk drive, listed for $1,565. *Byte* predicted that "most people" would opt for this minimal configuration, reflecting the PC's continuity with hobbyist machines. That wasn't true, though—most actual customers wanted disk storage and enough memory to use it. Installing another 32 KB of RAM on the main board, adding two internal disk drives and screwing in a floppy disk controller card, two 64 KB memory expansion cards (for a total of 176 KB), and a printer interface card more than doubled the cost to $3,405—not including a screen or printer. It would also fill all five expansion slots, leaving no room for the optional serial interface card, needed to connect a modem, or for the joystick interface card. As *Byte* observed, "you cannot put everything into the IBM microcomputer that you might want to" as "the expansion slots . . . will fill rather quickly."[28]

PC-DOS

Representatives of IBM had approached Bill Gates in the summer of 1980 to supply a version of BASIC that would run on the Intel 8088.[29] IBM expected to use a version of CP/M, which was the standard operating system for personal computers, just as Microsoft BASIC was the standard programming language. Digital Research was already promising a 16-bit extension, but when IBM visited Digital Research, Kildall was not there, and his wife, who handled the company's administrative work, refused to sign IBM's nondisclosure agreement.

Microsoft offered IBM an alternative 16-bit operating system. PC-DOS was based on 86-DOS, an operating system written by Tim Paterson of Seattle Computer Products. Paterson's work took about two months, and the code occupied about 6 KB. Seattle Computer Products referred to it internally as QDOS, for "quick and dirty operating system."[30] Its immediate inspiration was CP/M. Paterson did not have access to CP/M source code, but his DOS used the same function calls as CP/M and retained some command names. Like CP/M, PC-DOS reflected a minimalist aesthetic. Some commands were built in, and others were loaded from disk as needed. Users ran programs by typing their names. The commands could do things like copy files, display their contents, prepare (*format*) new disks for use, and customize the display. PC-DOS commands and error messages were overall less cryptic than those in CP/M or Apple's DOS. It was more forgiving to its users. By contrast, CP/M often needed to reboot the system if the wrong disk was inserted into a drive.

Here's what it was like to use an IBM PC in 1981. Turning it on, by flipping a big red switch on the back of the box, caused a satisfying clunk and a gentle whoosh from the power supply that would persist until it was turned off again. The screen flickered and flashed a cursor while the computer checked its RAM and other components for proper operation. This took ten to fifteen seconds; in later years a fully expanded PC needed more than a minute. Convinced that it was safe to proceed, the PC groaned as it briefly ground its floppy disk drives and then let out a loud beep. A flashing cursor appeared, and the first drive began to whir at 300 revolutions per minute. After the BIOS code recognized the PC-DOS 1.0 *boot disk*, the drive issued gentle clunks for about five seconds as its read head moved from track to track to read the system files. The entire operating system and a host of demonstration programs fitted on a single 160 KB floppy disk. (That included *Donkey*, a famously bad demonstration game cowritten by Bill Gates himself.)

DOS then took over, requiring the user to enter the current date, before printing a copyright message and displaying the A> command prompt. IBM offered an excellent full-sized detached keyboard with a layout that touch typists were used to. Making the computer do anything required typing commands, each key press producing a satisfying click remembered fondly by those who went on to use the cheaper, flimsier keyboards of later PCs. Replica and refurbished IBM keyboards now sell for hundreds of dollars each. The disk drive whirred and clunked frequently as DOS commands loaded—RAM was too precious to hold the code to change screen modes or format disks for an entire session. One of the benefits of having two drives was being able to leave the DOS disk in one while running applications or loading BASIC code from the other.

Besting Apple in the Business Market

The PC's initial advantage over the Apple was finesse rather than raw computing power. Its Intel 8088 processor, a cut-down version of the 16-bit 8086, was theoretically more powerful than the 8-bit processors used on earlier personal computers, but *Byte* observed that in practice "the necessity of funneling all data through an 8-bit path degrades the 8088's performance to the point where it is more like a fast 8-bit microprocessor with an extended instruction set than it is a 16-bit microprocessor."[31]

IBM's biggest success with the PC came in the business market. Larger companies, whose data processing departments were skeptical of Apple, had strong relationships with IBM. An IBM PC was comparable in cost to a video terminal. IBM even offered a special version of the PC configured with special hardware and software to mimic the 3270 series video terminals it sold for use with its mainframes.[32] But many corporate PC purchases were driven by individual initiatives, not official policies. Data processing staff, wedded to Wang word processors or IBM mainframes, were losing control. In December 1982, *Time* magazine named the computer its "Machine of the Year."[33]

The IBM PC faced little competition from Apple. The Apple II was four years old and optimized for computer hobbyists. For example, IBM's PC displayed a full eighty columns of upper- and lower-case text. Rather than add comparable features to the Apple II, Apple had incorporated them into its more expensive, and incompatible, Apple III, launched in 1980. Apple's rush to get the machine to market had combined with its desire to squeeze new capabilities onto a small motherboard to make its first batch of machines spectacularly unreliable. Apple finished up recalling and replacing the entire production run.[34]

This fiasco killed the Apple III's prospects, but Apple didn't release a major upgrade to the Apple II until the IIe arrived in 1983. Steve Wozniak later wrote, with a touch of overstatement, that "about the only salary Apple spent on the Apple II during that period—1980 to 1983—was on the guy who printed the price lists."[35] Over a ten-year lifespan the IIe sold millions of copies, many times the volume of earlier models. The new design lowered manufacturing costs enough to keep Apple afloat through the mid-1980s in its core home and education markets, whereas its useful but modest enhancements did little to challenge IBM for business use.

The 8087 and IEEE Floating Point

Doing engineering calculations or financial modeling cost a lot less with a personal computer, such as the Apple II, than with a mainframe or timesharing system. But only small jobs would fit into its limited memory and run at an acceptable speed. Complex models still required big computers. That began to change with the IBM PC. Even the original IBM PC could be expanded to much larger memory capacities than the Apple.

The other big difference was floating-point support. Since the 1950s, capable floating-point hardware support had been the defining characteristic of large scientifically oriented computers. The 8088 used in the original PC did not support floating point and its performance on technical calculations was mediocre. But every PC included an empty socket waiting for a new kind of chip, the 8087 *floating-point coprocessor*. The 8087 was the first chip to implement a new approach to floating point, proposed by William Kahan and later formalized in the standard IEEE 754. Its adoption by firms including DEC and IBM was a major advance for scientific computing, for which Kahan received a Turing Award. Code, even in a standard language like Fortran, had previously produced inconsistent floating-point results when run on different computers. According to Jerome Coonen, a student of Kahan's who managed software development for the original Macintosh, this standardization on robust mechanisms was a "huge step forward" from the previous "dismal situation . . . Kahan's achievement was having floating point taken for granted for 40 years."[36]

The 8087 was announced in 1980 but trickled into the market because it pushed the limits of Intel's production processes. Writing in *Byte*, Steven S. Fried called it "a full-blown 80-bit processor that performs numeric operations up to 100 times faster . . . at the same speed as a medium-sized minicomputer, while providing more accuracy than most mainframes."[37] The 8088 itself had only 29,000 transistors, but its coprocessor needed 45,000 to implement its own registers and stack.

Assembly language code and language compilers had to be rewritten to use special floating-point instructions, executed in parallel with whatever the main processor was doing. Scientific users quickly embraced the 8087, which made the PC a credible alternative to minicomputers. Fried had promised that "the 8087 can also work wonders with business applications," but software support was limited. Even Lotus-1-2-3, which existed only to crunch numbers, did not utilize it. Fried began a business selling patches to add coprocessor support to such packages.[38] Over time, IEEE-style floating point became a core part of every processor. By the time Intel launched the 80486 in 1989, its factories were just about able to manufacture this one million transistor chip with its built-in coprocessor. Software developers, particularly video game programmers, began to use floating-point instructions. By the late 1990s, PC processors competed largely on the strength of their floating-point capabilities.

The PC XT and the PCjr

The evolution of the PC reflected its success as a higher-end, business-oriented machine. In 1983 IBM launched the PC XT, which dropped the cassette interface and other features aimed at home users. Ambitious PC users had quickly run out of expansion slots. The XT added three more, plus sockets for a full 256 KB of RAM, which freed up more slots. Its biggest new feature was a standard 10 MB hard disk drive. Adding a hard disk drive transformed the experience of using a PC, bringing it closer to the feeling of using a minicomputer or workstation. New application programs were copied onto the hard drive to run. Even the slowest hard drives transferred data far more rapidly than floppy disk drives and, for most users, could hold a complete collection of programs and working data. As prices fell and capacities rose, hard disks soon became standard equipment on all but the cheapest PCs. By 1986, upgrade kits for older models were available for less than $500. Hard disks introduced new complexities into personal computing, requiring users to manage their directory structures, and opened new markets for hardware and software to back up their contents. The popular Norton Utilities package, created by Peter Norton, included programs to restore accidentally deleted files, navigate directory structures, and optimize hard disk performance.

Comparing a PC XT and an Apple IIe, both manufactured in 1983, gives an insight into the cultures that produced them. IBM's Color Graphics Adapter board ran the entire length of the case and contained 69 chips. The IIe followed Wozniak's tradition of efficient and idiosyncratic engineering. Apple's entire machine, including an expansion board that boosted memory and enabled high-resolution graphics comparable to IBM's, contained only 41 chips. Its color generation mechanism was a hack tied to the NCSA system used by American televisions, timing the generation of signal pulses to trigger color display. Apple's minimal floppy disk drive and controller card held a total of 12 chips, versus 51 for the IBM PC's. Tracy Kidder told a famous story about a Data General engineer sneaking a look inside the processor cabinet of a DEC VAX. He "imagined he saw a diagram of DEC's corporate organization" in its highly compartmentalized separation of functions onto twenty-seven circuit boards. The layout "expressed that phenomenally successful company's cautious, bureaucratic style."[39] IBM's PCs were likewise strikingly conservative designs, but their robust blandness and unrivalled expandability were more than enough to triumph over the quirky efficiency of Apple's aging design.

The XT set the template for office computing in the mid-1980s but IBM had not given up on the home market. About seven months after the XT, it announced the PCjr, a cut-down version of the PC in a smaller case with more equipment built onto the motherboard but less internal expandability. It had ROM cartridge ports for video games or programming languages and a wireless keyboard for use on the sofa. Before its launch, the PCjr was expected to dominate the home computing market, but it proved too limited for people who wanted a cheaper IBM PC and far too expensive to compete with the Commodore 64 for video gamers. The *New York Times* review dismissed its keyboard as "rather strange" with "all the grace of a rubber-kneed centipede." "Two pages into a manuscript," noted the review, "I was looking for a pad and pen with which to finish my story."[40]

The PCjr has become one of business history's most famously unsuccessful products, on a par with corporate disasters such as the Ford Edsel. It did, however, debut a new sixteen-color PC video mode intended to make video games more appealing. Software company Sierra On-Line worked with IBM to produce a game to show off the new graphics, delivering the first of the hugely popular *Kings Quest* "graphical adventures" designed by company cofounder Roberta Williams. Sierra's adventures, which also included the humorous science fiction *Space Quest* and risqué *Leisure Suit Larry* titles, used a hybrid of video-game controls and typed commands.[41] Unlike the Infocom games, which they quickly eclipsed, they relied primarily on pictures to describe the environments through which players traveled. The PCjr video mode is remembered as "Tandy graphics" because it was copied by Radio Shack for a popular line of low-cost PC compatible machines that replaced the aging TRS-80 series, succeeding where IBM had failed in bringing PC technology to the home computer market.

Lotus 1-2-3

The IBM PC opened up new opportunities for the rapidly developing microcomputer application software industry. In corporate offices, the little stream of personal computers that began with the Apple II became a flood of IBM PCs running Lotus 1-2-3, word processing software like Word Perfect, and database programs like dBase II.

Thanks to the huge growth in personal computer use, by the end of the 1980s the largest PC software firms had hundreds of millions of dollars in revenue and thousands of employees. Purchase prices were much smaller than for mainframe packages, but the most successful products sold millions rather than thousands. Rather than ziplock bags, the leading PC software packages sold for around $500 each and came in sturdy cardboard slipcase binders holding professionally produced manuals.

With the PC's announcement, IBM also announced the availability of word processing, accounting, games software, and a version of VisiCalc. Mitch Kapor, who had previously developed add-ins for VisiCalc and knew exactly how it could be improved, partnered with an experienced programmer, Jonathan Sachs, to start a rival firm, the Lotus Development Corporation.

Lotus 1-2-3 was similar to VisiCalc, but added graphing and database capabilities (indicated by the name's *2* and *3*) to its core spreadsheet features. It symbolized the shift of the PC software industry to a new, more professional business model. Because it was written specifically for the IBM PC, using assembly language to gain faster performance, it ran much faster than other spreadsheets including the PC version of VisiCalc. Kapor had obtained several million dollars from selling his earlier programs and from venture capitalists to underwrite the initial development and launch of the program. This let Lotus make a splash in January 1983 with a polished and complete program including a professional manual and a tutorial disk.[42]

By the end of its first year on the market, 1-2-3 had achieved $53 million in sales. One of its most compelling features was a *macro* system to record and replay sequences of commands. These sequences could be mixed with simple program code to figure out what actions to perform. According to Kapor, "That was one of the reasons we won in the marketplace, because it just totally unleashed the power of end-users."[43]

Lotus 1-2-3 was so popular that it inspired several clones, which copied the Lotus menu structure and macro command language. This raised a novel legal question: could copyright law be stretched to protect the *look and feel* of a program as well as its actual code? Lotus was initially successful when it sued the makers of a blatant clone, called The Twin, but eventually lost in another case (Lotus v. Borland) that established that command menus were not covered by copyright protection.[44]

WordPerfect

WordStar was the most popular word processing program for the IBM PC for the first few years. As with VisiCalc, it was a straight conversion, in this case from CP/M, which did not take full advantage of the capabilities of the PC. New features accumulated over the years only added to the awkwardness of its user interface. A 1983 review described WordStar 3.0 as "thousands of Rube Goldberg straps and ropes holding things together without any overall unifying concept (or even a dozen)."[45] Rivals copied the interfaces of popular word processing machines, which helped to lure their customers over to cheaper PC hardware. MultiMate copied the operation of Wang systems, making it easy for their users to migrate to cheaper PC systems. IBM's own package, Displaywrite, mimicked its Displaywriter word processing system.[46]

WordStar was eventually superseded by Satellite Software's WordPerfect, which first appeared for the PC in late 1982. Having begun life on a Data General minicomputer at Brigham Young University, it was better suited than WordStar to the powerful PC platform. It was fast (written, like 1-2-3, in assembly language and working directly with PC display hardware) and powerful, and it had a clean command structure based on combinations of the Shift, Control, and function keys. This took a while to learn, aided by a paper mask placed over the keyboard, but gave experienced users instant access to every function. They could switch between a clean text display and an alternative view that marked all formatting codes with special tags. Reaction to the program was captured by the title of one early review "Not Quite Perfect, But Certainly Superb."[47] *PC Magazine* called it "a solid contender for the word processing championship." The reviewer's declaration that "moving around within WordPerfect is a joy," because the Page Up, Page Down, and cursor keys all functioned intuitively, reminds us how low expectations were for usability at the dawn of the PC era.[48]

Later versions packed in more features, including support for many different printers and a built-in macro language to automate complex formatting and editing operations. WordPerfect release 4.2 in 1986 set the standard for the rest of the 1980s and finally overtook WordStar in sales. At its peak around 1990, WordPerfect controlled around half the market for word processing software.[49] Being based in Utah gave the firm a particular advantage in international sales and support, as many of the local residents had served on Mormon missions overseas.

dBASE

The last of the "big three" PC application areas was database software. This demanded the biggest tradeoffs. IBM PCs could provide a more responsive spreadsheet experience than any timesharing minicomputer and faithfully replicate the experience of using a dedicated word processing system. They lacked the storage capacities, reliability, networking capabilities and processing power to duplicate the database management systems running on mainframes and large minicomputers. PC databases were still useful, however, for small businesses that used PCs equipped with hard drives. A standard 20 MB hard drive could potentially store tens of thousands of records, enough to handle sales, customer and inventory records.

The most popular database package for the IBM PC, Ashton Tate's dBASE, was a conversion from CP/M. dBASE was a programming system and not a tool for the typical office worker, but it was an easier and more productive tool than a general-purpose language like Pascal and ran on hardware cheap enough for smaller businesses to afford. dBASE included tools to create on-screen forms for data entry and display and to manage structured data files. Its programming language was optimized for file operations, making it easy to search a file for records that matched specified criteria. A reporting module created templates for printed output. A booming community of contractors emerged to create custom dBase applications. Small software companies produced dBase applications for specific industries, such as auto mechanic shops, and sold them packaged with hardware.[50]

Like WordStar, dBase II was one of the launch titles for the IBM PC. Unlike WordStar, dBase retained its dominant position well into the PC era. dBASE III Plus, released in 1985, cemented its place as the standard PC database package. It retained that position until the launch of dBASE IV in 1988, which was full of bugs and eliminated some of the features developers relied on.[51] Those developers shifted to systems such as FoxBase and Clipper that used the same programming language. Ashton-Tate laid off many of its employees and sold itself to a rival. The dBASE language dwindled in the 1990s, but nothing that's widely deployed goes away quickly. Even today, a significant developer community continues to enhance and support dBASE applications.

Shareware

The PC helped turn the application software business into a fast-growing industry, but not every PC user wanted to spend $500 on Lotus 1-2-3 or WordPerfect. One option was simply to ignore copyright law and copy the installation disks from a friend. Software companies ran campaigns to discourage piracy. Some hoped that the hefty manuals supplied with their packages and the telephone support they provided to registered users would discourage piracy. A flood of independent guidebooks and the increasing ubiquity of photocopiers made that less of a problem. Lotus and several other leading firms turned to copy protection, introducing deliberate errors into floppy disks that users would be unable to reproduce with an ordinary disk drive. The floppy disk was needed even when the program was loaded from a hard drive. That was unpopular with users—the special disks didn't always work, and if they were lost or damaged the program would be useless. Software companies eventually abandoned these schemes in the face of complaints from large companies forced to manage thousands of key disks.[52]

Meanwhile, the increasing professionalism of the software industry made it harder to turn programs into money. Packaging a program with a nice box and manual, advertising it, and getting it into stores was dauntingly expensive. By 1983, a new business model had emerged: shareware. The term was popularized by a word processor, PC-Write, written and sold by Bob Wallis, a former Microsoft programmer. Wallis encouraged users to make copies for their friends, asking those who found it useful to mail $75 for a printed manual and technical support. Two slightly earlier and widely used releases used similar business models: the communications package PC-Talk and the database program PC-File.

All three packages were polished programs that got the job done for most people who tried them. One reviewer called PC Write "very good indeed," despite some quirks such as forcing users to quit one program and load another every time a document needed to be printed.[53] Not all users paid, but the ones who didn't at least helped to spread the program to others who might. Their authors received millions of dollars and were freed from most of the tasks, like telemarketing and sales, which conventional software companies had to support. It took only a small staff to answer questions from users and deposit the checks.

Another shareware product, the ZIP format for compressed files, is still widely used today. "PKZIP," named after its creator, Phil Katz, of Milwaukee, was introduced in 1989 as shareware package. It was free for "noncommercial" use, although Katz suggested a $25 donation from satisfied users. ZIP quickly replaced previous methods of compressing files for distribution, an increasingly important role as program files grew larger and strained the capacities for floppy disks and modem links.

Shareware programs spread as ZIP files, shared via bulletin boards, computer user groups, and public domain software libraries. The "libraries" were businesses that issued catalogs of disks that they prepared, for a few dollars, to duplicate and mail. Their catalogs also included truly public domain programs, whose authors had renounced their copyright. The largest, such as the Public Software Library in Houston, featured many thousands of disks.

Some shareware programs unlocked additional features, such as extra levels in video games, once the registration payment was made. The donation-based shareware model became less common after the late 1980s, but this variant, sometimes called "cripple-ware" or "freemium," proliferated with the spread of Internet access and the shift to mobile devices. Today most commercial software is downloaded freely, but users must pay a registration fee (sometimes from inside the application) for it to be fully usable.

THE IBM PC BECOMES AN INDUSTRY

Although almost all of today's personal computers and most servers are the direct descendants of the IBM PC, not a single one of the billion or so IBM-compatible machines sold from 2015 to 2019 was made by IBM. What began in 1981 as a single proprietary machine had by the late 1980s become the basis for a worldwide industry of thousands of companies that collectively produced millions of PCs every year. An unremarkable beige box had become an industry standard. In the remainder of this chapter, we explain that transformation and the process by which IBM lost control of its own creation.

Expansion Cards

IBM launched its PC with a small lineup of expansion cards, but it documented the interface so that other companies could produce add-in hardware. A thriving industry soon emerged to supply them—just as it had for the MITS Altair and Apple II. Smart consumers could buy a bare-bones PC from IBM and save money by purchasing the cards they needed elsewhere. In November 1983, *Byte* produced a list of 107 manufacturers of PC expansion cards. As well as lower prices, their advantages over IBM's own equipment included more memory capacity, squeezing several functions onto a single card (such as serial and parallel connectors), and capabilities that IBM itself had neglected entirely such as print spooling and speech synthesis.[54]

AST specialized in cramming multiple functions onto its expansion boards, to conserve the PC's limited number of slots. Its SixPak card became the industry's best-seller by squeezing most of what a PC user needed onto a single board: memory expansion up to the full 640 KB; serial, joystick, and parallel ports; and a battery-backed clock to maintain time and date settings.

The only other cards needed for a fully equipped PC were disk and display controllers. PC users needing crisp text chose IBM's green-screen monitor. Used with the official IBM display adapter, it could display nothing else. A company named Hercules produced a replacement that generated high-resolution graphics as well as text.

Hercules graphics became an unofficial part of the PC standard, the first to be added without IBM's backing.

With more slots free, PC card producers could offer entirely new capabilities. The PC's built-in speakers were able to do little more than beep or click. The AdLib sound card released in 1987 added musical capabilities. Its software interface was copied by another hardware producer, Creative Labs, whose SoundBlaster cards dominated the market through the 1990s. This set another unofficial standard.

MS-DOS Computers

Microsoft's agreement with IBM let it sell most of the same code to other computer makers, and eventually directly to consumers, as MS-DOS. It became so ubiquitous that people mostly just called it DOS (a term we use here to include both PC-DOS and MS-DOS). Even by the standards of the mid-1980s, DOS was not an ambitious operating system. It supported only one user and could run only one program at a time. Nevertheless, DOS transformed Microsoft from a company that mainly sold BASIC to one well on the road to dominating the market for personal computer software.

Most of the IBM PC, including the 8088 microprocessor, consisted of standard parts that could be ordered from a catalog. In 1982, seeing the success of the IBM PC, other companies began to order the similar parts, license MS-DOS, and build their own computers. This followed the model of CP/M: an operating system licensed by different hardware makers to use on a wide range of hardware.

Early versions of MS-DOS lacked some important features of PC-DOS, which meant that licenses had to write their own code to duplicate the additions IBM had made (such as the FDISK command to configure hard disk drives) and, as with CP/M, to adapt the source code to their hardware. The first producers of MS-DOS machines tried to improve on the IBM-PC and not just copy it. For example, the Victor 9000, designed by Chuck Peddle who had created the Commodore PET, was launched in 1982. *Byte*'s reviewer said he would "take the Victor" over a comparably equipped IBM PC, because it was "clearly superior in quality of display, amount of standard memory, standard number and versatility of I/O ports, and number of available expansion slots."[55]

Because no single company dominated the market for CP/M computers, programs like WordStar had to use software interfaces provided by the operating system to access hardware devices such as screens. The MS-DOS marketplace evolved differently. Both systems provided interfaces that programs could invoke to work with disk files, manage peripherals, and display text on the screen. But because the IBM PC was a huge hit, there was a de facto standard hardware platform for DOS computers even before Microsoft began to license MS-DOS. Relying on DOS to mediate interactions with the hardware meant foregoing access to most of the PC's advanced features, such as graphics. Even the IBM PC serial port, used to connect modems or printers, couldn't work to its full capacity when controlled by DOS. Programmers looking to create professional applications, tended to bypass DOS and work directly with the PC hardware. Programs such as Lotus 1-2-3 that manipulated PC hardware directly could greatly outperform those that did not.

Victor's was undoubtedly the better computer, but most customers considering an MS-DOS computer did so to run software written for the IBM PC. The same year DEC, which had begun to struggle at the lower end of the market in the face of new competition, introduced three personal computers. Not only were these incompatible with the VAX, but despite adopting MS-DOS, none of them were fully compatible with the IBM PC either. One of them, the Rainbow, was a modest success but could not slow down the IBM-compatible juggernaut.[56]

Cloning the IBM PC

The Victor 9000, DEC Rainbow, and other early MS-DOS computers weren't successful enough for software firms to tailor programs to their superior but incompatible hardware. In the long term, only MS-DOS computers that were fully compatible with the IBM PC could survive. Producing a compatible PC was harder than licensing MS-DOS. The core of what made a computer an IBM PC was the BIOS code stored on a ROM chip. IBM owned that code. It relied on copyright, which protects written works, rather than patents, which protect inventions, to prevent the duplication of its PC.

Around the time of the PC's announcement, three Texas Instruments employees left their jobs and started a company called Compaq. Legend has it that Rod Canion, Jim Harris, and Bill Murto sketched out a fully IBM-compatible PC on a napkin in a Houston restaurant. To get around IBM's control of the BIOS code, they used *reverse engineering*. A team probed the BIOS and wrote detailed functional specifications for its behavior. Those specifications were passed to an entirely separate *clean room* team, giving the "what" of the BIOS but not the "how." Then the clean room team wrote new code to do all the same things. This was an expensive but legal approach.

The Compaq computer, delivered in 1983, was portable in much the same way as the earlier Osborne—at 25 pounds, it "gave new meaning to the phrase pumping iron."[57] The real draw was compatibility with the IBM PC at a lower price. Compaq became the first start-up to exceed $100 million of sales in its first year in business.

Early in the development of compatible systems, *PC Magazine* ran a feature on the phenomenon, including tests on several machines. Its conclusion: compatibility ranged from "close but no cigar [the Compaq] to not even close."[58] Following the lead of Portia Isaacson, a computer scientist turned industry analyst, the trade press relied on two informal tests for compatibility: would the machine run Lotus 1-2-3? Would it run Bruce Artwick's *Flight Simulator*, licensed by Microsoft (figure 8.5), a game that exercised every nook and cranny of the IBM hardware to achieve what *Byte* called "fantastic graphics and realism"? If the answers were Yes and Yes, the machine was a true *clone*.[59] *Flight Simulator* was a particularly brutal test. To run it, a PC had to be rebooted, loading game code instead of DOS. While loading, it displayed weird graphics, because Atwick used the display as working memory to squeeze the game into just 64 KB of RAM.

The ready availability of PC video cards, disk controllers, and other components made it easier to get into the clone business as there was no need to design or commission what could be ordered from a catalog. AST took that step in 1986 by building motherboards to plug its SixPak multifunction cards into. It began to sell fully assembled computers, eventually becoming one of the top ten suppliers. AST wrote its own BIOS, but even that became unnecessary after Phoenix Technologies reverse engineered the IBM BIOS and started selling compatible chips as a standard part. The PC motherboard became just one more commodity available from a dozen different suppliers. The floodgates opened for PC clones. The shift was captured by *PC Magazine*, the industry's most successful periodical, when in 1986 it changed its subtitle from "the independent guide to IBM personal computers" to "the independent guide to IBM-standard personal computing."

The ever-falling price of PC compatibles extended their market from corporations and wealthy individuals to small businesses, schools, and hobbyists with middling incomes. This shift was most dramatic in Britain, where lower incomes and higher prices had limited the IBM PC's initial market to corporate buyers.

In 1986 Amstrad, building on the success of its word processors, introduced the PC 1512, shown in figure 8.6. Amstrad again applied the techniques that had brought down the cost of home computers. An engineer recalled opening up an IBM PC to

Figure 8.5

To redraw its 3D graphics rapidly, *Flight Simulator* displaced MS-DOS and bypassed the PC BIOS to work directly with the innards of IBM's color graphics hardware. This made it a powerful test of complete IBM compatibility.

Figure 8.6
Amstrad's IBM PC clones used large production runs and a highly integrated design to reduce manufacturing costs. Note the slogan: "Compatible with you know who. Priced as only we know how." The PC 1512 sold for little more than a home computer, packaged with a mouse, monitor, and the GEM graphical environment.

discover "these discrete devices all over the place, and not one custom device in the whole bloody thing. It had massive boards stuffed full of chips. So we thought we can save a lot of money here by designing our own gate arrays."[60] They integrated functions like graphics and printer ports onto the motherboard to eliminate expansion cards and slots. This integration brought down costs dramatically but sacrificed some of the customizability of a standard PC. Manufacturing a large production run in Asia created further economies.[61]

The PC 1512 was a little flimsy and lagged behind the capabilities of high-end machines, but it sold for £399 including a screen, keyboard, and one floppy drive. That put it into home computer territory, less than half the cost of the cheapest PC clone previously available in Britain. Reviewing the machine, Guy Kewney concluded "For the life of me, I can't suggest why you would want any other version of the standard PC . . . the machine is fast, easier to use than any other PC, cheaper than anything with the same specification."[62] Over the next two years, Amstrad sold far more PCs than IBM had ever managed in Britain, and in Germany where it was rebadged as a Schneider model. In the US, home computer producers Commodore and Atari both launched budget-priced PC compatible lines.[63]

The PC AT—A New Standard

Each new IBM PC model set a new de facto standard, which was quickly replicated by clones claiming to be "100 percent compatible." The most influential, and the point from which modern PCs evolved, was the IBM PC AT launched in 1984. IBM built the machine around a faster 16-bit Intel processor, the 80286. The price was higher than the original PC, $5,795 for an otherwise well-equipped model with no screen, but the AT included a lot more standard equipment. It had 16-bit expansion slots and floppy disks that stored more information, a full 1.2 MB, to work on larger files. IBM made many little tweaks, from changing the keyboard layout to adding a built-in clock and battery-powered memory to store time and configuration details when the computer was turned off.

AT stood for *advanced technology*. Its performance was closer to a minicomputer's than a hobbyist PC's. Because its new processor could handle up to 16 MB of RAM and a powerful new instruction mode, the AT was designed for multitasking, multiuser operating systems. In practice, however, its users mostly stuck with MS-DOS, which meant using the AT as little more than a fast version of the original PC.

The AT was rapturously received. *PC Magazine* called it an "ultra high performance technological marvel" that would leave the clone makers looking for new jobs. "This innovative machine is so good," it marveled, "that IBM is shooting fish in a barrel. With a cannon." Only AT&T and "perhaps management skilled Compaq" stood a chance against it.[64] Engineering an AT clone would indeed have been a challenging task for a single company, but spread across the emerging ecosystem of PC parts manufacturers, it was no more than a speedbump. Within a year every part of the AT had been duplicated by multiple suppliers, and there were more clone companies than ever.

The clone makers' "turbo ATs" were faster than the real thing. It turned out that the PC AT was designed to run at 8 MHz, but because of corporate worries about cutting into the market for bigger IBM computers had initially been hobbled with a 6 MHz timing crystal. Clone makers had no such concerns. Over the next few years they pushed performance further beyond the original IBM systems. For example, IBM offered an enhanced graphics card generating higher resolutions and more colors. The original card was slow and expensive, but by 1987 many PC users were upgrading their systems with cheap clone cards (including some of graphics specialist ATI's first products) providing improved performance, more flexibility, and higher resolutions.

Ethernet for Local Area Networks

By the mid-1980s, corporate computing managers had accepted that PCs were standard office equipment, whether they liked it or not. To make PCs more useful and connect them to larger systems, companies started to connect them together in *local area networks* (LANs). By 1984, over 20 LAN projects for the IBM PCs were being advertised.[65] Many relied on slower, cheaper serial connections, but the fastest and most effective used a connection mechanism called Ethernet.

Robert Metcalfe and David Boggs had conceived Ethernet in 1973. They worked at Xerox's Palo Alto Research Center (PARC) in Silicon Valley, about which we have more to say in the next chapter. Metcalfe was an alumnus of MIT's project MAC, having helped link its PDP-10 to the ARPANET. In 1972 he became "the networking guy at PARC," using his experience to connect its PDP-10 clone, MAXC, to the ARPANET.[66]

The real focus at Xerox was on networking inside the building, to connect single-user computers to each other and to a high-quality printer. The ARPANET model, with its expensive IMPs, was clearly not appropriate. PARC networked several Data General minicomputers in a star-shaped topology, but Metcalfe found this inappropriate for the office where machines are connected and disconnected frequently.[67] Instead, he recalled the ALOHAnet system that used radio signals to link computers on the Hawaiian islands.[68] For his office network, Metcalfe used the same approach but substituted a cheap coaxial cable for the "ether" that carried ALOHAnet's radio signals. A computer could be added to this *Ethernet* simply by tapping into the cable. In May 1973, he showed that such a system could handle a lot of traffic without becoming overloaded. Metcalfe worked with David Boggs to get a network running by the following year. He recalls that its speed, 3 million bits per second, was unheard of at the time, when "the 50-kilobit-per-second (Kbps) telephone circuits of the ARPANET were considered fast."[69] Ethernet's first commercial success came in 1979, when Digital Equipment Corporation, Intel, and Xerox joined to establish a faster, 10 million bits per second version, as a standard.

In the market for performance LANs, Ethernet's main opponent was IBM's *token ring* technology. Unlike Ethernet, this offered a guaranteed minimum level of performance even when network usage spiked. To prevent collisions, the channel was controlled by whichever computer held a virtual token, just as early railroads engineers had to hold physical tokens when proceeding onto unsignaled tracks.

The networked PCs of the late 1980s were not connected to the Internet. PC networks provided two basic services: file sharing and printer sharing. File sharing let everyone in a department access a shared hard drive, to collaborate and store their work in an accessible and frequently backed up location. Printer sharing let companies purchase and support a small number of fast, expensive printers instead of fitting one to each PC. These networked PCs gave their users a lot more autonomy and independence than the timeshared mainframes accessed through "dumb terminals" or "glass Teletypes" of the 1970s.

The company that emerged with over half the business by 1989 was Novell, located in the Salt Lake City area. Novell's Netware was a complex operating system able to turn a PC into the hub of an Ethernet network. It usually replaced MS-DOS entirely on that *file server* computer. Novell originally treated its software as an add-on to its business of selling Ethernet hardware but struggled against 3Com in that market. It took off after adopting a new strategy of focusing on the software. Netware 286, designed to exploit the power of PC AT compatibles, included a utility to compile a driver tailored to run efficiently with a particular network card and computer configuration. Competition soon turned Ethernet cards into cheap commodity items. Netware stayed expensive. To ensure a supply of competent network technicians, Novell launched a popular certification program. A Certified Novell Engineer would have little trouble finding work. Similar schemes were later introduced by other vendors, including Microsoft.

Although Netware is no longer used, Ethernet remains the dominant standard for wired local area networking. Ethernet connectors have been built onto most PC motherboards since the mid-1990s. Today the typical connection rate is a gigabyte a second, one hundred times faster than the version first commercialized back in the early 1980s.

THE PC AT THE END OF THE 1980S

IBM did not anticipate the IBM PC's huge success or the clone industry that would quickly develop. IBM could not demand royalties from the producers of PC clones. In the mid-1980s, as IBM planned a successor to its PC range, it was determined to remedy this mistake.

The PS/2: IBM Loses Its Grip

The new PS/2 line, launched in April 1987, was built around proprietary IBM technology, protected by patents. It was intended to sweep away the entire IBM PC family and its imitators, just as System/360 had replaced all IBM's existing computers and peripherals. Clone makers would have to pay for licenses.

The new machines were beautifully engineered, designed so that they could be taken apart without using a screwdriver. IBM built more capabilities, including graphics, onto their motherboards. It moved power switches to the fronts of the machines, changed their keyboard connectors, and added sockets for mice. Unlike earlier PCs, which used 5 1/4-inch floppy disks, all the PS/2 machines used 3 1/2-inch disks (which, being enclosed in a hard plastic shell, were no longer floppy and were known as *micro diskettes* in IBM speak). Whereas previous PC graphics were mediocre, the new video graphics array (VGA) output crisp text and detailed graphics with vivid and realistic colors. This perked up business presentations and charts around the corporate world and eventually established the PC as the video gaming platform of choice.

The most fundamental changes were made to the higher-end models. The 16-bit expansion slots in the PC AT had been coupled directly to the processor, creating compatibility problems as processors were sped up. IBM replaced this with a new, and meticulously patented, 32-bit alternative called MCA, for *micro channel architecture*. It was used in some bigger computers, too, and included powerful features recalling the channels built into classic IBM mainframes. Two devices, such as a network card and a disk drive controller, could exchange data without tying up the central processor, at speeds faster than the processor itself could handle. The new cards were supposed to be easier to configure. Setting up a conventional PC card took hours of fiddling with switches, drivers, and operating system files to determine settings. In contrast, MCA cards stored settings in a special kind of memory where they could be set automatically by an operating system.

IBM expected clone makers, and even expansion card producers, to start paying license fees to copy the MCA bus. This made sense. The clone industry had driven down costs ferociously, but during the early years of the PC its biggest innovation had been putting a handle on the box to make it portable. Surely its customers would continue to demand copies of IBM's latest models. A Gartner Group analyst was quoted saying "if the IBM 'clone' companies hope to keep their share of the corporate market, they'll have to match IBM's new personal computer architecture."[70] Exploiting this, IBM tried to extract punitive terms from clone makers. To gain access to the new technology they would have to give IBM up to 5 percent of the sales price of every PS/2 compatible computer. IBM was even reported to have demanded retroactive royalties on PCs already sold, which for a company like Tandy or Compaq would run to millions of dollars.[71]

OS/2

Like System/360, IBM's new range was launched with the announcement of a new operating system. OS/2 was developed jointly by IBM and Microsoft. It promised real benefits over MS-DOS, including the ability to multitask programs and access megabytes of memory. Reporting the launch, the *New York Times* quoted a Microsoft executive saying that PS/2 was "the most important introduction in the short history of personal computers," and had established "the computer architecture for the next decade."[72] OS/2 could be used on computers made by other companies, but IBM played up the benefits of its Extended Edition, which was full of connections to IBM databases and mainframe networks. Whereas the original PC had been a marginal project, developed with little input from IBM's senior management, PS/2 played a crucial part in IBM's grand *system application architecture* (SAA) strategy. No other firm was a leading supplier of mainframes, minicomputers, word processors, and personal computers. Justifying its higher prices, particularly for smaller computers where competition was tougher, would be easier if customers obtained tangible benefits from interconnecting IBM computers of all sizes. SAA was supposed to provide common user interfaces, programming interfaces, networking protocols, and office software across the entire IBM range.

Like OS/360 before it, OS/2 wasn't ready by the time the new hardware began delivery. The initial release, shipped eight months late, lacked key features including a promised graphical user interface called Presentation Manager. The first full version arrived in October 1988. IBM had expected memory prices to fall, but production shortfalls instead caused a worldwide spike in RAM prices in the late 1980s, meaning that the 3 MB of extra RAM needed to run OS/2 added about $1,000 to the price of a PC. This slowed the adoption of OS/2, even among purchasers of high-end PS/2 computers.

New Kinds of Computer Firm

Most clone makers were unwilling to pay IBM to use the patented elements of its PS/2 machines, although they were able to mimic some of their features. Case manufacturers, for example, began to feature bays for the new 3 1/2-inch disk drives. Graphics card makers produced video cards that duplicated, and improved upon, the capabilities of IBM's new VGA output. Instead of licensing IBM's micro channel architecture the PC industry followed an unexpected strategy: working together to evolve the PC AT into something competitive with PS/2. An initial step in this direction had taken place the year before PS/2 launched. Compaq got tired of waiting for IBM to make something new to copy and launched the Deskpro 386, which shoehorned Intel's newest processor into the existing PC architecture.[73] At the time this looked like a stop-gap solution, unable to harness the full power of the new chip. A review concluded that although "given an 80386 and unlimited freedom, a competent . . . engineer would not have come up with this design," Compaq had nevertheless extracted "better performance than expected out of existing standards."[74] *Newsweek* called it a "calculated risk" and warned that "customers may choose to wait for IBM, fearing that any other computer will be incompatible."[75] In fact the Deskpro 386 was a big hit. Compaq's approach was soon duplicated by other clone makers.

Compaq, part of a cluster of chip and PC companies based in Texas, exemplified the first wave of PC compatible producers. It did the same things as IBM's PC division, in the same way, but more aggressively at lower costs. Customers who chose Compaq over IBM knew they were getting a top-quality machine for a bit less money. Compaq prided itself on its engineering talent, which put it ahead of other clone companies in introducing new features. Once minicomputers fell on hard times, Compaq eventually purchased DEC to further enhance its technological capabilities. Like IBM, Compaq wrote its own BIOS and mass produced its machines in its own factory. Like IBM, it sold through networks of dealers who added a hefty markup to cover their own costs. In fact, Compaq decided to sell only through dealers that IBM had already authorized to carry its own PCs, even hiring the IBM executive who had put its PC dealer network together.[76] Starting a firm like Compaq would require a major investment to design a computer, buy a factory or set up contract manufacturing arrangements, and build a dealer network.

IBM's keyboard, case, motherboard, and expansion cards were custom built. So were Compaq's. As the PC components industry developed these items, like disks, memory chips, and processors, became standard items that could be ordered from a catalog. Moderately knowledgeable computer users saved money by building their own computers. This could be done in a couple of hours using no tool more exotic than a Phillips head screwdriver. The more commercially-minded built PCs to order for their friends and eventually set up computer businesses. Across the country, each small town or city neighborhood had at least one storefront PC "screwdriver shop," building new machines to order while servicing and supporting old ones. Some saved money by using pirated copies of DOS. The shift to assembly of standard components facilitated the exodus of component manufacturing to Asia, as suppliers could focus on low cost niches without needing to engineer or market a whole computer.

By the end of the 1980s the typical workhorse business PC was an IBM PC AT compatible assembled from a handful of standard parts. One was the motherboard, into which a processor and memory chips were inserted. This was screwed into a case, along with a power supply. A typical configuration filled three of its slots with a display adapter, a combined parallel and serial card to drive a printer and modem, and a disk controller card. Hard and floppy disks filled two drive bays of standard dimensions. A PC clone maker needed to procure only one custom part: a badge the same size and shape as the IBM badge, to be stuck into the standardized depression on the case.

The most successful of this new breed of computer companies was founded by Michael Dell in 1984 from his college dorm room in Austin, Texas. As demand grew he dropped out of college and began to advertise nationally. By 1988 the Dell Computer Corporation had made an initial public offering and was one of America's fastest growing businesses.[77] Firms like Dell rose and fell not on the basis of their engineering talent, of which they had little, but on the basis of their success in obtaining or directly importing components, the quality of their sales and service operations, and the efficiency of their distribution systems. Unlike Compaq, Dell sold direct to its customers, eliminating the overhead of working through dealers and distributors. That also reduced the costs associated with warehousing unsold computers. Dell assembled a computer only after it had been configured and paid for by a customer. It arrived within a few days. Each PC hardware component, except for the processor, could be chosen from dozens of hardware suppliers. Competition, economies of scale, and specialization drove down component prices at a startling rate. A computer that took a couple of months to go from production to sale would lose a significant fraction of its value on the way.

As users and software suppliers became ever more committed to the PC platform the costs of switching away from it rose, while the benefits of remaining within it increased. Even Digital Research eventually gave up on CP/M and tried to sell MS-DOS compatible operating systems. The only companies whose products were found in most PCs were Microsoft, maker of the standard operating system, and Intel, which was largely successful in stopping other firms from making viable clones of its processors. As PC prices plummeted, Microsoft stood to make more money on each PC sale than the company building the computer. Microsoft made its initial public offering of stock in 1986 when the company had about a thousand employees. Within eighteen months it had doubled, then doubled again. This exponential growth continued until 2001. Thanks to its practice of granting stock options to programmers, managers, sales staff, and other permanent employees Microsoft eventually created more millionaires than any other company in history. The practice of granting stock options had begun with Silicon Valley chip firms, but it was Microsoft that established the idea that a programming job could be a reliable pathway to wealth. After seven years, employees could use their windfalls to retire or start their own companies.

Portable PCs

By the late 1980s, portable PCs were becoming more practical. Early portables from Osborne and Compaq were full-sized computers with handles and tiny built-in monitors. They could be dragged from offices to homes or worksites. Programmable calculators were genuinely portable and could be used with batteries but had tiny screens and keyboards. The most portable computer with a real keyboard was Radio Shack's TRS-80 Model 100, developed by Kyocera of Japan (figure 8.7). It ran for about 20 hours off standard batteries and weighed only three pounds. Achieving those goals involved some significant compromises—no built-in disk drives, only 8 to 32 KB of memory, and a screen limited to eight lines of text. Its most enthusiastic users were journalists, who had previously dictated copy over telephone lines. They could now write stories in the field with the Model 100, filing them with its built-in modem. As well as a text editor, its ROM chips included programs to maintain to-do lists and address books and Microsoft BASIC, which enabled custom applications such as data logging and

Figure 8.7

RadioShack's TRS-80 Model 100 was the most successful portable computer of the early 1980s, despite its tiny screen. It was particularly popular with journalists. Courtesy of Computing History Photographs, Archives Center, National Museum of American History, Smithsonian Institute.

industrial control. It won InfoWorld's award for the best new hardware of 1983. More than six million were eventually sold.

Similar technologies soon led to battery powered PC compatibles. Toshiba's T1100, introduced in 1985, included a built-in 3.5-inch floppy disk drive and a full IBM-compatible monochrome display. It is often called the first laptop PC. (GRiD pioneered the format a few years earlier (figure 8.8), but its machines were not IBM compatible.) The T1100 and its gradually more capable successors kept their weight down to about nine pounds. By 1989, Toshiba had introduced higher-end options with hard disk drives, swappable battery packs, and 80286 processors.

It was possible to produce a smaller and lighter PC, but only by making deep compromises. In 1989, Atari unveiled its $400 Portfolio—a one-pound machine with capabilities similar to the original IBM PC. It included a low-power Intel-compatible processor, a Lotus 1-2-3–compatible spreadsheet, and an MS-DOS–compatible operating system

Figure 8.8
The GRiD Compass laptop pioneered the "clamshell" design. It used an orange electroluminescent display and bubble memory, with no moving parts. Photo by Eric Long, Smithsonian National Air and Space Museum (TMS A19890006000_PS01).

working with removable memory cards. Hailed as "the ultimate in yuppie accoutrements," the Portfolio was a feat of miniaturization, but its tiny, hard-to-read screen, limited storage, and fiddly little keyboard hurt its practicality as a business tool.[78]

The PC Triumphant

The PC's position at the end of the 1980s was unassailable. The IBM PC had evolved from a single model to the basis for a new kind of computing. To understand why, we need to consider changes in price as well as performance. Adjusted for inflation, in 1981 a PC with two floppy disks, a monochrome screen, and 176 KB of RAM had cost $5,000. What could the same budget buy in late 1989?

Nothing comparable to the original PC was still being sold. WarehouseData Products was liquidating PC/XT clones with twice the performance. $5,000 would buy five complete systems, including screens and hard drives—enough to run a small business. At Dell it would secure three turbo AT systems, each with close to ten times the processor power of the original PC. They had much better graphics, and their hard drives liberated users from the experience of endlessly swapping floppy disks. Blowing the whole sum on one machine would buy something spectacular. In the prime spot in the

THE DELL SYSTEM® 310
20 MHz 386.

The best combination of performance and value available in its class.

STANDARD FEATURES:
- Intel 80386 microprocessor running at 20 MHz.
- Choice of 1 MB, 2 MB, or 4 MB of RAM* expandable to 16 MB (using a dedicated high-speed 32-bit memory slot).
- Advanced Intel 82385 Cache Memory Controller with 32 KB of high speed static RAM cache.
- Page mode interleaved memory architecture.
- VGA systems include a high performance 16-bit video adapter.
- Socket for 20 MHz Intel 80387 or 20 MHz WEITEK 3167 math coprocessor.
- 5.25" 1.2 MB or 3.5" 1.44 MB diskette drive.
- Dual diskette and hard drive controller.
- Enhanced 101-key keyboard.
- 1 parallel and 2 serial ports.
- 200-watt power supply.
- 8 industry standard expansion slots (6 available).

**Lease for as low as $131/month.
△ Extended Service Plan pricing starts at $251.

40 MB TTL Monochrome System	$3,599
40 MB VGA Color Plus System	$4,099
100 MB VGA Color Plus System	$4,699
100 MB Super VGA Color System (800x600)	$4,799

Prices listed reflect 1 MB of RAM. 150 and 322 MB hard drive configurations also available.

*Performance Enhancements (Systems 325, 310, 316 and 220): within the first megabyte of memory, 384 KB of memory is reserved for use by the system to enhance performance. 4 MB configurations available on all systems. Call for pricing.

THE DELL SYSTEM® 316
16 MHz 386SX.

Expandable, affordable access to 386 architecture.

STANDARD FEATURES:
- Intel 80386SX microprocessor running at 16 MHz.
- Choice of 1 MB, 2 MB, or 4 MB of RAM* expandable to 16 MB (8 MB on the system board).
- Page mode interleaved memory architecture.
- VGA systems include a high performance 16-bit video adapter.
- LIM 4.0 support for memory over 1 MB.
- Socket for 16 MHz Intel 80387SX math coprocessor.
- 5.25" 1.2 MB or 3.5" 1.44 MB diskette drive.
- Integrated high performance hard disk drive interface and diskette controller on system board. (ESDI based systems include a hard disk controller.)
- Enhanced 101-key keyboard.
- 1 parallel and 2 serial ports.
- 200-watt power supply.
- 8 industry standard expansion slots (7 available).

**Lease for as low as $98/month.
△ Extended Service Plan pricing starts at $234.

40 MB TTL Monochrome System	$2,699
40 MB VGA Color Plus System	$3,199
100 MB VGA Color Plus System	$3,799
100 MB Super VGA Color System (800x600)	$3,899

Prices listed reflect 1 MB of RAM. 150 and 322 MB hard drive configurations also available.

THE DELL SYSTEM® 220
20 MHz 286.

It's faster than many 386 computers, and has a smaller footprint.

STANDARD FEATURES:
- 80286 microprocessor running at 20 MHz.
- Choice of 1 MB, 2 MB, or 4 MB of RAM* expandable to 16 MB (8 MB on system board).
- Page mode interleaved memory architecture.
- LIM 4.0 support for memory over 1 MB.
- Integrated diskette and VGA video controller on system board.
- Socket for Intel 80287 math coprocessor.
- One 3.5" 1.44 MB diskette drive.
- Integrated high performance hard disk interface on system board.
- Enhanced 101-key keyboard.
- 1 parallel and 2 serial ports (integrated on system board).
- 3 full-sized 16-bit AT expansion slots available.

**Lease for as low as $109/month.
△ Extended Service Plan pricing starts at $264.

40 MB VGA Monochrome System	$2,999
40 MB VGA Color Plus System	$3,299
100 MB VGA Monochrome System	$3,599
100 MB VGA Color Plus System	$3,899

Prices listed reflect 1 MB of RAM. External 5.25" 1.2 MB diskette drive available.

THE NEW DELL SYSTEM® 210
12.5 MHz 286.

The price says this is an entry-level system. The performance says it's a lot more.

STANDARD FEATURES:
- 80286 microprocessor running at 12.5 MHz.
- Choice of 512 KB, 640 KB,†† 1 MB, or 2 MB of RAM expandable to 16 MB (6 MB on system board).
- Page mode interleaved memory architecture.
- LIM 4.0 support for memory over 640 KB.
- Integrated diskette and high performance 16-bit VGA video controller on system board.
- Socket for Intel 80287 math coprocessor.
- 5.25" 1.2 MB or 3.5" 1.44 MB diskette drive.
- Integrated high performance hard disk interface on system board.
- Enhanced 101-key keyboard.
- 1 parallel and 2 serial ports.
- 3 full-sized 16-bit AT expansion slots available.

**Lease for as low as $64/month.
△ Extended Service Plan pricing starts at $190.

20 MB VGA Monochrome System	$1,699
20 MB VGA Color Plus System	$1,999
40 MB VGA Monochrome System	$1,899
40 MB VGA Color Plus System	$2,199

Prices listed reflect 512 KB of RAM.
††640 KB versions of the above systems are available for an additional $80. 100 MB hard drive configurations also available.

All prices and specifications are subject to change without notice. Dell cannot be responsible for errors in typography or photography. **Payments based on a 36-month open-end lease. †Leasing arranged by Leasing Group, Inc. In Canada, configurations and prices will vary. DELL SYSTEM is a registered trademark of Dell Computer Corporation. Microsoft, MS, MS-DOS and XENIX are registered trademarks owned by Microsoft Corp. Intel is a registered trademark; 386 and 386SX are trademarks of Intel Corporation. UNIX is a registered trademark of AT&T. Dell UNIX System V is based on INTERACTIVE Systems Corporation's 386/ix.™ ™Signifies trademarks of entities other than Dell Computer Corporation. △Service provided by Xerox Corporation. Service in remote locations will incur additional travel charges. ©1989 Dell Computer Corporation. All rights reserved.

SAVE NOW ON THE DELL SYSTEM® 325 25 MHz 386.
AN EVEN BETTER VALUE AT THESE NEW LOW PRICES.

STANDARD FEATURES:
- Intel 80386 microprocessor running at 25 MHz.
- Choice of 1 MB, 2 MB, or 4 MB of RAM* expandable to 16 MB (using a dedicated high-speed 32-bit memory slot).
- Advanced Intel 82385 Cache Memory Controller with 32 KB of high speed static RAM cache.
- Page mode interleaved memory architecture.
- VGA systems include a high performance 16-bit video adapter.
- Socket for 25 MHz Intel 80387 or 25 MHz WEITEK 3167 math coprocessor.
- 5.25" 1.2 MB or 3.5" 1.44 MB diskette drive.
- Dual diskette and hard drive controller.
- Enhanced 101-key keyboard.
- 1 parallel and 2 serial ports.
- 200-watt power supply.
- 8 industry standard expansion slots (6 available).

**Lease for as low as $178/month.
△ Extended Service Plan pricing starts at $370.

40 MB VGA Monochrome System	$4,899
100 MB VGA Color Plus System	$5,799
100 MB Super VGA Color System (800x600)	$5,899
150 MB Super VGA Color System (800x600)	$6,399

Prices listed reflect 1 MB of RAM. 322 MB hard drive configurations also available.

All systems are photographed with optional extras.

AD CODE NO. 11EL9

Figure 8.9

By the end of the 1980s, most PC companies purchased standard parts and screwed them together. This page, taken from a four-page Dell advertisement run on the inside cover of *Byte's* November 1989 issue, consisted almost entirely of technical specifications and prices in small print. Potential customers would compare the components and pricing offered by Dell with those of its many rivals. By the December issue, the base price of Dell's flagship System 325 had fallen another $800.

December 1989 *Byte*, just inside the front cover, Dell was advertising its flagship Dell System 325 (figure 8.9): a 386 running at 25 MHz with a 100 MB hard disk drive and a SuperVGA color screen. According to one widely used benchmark, that configuration would provide more than 25 times the performance of an original IBM PC.[79] *PC Magazine* reviewed it as "a complete package at a competitive price with an outstanding warranty," declaring it the "Editor's Choice" at half the price of the equivalent Compaq model despite some (literal) rough edges caused by Dell's use of a cheap commodity case.[80]

These remarkable improvements in power and affordability made the PC a standard part of the American office. Yet even $1,000 for a bargain-basement PC system, plus a few hundred for a printer and more for software and support, was a lot to spend for the ability to do word processing and spreadsheet work at home. Only 15 percent of American households owned a computer in 1990. Among African American households, the figure was 7 percent. Even among the richest 20 percent of households, two thirds had not yet made the purchase.[81]

By the end of the 1980s, DOS had picked up some new features, like a better BASIC and a full screen text editor, and limited support for larger memories, but its limitations held back the ever more powerful PC hardware from living up to its potential. The Dell System 325 had a processor able to support multiple users, virtual memory, and robust multitasking. Under DOS it could do none of those things, instead functioning as a faster version of IBM's original PC, which was itself an evolution of 1970s microcomputing technology. Microsoft's official answer was to use OS/2 instead of DOS, but almost nobody did. In chapter 10, we look at the further development of the PC over the 1990s and the eventual replacement of DOS—not by OS/2 but by Microsoft Windows. Before we tell that story, however, we set back the clock again to see what computing researchers, minicomputer producers, and Unix workstation companies had been up to while personal computers were evolving. You will see that the processors and operating systems that power our modern personal computers have as much to do with those developments as they do with the original IBM PC.

9 THE COMPUTER BECOMES A GRAPHICAL TOOL

On January 24, 1984, Steve Jobs, wearing a double-breasted navy blazer and garish green bow tie, took the stage at De Anza College (close to Apple's headquarters) and pulled a tiny Macintosh computer out of a bag. The computer sprang to life, proclaiming itself "insanely great" before running a slide show on its crisp monochrome screen to demonstrate its new graphical user interface (GUI). In the popular imagination this moment divides the history of personal computing into two eras: the dark ages of text-based computing, inherited from timesharing systems, versus the enlightened world of windows and graphics.

Conventional personal computers could display graphics as well as text, but although that power was exploited by individual programs such as video games and charting software, it was ignored by MS-DOS. Computing doesn't work that way anymore. We control computers by selecting things visually, using a mouse, a laptop touchpad, or a touch screen. Controls are often represented by little pictures as well as (or instead of) words. To rearrange items on a page or windows on a screen, we drag them into place. Text is graphically rendered in a variety of fonts and styles. In 1984, says the myth, the brilliance of a great innovator changed everything.

Having read this account so far, you are likely not surprised to learn that the reality is rather more complicated. The Macintosh as delivered was widely dismissed as an expensive toy, its sales dwarfed not just by the IBM PC and Apple IIe but even by that notorious flop, the PCjr. Eighteen months after introducing the Macintosh, Jobs lost the confidence of Apple's board and left the company. Most of the Mac's crucial new features were already available on another Apple machine, the Lisa, and much of the work behind that originated not at Apple but at Xerox. Throughout the 1980s, computers with graphical user interfaces had only a tiny share of the market and were much more expensive than mainstream personal computers, which is why we could tell the story of mainstream office computing through 1989 without mentioning them.

The most obvious new feature of the Macintosh was its graphical user interface (GUI). Its key elements were invented over the course of a few years in the mid-1970s by a small team working in a single research facility, Xerox's Palo Alto Research Center (PARC). But, less obviously, graphical computing as developed at PARC depended on new hardware capabilities—powerful processors, large memories (the lack of which crippled the first Macintosh), and high-resolution screens. To explain the diffusion of graphical user interfaces, we must understand the spread of those capabilities, initially to a new generation of microprocessor-based personal computers marketed as *graphics workstations*.

XEROX INVENTS GRAPHICAL COMPUTING

The Xerox Corporation was named after its breakout hit, office copy machines that used a process called xerography to produce fast, accurate photocopies on plain paper. By the early 1970s its original patent on the process had long since lapsed, and its efforts to keep out competitors with new patents had attracted an antitrust investigation that eventually led to a consent decree.[1] Xerox copiers remained the leading and most recognized brand, but with increasing competition in the market for copiers, the company was keen to identify and dominate future generations of high technology office products. In addition to its purchase of Scientific Data Systems, discussed previously, Xerox set up its Palo Alto Research Center in the Palo Alto foothills in 1970. Its mission was to anticipate the profound changes that the distribution of information on screens rather than through paper technology would bring to corporate offices.[2]

Xerox PARC

Two things made PARC's founding significant for computing. The first was the choice of Palo Alto: Jacob Goldman, director of corporate research at Xerox, had favored New Haven, Connecticut, but the person he hired to set up the lab, George Pake, favored Palo Alto and prevailed, even though it was far from Xerox's headquarters in the Northeast.

The second took place in the halls of Congress. As protests mounted on college campuses, a parallel debate raged in Congress over the US involvement in Vietnam, including war-related university research funded by the Department of Defense (DoD). Senator J. William Fulbright warned that science research was losing its independence in the face of a monolithic "military-industrial complex." In an amendment to the 1970 Military Procurement Authorization Bill, a committee chaired by Senator Mike Mansfield inserted language that "none of the funds authorized . . . may be used to carry out any research project or study unless such a study has a direct and apparent relationship to a specific military function or operation."[3] The committee intended to separate basic from applied research, but those doing advanced research on computing with ARPA funding felt that they were at risk.[4] Some members assumed that the National Science Foundation would take the DoD's place in funding basic research, but the NSF was never given the resources to take up the slack.

At that moment, George Pake was scouring the country's universities for people to staff Xerox PARC. He found a crop of talented and ambitious people willing to move to Palo Alto. ARPA funding was heavily concentrated at a few universities, including MIT, Carnegie-Mellon, Stanford, UC-Berkeley, and the University of Utah. Researchers from each of those ended up at PARC, including Alan Kay and Robert Taylor from Utah, and Jerome Elkind and Robert Metcalfe from MIT. (Metcalfe actually was getting his PhD from Harvard, but at the time he was recruited by PARC he had an ARPA-funded job at MIT.) Taylor had been head of ARPA's Information Processing Techniques Office. Chuck Thacker and Butler Lampson were among the alumni of the nearby Berkeley Computer Corporation who moved to PARC. PARC staff was encouraged to use SDS technology because SDS had just been purchased by Xerox. But the researchers there resisted and instead built a clone of the PDP-10, which they called the multiple access Xerox computer—MAXC for short, a pun on the name of Max Palevsky, the founder of SDS.[5]

Rather than perfect timesharing, the PARC team was determined to develop a new kind of interactive computing experience. Development of the hardware and software for a new computer, the Alto, was at the center of the lab's work from 1972 onward. Butler Lampson proposed its construction, with Charles P. Thacker as lead hardware designer. Thacker later won the Turing award for this work. Much of the architecture of personal computers powerful enough to support graphical user interfaces came from minicomputers such as the DEC VAX. But even when equipped with specialized graphics hardware, VAX machines were never intended for personal use. The Xerox PARC team had started by designing and building what was essentially a personal minicomputer. Each Alto coupled high-resolution graphics hardware directly to a powerful processor with, by the standards of the day, an absurdly large memory.

The Alto had a lot in common with the most advanced personal computers of the mid-1980s, but realizing those capabilities in the early 1970s required different technology. Altos used integrated circuits but, like minicomputers, their processors were too powerful to squeeze onto a single chip. In fact, the Alto had a novel architecture in which processor capabilities were spread around the machine rather than clustered on one circuit board. Each Alto had its own hard drive with a removable platter, like those used with IBM mainframes.

Each Alto was equipped with a mouse. Doug Engelbart did not join Xerox PARC, but many of his staff did, including Bill English (who did the detailed design of the original mouse). Researchers at PARC refined the mouse and coupled it with a unique high-resolution screen, arranged in portrait orientation to mimic a sheet of paper. This was bitmapped, so that its almost half-million pixels could be manipulated by flipping bits in memory. That was possible by the new availability of RAM chips, so that at least 128 KB of memory could be stuffed into an Alto.[6] With office work in mind, Bob Metcalf's ethernet system, discussed in the previous chapter, was invented to support the lab's Altos and let them share files and printers.

The first Alto went into use in 1973, and before long, Xerox had built enough to serve PARC's internal computing needs. An improved model was used more widely in the late 1970s, including a pilot project in the White House during Jimmy Carter's administration. Thanks to all this novel hardware, an Alto cost about $18,000 to build, far more than the personal computers that would begin to appear a few years later. PARC was using Xerox's riches to explore the potential shape of computing in the 1980s after costs had dropped enough to make the system viable.

Smalltalk and Object-Oriented Programming

The Smalltalk programming environment was developed by a small group led by Alan Kay, a charismatic young researcher. Adele Goldberg played a large part in its design. Kay saw the emerging Alto as a chance to create an "interim" version of something he first described in his 1968 PhD dissertation and later began to call a Dynabook. This was something like today's tablet computers: a thin, portable, battery-powered device with a high-resolution screen and (unlike most tablet computers) a keyboard. The *book* name reflected the idea that the device would be robust, portable, and central to education.[7] Kay was heavily influenced by the pedagogical theories of Seymour Papert, who stressed the need for children to learn through experimentation. Unlike an ordinary book, it would be *dynamic*, which to Kay meant it had to be highly interactive but easy and fun, unlike existing systems such as Doug Englebart's NLS.

Smalltalk was designed with flexibility and interactivity in mind, to put graphical objects of different kinds on screen and interact with them. Traditional programming languages assumed a text-based user interface. Applications coded with them were controlled with typed commands or selections from text menus. The program would print a list of options and wait for users to push a key to select one. Kay wanted the Dynabook to feel personal and interactive, displaying pictures for its users to interact with. Nobody yet knew what kind of graphical interface would prove most effective—ideas such as pull-down menus and icons had not yet been codified. During his PhD research, at the University of Utah, Kay worked with Ivan Sutherland, whose Sketchpad system was a major influence, through both its use of graphics and its support for clipping and scrolling windows so that users could work with objects too large to fully display on the screen.

As well as a new kind of user interface, Smalltalk codified and began to spread a new approach to programming languages called *object-oriented programming*. Kay took some ideas from Simula 67, a computer modeling and simulation language developed in Norway by Ole-Johan Dahl and Kristen Nygaard.[8] Traditional languages define data structures separately from the code that manipulates them. The new approach let programmers produce highly modularized code, in which data structures are defined together with the operations that programmers use to access their values or update their contents. These hybrid bundles of data and code were called *objects* by Kay. Each object was an *instance* of a standard *class*. New classes could be defined as special cases of existing ones, with additional capabilities or characteristics.

These capabilities had huge value for computer simulations, in which programmers needed to define representations of real-world objects together with the rules needed to define their interactions. They were also attractive for many of the people working around this time on new programming languages or development methodology features to make it easier for individuals or small teams to create complex, reliable software. We have already discussed the so-called software crisis and efforts to deal with it such as structured programming. Object orientation was an extension of ideas about variable scope and code modularity proposed by computer scientists such as Niklaus Wirth and Tony Hoare. Because all data was held inside objects, it could be manipulated only by using the *methods* explicitly provided in the code defining the corresponding classes. That enforced modularity and made it easier to reuse code between systems and to maintain systems. Smalltalk conceptualized the interactions between these objects as a kind of dialog achieved through the exchange of messages, an idea captured in the name Kay gave the language.

These new language features were at least as promising for graphical user interfaces as for simulation languages (see figure 9.1). Graphical user interfaces were much more complex than text-based menus, forcing programmers to manage a bitmapped screen festooned with buttons to click, panels to scroll, and menu options to select. Traditional interface methods were, to use a term popularized by Kay, modal. Users issued the desired command, which put the system into a mode. What it did in response to their next input would depend on the mode. For example, in *delete* mode, selecting a file would delete it. In *edit* mode, the same action would open it for editing. Kay favored a different interface style, in which users would first select the object they wanted to work on and then manipulate it to accomplish the desired operation. Providing that kind of open-ended interaction in a conventional programming language would be frustrating and inefficient—the program would have to be structured as a loop that constantly checked whether the user had just carried out each of a huge number of possible actions. In Smalltalk, the programmer could specify the code to be

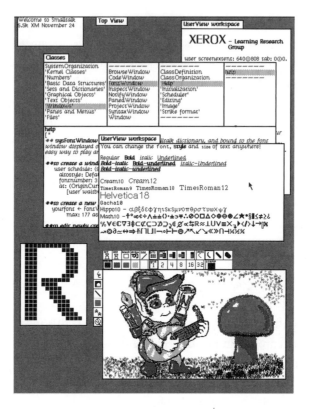

Figure 9.1

Xerox researchers pioneered the graphical user interface. This originated not as an operating system feature but as part of the Smalltalk programming environment, which included its own tools for drawing (bottom) and for editing event-driven code that defined the action taken when a user clicked on a graphical object. Image from Wikimedia user SUMIM.ST used under the Creative Commons Attribution-Share Alike 4.0 International license.

run when a particular region of the screen, button, or scroll bar was triggered, and then the system itself would figure out what objects should be alerted in response to a particular click. This was called *event-driven* code.

Kay has mentioned a long list of systems that influenced Smalltalk. One of the most important was the Lisp programming language. Kay said later that he "could hardly believe how beautiful and wonderful the *idea* of Lisp was."[9] Computer scientists tend to love languages that are minimalist and flexible, and almost invariably hate those (like PL/I, COBOL, and the defense-oriented Ada) that design committees have stuffed with features to meet the imagined needs of many different kinds of user. Smalltalk, like Lisp, had a very simple core but allowed users to extend this with their own code, layering objects on top of each other. Most of Smalltalk was itself implemented in Smalltalk, with the result that every aspect of the system could be manipulated with code even while a program was running. This was made possible by its structure: Dan Ingalls, who implemented the early versions, designed a virtual machine to underpin Smalltalk so that code could be efficiently interpreted.

Smalltalk went beyond Lisp by providing what was later called an integrated development environment (IDE), which included a text editor, a browser to explore the hierarchies of classes defined in code, and debugging tools to examine the current state of objects as programs executed. The IDE concept eventually became a standard part of software development for other languages. It was applied to Lisp itself by two firms, Symbolics and Lisp Machines, both founded in 1979 by rival factions of "hackers" from the MIT AI lab. Each sold small numbers of enormously expensive single-user computers, costing around $70,000 each. They had stack-oriented processor architectures optimized for Lisp. As general-purpose PCs grew more powerful, the already small market for such computers dwindled to nothing, but existing users stuck with them because of the huge productivity advantages their development environments provided to programmers.

Object-oriented programming was harder to grasp than some of the other novel features of Alto, such as mice and graphical controls, and spread more slowly. Some high-profile languages of the late 1970s, such as Niklaus Wirth's follow-up to Pascal, Modula-2, were designed to support increased modularity, but the full object-oriented approach was little known outside PARC until an article about it appeared in the August 1981 issue of *Byte* magazine.[10] Smalltalk itself found some enthusiastic users, for example, at Wall Street firms where the high cost of entry could be justified by the speed boost it provided to skilled application developers.

Object-oriented programming did not go mainstream until the 1990s with the addition of object-oriented features to established programming languages, most notably in the extensions to C described by Bell Labs computer scientist Bjarne Stroustrup in his 1985 book *The C++ Programming Language*.[11] C++ was a superset of C, meaning that existing C code continued to work and programmers could embrace or ignore the new object-oriented capabilities as they chose. This appalled Kay ("objects were a radical idea, then they got retrograded") and other purists, but it helped C++ quickly establish itself as a widely used language and a favorite for computer science instruction.[12]

What You See Is What You Get

Smalltalk was not an operating system. It had a defined interface for its development environment, which relied heavily on menus that popped up (appeared) when mouse buttons were clicked, but Smalltalk worked as a toolkit to allow the rapid prototyping of applications with graphical user interfaces rather than as a means of standardizing those interfaces. As the first major piece of graphical software running on the Alto, it was heavily influential on other programs developed for that machine.

One of these was the Gypsy text editor produced by Larry Tesler and Timothy Mott in 1976. Gypsy took the capabilities of a previous program, Bravo, developed by a group including Butler Lampson and Charles Simonyi, and reworked it with the first user interface to resemble that of now-standard systems such as Microsoft Word. For example, to add text, users simply used the mouse to set an insertion point and then typed. To copy text, one highlighted it with the mouse and then pushed the Copy key. Xerox researchers, following Kay, called this style of operation *modeless* because the results of triggering a function were consistent and did not depend on a previously selected command mode.[13]

Like Bravo, Gypsy exploited the graphical screen of the Alto to display text with different fonts, accurate spacing of letters, embedded graphics, and formatting features such as bold and italic text. Computerized publishing expert Jonathan Seybold dubbed this *what you see is what you get* (WYSIWYG), repurposing a catchphrase of Flip Wilson, the first African American comedian to make regular television appearances. Wilson used the phrase in character as Geraldine Jones, a brashly self-confident woman, as winking acknowledgement of the tension between his cross-gender performance and Geraldine's lack of pretense. The PARC staff borrowed it to define a simpler form of representational fidelity: the printed output would match the visual content of the screen as closely as possible.

This was made possible by another PARC invention, the laser printer. This merged the printing and paper handling mechanisms from a high-end Xerox copier with a powerful embedded computer able to draw high resolution images onto the copier drum with a laser, replacing the usual optical mechanism used to create an impression from the source document. The idea came from a Xerox engineer, Gary Starkweather, who in 1971 transferred to PARC. The computer control system was built by a team including Butler Lampson and relied on PARC's unique Ethernet infrastructure to quickly move large amounts of data from the Altos to the printer controller. By the late 1970s, IBM and Canon had introduced laser printers by licensing Xerox technology. Between income from licensing its patents and selling its own laser printers, Xerox recovered its entire investment in PARC many times over.[14]

Client-Server Applications

Xerox's invention of Ethernet, discussed in the previous chapter, provided an effective way of linking computers to one another in a local environment. PARC's use of the technology was a compelling counter argument to the idea, captured in Grosch's law, that a single large and expensive computer would be more cost effective than multiple smaller computers.[15] By the late 1970s, a new buzzword, *distributed computing*, had emerged to describe the idea of having big and little computers work together over computer networks—for example, using a minicomputer or personal computer to check data and update the user interface of an application and a mainframe or other large computer to maintain a secure, shared database.

The new approach would also require a great deal of innovation in software. One example of this at PARC was the repurposing of MAXC, the homebrew PDP-10, into a server for the Altos. MAXC was hooked up to the ARPANET, and incoming email for users at PARC was delivered there. By 1978, a program called Laurel had been developed for the Altos. This introduced what later became the standard way of working with email: users downloaded their messages to their personal computers to file and read them. Replies were uploaded back to the server. Laurel used a graphical interface, running on Alto, which split the screen into two window panes: one showing header information for the current message folder, so that the user could pick a message with the mouse, and the other showing the text of the selected message.

Approaches of this kind were called *client-server* computing—a program running on one computer (the client) made a request for a program running on another computer (the server) to do something, that is, to provide a service. IBM and other computer companies devised complicated networking strategies for distributed computing, including the OSI seven-layer model that we mentioned earlier. Many actual systems, however, relied on a simple but flexible mechanism to let a program running on one computer initiate a processing job on another computer: the remote procedure call (RPC). The idea was originated at PARC by Bruce Nelson around 1981 and was copied for other operating systems, becoming a standard part of Unix within a few years.[16]

Commercializing Alto

People only vaguely familiar with the PARC story often claim that Xerox's big mistake was not trying to commercialize its innovations. In fact, as Michael A. Hiltzik showed in his detailed history, *Dealers of Lightning*, Xerox made a serious attempt to bring Alto technology directly to market as a competitor to word processors and personal computers.[17] If anything, its mistake was to turn PARC's innovations too directly and too rapidly into commercial products, years before customers were ready for them or hardware existed to provide acceptable performance at a reasonable price.

PARC staff pitched their achievements to Xerox's top management as the foundation on which the firm could dominate the office of the future. Over the next five years, Xerox worked hard to turn PARC's prototypes into salable products. That required far more time, people (a team that grew to 280 developers), and money than their original invention. Xerox launched its Star computer (officially the 8010 Information System), a direct commercial refinement of the Alto, in 1981 (figure 9.2). A workstation cost $16,000, not including the servers, laser printer, and Ethernet networks needed to use it. Star was superior in every way to Alto. Its display was higher resolution, its Ethernet

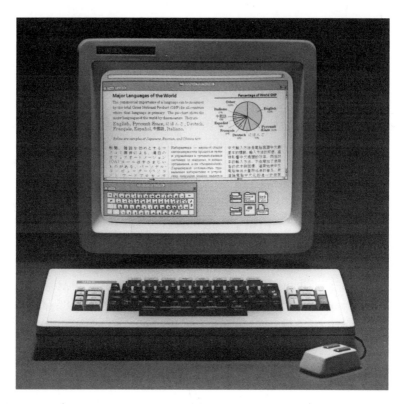

Figure 9.2

The Xerox Star was the first computer with a standardized graphical user interface based on a desktop metaphor (the printer and folder icons in the lower right). This image flaunted its unique ability to mix graphics and text in multiple fonts and languages, displaying the document on screen as it would look when printed. It took many years for Microsoft and Apple products to match these capabilities. Courtesy Xerox, image scanned by Digibarn.

was faster, and its new family of processors, based on an architecture proposed by Butler Lampson, was optimized for virtual memory and virtual machines. It came with powerful and polished office software, supporting compound documents (for example, embedding an editable spreadsheet chart inside a report).

Unlike Alto, Star's operating system had a standardized graphical user interface. This introduced the idea of a computer desktop.[18] As its designer, David Canfield Smith, described this, "Every user's initial view of Star is the Desktop, which resembles the top of an office desk, together with surrounding furniture and equipment. It represents a working environment, where current projects and accessible resources reside. On the screen are displayed pictures of familiar office objects, such as documents, folders, file drawers, in-baskets, and out-baskets. These objects are displayed as small pictures, or icons. . . . Star users are encouraged to think of the objects on the Desktop in physical terms. You can move the icons around to arrange your Desktop as you wish."[19]

Star was a technological marvel, but Xerox struggled to sell it against the affordable and flexible IBM PC systems being adopted by businesses for office work. PCs could not match its elegance or graphical capabilities, but they were much cheaper and faster at basic tasks like scrolling a page or loading a document. Powerful as the Star was, its complex operating system overwhelmed its hardware, particularly on the initial Dandelion processor. PCs were also more flexible. Xerox marketed the Star as an office automation machine, like the dedicated word processors of the 1970s, rather than as a personal computer. It never released tools to support the development of other applications for the Star.

Later versions had smaller price tags and faster processors, but Xerox sold only about 25,000 of its Star family. By the end of the 1980s, it had given up. Xerox had, as the title of a popular book described it, "fumbled the future."[20] Other firms brought the concept of windows, icons, a mouse, and pull-down menus (the so-called WIMP interface), and Ethernet to a mass market. Xerox proved less nimble than Apple or Microsoft, but out of fairness one should also compare Xerox with peer companies: DEC with its minicomputers and interactive operating systems, Honeywell with its Multics timesharing system, and Control Data with Plato. Eventually they all slipped or failed as businesses, but collectively they laid new foundations for computing.

BRINGING THE GRAPHICAL INTERFACE TO PERSONAL COMPUTERS

Alto-like computers were sold with more success to the kinds of people who had previously been using VAX terminals: engineers designing things like cars, planes, and computer chips, scientists running computationally intense calculations and simulations, and Wall Street firms with large technology budgets and a need for high-speed results. Physically, a *graphics workstation* was a small box with a single main board, like a personal computer; functionally, however, it was a powerful minicomputer for individual use. Most workstations used the Unix operating system, reflecting their minicomputer-like architectures. *Workstation* was a 1980s buzzword from office automation. *Graphics* simply meant pictures as well as words.

Graphics Workstations

Laboratories and engineering centers viewed workstations not as an expensive alternative to a PC but as a cheaper alternative to a VAX bundled with graphics hardware. The VAX brought the power of a scientific mainframe into the engineering division of a company. Workstations brought that power to the individual desktop.

Unlike the Star, which followed the Alto in building a custom processor from multiple chips, most workstations used a new Motorola microprocessor chip, the

68000, introduced in 1979. A series of articles in *Byte* by its designers began with an explanation that Motorola had chosen a clean-sheet design over compatibility with its older 8-bit chips, the path taken by Intel, to produce "the fastest, most flexible processor available," designed "for programmers, to make their job easier."[21] The 68000's instruction set, like the rest of its architecture, had more in common with a VAX minicomputer than with earlier microprocessors. The 68000 used 32-bit memory addresses, giving programmers the convenience of addressing up to 16 MB of memory as a single continuous area, unlike the convoluted segment system that Intel's processors of the era used. Its plentiful registers could hold and manipulate 32-bit numbers, even though the processor moved data more slowly in 16-bit chunks. It offered many addressing modes, and a larger instruction set made possible by extensive use of microcode. It had features to support multitasking. Later members of the 68000 family, such as the fully 32-bit 68020, provided virtual memory support and even larger address spaces.

First out of the blocks with a 68000-based workstation was Apollo, of Chelmsford, Massachusetts. Its founder, Bill Poduska, had previously cofounded Prime, the company that pioneered the 32-bit mini. In 1981, Apollo delivered a workstation with its own operating and networking systems. The price for a single workstation began at $40,000. It soon faced tough competition from Sun Microsystems, founded in early 1982, to commercialize a workstation developed for the Stanford University Network (hence the company's name). This continued the local tradition of shifting technology from publicly funded university projects to profit-making companies. Andy Bechtolsheim brought the Stanford hardware. Bill Joy brought expertise developed on the BSD Unix effort as another cofounder.[22] Silicon Graphics Incorporated, the other major workstation supplier, emphasized powerful graphics hardware and won a dominant position in film studios.

Workstation companies targeted small markets that would not support the cost of developing new technologies. Instead, they depended on what was called an open systems approach—using standard processors, memory chips, networking standards, peripheral connections, and so on. Combined with the inherent price-performance advantages of microprocessor-based systems over minicomputers, this gave them a huge price-performance advantage. The Sun-2, with virtual memory support, arrived in 1984. Instead of connecting a dozen expensive graphical terminals to a high-end VAX minicomputer, a lab or trading floor could now install a dozen workstations each with a powerful processor of its own. Prices were in the $20,000 range, a bargain for what *Byte* declared a "VAX-class machine."[23] Soon Unix-based servers were also replacing DEC equipment for other applications such as corporate email and databases.

Sun's slogan was "The network is the computer." Its workstations were usually found in networks, together with a more powerful server for file storage. Unix could store user profiles centrally, so that one's personal configuration would work on any computer on the network. A Unix program could run on another computer (such as a powerful server) but display its results in a local window. By the late 1980s, these local networks had been bridged onto Internet connections or corporate backbone networks to provide seamless access to resources far beyond the building.

Although DEC had strong research and development, its management style, commitment to its VAX product line, and determination to design and build as many of its own components as possible fatally hobbled its own efforts to sell workstations. By the late 1980s, the future market for VAX systems was starting to look shaky. DEC, still the second largest firm in the computer industry, struggled to mount an effective response. In 1992 it announced a quarterly loss of $2.8 billion and ousted its long-serving CEO, Ken Olsen. DEC would never return to sustained profitability.

The Apple Lisa

Seemingly running a controlled experiment to see whether the commercial failure of Xerox's Star was a failure of implementation or a doomed strategy, Apple tried the same gambit with its Lisa, introduced in 1983. Lisa had specifications similar to the graphical workstations of the era: a Motorola 68000 processor, crisp monochrome screen, hard disk drive, expansion slots, an operating system able to run several applications at once, and a full megabyte of memory. At launch it listed for $9,995—a relative bargain. Like the Star, it was pitched for administrative use and came with a suite of visually polished office applications. Unlike the Star, it was possible to develop software for Lisa, but Apple didn't make it easy. Lisa had to reboot into a different, text-based operating system to compile code.

Like Star, Lisa derived directly from the work of PARC. Apple hired Larry Tesler away from PARC to lead the development of system software for Lisa. This was far more polished than anything available on the Unix workstations of the era. Tesler developed the object-oriented Lisa Toolkit to help produce graphical applications, but the unfamiliarity of this style of programming was an additional challenge. Lisa copied features from Star, including its use of a simulated desktop to represent files, but Apple worked hard to refine and simplify the Xerox approach. For example, Apple had only a single button on its mouse and used actions within a desktop metaphor, such as dragging files from one folder to another to trigger complex tasks. Sun workstations attached different actions to each of the three buttons on its mouse, following the original Alto. Even the Star workstation, which had introduced the desktop metaphor, had two mouse buttons and many special keys for actions such as delete, copy, and move.

Byte enthused that in terms of its market impact "the Lisa system is the most important development in computers in the last five years, easily outpacing IBM's introduction of the Personal Computer."[24] Lisa had some unfortunate limitations. Its floppy drives were unreliable, and its operating system ran slowly and crashed frequently. Apple had copied Xerox's visual orientation but not yet its laser printer, so there was no way to properly print the beautiful documents it created. Despite those foibles, Lisa had exactly the core capabilities that would define the most powerful personal computers of the next decade: hard disks, networking, a graphical user interface, and slots for expansion. Users could load several applications simultaneously, cutting and pasting between their windows. That wasn't quite multitasking, as background applications were suspended, but the operating system did prevent applications from overwriting each other.

Lisa cost less than an Apollo workstation and sold much faster (around 100,000 Lisas versus a few hundred Apollo workstations during the first year each was available), but Apollo was a start-up company selling to a niche market.[25] Its performance registered with investors as a big success when the firm made its initial public offering in 1983. Apple had spent heavily to set the next standard for personal computing, targeting the huge market for document preparation. Against those high expectations, Lisa was off to a disappointing start.

The Macintosh

In an alternate universe one might imagine subsequent Lisa models gradually winning market share from IBM as they benefited from lower component costs, improved software, newly available laser printers, and faster processors. Apple's actual experience in the 1990s demonstrated that it is much harder to overcome the limitations of an established machine architecture than to improve hardware performance. But in our universe, history took a less predictable turn.

Apple had grown fast, but it was still a young company plagued by internal struggles. Back in 1979, Jef Raskin, a former computer science professor, had started a project to build a cheap, friendly personal computer with a built-in modem. Steve Jobs took it over, keeping little more than the idea of a cheap personal machine and the name: Macintosh. He dreamed of a self-contained little machine, used by individuals rather than companies. Feuding with the managers of the Lisa project, Jobs wanted to somehow borrow its user interface for this affordable personal computer. In pursuit of this impossible goal, he pushed the Macintosh team to perform miracles of coding and design.

The Macintosh was an approachable and diminutive box, almost a cube, with a gap at the back that worked like a handle (figure 9.3). Inside the box were a single floppy disk drive, using Sony's new 3 1/2-inch format, a high-resolution nine-inch black-on-white monitor, and just 128 KB of memory. Jobs personally mandated this absurdly small memory, one-eighth the size of the Lisa's, to keep the cost down. Much of the operating system was burned into ROM chips in order to leave RAM free, but the team

Figure 9.3

Although its user interface won many fans, the original Apple Macintosh had few compelling applications. Once the almost essential second floppy disk was added, this approachable little computer had a list price of nearly $3,000. Photo courtesy Division of Medicine and Science, National Museum of American History, Smithsonian Institution.

still had to leave out some core Lisa capabilities. Unlike the Apple II and IBM PC, the Macintosh was closed, that is, users could not add boards and were discouraged from even opening the case. With no expansion slots and no memory sockets, there was no official way to add a hard disk drive, connect to a network, or even expand its memory (although engineers did sneak past Jobs the extra lines needed to solder in higher capacity chips later).[26] As well as saving money, these constraints reflected Jobs's faith that he could make better decisions for its users than they could make for themselves, delivering a computer that worked like a self-contained home appliance.

Jobs had crippled the Macintosh with restrictions aimed to cut its price, but Apple nevertheless priced a system with two floppy drives at $3,000. Without the second drive, which wasn't available for several months following launch, users would be swapping disks constantly. Copying a floppy disk would take at least five swaps, sometimes twenty. Steven Levy, an early user, called this "a new high tech form of torture." Others joked about a new medical condition, "Macintosh elbow."[27]

Apple introduced the Macintosh with a legendary commercial that ran during the 1984 Super Bowl. This hyperbolically equated IBM and its drab PC with George Orwell's dictatorial Big Brother. Members of the Macintosh group called Jobs's ability to convince them of things that, outside his presence, seemed highly implausible as a "reality distortion field." Launched outside that field, the Macintosh landed with a thud. A Macintosh system with a tiny screen, tiny memory, and no expandability was barely cheaper than the improved Lisa 2 launched at the same time. After an initial flurry of sales, around a quarter of a million in 1984, Apple soon found itself delivering just five thousand Macintosh units a month.[28] Jobs responded by shutting down the Lisa division, ensuring that the Macintosh was Apple's only path forward.

The few people who did purchase a Macintosh during that first year tended to form a cult-like devotion, appreciating the obsessive attention that Jobs had paid to every aspect of the new machine's physical design and user interface. The whole system, including the detachable keyboard and mouse, was highly portable; early models came with a small carrying case. Its faster processor and stripped-down operating system meant that it responded to user inputs more rapidly than Lisa or the Xerox machines. The Mac's elegant user interface was its greatest accomplishment. When a file was opened or closed, its symbol expanded or contracted on the screen in little steps—to its fans, it just felt right.

The Macintosh shipped with simple word processing and painting programs, but additional applications were slow to appear. The first development system required purchase of a Lisa. There were few tools for the development of graphical programs, and available programming languages and methods were ill suited. In 1986, Tesler in consultation with Niklaus Wirth, the inventor of Pascal, unveiled a new language, Object Pascal, and an object framework called MacApp. These exploited the natural affinity of object-oriented programming with graphical user interfaces. Other tools and languages followed, but the transition was challenging for programmers.[29]

The one Macintosh market that showed great promise in 1984 was education. Drexel University quickly standardized on the Mac as the ideal vehicle for graphical, interactive educational computing. All freshmen were required to purchase one, at a negotiated cost less than half the retail price.[30]

In the spring of 1985, Apple's board of directors forced out Jobs. Macintosh computers then began to evolve back into expandable, Lisa-like systems. The first step had already been taken: when even Jobs had conceded that the original Macintosh was unusable, a 512 KB version was rushed into production. An easy-to-use networking system, AppleTalk, launched in 1985. The Macintosh Plus, in 1986, had a megabyte of memory and a hard drive interface, just like Lisa.

Desktop Publishing

By the time the Macintosh Plus arrived, a compelling use for it had already been discovered. The desktop publishing industry began in 1985 with the launch of Aldus Page-Maker, designed by Paul Brainerd (see figure 9.4). Brainerd had previously developed computerized production systems used by newspapers, and he recognized that a large potential market had opened up now that personal computers with the capabilities needed for page design were available.[31] PageMaker let amateurs tinker with fonts and graphics until their newsletters or posters looked just right (to them, if not to trained designers). Professionals could produce slick-looking pages more rapidly than ever before.

PageMaker worked with the new Apple LaserWriter printer. This cost $6,995, far more than the Macintosh it plugged into, yet still aggressively low by Apple's standards because rendering pages described in Adobe's new PostScript language required the printer to hold a more powerful processor and more memory than the computer did.[32]

As one reviewer concluded, "I can't count the number of times I've shown someone my Macintosh and they've said: 'But it's just a toy. . . .' Now at least I can show PageMaker to them and say 'Let's see your IBM do that.'"[33] Thanks to PageMaker, the

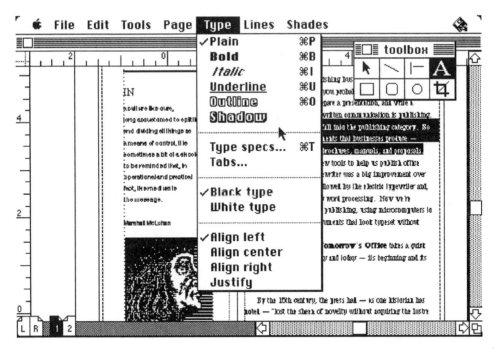

Figure 9.4

Aldus PageMaker was the first compelling application for the Macintosh, creating a new market for desktop publishing. Coupled with a laser printer, it could publish crisp text and graphics at a fraction of the cost of traditional typesetting technology. The Macintosh standardized on menu titles at the top of the screen, which when clicked, dropped down to present control options.

Macintosh, unlike Lisa or Star, offered a compelling business case to a small but well-defined group of users. Graphical computing was still too expensive for general office use, but for people who needed to produce high-quality printed output, it was a bargain if it eliminated the cost and delays of working with a traditional print shop.[34]

This market supported Apple through the mid-1980s, while Macintosh use spread through more schools and universities and the machine became a favorite with affluent individuals and successful writers. The Macintosh's dominance of the graphical design and publishing market was reinforced with the arrival of Adobe's Illustrator package for drawing and Photoshop for image manipulation.

Apple reacted by reorienting its products toward professional use. The Macintosh II series, launched in 1987, put back the expansion slots that Jobs had stripped out and, like Lisa, could be used with full-sized monitors. For the rest of the decade, high-end Macintosh hardware had specifications roughly equivalent to or slightly better than the most expensive PC-compatible systems. Macintoshes were pricier but utilized their hardware more effectively to give a far more polished experience.

Macintosh Competitors

The Macintosh was just one of the new platforms challenging the dominance of the IBM PC. It would be only a slight exaggeration to say that every design team of the mid-1980s that wasn't committed to IBM compatibility finished up building its computer around a processor from Motorola's 68000 range. Even Sinclair, looking to follow up its Spectrum home computer with a more powerful business-oriented model, chose one for its optimistically named Quantum Leap, launched in 1984. It was a dramatic flop.

Besides the Macintosh, the two most successful upstart platforms, both launched in 1985, were the Commodore Amiga and the Atari ST. Both were powerful computers with at least 512 KB of RAM, 3 1/2-inch floppy disk drives, and 68000 processors. Atari's slogan was "Power Without The Price." Its ST sold for half the price of a comparable Macintosh and added color graphics capabilities. Atari licensed the graphics environment manager (GEM) windowing system and a variant of CP/M from Digital Research. Its hardware and particularly its software were clumsy in comparison with Apple's, but *Byte* proclaimed it the "clear leader in price/performance" for the foreseeable future.[35] Amiga was initially far more expensive but included uniquely powerful sound and graphics chips, making it a favorite for video production work and game playing.

Amiga and ST sales in the mid-1980s compared well with those of the Macintosh and with all but the largest PC-compatible firms of the era, such as Compaq and IBM itself. They sold particularly well in Europe, to small businesses and computer enthusiasts for whom compatibility and polish were less important than value. Like earlier home computers, they were popular for video game players with limited funds, in part because disks were much easier to illegally copy than game cartridges.

Although portable and battery-powered DOS computers were becoming more practical during the late 1980s, the chips, screens, and hard disks needed by Macintosh-like computers consumed power too quickly for existing battery technologies to be workable. In 1989, Apple unveiled its $7,000 Macintosh Portable. Its specification sparked great excitement: a fully featured Macintosh with a hard disk drive that could run on battery power. It incorporated a relatively large (9.8-inch) screen that could, for the first time, display moving objects like mouse pointers without blurring horribly. A review called "trying to describe" the performance of the new "active matrix" display panel an "impossible task—you have to see it to believe it."[36] But its portability was severely limited by a 16-pound weight, mostly from the lead-acid battery Apple needed to ensure several hours of power.[37] One rival, a portable version of the Atari ST, was even less practical: it consumed twelve Duracell disposable batteries in half an hour.[38] Available technology forced so many tradeoffs that graphical user interfaces remained firmly desk bound for a few more years.

Commodore was eventually the most successful of the Apple's rivals, selling several million Amigas in the UK and Germany alone in the late 1980s, most of them low-cost A500 models for home use.[39] But neither Atari nor Commodore had the profit margins to duplicate Apple's polished operating system, or the market share to benefit from the economies of scale that drove down costs and pushed up performance for PC components. Both firms were slow to improve the processor speeds and graphical capabilities of their machines. By 1991, they no longer held clear advantages over the rapidly evolving PC platform, even for video gamers or graphics work.

NeXT, founded in 1985 by Steve Jobs after he was fired from Apple, offered a high-end line of graphical workstations also based on 68000 chips. Noting Apple's success in education, Jobs initially restricted sales to universities.[40] NeXT's commercial path was more consistent than Commodore's: its sales started small and remained tiny, despite introducing a second-generation model in 1990 and beginning to promote its work-stations outside the higher education market. It sold around 50,000 in all, many to Wall Street firms and national security agencies looking for rapid software development capabilities.

In 1993, the ST family was discontinued and NeXT gave up on trying to sell computers and shifted its focus to developing a version of its powerful operating system able to run on standard PC hardware. The next year Commodore International declared bankruptcy. The Macintosh was left as the only serious challenger to the dominance of the PC hardware standard. But by then, Microsoft had provided a credible graphical user interface to ordinary PCs with its newly popular Windows operating system. New capabilities were entering the mainstream of computing, and with them a further expansion in the practical universality of computers.

10 THE PC BECOMES A MINICOMPUTER

The decades since IBM tried and failed to set a new industry standard with the PS/2 range might seem a time of remarkable continuity: by 1990, Intel and Microsoft controlled the PC industry, and they still control it today. The personal computers of the 2010s, including Apple Macintosh models, were all direct descendants of the PC AT. They maintained compatibility through each incremental change, even as PCs evolved from desktop computers into servers, workstations, and laptops. People stopped talking about "IBM PC–compatible" computers and spoke instead just of PCs or, more revealingly, of Wintel computers, after their key features: Intel hardware and Windows operating systems.

By the late 1990s, the PC had killed the minicomputer and the graphics workstation. Yet from the viewpoint of technology and architecture, the situation is the reverse: the personal computer as we know it today was invented over the course of the 1990s, not in 1981 with IBM's first model or in 1977 by Apple. The PC architectures of the 2000s have more in common with those of 1980s minicomputers than they do with MS-DOS or CP/M. Since 2000, Windows has been based on an operating system designed by a former DEC engineer and patterned after a minicomputer system. From this perspective, the minicomputer never died. Rather, minicomputers shrank and replaced PCs without their users ever realizing it.

The original IBM PC was not so different from an Apple II. Even the typical MS-DOS computer of 1990 was essentially a much faster version of the same thing. Incremental hardware improvements continued apace through the 1990s, making PCs hundreds of times more powerful than their IBM progenitors. When the new millennium dawned, a high-end PC had one, or possibly two, 800 MHz Pentium III processors, an 80,000 MB hard disk drive, 512 MB of RAM, and a 21-inch screen running at 1600×1200 resolution. Laptops were almost as powerful and were well on their way to replacing desktop PCs as the most popular computing platform. The more fundamental changes, though, were in software and architecture. The new machine ran Windows 2000, a robust 32-bit multitasking operating system with a graphical user interface. It recognized and configured new hardware automatically. Video games and movies both played smoothly in high resolution. Forget the 640 KB limit imposed by MS-DOS—the new operating system could handle entire gigabytes of memory. It juggled dozens of programs, protecting them from each other so that if one crashed the others were unaffected.

Not all computers run Windows, but the minicomputer lineage is even stronger in Macintosh laptops and Linux servers, which both run software patterned after Unix. Unlike DOS, modern PC operating systems mediate every interaction between application programs and hardware. Security features protect the files and processes belonging to different users. Data from disks are streamed in and out on high-capacity channels without bogging down the main processor. When physical memory runs low, the hard drive jumps into action as virtual memory.

MOVING BEYOND DOS

The deficiencies of DOS were well known by the mid-1980s. Not even Microsoft thought it had a future. But efforts to replace it kept failing. Microsoft and IBM instead released new versions with incremental improvements—tools to automatically free up more usable memory, a mouse-driven but text-based control screen, a better BASIC, support for larger hard drives, and so on.

One obvious limitation of DOS was that it forced programmers who wanted to take advantage of the increasingly powerful graphical capabilities of PCs to bypass it to deal directly with the underlying hardware. A programmer who needed to print more than basic text output had to directly generate the series of control codes needed to produce formatted text or graphics on a particular printer. The producers of packages like WordPerfect and Lotus 1-2-3 invested huge amounts in writing device drivers for different printers and graphics cards. Like other DOS programs, they were primarily text based and ran with gratifying speed.

Memory handling was another frustrating aspect of PC use in 1990. DOS treated newer chips as faster versions of the 8088 processor used in the original IBM PC. This kept compatibility but forced application developers to write their programs in 64 KB chunks, known as *segments*. These all had to fit, along with DOS itself and any device drivers, into the first 640 KB of RAM, the *base memory*. Some programs with heavy memory needs, such as Lotus 1-2-3, were programmed to jump through hoops to store data elsewhere, but DOS hindered this more than it helped it. A computer with megabytes of unused RAM might still give an out-of-memory error if a program could not be given enough base memory. Ashton-Tate's dBase IV was a notorious offender. It was almost impossible to use under the newest DOS release (version 4), which gobbled more of the base memory for its own needs. Both products flopped, and Ashton-Tate foundered. Memory optimization was a major selling point of later versions of DOS, but users still spent hours fiddling with their AUTOEXEC.BAT and CONFIG.SYS files to free up base memory. The bafflement that DOS caused inspired Dan Gookin's 1991 book *DOS for Dummies*, full of cartoons and humorously reassuring explanations. Its spectacular sales launched an entire genre of "dummies" books.[1]

Early Alternatives

DOS's lack of support for multiple users or multitasking was perhaps the most fundamental of its weaknesses. Microsoft sold its own version of Unix, called Xenix, to computer builders as a more powerful alternative to MS-DOS. Running on a 286 PC, Xenix could power several attached terminals. Microsoft described Xenix as the future of personal computing and MS-DOS as a stopgap. When MS-DOS 2 launched, Microsoft promised that its Xenix-compatible features like hierarchical directories and pipes would make the transition easier.[2] Version 3 of DOS was originally supposed to be a multiuser, multitasking operating system with close ties to Xenix. That dream faded, in part because in 1984 the Bell System freed AT&T to sell Unix in competition with Microsoft, and in part because Xenix could not run the applications that attracted most users to the PC in the first place. Microsoft passed Xenix on to a smaller company, The Santa Cruz Operation.

Another shortcoming of DOS was its lack of a graphical user interface. Even before the Macintosh's announcement, companies were trying to provide Xerox-like interfaces for the IBM PC. In 1982, the creators of VisiCalc announced a product called VisiOn for the IBM PC, which failed badly enough in the marketplace to destroy their company. IBM developed a text-based windowing system for DOS called TopView, and in 1985 Digital Research launched a PC version of its Macintosh-like GEM interface. Microsoft's own equivalent was Windows, which ran on top of MS-DOS. Windows could multitask specially written graphical applications, although not smoothly. The first version, released in late 1985, was almost entirely useless. After playing with the surprisingly challenging Reversi game included with the release, most users found little reason to launch it again.

None of these products were widely adopted. A 1983 roundup of five windowing systems for the PC by journalist John Markoff concluded that they were "literally begging for faster hardware."[3] GEM, Windows, TopView, and VisiOn also suffered, in varying proportions, from lack of support or compatibility with popular applications, spurious but distracting legal challenges from Apple, and excessive complexity.[4] Windows and GEM were designed to work with special Macintosh-like applications, raising a further problem: because programs were written for DOS, few users ran Windows or GEM; but because few users ran Windows or GEM, most programs were written for DOS.

The limitations of DOS were particularly frustrating for users of high-end PCs built around Intel's 386 processor. It was a big step forward from the 286 chip, which had a compatibility mode run for old-style code and new mode to access more RAM and multitask programs but couldn't combine the two regimes. Yet DOS ignored the new features Intel had added to multitask programs written for the processor used in the original PC.

OS/2, which was supposed to replace both DOS and Windows, didn't fully solve the problem either. IBM insisted that OS/2 run on 286 models, which meant forgoing the advantages of the 386. By the time the first reasonably practical version of OS/2 was released in 1989, that already looked like a bad decision. OS/2 needed so much RAM and processor power that nobody without a 386 or better processor would attempt to use it. OS/2 versions of popular software like WordPerfect and Lotus 1-2-3 weren't yet available, but the decision to support the 286 meant that OS/2 could load only one MS-DOS program at a time. That made its advantages more theoretical than practical for most potential users.

Windows 3

On May 22, 1990, Microsoft issued Windows 3.0. Microsoft hyped it with a big launch party in New York City, but this was not, beyond computer industry insiders, a particularly newsworthy event. Previous versions of Windows had come and gone without winning over more than a tiny fraction of the MS-DOS user base. Macintosh users crowed about the superior stability, elegance, and performance of their platform. Microsoft itself was still publicly committed to OS/2, positioning Windows as a stopgap and solution for users with less powerful computers.

Instead, Windows 3.0 was a breakout hit, the product that finally shifted mainstream computer users into the age of the graphical user interface (figure 10.1). Windows was still not as elegant as the Macintosh system, but Apple charged a hefty premium. Someone looking for a new computer could get a bigger hard drive, larger screen, and more memory by choosing a Windows computer. Windows worked well enough to get work done with a growing number of powerful application programs that closely resembled their Macintosh counterparts. In 1991, those already included Microsoft's own Word for Windows and Excel, Corel Draw! (a graphics program similar to Macintosh favorite Adobe Illustrator), and Aldus PageMaker. Professional magazine publishers and graphical artists looking for high-fidelity reproduction stuck with the Macintosh, but office workers and home users with ordinary computers were now able to enjoy the key

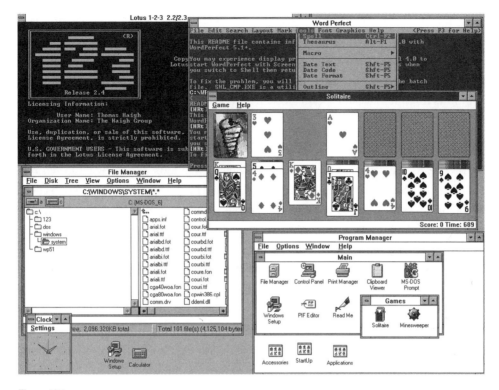

Figure 10.1

Windows 3 could multitask DOS applications such as Lotus 1-2-3 and Word Perfect, but the bundled Solitaire game was a threat to office productivity. Its interface split the capabilities of the Macintosh desktop between the Program Manager (used to launch programs) and the File Manager (to manipulate files via directory structures). Minimized windows appeared as icons on the desktop. Unlike the Macintosh, Windows placed the pull-down menus for each application at the top of its window, not at the top of the screen. In 1990, the 1024 × 768 resolution shown here was usable only with an expensive monitor connected to a recent graphics card.

benefits of graphical computing: mixing text and graphics in documents, cutting and pasting between applications, and seeing screen previews that matched printed output.

More importantly, a PC running Windows was still a PC. Thousands of DOS programs handled every possible need, including hobbyist packages such as astrological table generators, niches such as IBM terminal emulation, and untold numbers of applications written or customized for particular organizations. It would be years until Windows alternatives would be available for most of these, but the existing DOS code would usually run fine under Windows and could sometimes be smoothly multitasked to exploit the new capabilities of 386 processors. Aggressively coded MS-DOS programs that wouldn't run at all inside Windows could still be used, by exiting Windows to reach the standard DOS command line. Many users continued to spend most of their time in DOS, starting up Windows only to use a graphically intensive program or, perhaps more frequently, for a round of the perplexingly addictive card game Solitaire during the lunch break. For the first few months after its launch, Solitaire was the most commonly glimpsed native Windows 3 application. It familiarized users with the Windows user interface and persuaded them to keep Windows installed.

Windows Begins to Dominate

By mid-1992, Microsoft had sold more than ten million copies of Windows 3.[5] Spurred by this success, Microsoft issued a series of updates to Windows 3 and new DOS versions optimized for Windows. Together, these updates fixed bugs, smoothed some rough edges in the user interface, added networking capabilities and resizable fonts for screen and printer, and greatly enhanced multimedia capabilities. They also significantly improved performance; for example, Windows began to bypass DOS entirely when accessing disk drives. Combined with the flood of Windows applications, this fine-tuning made Windows increasingly attractive to large businesses. Within a few years, most corporate desktop computers were running Windows, and computer manufacturers were preinstalling both DOS and Windows on most new machines.

The success of Windows ended the already strained collaboration of IBM and Microsoft on OS/2. IBM poured a fortune into developing version 2.0 of OS/2, which shipped in 1992. Recognizing that most software had not yet been produced in OS/2 versions, IBM added support for multitasking DOS programs and for running Windows programs. Its marketing slogan was "A better DOS than DOS, and a better Windows than Windows." But neither this nor a push, in 1994, to boost performance with the OS/2 Warp release, came close to slowing the rise of Windows.

Microsoft aggressively challenged software firms who developed add-on products that boosted DOS capabilities, by incorporating those capabilities into new versions of MS-DOS or Windows. For example, in 1990 Stac introduced Stacker, which effectively doubled hard disk capacities by compressing their content. Windows took up a lot more hard drive space than DOS had, and Stacker was a well-timed hit. In 1993, Microsoft started giving away the same technology in MS-DOS 6.0, destroying the market for Stacker. Microsoft lost a patent infringement lawsuit and finished up paying royalties to Stac, but succeeded in preventing the emergence of a new software market.

As Windows grew in power and popularity, its reliance on MS-DOS started to seem more a handicap than an advantage. For example, someone installing a sound card and CD-ROM drive had to manually move tiny plastic jumpers around on the sound card to set the IRQ and memory range that it should use. The first few settings tried would probably leave the card mute, cause some other piece of hardware to stop working, or make the computer completely inoperable. It was then necessary to configure both DOS and Windows to use the new hardware, a complex process involving driver disks and configuration utilities. Making the CD-ROM drive work meant editing the CONFIG. SYS and AUTOEXEC.BAT files to insert apparent gibberish like "DEVICEHIGH=C:\ DOS\aspicd.sys /d:mscd0000" and "MSCDEX /D:mscd0000 /l:k."

Other limitations of Windows 3 were inherited from decisions made in the first version of Windows. The most constraining was a decision to store information on the user interface components of every currently running program in a single 64 KB memory segment. Users of computers with just a megabyte or two of RAM would never open enough programs for this to be a problem, but as the multitasking capabilities of Windows pushed users into paying for more memory, they were frustrated when their computer announced that it had run out of "system resources" and then refused to open more programs while most of its memory sat unused.

Microsoft Office

In its early years, Microsoft offered a wide range of products from programming languages such as Fortran, COBOL, and Pascal to a handful of games. However, its profits came mostly from two products: Microsoft's early success was built on BASIC, and by the mid-1980s most revenues came from DOS. Its spreadsheet package for MS-DOS sold poorly in comparison with competitors such as Lotus 1-2-3. Word, its DOS word processor, had a solid foundation and a clean user interface designed by Charles Simonyi, who had earlier produced the Bravo editor at PARC, but it posed little serious competition to WordPerfect.

The shift to Windows changed all that. By 1991, over 50 percent of Microsoft's revenues came from applications. Microsoft had produced capable Windows versions of its applications even before the launch of Windows 3.0, but with the exception of the Excel spreadsheet, these were little used. When Windows suddenly took off, Word (figure 10.2) and Excel began to sell in huge numbers. Competitors took years to produce comparably polished Windows software.

Microsoft's ability to field a powerful, mature, and easy-to-use set of office applications for Windows owed a lot to its earlier experiences with the Macintosh. Although Bill Gates and Steve Jobs are typically remembered as mortal enemies, their companies often worked together. When the Mac appeared in 1984, it had almost no applications software. For word processing, Apple offered only MacWrite, which showed off the Mac's graphical interface but lagged far behind the word processing capabilities of leading PC packages. It could not, for example, handle documents of more than ten pages.[6] Microsoft originally designed both Excel and the graphical version of Word for the Macintosh. Both were first-rate packages, giving Microsoft a significant head start when it was time to produce Windows applications.

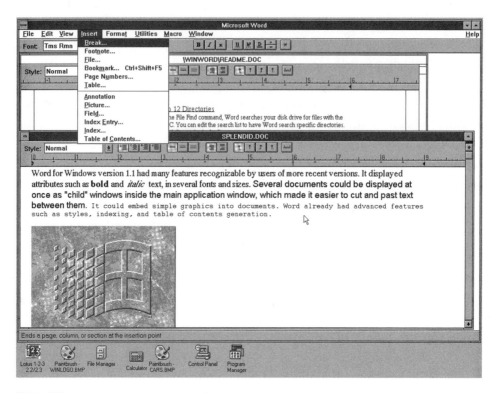

Figure 10.2

Word for Windows, based on a long-established Macintosh package, dominated the growing market for Windows-based word processing software created by the abrupt success of Windows 3.0. The broad resemblance of Windows to the Macintosh interface is clear here, including the rows of control icons, the pull-down menus, and the scroll bar.

Windows brought compelling advantages to office software users. Its more consistent user interface made it easier to pick up a new program. It was better at moving data or pictures between programs, for example, creating a graph in a spreadsheet and publishing it in a report. In DOS that meant opening one program, exporting the data, closing that program, and launching a different one. Windows let users run programs side by side, cutting from one and pasting into the other. Its WYSIWYG approach made it easy to manipulate graphics, fonts, and other formatting on-screen rather than messing with cryptic control codes and repeatedly printing test copies. This approach worked better than earlier efforts to produce "integrated" MS-DOS office software that combined spreadsheets, database, and word processing capabilities into a single program. Lotus offered such a package called Symphony for the IBM PC and was working on one for the Mac called Jazz. At Ashton-Tate, a Xerox PARC alumnus named Robert Carr developed a package called Framework. Neither was widely used. To satisfy the needs of home users, Microsoft and Apple also produced simple integrated suites, both called Works.

Microsoft had been experimenting since 1989 with bundles of its core Macintosh application programs Word, Mail, Excel (spreadsheet), and PowerPoint (presentations). By 1992 the bundle had been extended to Windows, branded the Microsoft Office Suite, and aggressively repriced to sell for around $500. That was about the price of a single program from a competitor, and thus someone who preferred Excel to Lotus 1-2-3 was likely to pick up the whole Office bundle, also depriving WordPerfect of a possible sale.[7] Large firms began to standardize on Office. Mike Maples, leader of the firm's application team, boasted that "My job is to get a fair share of the software application market, and to me that's 100 percent."[8] Later releases of Office, most notably Office 95, improved integration and offered a variety of other applications that came and went with different editions and versions of the bundle.

PowerPoint was initially the weakest of the Office products, based on a Macintosh application Microsoft had purchased in 1987. It therefore benefited particularly from the bundling strategy—why spend hundreds of dollars on a superior rival after having received a largely adequate package as free add-on? As PowerPoint became a ubiquitous vehicle for presentations, first at corporate meetings and eventually in elementary school classrooms, its influence on human thought and interaction has frequently been criticized. Google's Peter Norvig suggested that its use "makes it harder to have an open exchange between presenter and audience, to convey ideas that do not neatly fit into outline format, or to have a truly inspiring presentation." He illustrated the latter point with a parodic PowerPoint version of Abraham Lincoln's Gettysburg Address.[9]

As Microsoft expanded its range of Windows software, it eventually rolled out a rival to every notably successful application produced by a rival company, in areas such as small business accounting, drawing, and desktop publishing. Not all of these succeeded, but its rapid growth and the ubiquity of its products made Microsoft one of the most prominent American companies of the era. Bill Gates became a symbol of ruthless business competition and of the youthful obsessive arrogance of computing culture. Unix-loving computer science professors and design-oriented Apple fans looked down on DOS and Windows, but the financial results were hard to ignore. Microsoft's profits, and share price, rose year after year to create cohort after cohort of option-holding millionaires.

In 1994 Douglas Copeland, a young novelist whose reputation lay as an observer of the latest cultural trends, published what became the opening of his novel *Microserfs* in *Wired Magazine*. His memorable portrait described Microsoft as a monoculture of alienated Gates-fearing nerds living in rented houses, obsessing over the pop culture of the recent past, driving Mazda Miatas, working enormously hard, and attending "vesting parties" as colleagues earned their enormous payouts. "There's this eerie, science-fiction lack of anyone who doesn't look exactly 31.2," mused Copeland's protagonist. "It's oppressive.

It seems like only last week the entire Campus went through Gap ribbed-T mania together—and now they're all shopping for the same 3bdr/2bth dove-gray condo in Kirkland."[10] Accurate or not, the book testified to popular fascination with the firm's culture.

Windows rolled over IBM and Digital Research among the operating system suppliers, while Office challenged once dominant PC application providers such as Lotus, Ashton-Tate, and Word Perfect. By the mid-1990s, opposition to Microsoft had consolidated in two rival office suites: IBM bought Lotus and added some lesser-known packages to round out its bundle, and graphics software specialist Corel anchored a suite with WordPerfect. Neither could stem the steady migration of users to Microsoft. IBM made its last major release in 1999, although Corel has continued to deliver updates to the dwindling band of die-hard users of WordPerfect.

New Development Technologies

Changing technology had always challenged programmers to learn new technologies and approaches. During the 1990s, driven in large part by the transition to Microsoft Windows as the industry's dominant platform, the succession of new development tools and technologies was particularly rapid. In 1998, software developer Ellen Ullman wrote that in the two decades since taking her first programming job

> I have taught myself six higher level programming languages, three assemblers, two data-retrieval languages, eight job-processing languages, seventeen scripting languages, ten types of macros, two object-definition languages, sixty-eight programming-library interfaces, five varieties of networks and eight operating environments. . . . Given the rate of change in computing, anyone who's been around for a while could probably make a list like this. . . . UNIX programmers used to scoff at COBOL drones stuck year by year in the wasteland of corporate mainframes. Then, just last year, UNIX became old-fashioned. Windows NT is now the new environment, and it's time to move on again. Don't get comfortable, don't get too attached, don't get married.[11]

The shift toward Windows was associated with a transition among professional programmers toward object-oriented languages. By the mid-1990s object orientation had grown from its roots in Smalltalk into a powerful marketing slogan applied to the design of operating systems and their user interfaces, to database management systems, and to the partitioning of internal corporate applications into clusters of cooperating processes communicating over networks.

Behind the hype, change was more gradual. The most visible shift was that C++ replaced C as the preferred language for heavy-duty personal computer software development. Software development gradually shifted toward application frameworks and libraries accessed as objects. By the early 1990s, C++ was edging out Pascal as the most common language for computer science instruction in universities.

C++ was powerful, but like C it was baffling to novices and unforgiving to debug. BASIC remained a standard part of MS-DOS; in fact, Microsoft gave it a major upgrade with DOS 5 in 1991. Despite being the most the widely used programming language of the era, it was never favored by computer scientists. Even during the heyday of BASIC, commercial developers of microcomputer software had usually turned to assembly language, or later to C, to squeeze more speed and capabilities out of the little machines. A language less suitable for the development of Windows software was hard to imagine.

Yet it was thanks to Windows that BASIC enjoyed a surprising second life. Microsoft released Visual Basic in 1991. It became the COBOL of the 1990s, widely used inside businesses to develop custom applications. That success owed little to the inherent merits of BASIC, and a great deal to the convenience of its development environment, which was based on a Windows programming system that Alan Cooper, a freelance developer, had sold to Microsoft back in 1988.[12] Developers constructed applications by dragging interface elements such as scroll bars, control buttons, and database fields from a pallet to the places on screen where they were needed. Once the interface looked right, developers wrote snippets of code to specify what would happen when a button was clicked or the value in a text box updated. An object-oriented framework gave programmers the ability to programmatically modify the visual elements and control other aspects of the application such as pull-down menus. An integrated development environment made it easy to debug the program as it ran.

Microsoft even built Visual Basic into its core Office applications. That gave every user of Office access to a programming system, and multiplied the power of developers. For example, a database application written in Microsoft's Access database program could incorporate code to export data, launch Word, invoke Word's mail merge function to insert the data into a form letter, and then fax, email, or print the letters.

The increasing complexity of personal computer systems opened up new security issues. Computer *viruses* are self-replicating pieces of code that attach themselves to *infect* other files. (Self-contained malicious programs spread over networks are technically *worms*, although they too were often called viruses.) The first viruses to affect PCs were created in the mid-1980s. They spread by infecting the code run when a computer was booted from a floppy disk. This created a market for antivirus software to scan for infections. By building programming facilities into its Office applications, Microsoft opened up new vectors for infection, as code hidden in a Word document could take command of a computer to copy or erase files and to infect other documents. Once computers were connected to the Internet, viruses began to automatically dispatch infected emails. Some spread rapidly to infect millions of computers.

Microsoft's combination of operating systems, programming tools, development environments, and object frameworks could be powerful if, as Microsoft wanted, developers committed entirely to Microsoft tools, operating systems, and database management systems. Its tools were eventually bundled and integrated as the Visual Studio suite. They worked smoothly together, and Microsoft supported developers well. As Ullman described it, "an inundation of CD-ROMs continues. Quarterly, seasonally, monthly, whatever . . . UPS shows up at my door with a new stack of disks. 1234567890-New versions of operating systems, database software, developer libraries, development tools, device driver kits—everything you need to know to keep pace with Microsoft."[13]

Relational Database Management Systems

By the mid-1990s custom PC applications were being created using a new class of database management systems, based not on the simple dBASE technology inherited from the days of CP/M microcomputers but on complex minicomputer systems built in computing research laboratories. The leading database management systems of the 1970s and early 1980s were run on mainframes and designed for very high throughput for routine processes during an era of extremely high hardware costs. As we explained previously, most used what was later called the *network model* of data organization, also known as the CODASYL approach.

By the late 1980s, excitement in the software industry had shifted to a new kind of database management system based on the *relational model* of data organization. The relational approach was defined by IBM researcher Edgar F. Codd in his 1970 paper "A Relational Model of Data for Large Shared Databanks."[14] Codd won a Turing award for this model, which organized data in a simple and conceptually elegant way. Its key advantage was flexibility: the connections between different data tables (for example, between customers, orders, and products) were specified each time data was retrieved, rather than specified when the tables were created and hard coded into the way the data was stored on disk. Its primary disadvantage was implementational complexity: figuring out an efficient way to execute any query the user happened to enter was far more complicated than following fixed paths.

Two research prototypes proved the relational approach was viable. IBM's System R, on a large mainframe, was a feat of software engineering. Its field tests in 1978–9 made a strong case for the new approach.[15] Because of the huge investment IBM and its customers had made in existing systems, it initially marketed the commercial version, SQL/DSS (DSS stood for decision support system) to use for analytical querying rather than routine processes. Only in the 1990s, when it was rebranded DB/2 (data base 2) did IBM position relational technology as a replacement for its earlier mainframe database software.[16]

The other early implementation had a greater initial impact. CODASYL-style systems had not been as widely used on DEC machines or other Unix-based minicomputers.

Michael Stonebraker of the University of California at Berkeley led the development during the mid-1970s of a robust Unix-based relational database management system called INGRES. Much of the development work was done by graduate students, who took their experience (and in many cases, the code itself) to a cluster of start-up companies producing database management systems. One was Sybase, whose system was eventually licensed by Microsoft as SQL Server, for use with Windows NT.[17] Another firm, Relational Technology, was founded in 1980 by Stonebraker himself to commercialize INGRES.[18] Oracle started a few years later, but was leading the market by the end of the 1980s. That owed a lot to its aggressive sales culture, set by its founder Larry Ellison.[19] By the end of the century it had marginalized other producers of commercial database management systems and was entrenched, along with fellow software producers Microsoft and SAP, as one of the world's largest and most valuable companies.

Client Server Applications

Perhaps the most important milestone in Oracle's history was the release in 1992 of version 7.0 of its database management system. It capitalized on booming interest in client server applications.[20] Few buzzwords were harder to escape in the computer industry of the late 1990s than *client server*, whose origins at Xerox PARC were discussed in the previous chapter. The newly acquired graphical user interface and networking capabilities of Windows PCs made them the obvious interface points for applications built around corporate databases.

In the business computing world, client server development was associated particularly with custom corporate applications running on personal computers that stored their data in a relational database management system running on a server. This combined the best features of personal computing and traditional timesharing systems. Windows applications were easier to use and more convenient than grappling with a terminal emulator to log into a mainframe or minicomputer to use a text-based application. Building the application around a database management system, rather than file storage, made it easier to back up and secure the data and to let applications pool their data. It also allows developers to code rules into the DBMS to validate changes to stored data.

Making client server work in practice required coaxing immature and rapidly changing technologies to function together. Ullman described a typical moment in IT work—trying to configure a system. Three people, all former employees of the company that produced it, were stumped for hours by its refusal to work as specified.

> For a moment: guilt. Three meters running by the hour: experts are expensive. We should know this, I thought. I imagined other, better experts who could do the job more quickly. But the guilt passed. I'd been doing this long enough to know that, pitiful as we were, we were it. The three of us struggling through the running of this database on a new operating system,

flipping through incomplete documentation, using trial-and-error settings of options—this was what passed for expertise in our business.[21]

Companies had computerized themselves over decades, one function or department at a time. The result was a messy patchwork of systems and technologies, some produced internally and others purchased from software companies. Getting anything done meant writing programs to transfer data from one system to another. According to a 1996 article, "The Integration Nightmare," "Those links are usually cobbled together ad hoc in response to the needs of the moment—the chewing gum and bailing wire approach to systems architecture," yielding "a tangled web of customized connections that must be torn apart and rebuilt each time the system is changed." An estimated "35 to 40 percent" of programmer time in corporate IT departments was spent keeping data in files and databases consistent. This could deter companies from changing the way they did business, upgrading their systems, or exploring new opportunities.[22]

Enterprise software packages could replace this mess with a single, pre-integrated system with modules for almost any conceivable need. SAP, the integrated administrative software company we discussed in chapter 3, took the lion's share of this market. By the 1990s, more and more large companies were throwing out their custom-written or single-purpose packages for business processes like payroll, accounting, and production scheduling. The IT services industry boomed as firms signed contracts for tens or hundreds of millions of dollars to handle the massive conversion efforts.

This growth was made possible by a technological shift from mainframes to PC networks. Rather than adopting the original centralized mainframe model, these new customers were flocking to SAP R/3 (introduced in 1992), which used a *three tier* model. In the most common configuration, users worked with Windows client programs running on a desktop computer. This communicated with the SAP *business logic*, typically hosted on a powerful Unix server. The third tier was the database management software, typically Oracle, used to hold the massive integrated database that underpinned the firm's operations.

INVENTING THE WINTEL PC

The PC ecosystem was bigger than any single company. When IBM abandoned the PC for the incompatible PS/2 line, it was left without a single dominant actor. Although earlier products such as the Hercules graphics card and Ad Lib sound card had extended the PC platform without IBM's involvement, they did not challenge IBM's control of its core elements—standards for the processor, motherboard, and expansion slots.

The PC Standard after IBM

The norm for the PC industry during the late 1980s and 1990s was to make rapid progress in tiny increments. If a Korean firm offered a new video card 20 percent faster and 10 percent cheaper than the Taiwanese model that a PC maker was currently screwing into its machines, then there was no reason not to order a few thousand copies to cover the next couple of months of production. Both cards would be compatible with the standard set by IBM, and most customers would never notice the difference.

The arrival of Windows exposed the limitations of this approach. Even on the most powerful PC money could buy in 1990, Windows 3.0 was slow, prone to crashing, and generally clunky. But its rapid adoption spurred fundamental changes to PC hardware and software, which remade both the PC and the PC industry during the mid-1990s.

The extreme decentralization that made it easy to adopt improved components made it much harder to direct the platform as a whole. The PC wasn't defined by any of its components. It was defined by the connectors that joined them together. The only way that all those parts could fit together is if each followed the design of the original PC AT. Improving the PC in more fundamental ways required firms to work together to agree on new connectors.

There was a precedent for this, in a collaborative effort to expand the maximum memory capacity of PC without the involvement of IBM. Spreadsheet users were building larger models and running out of RAM to hold them. New, higher capacity RAM chips made it feasible to squeeze more memory into a PC. But the PC's original design imposed a 640 KB memory limit. In 1985 Lotus, Intel, and Microsoft had jointly announced a joint standard for *expanded memory* so that hardware suppliers could produce higher capacity add-in boards. It relied on a paging technology, switching blocks from the larger memory pool into the address range accessible by the processor.

After IBM's shift to PS/2 orphaned the PC, these collaborative efforts became the only way to advance its architecture. The IBM PC AT used a complex hard disk controller. This set a de facto standard (known as ST-506), and many firms began to produce better and cheaper compatible drives and controllers. However, by the end of the 1980s, PC clones were shifting to the new integrated drive electronics (IDE) connector, which put the control electronics on the drive itself. This lowered costs, improved performance, and supported much higher storage capacities. Drive manufacturers cooperated with motherboard producers (and the producers of chipset and BIOS components who supplied them), computer assemblers, and of course Microsoft to incorporate the necessary changes.

The biggest challenge facing PC producers was IBM's new micro channel architecture. Keeping the standard PC architecture alive would eventually require them to match its capabilities. The first effort to do this, the EISA consortium assembled by

Compaq, signaled the new dynamics of the PC industry. The group designed an improved but backwardly compatible version of the AT expansion bus (which it also standardized) to be called the enhanced industry standard architecture (EISA). EISA remained rare, used mostly to connect high-performance disk drives and network cards to servers, but it signaled the determination of the PC industry to evolve its platform independently of IBM.[23]

Another industry consortium, the Video Electronics Standards Association (VESA), was set up in 1989. Dozens of companies were producing low-priced cards almost perfectly compatible with IBM's own VGA graphics chips, which offered resolutions of 640×480 pixels in sixteen colors. Banding together, they agreed on standards for higher resolution modes with more colors, including the widely used super VGA (SVGA) mode relied upon by many users of Windows 3. Redefining VGA as an open standard rather than the imitation of an obsolete IBM design allowed the creators of monitors and graphics cards markers to greatly improve their resolutions during the 1990s. The PC standard abandoned by IBM was beginning to look like something with an independent future. No single company could hope to dictate changes, but ad hoc groups were beginning to take charge of different parts of the platform and agree on standards for their connections. Today, VESA is best known for another important act of standardization: fixing the mounting holes on the back of flat screen televisions.

Other weaknesses exposed by Windows required more fundamental changes. One problem was the woefully slow rate at which existing graphics hardware could redraw the screen. The modular nature of the PC was a curse as well as a blessing. Imagine someone who purchased a fast new PC to run Windows 3.0 when it was launched. She could recalculate complex spreadsheets in the blink of an eye, but would be frustrated to see Word for Windows scroll through a document so slowly that the screen visibly rippled from top to bottom. With each click of the scroll arrow, the processor had to funnel the entire contents of the window back into the graphics card though a low-capacity connector designed in 1984 for the PC AT.

Graphics board producers made better Windows performance their top priority. In 1991, the 911 graphics chip, the first widely used hardware designed specifically as a windows *accelerator*, was announced by fledgling hardware firm S3. A specialized processor on the graphics card handled the grunt work required to draw straight lines or scroll a window, reducing the amount of data being pushed through the connector bottleneck. The agonizingly slow graphics performance of Windows also gave the industry a strong motivation to remove the bottleneck itself. That took coordinated changes by companies including the makers of motherboards, of expansion cards, and of the chipsets that controlled the slots. The VESA consortium followed up its work on VGA with a faster connector for graphics cards, which improved performance spectacularly. It matched the height and spacing of the old slots and therefore worked with existing cases.[24]

Intel Takes Charge

Of all the features inherited from the original PC AT, the hardest to change was something that would have been trivial to modify if a single company had controlled the PC standard: the case design. The case and motherboard of an early 1990s PC were produced by different companies. A decade later, this design was making PCs look clunky, as much more powerful workstation rivals shrank into fashionable "pizza box" cases. The motherboard couldn't shrink too much as it had to reach the mounting holes used by the original IBM board and align its expansion slots and keyboard connector with the cutouts in the case. That made the case wide. Expansion cards followed the dimensions set by the original PC in 1981. That made the case high. The power supply occupied the right rear corner, as in the IBM original, with an inconveniently located switch protruding through another standard hole. Even the mini-tower format that became popular in the early 1990s just shortened the traditional desktop case and turned it sideways.

The constraints were more than just aesthetic. They hurt the ability of PC producers to exploit new chip technology. Graphics, network, and sound controllers were rarely integrated onto PC motherboards, as they were on Macintosh and workstation computers. Why? Not because of any technological limitation, but simply because standard cases provided no holes for additional motherboard-mounted connectors to poke through. A hole cannot be emulated for backward compatibility or included as a different mode. It is either there or it is not.

In the 1980s Intel had left other firms to design the motherboard used with its processors, and to create all-important support chips (chipsets) that handled tasks such as interfacing with memory chips and expansion cards. This changed with the launch of its Pentium processor in 1993 (to which we return subsequently in this chapter). The new processor was paired with high-speed expansion slots, PCI, also designed by Intel. PCI was first used for graphics cards, which most urgently required the additional bandwidth, but as its use spread to other components such as network and sound cards, it gradually replaced the old-style slots inherited from the PC AT.

Intel began to sell entire motherboards, leveraging its increasing control of PC designs to become the leading supplier of chipsets to other motherboard producers. By 1996, it had the clout to propose a new standard format and have the producers of motherboards and cases go along. Intel's introduction of the ATX motherboard format in 1996 marked the disappearance of the last significant element of the PC standard to be derived directly from an actual IBM machine. ATX revised the case design, power supply connections, and system board layout design. It added a large case opening, with room for lots of new sockets on the motherboard itself. Sound and networking capabilities could then be provided by the Intel chipsets, eliminating the need for separate cards.

In 1998, Intel tried selling graphics cards. That failed, but the following year it began building graphics capabilities into the chipsets it was selling to motherboard makers. This worked well enough to kill the market for separate graphics cards on most laptops and on cheaper desktop computers. Customers paid less for their PCs, but more of the money went to Intel.

By the end of the decade, Intel controlled the evolution of the PC hardware platform almost as completely as IBM had controlled it in the mid-1980s. Intel used its new dominance to speed the adoption of some new technologies, such as the universal serial bus (USB), by building them into its chipsets. USB was a boon to computer users, replacing custom connectors and controllers for peripherals such as scanners, printers, keyboards, mice, and external disk drives with a single compact and flexible socket. Intel used the same power to derail the adoption of other technologies, such as high-speed IEEE 1394 (FireWire) peripheral connections.

Graphical Laptops

The transition to Windows was challenging for laptop users. A Windows machine needed more memory, more hard disk storage, and responsive high-resolution color screens. In 1990 those requirements strained most of the available desktop computers and overwhelmed laptops that had been perfectly usable with DOS.

Within a few years, however, hardware had jumped forward to provide the full power of a capable Windows computer in a briefcase-friendly seven-pound package. That was made possible by greater miniaturization and integration. Integrating ever more functions onto a handful of support chips was helpful for desktop computers, which could be produced with smaller motherboards and fewer expansion slots. For laptops it was essential to be small, light, and power efficient. Laptops could accommodate just one or two small, credit card-sized add-ons, such as Ethernet adapters. Features like powerful graphics, sound, networking ports, Bluetooth wireless, and infrared data transfer all had to be integrated onto a motherboard far smaller than that of a desktop computer.

By 1996, for example, consulting and accounting firm Price Waterhouse was issuing its road-weary employees Toshiba's Satellite Pro 420CDT models boasting an 11-inch color screen with an 800×600 resolution, integrated sound and CD-ROM drive, a built-in pointing stick to substitute for a mouse, a lithium-ion battery able to power three hours of use, and a full array of ports to connect full-sized screens, keyboards, or printers when the consultants were back in the office. Apple's line of PowerBooks (figure 10.3) packed the Macintosh into a similarly attractive format, complete with a trackball underneath the keyboard.

Figure 10.3

Apple's PowerBooks, offered from 1991, were the first laptops to provide a practical mobile experience using a graphical user interface. They captured a large part of the laptop market. Note the mechanical trackball used as a mobile alternative to the mouse. The PowerBook 180c shown here, from 1993, was a top-of-the-range model with an 8.4-inch color screen. Courtesy diskdepot.co.uk, used under license Creative Commons Attribution-Share Alike 3.0 Unported.

IBM's ThinkPad line of laptops, introduced in 1992, successfully challenged Toshiba for the corporate market. Simple black boxes when closed, they opened to reveal big screens and striking splashes of color, including a red pointing stick. The most striking design element on one early model was the butterfly keyboard, which unfolded to its full width when the screen was raised. That approach was retired when bigger screens meant that the keyboard no longer determined the width of a folded laptop.

From the mid-1990s to the mid-2000s, laptops advanced rapidly but incrementally. Displays grew thinner and brighter, battery life longer, and built-in features more extensive. There were still trade-offs to be made—manufacturers typically had a slim and light range without built-in optical disks for users favoring portability, a mainstream range, and a larger, heavier, and more capable *desktop-replacement* range for those willing to carry extra weight. These changes in quantity eventually added up to a change in kind. As the premium paid for choosing a laptop over a desktop shrank and the performance gap narrowed, laptops shifted from luxury products for traveling professionals to the default purchase for home users unwilling to rearrange furniture to accommodate a bulky desktop PC.

THE RISC CHALLENGE TO WINTEL

At the start of the 1990s, workstation manufacturers hoped to soon reduce the costs of their products and broaden their appeal enough to displace PCs as the default corporate purchase. PCs could not match their sophisticated operating systems, big screens, integrated networking, and windowing systems. These features would surely become ubiquitous—but how? Would workstations get cheap and user friendly enough to take over the market for PCs? Or would PCs develop the power and stability needed to challenge workstations? Most workstations of the 1980s used processors from the Motorola 68000 series, more powerful versions of those found in personal computers like the Apple Macintosh series. By 1990, however, workstations had adopted higher performance processor architectures, designed using the reduced instruction set computer (RISC) principle.

Origins of RISC

To understand that challenge, we must move back in time to the early 1980s when the RISC design philosophy was being developed. The first commercially produced electronic computers had very simple instruction sets, so it took many instructions to accomplish anything. With each new generation, designers made individual instructions more powerful by adding more complex addressing modes and instruction formats. When each instruction accomplished more, any given program could be written in fewer instructions. This saved memory space, reduced the complexity of compilers, and made assembly language programming less tedious. DEC's VAX, designed in the mid-1970s, provides a classic example. The VAX presented its programmers with more than 250 instructions to manipulate data in almost every conceivable way. There were instructions for integers, floating-point numbers, packed decimal numbers, and character strings, operating in a variety of modes.[25] Even a simple addition could specify any combination of three register or main memory locations (two sources for the numbers to be added and one destination to store the result). It offered a floating-point instruction to evaluate a polynomial using a table of coefficients. Since the 1960s, it had been common practice for the processor to implement complex instructions with microcode. When a programmer asked the VAX to evaluate a polynomial, that triggered a long series of simpler internal steps.[26] Above all, complex instruction sets were supposed to make computers run faster, in part by reducing the number of times the computer had to fetch and decode new commands.

Those assumptions had been long accepted, but in the mid-1970s John Cocke of IBM argued that a computer using more and simpler instructions to complete a given task would outperform one with fewer and more complex instructions. Magnetic core had been replaced by semiconductor memory, which was faster and thus reduced the time cost of fetching new instructions from memory. Thomas Watson Jr. once argued that IBM needed a few "wild ducks" in its ranks—people who were not content to accept conventional wisdom. Cocke fit that description.[27] His ideas led to an experimental machine called the IBM 801, completed under the direction of George Radin in 1979.[28]

The idea behind RISC was that in an age in which code was produced by compilers and memories were large and fast, it made sense to have smarter compilers and simpler hardware. The compiler, rather than the chip itself, would handle the job of optimizing the conversion of complex instructions into simple operations. Paring the instruction set down to its essentials made the compiler's job easier and hardware optimization simpler. Chips would be equipped with many registers, into which the data to be worked on was loaded. Removing addresses from arithmetic instructions reduced the need to wait for memory accesses. The simpler, smaller chip was not just cheaper to produce, but it could also be run reliably at a higher clock speed.

Commercialization of RISC

In 1980, a group at Berkeley led by David Patterson, after hearing "rumors of the 801," started a similar project. That project originated the name RISC. Another project, called MIPS, began in 1981 at Stanford under the leadership of John Hennessy.[29]

Hennessy and Patterson became evangelists for RISC, supported by quantitative measurements proving that a RISC design could squeeze much more processing power out of a piece of silicon. The humble microprocessor, born of a pocket calculator, would soon surpass minicomputers, mainframes, and even supercomputers in performance. As they publicized their work they were met with skepticism, but before long, commercial chips were beginning to prove RISC's power.

Commercial RISC chips did more than just shrink their instruction sets. They borrowed two techniques already proven in scientific computing. *Superscalar* capabilities duplicated logic units, so that a single chip held two or more arithmetic engines working in parallel. This was much easier to accomplish with a RISC design, because each unit was so compact that several could be squeezed onto a chip. *Pipelining*, pioneered by IBM's Stretch, redesigned the processor so that the circuits used to read and recode instructions were kept busy on future instructions while the current one was executed. This too was more practical in RISC designs, because most operations took place in a fixed time interval (like tasks on a factory assembly line). Both approaches required instructions to be optimally ordered, reinforcing reliance of RISC on advanced compilers.

Patterson worked with Sun to develop the scalable processor architecture (SPARC) processor based on his research at Berkeley. Sun licensed the design to other companies with the hope of making SPARC a standard.[30] Hennessy's project also spawned a commercial venture, MIPS Computer Systems. Silicon Graphics reworked its line of workstations around MIPS chips.[31] Hewlett-Packard went with a RISC design called *precision architecture*.

Most computer industry commentators believed during the early 1990s that this new technology would inevitably supplant traditional processor designs, such as the Intel processors used in PCs. The 386 chip was well engineered and could support modern operating systems. But Intel was struggling to boost its performance. In 1985, when the 386 chip was launched, the fastest ran at 16 MHz. By 1992, Intel had barely doubled that clock speed (to 33 MHz), whereas the fastest RISC chips were running at more than 150 MHz.

When the 486 was introduced in 1989 it seemed little more than a stopgap. Cache memory had been one of the most important features of the 386 processor. The original PC AT had run its RAM chips and expansion slots at the same clock rate as the processor. Processors were speeding up faster than affordable RAM chips, yet there would be little benefit to processing an instruction faster if the computer just had to wait longer for the next instructions and data to arrive from RAM. The solution was a small high-speed *cache* memory running at the full speed of the processor, holding a copy of the areas of memory the computer was currently working with. Most memory reads came from the cache, greatly reducing the performance penalty caused by the slower main memory. Alongside the 386 processor itself, Intel sold a specially designed cache controller chip and the optional 387 coprocessor.

With the 486, it combined all three chips along with the cache memory into a single, very expensive package. This integration roughly doubled performance over the 386. The final DX2 and DX4 models of the 486 ran the chip's internals at two or three times the speed of its connection to the rest of the computer. That seemed about as far as the Intel architecture could be pushed.

Meanwhile, RISC systems were becoming cheap enough to compete directly against the most powerful PCs. Sun took a major step in 1989 with the launch of its mass-produced SPARCstation 1 workstation, housed in a compact case. Prices started around $10,000. *Byte* noted that Sun's SPARCstation provided around ten times the floating-point performance of a conventional workstation equipped with a coprocessor chip.[32] Sun even started to attract conversions of office software such as Lotus 1-2-3 and WordPerfect. PC-based workstations weren't always a better value. In December 1990, for example, *Byte* reviewed one of the first high-performance 486-based computers from NCR, priced at a staggering $31,600 as tested. The cover of the same issue, "PC-Priced Workstations," teased a story profiling budget-priced Sun-compatible SPARC machines for as little as $6,000.[33]

The biggest RISC challenge to Intel's position in the personal computer market came from the PowerPC architecture. It was a miniaturized, single-chip version of the POWER architecture that IBM had introduced on its powerful RS/6000 servers and workstations in 1990. IBM, Apple, and Motorola worked together to develop this and promote it as the next hardware standard for the computer industry. It was supposed to replace both Intel PCs and Motorola-powered Macintosh computers.

The first PowerPC processor, the 601, was ready by 1992. IBM held off trying to sell its own hardware until it had an operating system ready, leaving Apple to introduce the would-be killer of the Intel PC with its first Power Macintosh model in 1994. Apple had rewritten only the most critical parts of its aging MacOS operating system, retaining most of the old code to run slowly under emulation. This let Apple shift over to the PowerPC quickly, but it also ate up most of the performance gains. Application software producers, such as Microsoft, similarly relied on emulation to run old code rather than completely rewriting their applications. That gave Intel a crucial window in which to boost the performance of its own chips before optimized software could fully realize the benefits of PowerPC.

Microsoft's Response: Windows NT

DOS and Windows were entwined with Intel hardware. If RISC chips were going to deliver higher performance and lower costs, this was not good news for Microsoft. In 1991, Microsoft announced its next-generation operating system, and bid for long-term survival, Windows NT. Nobody yet knew which, if any, of several rival RISC designs might set any future standard. NT compartmentalized all the code tied to one or another processor design into a *hardware abstraction layer*. The rest of the operating system code could be reused for different processors. Windows NT application programs were forbidden from communicating directly with the hardware. They interacted with disks, screens, and other devices by using software interfaces to ask Windows to carry out operations. As those interfaces were standard across all versions of NT, applications would need to be recompiled for new processors but not rewritten substantially.

Windows NT had little in common with MS-DOS but a great deal in common with DEC's minicomputer operating systems. That was not a coincidence. Its development was led by Dave Cutler, one of DEC's lead operating system designers during development of the VAX in the 1970s. In the mid-1980s, he led the creation of a RISC processor at DEC. It was canceled in 1988, giving Microsoft a chance to lure away Cutler and his team. Early in its development, when Microsoft and IBM were still supposed to be cooperating, the project was known as OS/2 3.0, but with the success of Windows, the connection to OS/2 was dropped. On the surface NT looked just like Windows 3.1, and it reimplemented the software interfaces used by conventional Windows applications so that many would run without modification. Under the hood, however, it was a next-generation version of DEC's VMS operating system rather than a reworking of the shaky pile of code Microsoft had built on top of MS-DOS. Even its name suggested this: jokers noticed that incrementing each letter in VMS by one position gave W(indows) NT. (NT officially stood for *new technology*.)

To make sure that no old code or Intel-centered design assumptions crept into NT, Cutler's team developed and used it on RISC chips first and tested it on Intel chips later. On its launch in 1993, Windows NT included ready-to-run versions for Intel, MIPS, and DEC's new Alpha processor. Alpha, based in part on the chip Cutler was developing, was DEC's attempt to move beyond the VAX architecture to make up lost ground in the market. Intel's faster 486 chips lagged far behind the best RISC offerings when tested with Windows NT.

As a 32-bit operating system, NT could handle programs with very large memory needs. It supported Unix-like capabilities like robust multitasking, remote procedure calls, and a huge variety of networking protocols. Cutler was looking to beat Unix and not just match it. NT was better at managing the rights of users to access resources, keeping files safe from corruption, and being quickly and easily set up.

Microsoft programmers were required to submit their work to a central file at the end of each day, where overnight all programmers' work was compiled into a daily *build*. If your contribution caused the system to crash, you were responsible for fixing it. That build then became the basis for the next day's work.[34] Furthermore, as soon as the build became marginally functional, members of the programming team were required to use it, regardless of how inefficient that might be. This requirement made life difficult, especially when the software was in an early stage and little of it worked well, but it kept the programmers focused on shipping a finished product of high quality. This process, too, had an evocative name: "eating your own dog food."[35]

Windows NT was a solid foundation for network servers as well as personal computers. This was a new development for the PC. Novell Netware was a server-only operating system optimized for file and printer sharing. NT was much better than Novell for client server applications, for which servers had to run database software or custom code. By 1997, Microsoft offered several editions of NT, from the $4,000 Enterprise Edition (eight-processor support and features to balance work within a server cluster) all the way down to a single-processor Workstation Edition ($319 at launch). All editions actually used the same code, with features like support for very large memories disabled in the cheaper packages.

Intel Battles RISC

Intel's processors did not have the architectural elegance of the newer rivals, such as DEC's Alpha chip, but the firm did boast an excellent set of production capabilities and a development budget larger than the entire revenues of many of its competitors. It was not prepared to surrender its dominance without a fight. In 1993 it introduced the Pentium processor. This chip was given a name rather than a number, after legal rulings that numbers, such as 586, could not be protected as trademarks.

The first major redesign of the Intel core since 1985, the Pentium incorporated the pipelining and superscalar techniques common in its RISC competitors. The chip's floating-point performance was much improved, addressing Intel's biggest single weakness. It was even possible to harness two Pentium processors to run in parallel. Yet the first Pentiums were not well received. Reviewers noted that they were expensive, prone to overheating, and for many tasks provided only modest performance gains. *Byte*'s first round-up of Pentium-based PCs, costing at least $8,500 each, featured "roaring fans, special cooling hardware, and sturdy tank-like cases that *must* be kept closed, or the electronics will surely melt."[36] In 1994, a design flaw was revealed that could cause incorrect results in dividing. Intel was slow to promise free replacements, causing outrage.

Intel saved the Pentium by releasing faster versions and lowering prices far more rapidly than it had with earlier chips. That included a lower-power version that could be packaged into a small, power-efficient laptop. As a mainstream processor the Pentium was a hit, but as a workstation processor it still underwhelmed. RISC's simplicity made it possible to drive chips at a higher clock speed. In March 1995, *Byte* magazine reviewed a collection of Windows NT workstations. The Pentium ran at only 90 MHz, whereas the simpler RISC design of the Alpha zoomed along at 275 MHz. Because of this handicap, even dual Pentium workstations delivered only about half the performance of the MIPS-based machines and fell still further behind DEC's Alpha.[37]

Intel countered with the Pentium Pro, released later that year. This deployed a RISC-like core as the high-performance heart of a complex chip able to interpret existing code. It pushed Intel's engineering advantages to their limits by squeezing up to 1,024 KB of high-speed cache memory into one or two additional silicon chips placed within the same ceramic package as the processor itself. This could be accessed much more rapidly than memory on the motherboard, which greatly improved its performance with 32-bit operating systems such as Windows NT. The new chip remained an expensive option, owing to the complexity of this manufacturing process, but it put Intel back into contention for workstations and servers. Intel-based workstations still lagged Windows NT performance of Alpha-based alternatives, but the gap was clearly—and unexpectedly—narrowing.

The Pentium Pro got a public relations boost in 1996 when the ASCI Red super-computer being built at Sandia National Laboratories became the first computer to clock in above one teraflop (a thousand billion floating-point operations) per second on a standard benchmark. Whereas Cray computers had relied on exotic materials and custom construction, the new generation of supercomputers used large numbers of standard processors and other mass-market components running in parallel. The same processors now powered desktop workstations and world-beating supercomputers. Running 9,298 processors in parallel, ASCI Red retained the title *world's fastest computer* for about four years.

Intel's new architecture went mainstream with the Pentium II series, introduced in 1997. This tweaked the high-performance core of the Pentium Pro, moved the cache memory onto the same package as the rest of the processor, and improved performance with the 16-bit code still common in conventional versions of Windows. "You can't ignore the dramatic shift occurring in the PC industry now," noted *Byte*. "Traditional Intel-based Windows systems are aggressively moving into markets previously dominated by Unix workstations."[38] The Pentium II, and slightly improved Pentium III, formed the backbone of Intel's product line until 2003. As *Byte* had promised, they brought workstation-level performance to ordinary personal computers without sacrificing compatibility with standard applications and operating systems.

Windows 95

Most of those fast new Intel processors were running Microsoft's new mainstream operating system. Windows 95 ended the five-year run of the Window 3 series and marked the end of MS-DOS as a separate product. It was launched in 1995 with a marketing push estimated to cost $300 million. According to the box, Windows 95 needed only 4 MB of RAM, a third of the minimum for NT, although anyone who took this seriously was likely to grow old waiting for it to switch between programs.

As an operating system, Windows 95 was less advanced than Windows NT. As a product, its goal was even bolder: provide most of the advantages of NT to users of less powerful computers, throw in powerful multimedia features, and retain almost perfect compatibility with existing applications. Network performance was much improved, including built-in support for the Internet's TCP/IP protocol, for dial-up Internet access, and for Internet applications such as telnet and ftp. Like NT, Windows 95 offered *pre-emptive* multitasking, allowing smooth simultaneous operation of several programs and letting the user terminate an errant program without restarting the computer. From a purist's point of view, its implementation of these features left a great deal to be desired. The need to run old applications and work on less powerful computers resulted in a complex hybrid of NT-style 32-bit features with old-fashioned

16-bit Windows and residual chunks of DOS. That sacrificed stability for compatibility: an old or badly written driver for something like a network card could still crash the entire system. Microsoft was clear that Windows NT was the "industrial strength" option while Windows 95 was for consumers and businesses with shallower pockets.

The new user-interface design was more attractive, more consistent, and more powerful than its predecessors. It introduced the Start menu, system tray, and icons for running programs that have adorned the edge of the screen ever since. Millions of dollars were spent just to license part of the song "Start Me Up" from the Rolling Stones for television ads promoting the new system. Skeptics quibbled that only Microsoft could make an operating system in which you used the Start button to shut down the computer. Windows 95 combined an almost Unix-like power with a decidedly non-Unix-like commitment to ease automatic self-configuration. New *plug and play* capabilities attempted to automatically recognize new hardware and install the appropriate drivers. At first this was rather hit and miss. After the redesign of motherboards and expansion hardware to accommodate automatic configuration, it eventually began to work as promised. Windows 95 also improved support for laptops.

Microsoft took just four days to sell its first million upgrade packages, many of them to eager customers lining up at midnight launch events.[39] Within five weeks, it had sold seven million. Microsoft's true dominance, however, came from its lock on the operating systems pre-installed on new computers. Windows 95 was powerful enough that there was little reason for most users to consider switching to IBM's OS/2, which retained some theoretical advantages but had its own quirks and limitations. IBM gave up on OS/2, leaving it as one player among many in the personal computer market. A few years later, as part of a shift from hardware to computer services, IBM withdrew from selling desktop computers. In 2004, when it disposed of its highly successful ThinkPad line of laptop computers to upstart Chinese company Lenovo, IBM left the PC business it had created 23 years earlier.[40]

With minor upgrades, Windows 95 remained the standard PC operating system for the rest of the decade. A succession of updates and improved versions kept it fresh. In April 2002, according to the Garner group, the billionth personal computer was shipped.[41] Most of those computers used a variant of Windows 95, and a large portion of the computer users in the world had no experience with other operating systems.

Intel and Microsoft Victorious

By the time Windows 95 launched, the RISC challenge to Intel was already beginning to collapse. Microsoft dropped Windows NT support for MIPS in 1996, after Silicon Graphics failed to persuade other firms to build workstations with its processors; and for PowerPC in 1997, when it became clear that other computer manufacturers were not going to follow Apple's lead. DEC's Alpha chips earned a loyal following among workstation users in the late 1990s, but the firm couldn't speed them up quickly enough to maintain a compelling speed advantage over Intel-based competitors. Compaq took over DEC in 1998. The next year its managers closed the group working to support Windows NT on Alpha chips. The RISC challenge to Intel for workstations and personal computers was officially over.

Apple's reliance on PowerPC then became a weakness rather than a strength. Even when reimplemented in PowerPC code, its MacOS operating system lacked crucial features such as robust multitasking and support for multiple processors. It was supposed to be only a stopgap. Apple's alliance with IBM also included development of a shared object-oriented operating system Taligent, but that effort quickly fell victim to corporate politics and technological challenges. Apple's promises shifted to a new project, Copeland, to produce a modern operating system able to run old programs. That plan was abandoned, too. The existing MacOS grew ever less stable as new demands were placed on its rickety foundation. Apple struggled: between 1993 and 1997 Apple fired three chief executive officers. The last of those, Gill Amelio, lasted little more than a year. His fascinating book, *On the Firing Line*, laid bare the company's ingrained dysfunction and the ineffectiveness of conventional managerial remedies in fixing it.[42] Macintosh's market share continued to shrink, from a high of 12 percent in 1991 down to 2.7 percent in 1998.

In October 1998, Microsoft announced that its forthcoming Windows NT 5.0 would support consumer hardware and software well enough to work for everyone. It renamed it Windows 2000. In the end Microsoft pushed the convergence back by one more release. In October 2001, Microsoft replaced both Windows 2000 and Windows Millennium Edition, the last of the Windows 95 series, with Windows XP, one of its most solid and longest-lived operating systems. Typing *ver* into a command window revealed that Windows XP was really NT version 5.2. Since then, Microsoft has launched many new versions of Windows, some more successful than others, without ever settling on a coherent numbering scheme. There's a joke that someone who learned to count from 1 to 10 with Microsoft would struggle in life, proceeding through 2, 286, 386, 3.0, 3.1, 3.11, 3.5, 95, 4.0, 98, ME, 2000, XP, Vista, 7, 8, and 8.2 before skipping over 9, to finally reach 10. Under the hood, though, every new version of Windows since 2001 has been an evolution of Windows NT.

Apple repeatedly failed to match Microsoft's trick of producing a robust new operating system with backward compatibility. Given that its tight control over Macintosh hardware made this job easier, this was a sad reflection on its management and engineering capabilities. Only after Steve Jobs returned to the helm of Apple in 1997 did the company finally stick with a strategy long enough to finally deliver a new operating system. OS X, released in 2001 to the enormous relief of Apple's remaining customers, rebuilt the familiar Macintosh interface on a new foundation borrowed from Jobs's failed NeXT workstations (and therefore from the BSD derivative of Unix). Coupled with Microsoft's launch of Windows XP, it completed the migration of the personal computer industry to robust and powerful minicomputer-like operating systems.

Four years later, Apple announced that all future Macintosh models would be based on standard Intel processors. Since then, the vast majority of personal computers have used Intel chips. Most of the rest use compatible chips from other firms, a challenge that has ebbed and flowed over the years. The AMD Athlon, introduced in 1999 often beat Intel's most powerful chips in performance as well as price. On this and other occasions when Intel looked vulnerable, it recaptured the initiative within a year or two, drawing on its formidable engineering and manufacturing advantages.

In August 2000, for the first time, a majority of American households reported owning a personal computer. For households earning more than $75,000 a year a computer was practically a necessity, with 88 percent ownership.[43] The 1990s had been a time of great excitement in the personal computer industry, as new architectures and operating systems were introduced with formidable alliances behind them. Yet the decade ended with Microsoft and Intel more dominant than ever. That sounds like an anticlimax, but to overcome the threat, both had remade their technologies from the inside out. Microsoft had thrown out every piece of DOS and Windows 3.0, replacing them with a minicomputer operating system so carefully that most users never noticed. Intel had put the heart of a workstation-style RISC chip inside its new processors, using it to run old code faster than ever before. In both cases the combination of high performance and backward compatibility proved unassailable.

Intel's success with desktop, laptop, and server computers was not repeated with every kind of computer. As RISC chips vanished from the desktop, they spread elsewhere. MIPS scored an early win as the supplier of processors for the Nintendo 64 consoles of the late 1990s. From 2006 to 2012, custom chips derived from IBM's PowerPC architecture powered the flagship models from all three dominant producers of video game consoles: Microsoft (for the Xbox 360), Nintendo (the Wii), and Sony (the PlayStation 3). PowerPC chips were also used widely in cars. Today's most widely used processor architecture, ARM, is a RISC design. It powers the vast majority of smartphones and tablets and now threatens Intel in the market for personal computers. We tell those stories in subsequent chapters.

11 THE COMPUTER BECOMES A UNIVERSAL MEDIA DEVICE

This is the part of our story in which defining what a *computer* is gets a lot harder. If an IBM 650, an Altair, a PDP-8, or an IBM PC materialized unexpectedly in your bedroom, you would probably shout something like, "How did that weird old computer get in here?" Each of those fits our idea of a computer: a metal box studded with lights, switches, and connectors. We've already talked about computers being embedded inside other systems. In the 1960s, minicomputers were built into missiles, space probes, lighting systems, aircraft, and even tractors. They started to shed their recognizable boxes and merge with other kinds of equipment. The process accelerated in the 1970s with the arrival of cheap microprocessors, powering pocket calculators and Atari games consoles.

From the 1980s to the early 2000s, two processes ran in parallel. On one track, the personal computer gained new capabilities. With them, it inched closer to becoming a universal media device—making telephone calls, playing and storing audio files, playing movies, storing and editing photographs, and playing games. On the other, less visible track, computers were making their way inside music players, televisions, cameras, and musical instruments. They dissolved the technologies inside but left the husk intact.

Following the trajectory sketched out by Gordon Moore, the amount of processing power, RAM, and storage that could be built onto a circuit board that was small and cheap enough to build into a consumer electronic device rapidly increased over the 1980s and 1990s. Cheap microprocessors underpinned a wave of new consumer devices, from microwave ovens to digital cameras, designed around tiny computers running fixed software. We tend not to think of these things as computers now, but their creators often marketed them as being, or containing, computers. In his 1979 book, *The Micro Millennium*, computer scientist Christopher Evans explored the likely results of computerization on things like cars, credit cards, games, and work. He promised that "at home there will be speaking bathroom scales, freezers that tell you to restock them, cookers that tell you how the meat is coming along, telephones that tell you how many people have called in your absence . . ."[1] In the longer term, he predicted the "death of the printed word," the end of paper money, and a life of leisure as intelligent machines took over the responsibilities of workers. A decade later, Mark Weiser of Xerox PARC coined the term "ubiquitous computing" to describe the potential of new formats such as smartboards, ID badges, and tablet computers as an alternative to the then-dominant model of personal computing.[2]

ORIGINS OF DIGITAL MEDIA

Embedded computers were particularly important in changing the way people experienced music, films, and television. Advances like MP3 music players and high definition televisions were made possible by advances in digital signal processing. Signals may originate in a variety of forms: voice, music, video, radar signals, scientific data from ground-based instruments, signals from deep space probes, and others. In the mid-twentieth century, signals were processed with complex analog circuits, each tailored to the specifics of the incoming signals. The transition away from analog processing required advances in computing and information theory.

A circuit called an analog/digital converter samples the data and converts it into a stream of digital bits. Sampling drew on the work of Bell Labs' engineers Harry Nyquist, who published an analysis in 1928, and by Claude Shannon in papers published in the late 1940s. A theoretical breakthrough came in 1965, with the publication by James Cooley and John Tukey of a method of carrying out a Fourier transform of a signal that was much faster and thus more practical than classic methods.[3] In the words of computer scientist Allen Newell, the discovery of the fast Fourier transform (FFT) "created the field of digital signal processing and thus penetrated the major bastion of analog computation."[4] The FFT allowed one to decompose a complex signal into combinations of basic periodic frequencies, just as a musical chord played on a piano is the result of hammers hitting several strings, plus their harmonics. Once decomposed, a computer can process the signal in any number of ways. Over time, these techniques migrated from large, expensive computers, like those used to handle communications with space probes, into cheap personal computers and consumer electronics.

During the 1990s, personal computers got much better at integrating graphics and sound. By the end of the decade, a reasonably well-equipped PC could play high quality movies from DVDs, store the contents of hundreds of music CDs on its hard drive, and stream radio programs from around the world over the Internet.

Early Multimedia

The term given to these new capabilities was *multimedia*. Neither the word nor the idea was new. Artists had been combing different media for decades. The most influential group exploring the possibilities of new technology for creative purposes was the Media Lab at MIT, established in 1985. It grew not from MIT's computer science teams, but from the Architecture Machine Group, founded by Nicholas Negroponte. Negroponte had trained as an architect, but his focus was on the interaction of humans with information and media technologies. One of its early projects was the Aspen Movie Map, created in 1978–9 (see figure 11.1). It was created by driving a car fitted with movie cameras around Aspen, Colorado. Sequences were transferred to laser disk (an analog video system) controlled by a minicomputer. Users could zoom in and out of a map and travel virtually through the streets, changing viewing angles and direction of travel. Some buildings could be entered and explored. The sizable tab was picked up by ARPA, which was interested in virtual training environments.

The minicomputer wasn't yet powerful enough to store video sequences digitally—it just directed the disk to play sequences. Video players reading LaserDiscs, optical disks the size of LP records, had been on sale since 1978, but because they were more expensive than video cassette recorders and could not record broadcasts, they

Figure 11.1

Left: Andrew Lippman (project leader) and John Boren (who designed the rig) balance on top of a truck to adjust the four-way camera assembly used by a precursor of MIT's Media Lab in fall 1978 to capture images at ten foot intervals for the Aspen Movie Map. Courtesy Andrew Lippman. Right: A similar camera assemblage on a Street View car, photographed on the Google campus in 2010 by Wikimedia user Kowloonese.

never appealed to the mass market. The discs stored analog video, not digital, though later versions coupled it with digital audio.

The group also experimented with producing a digital model of the town and using it to create realistic 3D views, but this could not yet be done in real time. Relatively realistic computer graphics sequences, with solid objects, lighting, and shadows, first appeared in movies like *Tron* (1982) (see figure 11.2), *Star Trek II: The Wrath of Khan* (1982), and *The Last Starfighter* (1984). This required massive amounts of computer power. The specialist company Digital Productions needed ten seconds of time on its Cray X-MP supercomputer to render each frame and output it to film. The twenty-seven minutes of digital footage for *The Last Starfighter* took months of preparatory work and tied up the Cray for so long that Digital Productions had to temporarily abandon its lucrative trade in animated commercials.[5]

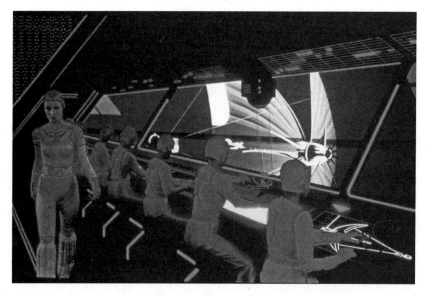

Figure 11.2
An image from the 1982 Disney movie *Tron*, which broke new ground by including more than fifteen minutes of purely computer-generated animation and, as here, by mixing filmed and generated elements in the same shots. Computer scientists Alan Kay advised Disney on the film, which required cutting-edge computer power and expertise. (Kay later married writer Bonnie MacBird, who edited her original script for *Tron* on an Alto at Xerox PARC.) The Solar Sailer visible through the window was animated by Information International, Inc., founded in 1962 by Ed Fredkin, an early adopter of DEC computers and later the director of MIT's Laboratory for Computer Science. Information International had the most advanced capabilities of the four companies hired to produce graphics for the movie. It relied on the unique Foonly F1, the most powerful PDP-10 compatible computer, custom built in the mid-1970s by former members of the Stanford AI Lab. Photo: Moviestore Collection Ltd / Alamy Stock Photo.

Computerizing Music and Speech

Generating speech and music electronically was much easier than creating realistic images. Speech consists of a small number of distinct sounds, called phonemes. English has fewer than fifty of these. The idea of producing those sounds electronically goes back to a Bell Labs patent issued in 1928. The original motivation was to compress and encrypt speech as digital data for secure transmission. Homer Dudley demonstrated a keyboard-driven speech synthesizer at the 1939 World's Fair. By the 1960s, researchers had built computerized speech synthesizers able to automatically turn text into recognizable speech. In 1978, Texas Instruments introduced a toy called the Speak & Spell which taught children to spell by pronouncing words as they keyed the letters. One Bell Labs engineer remarked that the toy generated as much excitement among electrical engineers as it did among children.[6] The $50 toy marked the emergence of digital signal processing out of the laboratory and into public consciousness. It was made possible by a new chip, the TMS5100, which miniaturized all the circuits needed to synthesize speech. The chip, and its successors, made their way into video games, computers, and other electronic devices.

Music could be synthesized using similar techniques, although the first synthesizers used complex analog electronics. They were created as laboratory projects or manufactured in small batches. Their clean, otherworldly sounds were the foundation of several 1960s hits, including the single "Telstar," inspired by the first telecommunications satellite, and the 1968 album *Switched-On Bach* by Wendy Carlos.[7] Analog equipment such as the thunderous Mellotron or sweet Hammond B-3 organ coupled with mechanical Leslie rotating speakers needed constant readjustment by electronics experts and took semitrucks and large crews to haul from one gig to another. By the 1970s, analog synthesizers were a fixture of popular progressive rock bands such as Yes, Genesis, and Emerson, Lake & Palmer, and defined the aesthetic of electronic music pioneers Kraftwerk. In 1977, Donna Summer's breathtaking "I Feel Love" brought the futuristic Moog synthesizer sound to discotheques where it inspired entire genres of electronic dance music.[8]

Digital synthesizers, built around standard chips able to turn input data into sounds, were easier to play and needed less tinkering. The earliest used a technique called FM synthesis to turn a handful of input parameters into a musical note. Its bright, crystalline sound evoked an alien world free from the entanglements of history. Cheap FM synthesizers came with many predefined *voices*, optimistically labeled with the names of musical instruments to which they sounded almost, but not quite, entirely different.

By the 1980s, digital synthesizers were cheap enough to be purchased by aspiring musicians and teenage amateurs. After its 1983 introduction, the Yamaha DX7 sold more than 160,000 in four years. It exploited a new interface, the musical instrument digital interface (MIDI), which allowed keyboards, sequencers (dedicated control computers), or regular computers to automate the control of electronic instruments. The Atari ST, one of the few computers to include MIDI sockets as standard, was a common sight on stage during the late 1980s. Synthesizers defined much of the popular music of the era, particularly in Britain where their possibilities were variously explored by the perky Human League, mopey Depeche Mode, camp Soft Cell, orientalist Japan, and wonderfully arch Pet Shop Boys. Not everyone liked the new sound—one disgruntled writer coined the term "Casio effect," after a leading producer of cheap keyboards, to describe the displacement of a technology by a convenient but inferior rival.[9]

Keyboard synthesizers were coupled with electronic drum machines, grinding out robotic rhythms that no human could hope to replicate. The iconic drum machine, the Roland TR-808, was labeled "Rhythm Composer. Computer Controlled." It combined analog drum synthesis with digital sequencing, driven by an embedded microcontroller (the computer doing the controlling), running a program burned into ROM. Microcontrollers account for most of computer technology's power as a universal technological solvent. As of this writing, they cost just a few cents each when ordered in bulk, making them cheaper and more flexible than the huge range of control mechanisms they've dissolved, from rotating cylinders in the mechanical programmers of washing machines to hardwired digital logic in burglar alarms.

The noises produced by the TR-808, particularly the bass effects, were nothing like real drums, but that ultimately made them more interesting—and more immediately recognizable—than expensive rivals better able to pass for human drummers. On

Figure 11.3

The Roland TR-808 drum machine flaunted its digital sequencing ability in the label "Rhythm Composer. Computer Controlled." (Image by Wikimedia user Brandon Daniel used under Creative Commons Attribution-Share Alike 2.0 Generic.)

"Psycho Killer," Talking Heads exploited their resemblance to gun shots to evoke a twitchy psychopath.[10] Decades after the end of its short production run, the 808 maintains a cult following, testified to by Kanye West's 2008 album *808s & Heartbreak*.

Electronic drum rhythms became the backbone for many new musical genres, including much hip-hop and the approaches subsumed under the broad label "electronic dance music." As media scholar Jack Hamilton wrote, the popular music of the late twentieth century was characterized by "the move from thinking of music in terms of harmonic progressions and conventional song structure to thinking of music in terms of *sequences*, discrete passages of sound and time to be repeated and revised *ad infinitum*." The task of musical production had, by this definition, become a kind of computer programming. "More than any other instrument of its time," argued Hamilton, "the 808 both enabled and catered to this shift."[11]

Sampling and Digital Recording

Making digital instruments sound realistic would take a different approach, based on digitally recording real instruments. Music recording and playback captures only one property: loudness. The different tones our ears hear in speech or music come from vibrations at different frequencies, from 20 hertz, that is, vibrations repeated 20 times a second, to 20,000 hertz.

The more accurately a recording medium can capture changes in volume, the more faithfully it can reproduce these vibrations. Analog media represent changes in volume with a proportional change in some other quantity. A microphone turned changes in volume into changes in resistance. Hooked up to a tape recorder, these became changes in the magnetism of the tape. The needle in a turntable followed a groove in the record. As it moved up and down, the electrical current generated by a crystal changed in proportion to the loudness of the music.

Digital media encoded audio differently. Computers had always used digital representations to read programs and data. Using bits to represent sound was conceptually simple, but practically and economically challenging. The digitization of sound had been pioneered by Bell Labs in the late 1940s, with a view to transmitting it digitally in encrypted form (see figure 11.4). Some of the first experiments in storing and manipulating audio as computer data were carried out by Thomas Stockham using the TX-0 at MIT. He went on to start a company, Soundstream, to commercialize digital recording and remastering. Its first public release in 1976 was a set of recordings by long dead opera tenor Enrico Caruso with the noise and distortions of the old acoustic recording process algorithmically corrected.[12]

Speech synthesis, including that used in the Speak & Spell described previously, was one of the first applications of digitized audio: each phoneme lasted for a fraction

Figure 11.4
A SIGSALY installation, c. 1943. The device was used to encrypt voice communications between the US and UK during World War II. The US National Security Agency claims that this was the beginning of the digital revolution in sound. Note the twin turntables, suggestive of the hip-hop decks of the 1980s.
Source: National Security Agency.

of a second, so digitizing less than a minute of audio would give all the sounds needed to produce speech. In 1982, newsreader Kenneth Kendell had his voice sampled as the basis for an optional speech synthesis upgrade for the BBC microcomputer. It consisted of two chips: a Texas Instruments synthesizer and a ROM holding his elegant phonemes and a selection of common words.

Wave table synthesizers took a library of recorded sounds, analogous to phonemes, and manipulated them as needed to create the particular notes and effects needed when a key was played. The technique was commercialized by Fairlight, whose first Computer Musical Instruments introduced in 1979 cost tens of thousands of dollars (figure 11.5). The *computer* tag was earned: each instrument included two Motorola processors, and waveforms could be edited by touching a screen with a light pen.

Sampling was not always used to mimic conventional instruments. Any sound, from the breaking of a bottle to a baby's wail, could be sampled and used as a new instrument. Some drum sounds and vocal elements from classic recordings, particularly soul music, were widely sampled and reappeared in hundreds of pieces of music. Musicians used sampling to incorporate snippets from movies or political speeches into their recordings.

By the early 2000s, digital pianos, which never need tuning, had largely displaced traditional home pianos. A digital piano has a full-size piano keyboard but makes music with wave table synthesis rather than with strings and hammers. Only the finest

Figure 11.5
The Fairlight Computer Musical Instrument, 1979, was a hybrid of an electronic organ and a computer work-station, with dual processors, a screen, two disk drives, and both computer and musical keyboards. Courtesy Blackmagic Design.

real pianos retained significant resale value; others went to the scrap heap at their own-ers' expense.

DIGITIZING MUSIC

Digitized sound took a lot of very expensive memory to store. The first sampler priced under $10,000, the E-mu Emulator released in 1981, had just 128 KB of memory to hold the sound being manipulated, which it could copy onto a floppy disk for later use.

That was not nearly big enough to hold a full song in high-quality stereo. Even a personal computer hard drive of the early 1980s would struggle to do that.

Turning sound into microscopic variations in tape magnetism and back again inevitably introduced noise and distortion. The more often an analog recording was copied during the editing process, the worse the quality became. To preserve quality when mixing music, it made sense for a professional recording studio to invest in a minicomputer-based digital editing system, with large hard drives and digital backup tapes. But digital editing remained the exception to the rule throughout the 1970s.

Compact Discs

Selling digital music recordings to the public would require a new medium that was able to store as much data as several dozen IBM PC AT hard disk drives but cheap enough to sell for a modest premium over a vinyl record. That unlikely combination of attributes defined the compact disc, the first digital media device to reach ordinary households. Each shiny plastic disc contained millions of tiny pits, encoding more than an hour of music. A laser beam scanned them, tracking a spiral outward as the disc rotated. These pits encode sound as a series of 16-bit numbers, giving more than 65,000 possible values. Its volume was sampled 44,100 times a second and so could reproduce audio frequencies up to 22,000 hertz—matching the typical limit of human hearing. Catalog data coded the position of tracks, so that listeners could skip almost instantly to a particular song.

The first CD player, a Sony model, cost around $900. Initial sales were to technology enthusiasts and hi-fi obsessives. Hi-fi enthusiasts invested huge sums in improved technology to minimize the distortions inherent in sound reproduction: turntables with better magnets and smoother motors, amplifiers with superior electronics, tapes with superior magnetic materials, and even luxury cables to run current into speakers. Compared to those, a CD player was a very cost-effective upgrade. During 1983, Americans purchased around 30,000 CD players. As *Stereophile* put it in its review of the player, "The remote-controlled model which we received will sell for $1000 when it starts appearing in stores in March '83. This is a staggering cost for your John Q Public type record buyer, but a modest outlay for any audiophile accustomed to the idea of shelling out $1000 for a cartridge to put in a $1000 arm on a $1000 turntable and feed to a $1000 preamp."[13] The reviewer marveled at its lack of background noise, noting that "every serious music listener who heard this system fell instantly in love with the sound of it."

The most novel part of the machine was its digital-to-analog conversion, handled by a set of four custom Sony chips. The operations of most CD players were controlled by microprocessors—chips that combined processors with small amounts of memory and other necessary capabilities into a single package—running control programs burned onto ROM chips.

The first disc to sell a million copies, in 1985, was *Brothers in Arms* by Dire Straits. As CD players became a fixture of home entertainment centers, their design shifted from the wood and metal trim of the 1970s to a new aesthetic of black plastic and flickering LED bars. Bold letters and decorative block diagrams appeared on stereo systems, giving them the look of props from a low-budget science fiction show. By 1988, CDs were outselling vinyl records.

Digital Encoding

Thanks to their digital nature, compact discs were promoted as immune to the scratches, dust, and erosion that gradually degraded the quality of vinyl records. In 1981, a presenter of the BBC program *Tomorrow's World* introducing them to Britain, memorably attacked one with a sharp stone to demonstrate its resilience. A breakfast television presenter spread honey on a disc and poured coffee over it. These claims were exaggerated—in fact a fingerprint or small scratch could prevent a disk from playing.

Yet CDs did fail differently from analog media, for two reasons. First, *thresholding*: the laser sensing mechanisms needs only to determine whether a pit is present or absent (leaving a *land*) in each position. Even if the pit is a little deeper or shallower than intended or the laser is slightly off frequency, the disc can nevertheless be read perfectly if each pit can be distinguished from a land. In contrast, even the most expensive turntable will introduce some distortion when turning a record's groove into electrical impulses.

The other reason is *redundancy*. There is nothing magical about digitality to protect from scratches, manufacturing defects, or dust particles. Some bits will be read incorrectly or not at all. Those errors must be compensated for. Audio CDs produced strictly according to the standard could record up to 72 minutes of digital music data. If written directly to the disk as a series of bits, that data would fill about one third of its capacity. The other two thirds provide redundancy. Music data is grouped into blocks of 24 bytes, each padded to 33 bytes using a scheme called cross-interleaved Reed-Solomon code (CIRC). The extra information allows for the detection of errors and the automatic correction of misread bits. More redundancy is added as the data is turned into patterns of pits so that it can be read more reliably: each byte (eight bits of data) is encoded as fourteen bits (plus three spacing bits). The net result is that 192 bits of music takes 588 pits and lands to store. Similar techniques are used with most other digital media and explain its general reliability.

Musicians against Digital

The branding for CDs flaunted their digital technology. Each player was required to display a logo reading "Compact Disc Digital Audio." Early discs, particularly for classical music, were often stamped with the letters "AAD," "ADD," or "DDD." The first letter specified whether the music had initially been recorded in an analog or digital format, and the second how the recording had been mixed. The assumption was that more digitality meant better music, as each analog operation would introduce hiss and distortion.[14]

Many early CDs were poorly mixed and encoded, falling far short of the theoretical limitations on their audio quality and, more subjectively, lacking qualities such as "warmth" or "depth." Neil Young, one of the most successful performers of the baby boomer generation, was particularly critical. He is sometimes credited with organizing a group called Musicians Against Digital (an apparent parody of Neil Diamond's Musicians Against Drugs). Young complained that "Shit doesn't sound right. . . . The first time, 'Hey—no hiss, wow, great. You didn't realize there was no *sound* until a little while later."[15] The clash is one of science versus romance. As Young put it, "The mind has been tricked, but the heart is sad." Higher resolution digital audio disc formats were introduced at the end of the 1990s but found few takers. Neither did Young's own compression-free digital audio player, launched in 2014.

Instead, as we discuss next, mainstream music consumption in the 2000s shifted toward highly compressed data files that sounded significantly worse than CDs. Audiophiles, who continue to spend thousands of dollars on speaker wire, were naturally appalled. Many still fetishize analog audio. The most expensive pieces of modern hi-fi equipment are based on antique technology, from minimalist turntables engineered with microscopic precision to amplifiers that eschew transistors in favor of vacuum tubes flaunted like precious stones in elaborate metal cages.

CD-ROM

The format for storing data on CDs, *CD-ROM*, took longer to commercialize than originally expected. CDs store digital data, but it wasn't enough simply to install a digital output alongside the stereo connectors. Each byte of computer data must be retrieved correctly or not at all. In contrast, music CD players facing a momentary read error made up (*interpolated*) data to fill in the gap on the basis of the sound stored before and after it. CD-ROMs needed additional levels of redundancy and error-correction data. When finally standardized in 1988, CD-ROM allowed for a maximum data storage of around 680 MB, at a time when PC hard drives could rarely store even a tenth of that much data.[16]

The technology improved fast. Early CD-ROM players were bulky external units hooked up to special interface cards. They were aimed at specialist users: schools, libraries, and corporations needing to access large databases. By the mid-1990s, internal drives were becoming standard hardware that added only a small amount to the cost of a desktop computer. By the late 1990s they had shrunk enough to be fitted inside laptop computers, too. The first CD-ROM drives spun discs at the same speed as audio CD players, which made them significantly slower than hard drives to retrieve data. There wasn't much point in spinning faster to play a symphony in ten minutes, but with data, faster retrieval was usually better. Drives reached 4X speed by 1994.

Databases and bulky reference works were an obvious use for the new disks. The Encyclopedia Britannica filled 32 hefty volumes with tiny text, weighing over a hundred thirty pounds. It was an aspirational purchase, heavily promoted by door-to-door sales people, but few families could rationally commit $1,400 and five feet of shelf space to a reference work. In 1989, Britannica launched a multimedia CD-ROM version of its less prestigious Compton's encyclopedia. The text and pictures of all 26 volumes fitted on one disc, together with hyperlinks between entries, animations, and sounds. Microsoft's rival Encarta, a flashy update of the obscure Funk & Wagnall's encyclopedia, listed for $395 on its debut in 1993 as part of the new Microsoft Home lineup of CD-ROM products. Competition was fierce, and as discs cost almost nothing to produce, unlike printed books, it quickly drove down prices. Within two years Encarta was being given away with CD-ROM drives and new PCs.

The other obvious use for all that disc space was video. Video decompression and playback was built into Windows 95, and to show it off the CD included high-quality music videos from Weezer, Sheryl Crow, and Edie Brickell. Digitized video also began to appear in video games, on both PCs and game consoles. Sega led the way with an add-on CD player for its popular Genesis console, launched in 1992 in the US. By 1995, consoles included CD-ROM drives as standard features so that games could be supplied on cheap, voluminous discs rather than expensive and limited cartridges. Captured video was used for intermissions in the game action. The epic space combat game *Wing Commander III: Heart of the Tiger*, released at the end of 1994, included more than three and a half hours of video spread over four discs. *Entertainment Weekly* said it made "as great a leap over previous CD-ROM games as Star Wars did over previous science-fiction movies," in a nod to Mark Hamill's starring role in both.[17] The other popular use for digitized video on CD was pornography.

Once CD drives became common they became the default medium for program distribution, replacing the ever larger stacks of floppy disks needed to hold complex software. Floppy disk drives were less and less used, disappearing from new computers in the early 2000s. Windows 95 was the first version to be distributed primarily on CD. Those requesting a copy on floppy disk would receive thirteen disks recorded in a special high-capacity format. High-capacity alternatives, primarily the Iomega ZIP drive, were popular in the late 1990s but were killed by newer CD drives able to write (*burn*) data onto disks as well as read it.

DIGITIZING IMAGES

Computers with bitmapped screens used numbers stored in a special memory area, the frame buffer, to drive graphics hardware that turned them into pictures on a screen. Programs changed those numbers to update the image. Cameras and scanners go the other way: turning a captured image into a collection of numbers.

Document Transmission by Fax

Today we usually transmit documents as computer files. Electronic document transmission was technically possible back in the 1980s, uploading and downloading files using corporate internal networks or public online services. But in practice, a recipient was unlikely to have the hardware, software, and skills needed to retrieve the file.

Instead, businesses adopted fax (short for *facsimile*) machines: cheap special purpose computers combining a scanner, modem, and printer. These worked like copiers, but the paper on which the copy was printed might be on the other side of the world. Fax machines took advantage of the existing telephone infrastructure and the familiarity of dialing a phone number. Most businesses ordered an extra line and left the fax machine turned on and set to answer calls. Users paid for the call, but to send a letter or sales order a fax was much faster than a postal service and much cheaper than overnight delivery with a service like FedEx. Fax was particularly important to the expansion of international trade, for example, by letting small PC assemblers order components directly from Taiwan.

There was nothing new about the electronic transmission of pictures. Since the 1920s, photojournalists had used *wire transmission*, over public telephone lines, to rush images from and to newspaper offices. Those analog machines fixed the photograph to a drum and scanned it in a spiral pattern. As historian Jonathan Coopersmith has shown, entrepreneurs had been trying just as long to turn facsimile transmission into a general-purpose method for delivering business documents. By the 1960s Xerox had a viable service, but because its analog machines were built around high-precision components, they remained too expensive to really take off.[18]

In contrast, the mass market fax machines of the 1980s were built with the same blocks as home computers and video game consoles: microprocessors, ROM chips with software burned on them, and RAM chips. These components were packaged to appear as a hybrid of a telephone (with a keypad and handset) and a copier. Most of the new machines used cheap, quiet, thermal printing mechanisms to burn images onto a roll of shiny paper.

The fax machine dialed a number and tried to negotiate with the machine that answered, sharing its first shrieks and beeps so that its user could verify that the call was not answered by an irate human. Then the noises stopped and it began dragging the document slowly past its scanner head. The machine used standard methods defined by the International Telecommunications Union to encode and transmit pages, ensuring that the receiving machine would be able to understand the transmission. The Group 3 digital coding scheme was devised in 1977 in Japan around the potential of cheap microprocessors, in what Coopersmith called "the most important event in fax history since 1843." Group 3 compressed each scanned page to transmit digitally in as little as fifteen seconds, much faster than earlier analog fax machines taking up to six minutes. When it was formally accepted in 1980, about 250,000 fax machines were in use in the United States. By 1990 there were five million.[19] In Japan, fax was even more widely used, as written Japanese was hard to represent in telegram, telex, or email but easy to transmit as an image.

As the machines were mechanically simple, their pricing depended on the costs of computer chips and the electronic scanner, which fell rapidly. Electronic innovations, competition based on standards, and economies of scale drove down the cost of a basic machine from more than $7,000 in 1983 to less than $1,000 by the end of 1985.[20]

Fax capabilities got even cheaper when they were built into computers and modems. PC fax boards cost hundreds of dollars when they first appeared in the late 1980s, but within a couple of years the chipsets purchased by modem producers included fax capabilities as a standard feature. WinFax (introduced in 1990) added a virtual fax machine as a printer option available to any Windows program. It sold well, until Microsoft made that capability part of Windows 95 as a standard feature. Corporations installed fax servers, so that an entire office could share one or two fax modems via the local area network. Once the Internet became popular, similar services were offered to the public, so that, for example, the list of recipients for a document could include both email addresses and fax numbers. Fax capabilities were also added to high-speed digital copiers which, like consumer fax machines, were computers coupled with printing and scanning capabilities.

In the early 2000s, email attachments and file uploads began to displace fax usage, but it has still not vanished completely. In 2018, the British government announced a ban on fax machines in its National Health Service, effective 2020. Not all of its nine thousand remaining machines were monitored, a problem for anyone sending urgent medical information, and the machines could not be counted on to protect patient privacy.[21]

Scanners

The same technologies that made cheap fax machines possible made their way into cheap scanners. Some models were handheld and dragged over paper like an oversized mouse, and others worked with a drum like the scanners in fax machines. Most, however, were flatbed scanners—the paper remained still on a piece of glass while the scanning mechanism moved below it.

Scanners were a popular, but expensive, part of desktop publishing operations in the late 1980s, along with Macintosh computers, PageMaker software, and laser printers. By the mid-1990s, the price of color scanners had dropped to a few hundred dollars and hard drives were big enough to hold large image collections. Scanners became popular consumer add-ons, and families began to digitize their photo collections.

Industrial grade scanners let businesses scan and destroy incoming paperwork, converting it to electronic images. Specialist scanners, fitted with devices to turn pages, were used to digitize entire library collections by groups such as the Internet Archive and Google's Books project.

The direct result of scanning is a digital picture of the document. Using software for *optical character recognition* (OCR) makes it possible to turn this into searchable or editable text. Since the 1960s, bank machines had routinely been reading account and sorting codes on checks, but only because they were printed in magnetic ink using a special typeface that exaggerated the differences between numbers.

Reading normal print in different typefaces was a much harder challenge. It was first done commercially in the mid-1970s by the reading machine invented by Raymond Kurzweil. This hooked a computer to a flatbed scanner to a speech synthesizer. The equipment was bulky and very expensive but promised visually impaired users an independence previously unavailable at any price. This was just one example of the uptake of computer technology to benefit disabled users. As historian Elizabeth R. Petrick has shown, the culture of tinkering that surrounded early personal computers included their adaptation, for example, to use alternative control methods for users unable to work with conventional keyboards. Computers could also be programmed to control other devices, such as wheelchairs.[22]

Like many other technologies first adopted by disabled users, OCR eventually proved useful to broader populations. OCR software of the late 1980s, such as OmniPage for the Apple Macintosh, remained slow and fiddly, requiring many manual corrections.[23] As computers grew more powerful, the capability became more useful, and it was eventually built into Adobe's popular Acrobat software so that scanned documents could easily be converted into searchable PDF files.

Digital Cameras

There was nothing new about the idea of representing a picture as a grid of dots. Early television cameras captured images by focusing light onto a grid (or mosaic) of light-collecting elements that was then scanned by an electron beam. In 1945, working on the *First Draft* EDVAC design, John von Neumann was fascinated by the potential of the iconoscope, an electronic tube then used in television cameras, as a storage device. Even the term *pixel*, introduced with the transition to digital images, was a contraction of *picture element*, a term used since the early days of experimental television.

In television, however, intensity of each point was transmitted as an analog value and reproduced, fuzzily, as part of a continuous line. Manipulating video with a computer required turning those intensities into numbers, the job of an exotic device called a frame grabber. This captured a single frame from a video input and turned it into a bitmap image. Frame grabbers were used for video production work and were built into specialist video manipulation hardware to create special effects. These devices were expensive, purchased mostly by video production companies to enliven music videos, advertisements, and wedding footage with titles and special effects. A related piece of hardware, the gen lock, synchronized computer displays with other video feeds so that computer-generated titles and graphics could be added.[24]

Today digital video sensors are everywhere. The crucial development was the charged coupled device (CCD), which combined a semiconductor with a light-sensitive layer. Fairchild Semiconductor began to sell a 100×100 light sensor in 1974. That provided the basis for an experimental digital camera at Kodak. When light was focused onto the sensor matrix, numbers could be read off the chip. Space missions had a particular need for tiny and reliable digital imaging technologies, creating pictures that could be beamed back to Earth. Techniques had been developed in the 1960s, originally for spy satellites, to expose film and then scan it and transmit images digitally back to earth. Being able to take high-quality digital still images directly was much simpler and faster. By 1978, a KH-11 spy satellite was using a CCD that reportedly had an 800×800 resolution.[25] The Hubble Space Telescope, launched in 1986, used a similarly sized mirror but gave much higher resolution CCD sensors a very public showcase.[26]

Back on earth, the first big market was for cheaper one-dimensional sensors able to scan a single line. Flatbed scanners and fax machines moved the scanner across the page to capture the entire image gradually. (A similar digital scanning approach had been pioneered with the photo diode cameras of the Viking Mars landers. It worked well, albeit slowly, as neither the platform nor the landscape was moving.)

CCDs were used in some of the analog camcorders of the 1980s, bulky devices that combined a video cassette recorder and a television camera into a single box. The technology made a brief consumer appearance in 1987, in the PXL-2000 PixelVision camera produced by toy company Fisher Price. It recorded highly pixelated video onto standard audio cassettes, later becoming a favorite of hipster artists.[27]

Commercializing digital cameras took longer, because many sensor elements were needed to capture an entire image at once. By the mid-1990s, higher resolution sensors and the chips and memories to deal with the large files they produced were becoming affordable. They made their way into two related kinds of product. Digital video cameras could store one hour of crisp, high-resolution footage on special tapes as 13 gigabytes of computer data. Computers fitted with a FireWire connection could extract digital video, edit it, and write the results back to the tape without any loss of quality.

The other kind of digital camera was patterned after traditional cameras. Camera manufacturers competed on *megapixels*—how many millions of pixels the sensor element handled. At the end of the 1990s, most had just one or two megapixels, capturing images that looked good on screen but would appear jagged when printed out.

Because they were optimized for still images, which took less space than video, most cameras used chip-based flash memory cards rather than tape (though some early models used floppy disks or CDs). Flash retained data when power was turned off but could be quickly and selectively overwritten. It was introduced by Toshiba in 1987, finding early applications in computers to store configuration settings for computers and to hold BIOS code in an easily updatable form. The cards used in early digital cameras could store only a few megabytes but, as with other memory chips, their capacities rose into the gigabytes as transistors shrank. Because it was very compact and power efficient, high-capacity flash memory was a crucial enabling technology for the creation of new portable devices. Semiconductor memories able to store hundreds of gigabytes ultimately replaced hard disk storage in most PCs, although this took longer than expected because magnetic disk capacities increased even faster than chip densities during the 1990s and early 2000s.

The digital cameras of the late 1990s were bulky, had small screens, and would deplete their batteries and fill their memory cards after taking just a few dozen images. Compared to the models available even a few years later, they were terrible, but the relevant comparison was with consumer film cameras. Conventional film cartridges held only 24 or 36 pictures. Seeing those pictures cost at least ten dollars and usually took three trips to a drugstore—to buy the film, to drop it off for processing, and to collect the prints. Pocket-sized cameras forced users to squint through a plastic window, giving a vague idea of what might appear in a photograph. Larger, more expensive single lens reflex cameras took better pictures and showed whether an image was in focus. Little wonder that most people took out their camera only for vacation trips and special occasions.

Even the most primitive digital cameras enabled new photographic practices. Digital cameras caught on fastest for business that needed to shoot images and use them immediately, for real estate sales, corporate newsletters, or identity cards. Their direct competition was Polaroid instant cameras, which had high running costs and mostly took small pictures. As prices dropped and picture quality improved, consumers began to buy digital cameras and to take far more pictures than ever before. Vacations were now captured with hundreds of pictures, not just one or two rolls of film. Teenagers could mimic the practices of fashion photographers by taking a few dozen shots of a friend and using the best one. Since the early 2000s, daily life has been visually recorded on a scale unmatched in earlier history, a phenomenon known as *ubiquitous photography*.[28]

Early memory cards held only a few megabytes, needing aggressive compression to hold even a dozen images. That was provided by a new image format, the JPEG (named for the Joint Photographic Experts Group). In 1991, when libjpeg, a widely used open source code module for JPEG compression, was released, it took a powerful PC to create these files. By the late 1990s, the necessary computer power could be put into a camera, although early models would be tied up for several seconds processing each image. Once the memory card was full, users moved the files onto a computer. Digital photography was another of the practices made possible by the arrival of PCs with voluminous hard drives as a standard feature of middle-class households. People who wanted to print out their photographs could still go to the drug store or purchase an affordable little color printer, but photographs were viewed more and more on screens. They were shared with friends and family by email, or by copying them onto a Zip disk or burning them onto a CD, rather than by handing over an envelope full of duplicate prints.

Screens got bigger, images sharper, battery life longer, camera bodies smaller, and sensors better. By the early 2000s, the adoption of sensors with a dozen megapixels meant that image quality was limited primarily by the quality of the camera's optics. Cameras began to use a different sensor technology, called CMOS after the chip technology on which it is based. CMOS imaging was prototyped at the Jet Propulsion Laboratory, for use on deep space probes. The new technology produced camera sensors that were cheaper, smaller, and lower powered than those based on CCDs. By 2006, a camera costing a few hundred dollars would fit in a trouser pocket, take hundreds of images without changing a battery or a memory card, and offer better image quality than any compact film-based consumer camera. Improvements under low light conditions—taking photographs at night or indoors without a flash—were particularly dramatic.

Digital Video

Cameras weren't the only piece of consumer electronics to be stealthily turned into computers on the inside. The same thing happened to televisions and video players. Since the 1950s, television pictures in the US had been transmitted as images broken into 483 scan lines arranged horizontally across the screen. At the same time, televisions were getting bigger and bigger. In the 1960s, a 25-inch color screen would have been the proud centerpiece of a living room. By the mid-1990s, screen sizes had crept up to 40 inches. People looking to share sports events with their friends were opting for even larger back projection models, weighing hundreds of pounds. Every detail was visible from the far side of a McMansion's living room—good for a Super Bowl party but jarringly fuzzy if one sat any closer.

Recorded video looked particularly bad, because VHS recorders captured only half that resolution. When digital video disc (DVD) players arrived in 1997, initially priced around a $1,000, they became the fastest-adopted consumer devices in American history. By 2003, half the homes in the United States had a DVD player, and players could be purchased for as little as $50. DVD was, in effect, the extension of CD technology to play digital video as well as audio. The discs were the same size, and DVD players could also handle CDs.

DVDs gave much better video quality than VHS tapes. Each disc held about as much data as six CDs. That wasn't nearly enough to store an uncompressed movie at resolutions of up to 720×480 pixels, so players were built around custom chips able to reconstruct a crisp and smoothly moving image from a trickle of data. Cheap players did a worse job, particularly with subtle variations in dark areas of the screen. As the players were disguised computers, discs included interactive menus and even simple games.

Discs were cheaper to manufacture than tapes. After a few years, the industry dropped its prices to encourage people to build collections as with CDs and audio tapes rather than rent movies from video stores. In 2004, VHS players abruptly vanished from the shelves of electronics retailers after sales collapsed, although production did not cease entirely until 2016.[29]

DVD drives quickly replaced CD-ROM drives in higher end personal computers. For users with large monitors this was an attractive alternative to hooking up a player to an ordinary TV. They were also attractive for video games—the seemingly vast storage of data CDs had quickly been swallowed up by video content, resulting in frequent disc swaps during play.

Digital Television

The adoption of DVDs made the poor image quality of conventional televisions painfully apparent. An international transition to high definition had been agreed upon in the 1980s, but to make it a reality the entire production chain, from filming to broadcasting, had to be remade. Broadcasting frames of up to 1,080 lines, each 1,920 pixels, required aggressive video compression. North America and Europe both adopted MPEG-2 compression, closely related to the MP3 and JPEG formats. In August 1998, the *New York Times* reported that the first true high-definition televisions would cost at least $8,000, with $2,000 of that for the computer chips needed to decompress video.[30] In practice, most early HD sets could handle only a lower resolution.

In the US, the official changeover took place on June 12, 2009, with the end of broadcasting of regular analog signals. This was mandated by the Telecommunications Act of 1996, although Congress pushed the deadline back several times after high-definition television was adopted more slowly than anticipated. To help the mostly poorer consumers who wished to watch broadcasts on older televisions, the act subsidized digital converter boxes. Their cost was a fraction of the $20 billion the government received by auctioning off the radio frequencies formerly occupied by television stations. They were snapped up by companies like Verizon Wireless and AT&T Mobility to be used for high-speed data services.

High definition images had a wider aspect ratio, modelled on movie screens. By the time HDTV started to become a reality, television display technology was also changing. Traditional televisions, including the first high definition models, blasted electrons down a tube to make the screen at the end glow. A bigger screen meant a longer tube, and hence a bigger and heavier cabinet. Sony's biggest model had a forty-inch screen and weighed more than 300 pounds.

In the 1970s, Evans had predicted that "the flat, super high definition screens of the 90's" would be "wall sized displays offering vivid and compelling realistic images."[31] Two newer technologies, both borrowed from computer displays, enabled the creation of thin, lightweight large-screen televisions. One technology, plasma, were first available in the US in 1997. The technology worked well for larger television screens and gave excellent color reproduction. By the 2010s, the other, liquid crystal displays (LCDs), began to dominate. These televisions were, essentially, giant versions of the screens used since the 1990s for laptop computers and since the early 2000s as desktop computer monitors.

The convergence of computer and television technology was complete. Televisions had the same range of digital inputs as computer monitors, displayed similar resolutions, and were built from the same technologies. In fact, televisions were themselves computers. As the cost of powerful computer chips fell, even affordable televisions began to incorporate *smart TV* features. They had USB ports to play videos and music from hard disk drives, Ethernet ports, and Wi-Fi connections to access computer networks and they let users download and run applications.

Even someone with a DVD player might revert to videotape to record television broadcasts. That task fell to another kind of stealth computer, the digital video recorder (DVR). Writing in 2000, Michael Lewis predicted that this would spell the end of traditional mass-market television advertising, as users skipped advertisements and decoupled their viewing from broadcast schedules. Each user "would create, in essence, his own private television channel, stored on a hard drive in the black box, tailored with great precision to his interests."[32] The most successful was the TiVo, introduced in 1999, a PowerPC-based computer with a hard drive of at least 14 GB to store video, a modem to retrieve program schedules, and a television tuner and MPEG-2 digitizer chip to extract video from a cable TV input. TiVos could be asked to record individual programs or entire seasons of a show. On the basis of those selections, they would guess what shows their masters might enjoy and record those, too. This led to a spate of comedy routines with the punch line "My TiVo thinks I'm gay." This combination of technophilia and fragile masculinity captured a cultural moment. TiVo's moment in the sun was brief, as services that streamed video over the Internet began to displace traditional models of broadcast television.

Analog set-top cable boxes gave way to computer-based digital replacements, sometimes with built-in Tivo-like capabilities. Analog cable TV had delivered around sixty fuzzy channels, but their replacement boxes used digital compression to provide several hundred channels in higher definition, plus video on-demand services. A media enthusiast of the early 2000s might have half a dozen powerful computers hiding in one corner of her living room: a HDTV, a DVD player, a Tivo, a digital set-top box, and a couple of games consoles.

Since the initial adoption of digital video technology, resolutions have continued to improve. Blu-ray discs were essentially higher capacity DVDs, with 50 gigabyte capacities able to play movies in full high definition without losing any image quality to compression. 4K televisions and UHD Blu-ray disks used four times the pixels of regular high-definition sets. The extra detail of 4K was visible only to people sitting close to very large televisions, although the new formats were accompanied by improvements to the encoding of color that let televisions display more vivid and realistic images.

DOWNLOADING MUSIC

One of the things that the music industry liked about CDs was that they could not initially be copied. Users could record the music onto tape, but that hurt the quality and eliminated the special features of CDs. CD-ROM readers could extract flawless digital audio data, but that did not threaten music sales. Even in 1997, a large hard drive of 4 GB would be filled by eight albums. It was much cheaper to buy them on CD.

The MP3 Format

What made it practical for users to start building up music libraries was the spread of effective compression technology. The MP3 file format could compress a music CD to perhaps 20 MB. That sacrificed audio quality, but it still sounded better than a tape copy. The MP3 filename extension was a contraction of MPEG Audio Layer III. MPEG itself stood for Motion Pictures Experts Group, an organization supported by the Frauhofer Institute for Digital Media Technology in Ilmenau, Germany. In the early 1990s, ISDN digital telephone lines were expected to replace traditional phone lines and modems. They could carry 128 kilobits (16 KB) a second, which was enough to transmit video digitally if, and only if, very efficient compression techniques could be developed. Playing back high-quality movies from CDs was another goal motivating the Frauhofer researchers.[33]

The video compression techniques devised by MPEG during the early 1990s were eventually widely used, but at the beginning, the world seized on the audio compression provided by the most complex (layer 3) of its proposed algorithms. Hard drives could hold hundreds of MP3 songs with room left for programs and other data. Rapid increases in processor power meant that personal computers had no trouble handling the computationally demanding decompression work needed to play the files. Windows 95 could play music smoothly as a background task. An attractive shareware player called WinAmp, first released in 1997, made it easy to pick songs from a large collection and play them. WinAmp was highly modular, letting users download visualizations and plug-ins to play other audio formats.

The music industry had rejected the new technology, so MP3s caught on first with college students and commercial distributors of pirated music.[34] Students were heavy consumers of pop music, but typically had little money to buy recordings or space to store them. But they had squeezed PCs into their dorm rooms, which were wired with high-speed Ethernet connections to campus resources and to the broader Internet.

Students were sharing files with their friends over ad hoc local networks. In 1999 Sean Parker, an entrepreneurial nineteen-year-old, and Shawn Fanning, an undergraduate at Northeastern University, created a program called Napster to automate the process of sharing music files. Within a year, tens of millions of people were using it. The program scanned music folders, adding their contents to a catalog hosted by Napster. When someone searching the catalog clicked to download a song, their computer downloaded it directly from the computer of a user (see figure 11.6). Parker hoped that this *peer-to-peer* mechanism would shield Napster from copyright infringement claims.[35]

After lawsuits from the music industry nevertheless wiped out Napster, its place was taken by other, truly decentralized, systems. Starting in 2005 the Recording Industry Association of America brought a series of widely publicized lawsuits against their users. In one case, a woman was fined $1.92 million for making twenty-four songs available for download.[36] This failed to halt the growth of filesharing systems. They were not just cheaper than purchasing music legally, but also faster and more convenient.

Music Players
Although a music collection built from MP3 files was more convenient than a shelf full of discs, it took a high-end PC to store and replay it. That worked for listening to music

Figure 11.6

The Napster client running in 2001. Search results (top panel) came from Napster's own servers, but the seven simultaneous downloads taking place in the bottom window were copying songs directly from the hard drives of other Napster users. The process often slowed down or failed, which may be why the user is downloading three copies of "You Can Call Me Al." Image from Wikimedia user Njahnke, shared under a Creative Commons Attribution-Share Alike 4.0 International license.

while working in a dorm room, or, for people who traveled with laptops, in a hotel room. It was not so great for listening to music when walking down the street or driving in a car. Those were important markets. Back in the 1950s, small battery-powered portable radios had been the first big consumer application for transistors. In the 1980s, Sony's Walkman tape player found a place on the belts of teenagers around the world.

By 2000 cheap, energy-efficient processors were powerful enough to tackle MP3 decoding. The first pocket-sized players, such as the Diamond Rio and Creative Labs Nomad series, stored around 64 MB—maybe three albums of highly compressed music. Their users had to choose and download songs to fit their mood and then fiddle with clumsy controls to play them. Sony, reflecting its position as the premium consumer electronics firm, countered with a slicker Memory Stick Walkman that cost twice as much. As owner of CBS Records, Sony wanted to protect music sales. Music labels were initially reluctant to use the MP3 format because the files were so easy to share. Sony therefore chose a proprietary file format and memory card design so that the ability of users to share music purchased online could be controlled.

These players were quickly quashed by a new entry, Apple's iPod (figure 11.7). The first version, launched by Steve Jobs in 2001, was promoted with the slogan of "a

Figure 11.7
Apple's original iPod was controlled by four buttons, and a mechanical control wheel turned to move through menus and song lists. It held up to three thousand songs on a miniature hard disk drive with more than a thousand times the capacity of IBM's original RAMAC unit. Image created by Wikimedia user Miguelon756–5303, used under the Creative Commons Attribution-Share Alike 4.0 International license.

thousand songs in your pocket." They fit because Apple had built it around a new, miniaturized, 1.8-inch hard drive able to store 5 GB. The user interface was stripped down to its essentials, centered on a spinning wheel. Four buttons let users start/stop playback, jump forward or backward with songs, and return to the previous screen. As with the original Macintosh, back in 1984, the interface broke new ground. Unlike the original Macintosh, the iPod was ideally suited to its intended task.

Other companies had already built players around hard drives. What Apple did have was first rate industrial design and an aesthetic of elegant simplicity. Reviewing the device, Walter Mossberg of the *Wall Street Journal* called it a "terrific digital music player that solves all [the] problems" with rival hard drive based players like Creative's Nomad Jukebox, launched the previous year, which looked like a bloated, budget priced CD player.[37] Steven Levy devoted an entire book, *The Perfect Thing*, to the iPod, calling it "the most familiar, and certainly the most desirable, new object of the twenty-first century."[38]

It felt just right in the hand, small but reassuringly heavy, and it vibrated like a living thing, issuing a gentle whine and a barely perceptible shudder as its hard drive spun up.

Perfection can always be improved upon. Every year Apple released a new version of the iPod. It added support for Windows PCs and USB connections. The buttons moved around and then vanished into the wheel itself. The wheel changed from a spinnable disk to a strokeable surface. The screen grew large and crisper, changed from black and white to color, and showed first photograph collections and then movies. Battery life improved. Drive capacities grew, peaking at 160 GB for the final generation in September 2007. Launching that model, Steve Jobs was able to boast that Apple had sold 110 million iPods.[39] It was the firm's most popular computer, outselling the combined sales of all Macintosh models more than ten times over. Their telltale white earbuds were inescapable in affluent neighborhoods, and clock radios with docking connectors for iPods became a standard fixture of chain hotels.

The iPod had changed the way people purchased and consumed music, ending the century-old practice of visiting a store and purchasing a disc or cylinder holding a recording. The first appeal of MP3 players was for those whose hard drives were already full of music files. Newer software made it easier to "rip" audio from CDs into MP3 files, but illegally downloading a music file remained the fastest and most convenient way to get new music. That changed when Apple launched its iTunes online music store in 2003. It wasn't the first online music marketplace, but it was the first to combine affordability with convenience and a large selection. Songs were priced at just 99 cents each. By 2010, iTunes was the largest music retailer in the world, driving a shift away from albums and back toward free-standing hit songs. iTunes also distributed independent radio shows, leading to an enduring boom in what are still called *podcasts*.

Apple introduced a range of other iPod models, none with the iconic stability of the regular iPod, which had a fixed basic form growing just a little thicker or lighter in each annual rebirth. Its Nano models, first offered in 2005, were built around flash memory rather than hard drives. Early models held as little as 1 GB of music. Their shapes mutated dramatically from year to year. Some vintages were wide like business cards, other tall and thin. One was square and could be worn as a watch. Cameras, microphones, and radios came and went with dizzying rapidity.

The iPod did nothing that a desktop or laptop computer could not already do, but it could do it gracefully, from a pocket, for hours at a time during walking or jogging or flying. Users preferred a specialized box, hiding the universal technology of computing, over the physical constraints of a flexible personal computer. Albums had vanished, but music collections retained a reassuring physical heft even in digital form. An old iPod is a snapshot of one's past self. That emotional bond drove the plot of the 2017 movie *Baby Driver*, whose titular car thief listened almost continually to a collection of purloined players. Apple was so strongly associated with the iPod that in 2009 *The Onion* ran a satirical report about a "MacBook Wheel" laptop which used a control wheel instead of a keyboard. "Everything," a fake Apple spokesman boasted, "is just a few hundred clicks away." A customer noted, "I'll buy almost anything if it's shiny and made by Apple."[40]

The ARM Processor Architecture

The double core PP 5002 chip at the heart of the iPod gave it the processing capability of a mid-1990s desktop PC, but the iPod could run for hours on a tiny battery. Portal Player, a small firm, had created the chip specially for Apple. Like most other companies creating similar products, it did not start its design from scratch. Instead it licensed a processor design from a British firm called ARM. Traditional microcontrollers for simple control applications were sold as standard chips, often based on old, 8-bit processors. The ARM approach provided more computing power using a modern RISC architecture. ARM customers reduced the cost and size of their control boards by merging the processor and other needed capabilities to produce what was called a System on a Chip (SoC). Portal Player did not own a factory. It used computer design tools to integrate the components of the SoC, sending the resulting data file to a chip factory for production.

ARM began in 1983 at Acorn Computers, producer of the BBC Micro. The firm had strong connections to computer science research. Influenced by academic work on RISC two of its engineers, Steve Furber and Sophie Wilson, were inspired to create their own fast and simple processor. The Acorn RISC Machine (ARM) processor appeared first in Acorn's 1987 Archimedes computer. It was hailed as "the fastest computer I have ever used, by a considerable margin . . . just about everything you do happens instantly," but Acorn lacked the resources to compete against the increasingly dominant IBM PC and Macintosh platforms.[41] Recognizing that the strikingly efficient processor had a larger potential market than the computers it powered, in 1990 Acorn spun off ARM Holdings to sell ARM technology for other devices. ARM designs began to show up in all kinds of products, from laser printers and disk drives to mobile phones and network routers. By the early 2000s, ARM dominated the market for embedded 32-bit processors, becoming Britain's biggest computer hardware success story. ARM designs have even been used to produce supercomputers.

The creation of mobile devices, many built around licensed ARM processor cores, was made easier by the maturation of another technology: general-purpose field programmable gate array (FPGA) chips that could be programmed electronically for particular applications. This was a much cheaper process than producing custom silicon and was ideal for prototype devices or equipment with small production runs, for which conventional ASIC chips would not be viable.

3D GRAPHICS AND GAMES

We have already mentioned Ivan Sutherland's pioneering Sketchpad computer drawing systems. He later partnered with David Evans to found computer graphics business Evans & Sutherland, which launched its business by building custom flight simulator installations around exotic hardware. If a million-dollar simulator saved one plane crash it would pay for itself many times over.[42] By the 1970s it was supplying general-purpose *graphics boxes* for DEC minicomputers—cabinets full of electronics that coupled with minicomputers to drive high-quality interactive vector graphics displays. Hardware in the boxes relieved the main processor of the complex matrix calculations needed to simulate the appearance of objects when viewed from a particular perspective. This was called 3D graphics, even though the images were not literally presented in three dimensions. At that point such processors cost in excess of $125,000 and fewer than 200 were estimated to have been installed, most of them for research, modeling, and computer-aided design. With them computers could rotate molecule visualizations or simulate landing on an aircraft carrier by redrawing in real time.

The success of the VAX in the late 1970s expanded the market for graphics boxes. As well as wireframe graphics of the kind produced by Evans & Sutherland's Picture System series, VAX users could choose raster graphics boxes that output television-style bitmapped images. As well as simulating perspective, these could color objects and shade them realistically on the basis of virtual lighting conditions. Advanced graphics boxes greatly speeded the rendering of simulated 3D views. They are the ancestors of the graphics hardware that began to appear in personal computers and games consoles from the mid-1990s onward.[43]

Virtual Reality

In the late 1980s, the prospect of realistic interactive graphics was bound up with enthusiasm for *virtual reality*. Users put on headsets showing a different view for each eye, creating a true 3D experience. Turning their heads would change the view, and special gloves would control virtual hands to pick up and work with objects (figure 11.8). The technology had been developed by NASA for space simulation. In the longer term, enthusiasts looked forward to full body sensory immersion.[44] They were hoping to realize the exciting but technologically vague promises of interaction in *cyberspace* made in the cyberpunk worlds of authors like William Gibson.[45]

Work on virtual reality was underpinned by the advanced graphical capability of Unix workstations, which PCs could not rival. Silicon Graphics, the fastest growing workstation company of the early 1990s, led the charge. Silicon Graphics computers powered early VR headsets and were featured prominently in the 1993 movie *Jurassic Park*, a nod to the firm's part in bringing its unprecedentedly realistic dinosaurs to life.

Figure 11.8
NASA's Ames Research Center in Silicon Valley was the most important venue for early work on virtual reality in the mid-1980s. Note the data gloves, stereo headset, and head-mounted motion sensors. NASA photograph.

Its products were standard equipment for movie production and also for engineering and architecture applications needing to produce interactive 3D visualizations.

The Silicon Graphics RISC-based Indigo workstations, introduced in 1991, supported *geometry engine* graphics processors to speed the rendering of 3D graphics. The firm also created the standard OpenGL programming interface to describe 3D worlds. This provided abstraction and portability: applications produced OpenGL descriptions of environments, leaving questions about how to realistically render them to the interplay of the operating system and the graphics system. In 1996, at the peak of its success, Silicon Graphics purchased Cray Research. That reflected the amount of computing power needed for very realistic graphics and its determination to become a leader in other areas of high-performance computing.

This attitude turned out to be hubris, as the purchase did not yield the desired profits. Nor did the technology deliver on the virtual reality hype. Some consumer products did come to market from firms like Nintendo and Sega, but none were successful. Jerky displays make people feel sick because changes in perception did not keep up with movement of the head. More fundamentally, immersive interfaces turned out to be badly suited for practical tasks. The idea of flying through a database, walking around an account ledger, or manipulating a financial model by reaching out and rearranging its components may seem exciting. In practice, clicking and typing has so far proved more effective.[46]

3D Games

3D techniques drawn from Silicon Graphics machines did transform video gaming, but without special headsets or data gloves. In video gaming, "3D" usually meant that instead of a cartoon-like flat view or a fixed overhead perspective, the player appeared to be interacting with a world drawn with realistic perspectives. The techniques were first used extensively in simulation games, like the mid-1980s hit *Elite*. These drew only wireframe outlines of objects, but the calculations required to animate them smoothly required programming feats. The 16-bit processors of the late 1980s could handle more complex shapes, colored in to simulate solid objects. Even so, they typically animated only a small part of the screen, filling the rest with flight controls, maps, or steering wheels.

By the early 1990s, fast PCs could fill whole screens with relatively elaborate graphics. *Doom*, originally released as freeware in 1993, did more than any other game to establish the PC as a favored platform for action games. A space marine blasts through a series of maze-like levels, unlocking doors, collecting ever more powerful weapons, and destroying increasingly deadly clusters of marauding demons. Its graphics were astonishing. Although its shapes remained quite simple, *Doom* wrapped everything in textures, like brick patterns for a wall or red scales for a demon, to give the illusion of detail. Its simulated areas of light and darkness added to the mood of dread. Its fans began to organize LAN parties, temporarily networking their computers to fight against each other.[47]

Doom introduced the concept of the *game engine*, by separating the code needed to manage events in the game world and present them to players from the "assets" such as objects, monsters, and tunnels stored in data files.[48] Infocom and Sierra On-Line had taken a similar approach to adventure games, but the high-performance action games had previously integrated the functions closely. *Doom* required elaborate and highly reusable graphics code, making the engine approach to software engineering (already established in areas such as expert systems, databases, and graphics rendering) highly effective. Users exploited this to create data files holding their own maps and variants that kept players hooked long after they beat the original game. *Doom's* creator, id Software, was able to license the engine to other companies for their own games. Today's game engines are among the most complex and expensive to develop of all software technologies.

3D Graphics Accelerators

The rapidly growing market for PC gaming encouraged hardware designers to duplicate the 3D rendering technology used by exotic workstations. This was pioneered by 3dfx, a start-up company founded by veterans of Silicon Graphics. Its Voodoo Graphics chipset handled the most laborious tasks, such as mapping textures over geometric shapes, without tying up the main processor. The Voodoo expansion cards, launched in 1996, ran alongside conventional VGA cards to boost 3D performance.

Quake, the blockbuster sequel to *Doom,* provided gamers with a powerful case to upgrade. An early review noted that it was "set in a true 3-D world" so that buildings could be viewed, and enemies attacked, from any angle. Its monsters were "sick, twisted and perverse" with the "bloody grimaces and entropic bioforms" intensified by "utterly convincing animation." According to the reviewer, "The first time I was attacked by a hook-wielding enemy, I actually dropped the mouse and backed away from the computer."[49]

In February 1997, an OpenGL version of the *Quake* engine was released, letting users benefit from 3D accelerators. It became the de facto benchmark for graphics hardware. Without extra hardware even the most powerful Pentium chips portrayed *Quake*'s monsters at a chunky resolution of 320×200 pixels and painted them with an unrealistic palette of 256 colors. Microsoft Windows might use seven times more pixels on the same computer system, but 3D graphics were more demanding, so that was all the processor could animate smoothly enough to maintain the illusion. With a 3Dfx card, the resolution quadrupled and the scary creatures were displayed in 65,536 shades of gore. This was the most jarring jump in video quality in gaming history. A reviewer noted, "A *Quake* player that hasn't seen *GLQuake* on the Voodoo chip hasn't really played *Quake*. It is almost unreal how much better it looks."[50]

A race began to produce more powerful accelerators, delivering even higher resolutions and letting game designers create more intricate worlds, more realistic textures, and more complex shapes. For a few years 3dfx chips dominated the market for serious gamers, and rivals focused on those unwilling to purchase two graphics cards. By the end of the 1990s, however, video gamers were defecting to the products of two other companies: Nvidia (which purchased the remains of 3dfx in 2000) and ATI.

Over time, graphics chips relieved the main processor of ever more complex tasks. For example, Nvidia's GeForce chips, released in 1999, relieved the CPU of the need to calculate geometrical transformations and lighting conditions. The new chips were promoted as graphics processing units (GPUs). High-end models came with huge cooling systems and noisy fans, and computer makers started including more rugged power supplies to cater to their needs. Even Intel's new PCI connectors weren't fast enough to keep up with the appetite for data. In 1997, motherboard makers added a new connector, the accelerated graphics port (AGP).

As computing power grew rapidly during the early 2000s, ordinary computer users had less need to upgrade frequently. Older computers could handle word processing and Web browsing without slowing down. Gamers, on the other hand, always needed more computer power. Last year's graphics card would not do justice to this year's game. Communities grew up to tinker with gaming PCs, fitting them with performance tweaks like cryogenic cooling systems to keep chips running far beyond their rated speeds from melting. Manufacturers responded with premium cards and motherboards designed for tweaking. Like special editions of sports cars, they offered a little more performance for a lot more money. The visual aesthetics of high-end PC components changed, too, with brightly colored circuit boards, lights, and transparent cases to attract the envy of other gamers.

While the overall costs of high-end PCs were falling dramatically, the price of high-end graphics cards continued to rise. Cards based on Nvidia's GEForce 8800 GTX, launched in 2006, cost around $600. That made them the most expensive part of most of the computers graced by their presence, but they also held most of the processing capability. These graphics cards could burn more power than the rest of the computer combined. Running two side by side, for top performance, required a PC with an 800-watt power supply (about twelve times more than IBM's original model).[51]

More casual gamers, willing to play older titles or tolerate lower resolutions, were able to use the increasingly powerful capabilities built into ordinary graphics cards, games consoles, and eventually laptop computers. From Windows 95 onward, Microsoft included extensive support for 3D graphics with its DirectX interface, meaning that video gamers no longer had to exit to DOS and fiddle with command line settings for high performance gaming.

The economies of scale enjoyed by PC graphics card companies put the final nail in the coffin of the specialist workstation companies. Silicon Graphics liquidated in 2009. Sun smartly shifted its focus to Internet servers. But there, too, competition from PC platforms eventually caught up with it. Oracle absorbed Sun in 2010.

3D Comes to Games Consoles

Games consoles made 3D graphics ubiquitous. Individual consoles have life cycles of at least five years, as companies invest huge sums to design and manufacture millions to recoup those costs. The first generation of consoles with 3D graphics capabilities included the Nintendo 64, launched in 1996, which was based on Silicon Graphics processor and graphics chips. Familiar Nintendo characters like Mario and Link (hero of the *Zelda* series) were suddenly moving around in 3D landscapes, requiring a complete rethinking of game mechanics and the control systems used to jump around platforms or target monsters. Sony's first console, the PlayStation, was equally influential.

Hit games from this period established the genres, conventions, and brands that dominate the video game industry to this day. *Tomb Raider* (1996) dressed up logic problems in a gender-switched pastiche of the Indiana Jones movies, inaugurating a series of sequels and big-budget movies. These told the story of Lara Croft, whose physique implausibly and controversially resembled a Barbie doll. *Resident Evil (*1996) mixed zombie killing with puzzle solving and scares to establish the "survival horror" genre. It spawned a series of determinedly mediocre movies that have collectively earned well over a billion dollars at the box office. *Half Life* (1998) melded the first-person shooter mechanics of *Doom* and *Quake* with more complex puzzles and a strong narrative, inspiring many imitators. Similar technology was applied to realistic tactical combat in the *Rainbow Six* series of games (also 1998), inspiring the long-running *Battlefield* (2002) and *Call of Duty* (2003) franchises. *Grand Theft Auto III* (2001) set players loose in a beautifully detailed fictional city modeled on New York. There were missions and a strong storyline, but what players remembered was the freedom to explore a busy city as carefree criminals, running over innocents, stealing cars, or shooting prostitutes as they wished.

Not all the hit games of the period were 3D based. Blizzard Entertainment created the hugely successful *WarCraft* (1994) and *StarCraft* (1998) series of games. These combined strategic elements—researching technology, gathering resources, building facilities—with real-time combat. *StarCraft* was the PC smash hit of the year on its initial release, but unusually for a video game sold even more copies over the next decade. *StarCraft* had a strong scripted storyline, but its unique success was as a multiplayer game, thanks to the new availability of Internet-based contests as well as the continuing popularity of LAN parties. In South Korea it established the first community of professional video gamers, earning their livings by playing televised tournaments.

Over time, consoles evolved closer to PCs. They shifted from cartridges to optical discs for game distribution, and adopted the same graphics chips as PCs. Microsoft's Xbox, launched in 2001, exemplifies this (figure 11.9). It was based on a Pentium III processor, a hard disk drive, an Nvidia graphics unit, and a DVD-ROM player—typical equipment for a gaming PC of the same era. It ran a customized operating system closely based on elements of Windows. Even the name, a contraction of DirectX Box, promoted a Windows gaming technology. Many users found a black box, ready to be placed under a TV and controlled from the sofa with video game handsets, more convenient and relaxing than fiddling around with a PC and sitting up at a monitor to play games.

Those willing to invest enough time and money could run most games on a regular PC with better graphics, but the Xbox provided a standard platform for developers that they could confidently push to its limits. The most successful Xbox game, *Halo: Combat Evolved,* was one of the first to successfully reinvent the precise mouse control

Figure 11.9

Under its lid, the Xbox was built mostly with standard desktop PC components, including this Toshiba DVD-ROM and Western Digital hard disk drive. Its price was initially subsidized by Microsoft, making it a tempting target for hackers looking for a cheap but powerful computer to repurpose. Image by Wikimedia user Evan-Amos.

of first-person shooters like *Quake* for the control sticks of games consoles. Its unsubtle mix of frenetic combat and drivable vehicles has sustained a long series of sequels.

The Xbox was neither a smash hit nor a flop, but Microsoft's cash flows from Office and Windows let it invest in the system without immediate payback. Microsoft was willing, early in its life cycle, to sell the Xbox for less than the cost of its parts, recouping the losses in subsequent years as component prices fell. High profit margins on games that were sold for up to $60 each further offset the subsidized hardware. To prevent users from buying the box and wiping it clean for use as a cheap computer, Microsoft installed a digital signature enforcement chip. This was intended to make it illegal for users to customize their own hardware, which caused controversy and lawsuits around the Xbox and subsequent generations of consoles. Linux hackers eventually defeated the protection.[52]

By 2006, when Microsoft replaced the original Xbox with the improved Xbox 360, it had fallen well behind the power of current PC models. The cycle has since repeated itself, with four generations of Xboxes and PlayStations competing for the affection of hardcore gamers. Nintendo went in a different direction, packaging less powerful technology in more innovative ways. Its Wii console, introduced in 2006, expanded the market by offering unique motion-sensitive wireless controllers. These could be swung like a baseball bat, pointed like a gun, or turned like a steering wheel. Many of its most popular titles were sport and fitness based, inviting users to get up and dance around their living rooms rather than slump on the couch.[53]

DIGITAL MEDIA TRIUMPHANT

More disruption was to follow over the next decade. To explain the next steps in digital convergence, we must first explore the parallel story of the commercialization of the Internet around a new online communications system, the World Wide Web. Together the Internet and Apple's iPhone, launched in 2007, began to replace the many different computer-based digital devices described in this chapter, including music players, video disk players, and cameras, into a single platform designed around the possibilities of constant connection.

Even before the iPhone launched, digital transitions had shaken established industries to pieces. The music industry, for example, had to shift from selling huge numbers of full albums with heavy profit margins to selling smaller numbers of individual songs through Apple's iTunes service. Photography was hit even harder. Kodak, the leading American supplier of photographic film and related products, invested heavily in new technologies. In the 1970s its research staff had pioneered digital photography, and in the 1990s it was one of the first companies to sell digital cameras for personal use. From 1992 it made a big push to promote the PhotoCD format: consumers would drop their films off in drugstores and receive high-definition scanned copies as well as traditional prints. Kodak even sold special CD players to display the pictures to television sets. In 1996 it introduced a new hybrid film format, recording exposure details on a magnetic strip alongside the analog image. It built a new business selling photo printers to home computer users. Nothing worked. Consumers wanted to avoid the cost and inconvenience of dealing with film, the only business in which Kodak had unique advantages. In 2012, Kodak declared bankruptcy, announcing a plan to reemerge as a smaller firm focused on digital imaging services for businesses.

It was a new world. People who brought home one of the first Apple II or TRS-80 computers in 1977 were self-consciously introducing a computer into their lives. These devices were digital outliers in a personal technology world dominated by analog media machines such as televisions, record players, telephones, and cameras. Thirty years later, all those analog devices had been largely replaced by hidden computers manipulating digital data.

12 THE COMPUTER BECOMES A PUBLISHING PLATFORM

In 1995 the World Wide Web emerged from academic obscurity to make front page news and become the object of a new technological gold rush as companies scrambled to build browsers, set up Websites, and sell services. It was the new "killer application," so desirable that users purchased entire new computer systems just to access it. Bill Gates suddenly realized that it might kill Microsoft, too, in a technological "tidal wave" that overwhelmed its desktop computing monopoly. For once, the hype about a computer revolution was justified, unlike the business plans of most of the thousands of companies aiming to get rich from it. By 2001, more than half of American households had subscribed to an Internet service. Many of them purchased a computer primarily to access Web and email services.[1]

Unlike much we have written about so far, the Web's importance to the world's development over the last few decades is easily observed and widely discussed. Scholars writing historically about almost any industry, from popular music to sex work, will reach a point in the 1990s or early 2000s where they must grapple with the role of the Internet. Our aim is not so much to uncover unknown aspects of the history of the Web as to situate that history firmly within the broader stories of the Internet and of computing.

When the Web began in the early 1990s, it was an electronic publishing system with some similarities to the systems discussed previously, such as Minitel, Prestel, and Plato. So, it makes sense for us to begin by looking at the development of commercial online systems, to understand how people were getting online and what made the Web and the Internet such an attractive alternative.

Ordinary people were going online in the early 1990s, but not by connecting to the Internet. The most widely used system of the era was America Online (AOL). Its roots lay in a set of online services launched from 1985 onward, most notably Quantum Link for the Commodore 64. These services offered a richer experience than terminal-based competitors by including graphics and interactive features in the client software used to access them.[2]

In 1991, Quantum was renamed America Online and began to merge its machine-specific services under the new brand. The focus shifted to PC clients, first for DOS and, from 1993, for Windows. Users paid a small subscription fee plus an additional charge for each hour of connection time. This gave access to many services hosted on AOL's servers, including news, games, shopping, travel reservations, and chat rooms. Like other online services, AOL had an email system but could send messages only to other AOL accounts in 1992, AOL added a gateway to exchange messages with Internet users. These interconnections changed the character of the Internet. One grumpy journalist complained that "sporting an email address ending in aol.com is an instant sign that you're probably too stupid to be taken seriously." When Usenet newsgroups were first made available to users of the Delphi online service, in 1993, existing users named it "the year September never ended." Every September had seen an influx of new undergraduate users, who had to be tutored in appropriate *netiquette*. More new users of other commercial online services soon followed, overwhelming established online cultures and communities.[3]

Much of the material on AOL came from other companies, which received a share of the money that users paid to access their contributions. For example, the Motley Fool investment site began as an AOL publication before shifting to the Web. From 1993, AOL promoted its service aggressively, mailing out many millions of discs holding the AOL software and promoting a free one-month subscription. AOL grew very rapidly, reaching one million users by August 1994 (figure 12.1).[4]

THE WORLD WIDE WEB

AOL's approach seemed like it would finally get ordinary people consuming and communicating electronically. As Microsoft put together Windows 95, it offered its own similar online service, the Microsoft Network (MSN), as one of its highest profile features. However, by the time Windows 95 was launched, the public was more excited by the prospect of connecting to the Internet. This unexpected twist was driven by the enormous popularity of a new Internet application, the World Wide Web, invented in an equally unexpected place: the high-energy physics laboratory CERN on the Swiss-French border.

The Web incorporated hypertext techniques, but those had already been implemented in several widely used systems. Its primary strength was simply that it ran on the Internet. When we last talked about the Internet, it had been opened to commercial use but was still primarily a system for scientists and researchers. Without existing Internet protocols, services, and infrastructure to build on, there could have been no

Figure 12.1
The website NoMoreAOLCDs.com. In 2001, as AOL continued to distribute vast numbers of discs holding software and trial offers, Californians Jim McKenna and John Lieberman began to collect unwanted CDs with the aim of eventually dumping a million of them in front of AOL's head office. The tally had reached 410,176 when the campaign ended in 2007.

Web. But without the Web the millions of ordinary people who began to connect to the Internet in 1994 might have favored other networks.

Hypertext

The Web's fundamental concept of structuring information as *hypertext* goes back to Vannevar Bush, whose seminal 1945 essay discussed the potential of microfilm and electronics to build huge personal libraries of documents annotated with navigable links.[5] This had a major influence on Ted Nelson.[6] His self-published manifesto *Computer Lib/Dream Machines* defined hypertext as "forms of writing which branch or perform on request; they are best presented on computer display screens."[7] Nelson also praised Engelbart's On-Line System (NLS), which was structured primarily around hierarchical outlines but also included a linking mechanism.[8] Nelson worked unsuccessfully through the 1970s and 1980s to make an ambitious online hypertext publishing system Xanadu come to life. He remained influential on, but always outside, the academic community, inspiring an active community of hypertext researchers.[9]

Hypertext publishing was adopted for technical documentation including help systems and reference manuals. Creating hypertext works was difficult, however. In 1987, Apple's Bill Atkinson opened up the technology to a much broader range of users and applications with Hypercard for the Macintosh. Hypercard was easy to grasp and let even a computer novice program simple games or educational materials. More people were exposed to hypertext via the Windows online help system introduced with Windows 3.0 in 1990, which provided electronic documentation for the operating system and its applications. So did CD-ROM reference books, including the popular Encarta encyclopedia. Being able to cross-link topics was vital for these applications, but unlike links on the Web, their links extended only to other pages within the same electronic book.

Gopher and WAIS

The Internet of the early 1990s was growing so rapidly that even experts found it hard to stay on top of its expanding collection of tools, protocols, file transfer sites, telnet services, and newsgroups. It began to feel like a large library that had no card catalog. Programmers at the University of Minnesota responded by creating Gopher.[10] Gopher allowed students and faculty to easily locate information such as class schedules, administrative policy statements, and sporting events. The name was a pun, applying the University mascot to a program that would "go fer" the data and retrieve it. Gopher displayed information as menus of menus in a hierarchical tree. When it spread to other organizations, users could follow links from one Gopher system to another. Gopher was released in April 1991. By April 1994 there were around seven thousand Gopher servers online.

At the supercomputer company Thinking Machines, Brewster Kahle and his colleagues developed a system called wide area information service (WAIS) that allowed the user to search the contents of files. WAIS scoured the Internet for documents and indexed their content, showcasing the heavy-duty processing power of the Thinking Machines supercomputers. WAIS was not easy to learn, but when it worked and retrieved exactly what a person wanted to find, it took one's breath away. Gopher's growth had accelerated after the establishment of Veronica, a similar service that indexed "gopherspace" so that users could search for information by terms as well as navigating through hierarchies.

Origins of the Web

Gopher and WAIS were rendered obsolete by a new interface for Internet resources: a hypertext system called the World Wide Web. The Web fell far short of Nelson's vision for hypertext, but it took one great step forward in comparison with other hypertext systems of the era: because it ran on the Internet, a link could point to a page on another computer just as easily as one on the same computer. Berners-Lee (figure 12.2), who wrote the original Web prototype in late 1990, stated that "[t]he Web's major goal was to be a shared information space through which people and machines could communicate."[11] It organized existing Internet resources, such as newsgroups and file transfer sites.

Berners-Lee worked with Robert Cailliau, leader of CERN's Office Computing Systems group, to develop and promote three simple but effective standards that defined the Web. The *universal resource identifier*, later called the *uniform resource locator* (URL), could "point to any document (or any other type of resource) in the universe of information."[12] Web links were a simple combination of two existing ways of specifying a location on the Internet: a computer name (expressed with the conventions of DNS) and a file location (expressed as a Unix path). To transfer information from Web servers to Web browsers Berners-Lee and Cailliau created the simple *hypertext transfer*

Figure 12.2
Tim Berners-Lee (right), with email pioneer Ray Tomlinson, at the American Computer Museum, April 2000. Photo by Paul Ceruzzi.

protocol (HTTP), which relied on existing Internet protocols. Finally, they defined a *hypertext markup language* (HTML) to tag documents with hypertext features like links and headings. That too was based on an existing technology, the *standard generalized markup language* (SGML) approach to tagging text with metadata.

The World Wide Web grew slowly at first, as Gopher had a head start and initially remained more popular. To view Web materials, one used a program called a browser. Berners-Lee's own browser was a mouse-based application able to display graphics. Its users could create and update Web pages as well as view them, a central part of his original vision for the Web. He had developed it with a NeXT workstation, taking advantage of its object-oriented tools for rapid application development. Relying on the NeXT software let him produce a prototype browser quickly, but it also limited that browser to a few thousand potential users. The most widely used early Web browsers, including a text-mode program called Lynx, instead presented screens similar to Gopher's, which had numbered menu selections.[13] These were based on code written at CERN and distributed to spread Web access.

Graphical Web Browsers

The small community of Web enthusiasts knew that it would need robust graphical browsers for widely used platforms. One of them was Marc Andreesen, a student at the University of Illinois with a job at the National Center for Supercomputing Applications, funded by the NSF to make supercomputers accessible to Internet users. By June 1993, Andreesen and Eric Bina, a Unix staff specialist at the center, had released a test version of a browser that they later named Mosaic. Mosaic's seamless integration of text and images made the potential of the Web instantly apparent (see figure 12.3).[14] The first Mosaic users were people who already had powerful Unix workstations and fast Internet connections, found mostly in universities and research labs. Its availability accelerated the Web's growth. A web crawler program created by an MIT student discovered only 130 active Web servers in mid-1993, but 623 when it was run again at the end of that year.[15]

Larry Smarr, who led NCSA, used its resources to promote Mosaic and rapidly produce versions for Windows and Macintosh computers, greatly broadening its potential user base. *Wired Magazine* enthused that "Prodigy, AOL, and CompuServe are all suddenly obsolete—and Mosaic is well on its way to becoming the world's standard interface. . . . The global network of hypertext is no longer just a very cool idea."[16]

In early 1994, Andreesen was approached by Jim Clark, the recently departed founder of Silicon Graphics. Together they founded what became the Netscape Communications Corporation and decided on a Mosaic-like browser as their first product. Clark and Andreesen hired many of the programmers who had worked on the software at the university. Netscape introduced its Navigator browser in September 1994. It

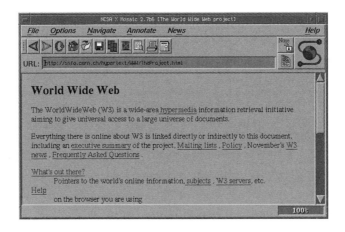

Figure 12.3

The first widely used graphical Web browser was xMosaic in early 1993, seen here visiting CERN's website. It ran on Unix workstations using the X Window system. Features such as the forward and backward buttons, the URL bar, and the underlining of hyperlinks remain common in today's browsers. The large globe spun when data was being received.

quickly supplanted Mosaic as the most used Web browser, taking advantage of the Web itself as a new channel for commercial software distribution.[17]

Netscape helped to sustain the exponential growth of Web publishing to more than 10,000 servers by the end of 1994. That doubled again in the next six months, reaching 23,500 in June 1995. Then along came Windows 95 with its built-in TCP/IP stack, dial-up networking, and multitasking capabilities. Affordable personal computers could now easily run Netscape on a big color screen, delivering tolerable speeds over a telephone line. The Internet became something else. It was no longer only a facet of computing technology but a part of entertainment, consumer spending, and popular culture. The millionth Web server came online around March 1997, and the 10 millionth by December 2000.[18]

Internet Service Providers

As the Web drew ever more users to the Internet the number of Internet service providers grew almost as rapidly. AT&T's Worldnet service, launched in 1995, set a de facto standard for Internet access—unlimited modem access for $20 a month. It opened up access points across the country, so that most potential customers could access the service with a local phone call. AOL and MSN hurriedly remade themselves as Internet service companies, offering Web browsing and other Internet services as well as their proprietary content.

AOL reached 6.2 million users by June 1996, making it the largest online service provider, but its growth was beginning to slow in the face of competition from lower priced Internet firms. At the end of that year AOL switched to a flat monthly fee of $20 to match the pricing model of many Internet service providers. A surge in usage overwhelmed its access points for a while, but the firm's growth rate jumped again.[19] By June 1999, it had twenty million active customers.[20] Many of its users barely understood the difference between AOL and the Internet.

Although most customers chose large ISPs, such as AT&T or AOL, most ISPs were much smaller operations, serving local communities. Some were converted bulletin board services. At the peak, in March 2000, *Boardwatch* magazine listed more than seven thousand firms selling Internet access to the American public.[21]

Other Applications

Most of the new Internet users of the mid-1990s were so centered on the Web that they had little sense of the difference between this new application and the Internet it ran on. Internet companies typically pointed users toward sites from which they could download the other tools of the mid-1990s Internet: a Usenet newsreader, an email client, a chat program, a telnet terminal-style client, and so on. Although the established Internet Relay Chat system, for exchanging real-time messages in *chat rooms* proved popular, other consumer-oriented messaging systems sprang up.

These delivered short messages more quickly than traditional email, ideal for exchanging messages in a conversational style. The services were particularly popular with teenagers as an alternative to chatting on the telephone. The most popular of these, AOL Instant Messenger (AIM) won quick adoption thanks to its integration into AOL's software. AOL produced free messenger software for a variety of platforms, including early handheld computers, for those who got their Internet access elsewhere. Yahoo and Microsoft both ran their own successful messaging services, though AOL bought out the most successful independent rival, the Israeli platform ICQ ("I seek you"), in 1998.

Game companies were also quick to exploit the potential of rapidly growing base of Internet users. Although some 1980s online services had offered online games, playing regular PC games against other users had required hooking computers together with serial cables or as we mentioned earlier in the case of *Doom*, ad hoc local area networks. Games running on Internet connected PCs let players share game worlds with hundreds of thousands of other players. The new model was launched by *Ultima Online* in 1997, a spin-off from a series of computer roleplaying games going back to the Apple II. Because the game world was maintained in a data center, players could interact with each other to trade or collaborate on quests, adding the social elements of a face-to-face roleplaying game. The game progressed whether they were logged in or not, and the actions their characters took, like building a castle, became a permanent part of the game world.

At its peak, the game had around a quarter million players, each paying a monthly subscription fee. Items in the game's online economy were traded by players for millions of dollars of real money, and low-paid Chinese "gold farmers" spent long days clicking through mind-numbingly boring chores to generate in-game currency to be sold to Western players on eBay. Even people playing the games for fun tended to settle into an all-consuming worklike routine, spending dozens of hours each week to build the virtual power and wealth of their online characters. As journalist Julian Dibbell, a veteran of early text-based multiplayer online games, described the life of one player determined to generate enough in-game cash to buy a large plot of imaginary land, "he had to bring Nils Hansen's blacksmithing skills up to Grandmaster. To reach that level, Stolle spent six months doing nothing but smithing: He clicked on hillsides to mine ore, headed to a forge to click the ore into ingots, clicked again to turn the ingots into weapons and armor, and then headed back to the hills to start all over again, each time raising Nils' skill level some tiny fraction of a percentage point. . . . Every day, month after month, a man was coming home from a full day of bone-jarringly repetitive work with hammer and nails to put in a full night of finger-numbingly repetitive work with 'hammer' and 'anvil'—and paying $9.95 per month for the privilege."[22]

WEB PUBLISHING

As the Web rapidly developed, many enthusiasts saw it as a utopian innovation with the potential to transform human life for the better, if only governments would stay out of its way. John Perry Barlow was a user of the WELL and lyricist for the Grateful Dead. With Mitch Kapor, of Lotus fame, and free software enthusiast John Gilmore he had cofounded the Electronic Frontier Foundation in 1990. In 1996, Barlow circulated on the Internet a "Declaration of the Independence of Cyberspace." It opened with the words, "Governments of the Industrial World, you weary giants of flesh and steel, I come from Cyberspace, the new home of Mind. On behalf of the future, I ask you of the past to leave us alone. You are not welcome among us. You have no sovereignty where we gather."[23]

In William Gibson's stories, cyberspace was an immersive virtual environment so realistic that traumas suffered when someone was "jacked in" could lead to real-life death. That didn't have much to do with the reality of the early Web, but the cyberspace metaphor suggested that the Internet was a place of its own, existing outside physical geography. It was not only a communications medium, nor even merely a lifestyle, but an actual place to live and work.

Barlow's shrill tone captured the faith of Internet enthusiasts that technology was about to make traditional political institutions obsolete. It built on a broader, largely Californian network of subcultures celebrating the application of technology to projects such as life extension, cryogenic freezing, and the uploading of human bodies to machines. A glossy new magazine, *Wired*, founded in 1993 was helping to bring some of those ideas into the cultural mainstream.[24]

The end of the cold war and the collapse in 1991 of the Soviet Union were sometimes attributed to socialism's inability to match the West's mastery of microelectronics, personal computers, and networking. Now technology was about to render Western governments equally irrelevant. Enthusiasts opened *cybercafes* with Internet-connected computers, an *Atlas of Cyberspace* was published, and the US government agreed not to force the collection of sales tax on online purchases made across state lines.[25] Excitement about the utopian potential of the Internet to transform education and create new business opportunities led to worry in the Clinton administration that a *digital divide* in Internet access would compound existing patterns of racial and geographical inequality.[26] Its initiatives to provide public access via schools and libraries paled against enormous structural disparities in opportunities and school funding.

The Limitations of the Web

The Web began as a very simple system, which was inevitable given the limited resources available for the project at CERN and its initial use to catalog the Internet's existing resources. Unlike Ted Nelson, Berners-Lee wasn't trying to produce a commercial system that could make online publishing economically viable. Unlike the community of academic hypertext researchers, he wasn't trying to solve the knotty problems involved in keeping hyperlinks current and accurate as time goes by.

The Web was just a thin layer on top of the Internet's existing infrastructure. Because there was no central database of hyperlinks, users could follow links out from a page but not go the other way to see everything that linked to a page. Between the time a link was created and clicked, the page to which it pointed might have been edited to remove relevant information or deleted completely. Most of the external links on Web pages eventually stop working. Ted Nelson and Doug Engelbart were among the Web's harshest critics. Nelson didn't even consider the Web to be true hypertext. Xanadu was supposed to hold old versions of a page forever, so that the linked material would always be available. Even Tim Berners-Lee complained that only half of his vision had come true with commercial browsers like Netscape. He initially wanted a Web that was as easy to write to as it was to surf.[27]

The Web also lacked a built-in search capability. Earlier electronic publishing platforms, like the LEXIS-NEXIS database service and AOL, used centralized servers. Publishers had to tag information correctly and provide it in a standardized format. This made online publishing hard, but searching and indexing easy. With the Web, that was reversed. Publishing was trivially easy—just start a small Web server *daemon* program on an Internet-connected computer and throw some HTML files in a directory. For that reason, searching and indexing the Web was enormously challenging.

Just as important was the lack of a payment system. Nelson had imagined authors and publishers shifting from books and newspapers to his Xanadu system. To make this economically viable, Nelson mandated a mechanism whereby readers made small micropayments for each page read, a portion of which would be passed on to authors as payment. Online services like AOL and Minitel charged users for connect time and premium services, splitting the proceeds with the outside companies publishing information through their platforms. The Web made publication easy but payment hard. CERN and other physics research centers gave away their publications, which had been produced with public funds. News publishers like the *New York Times* and CNN were attracted by the Web's popularity to start giving away their stories, but in the long term this practice threatened to destroy their businesses.

Many different initiatives were proposed to tackle these shortcomings but because the Web was built around existing Internet technologies, it would have taken a full-scale redesign of the entire Internet, not just the Web, to fix them. The solutions that emerged were complex and partial. For example, the problem of links going bad or pointing to pages that had long-since changed was targeted by Brewster Kahle, the creator of WAIS. In 2001, his Internet Archive group launched a Wayback Machine that provided access to saved static copies of Web sites. Preserving the Web is a remarkable feat of engineering, and by entering an old link one can sometimes retrieve the page it used to go to. But the archive is fragmentary: it holds some sites but not others, and some pages but not others, even within a website. For less popular sites, months or sometimes years go by between snapshots.

Search—The Web's Missing Links

The word *browser* captures the way that Berners Lee expected people to use his creation: clicking links from one page to another was like browsing a market from one stall to the next. Directory pages used hypertext to build ad hoc catalogs. As the Web expanded, this became less viable. Any foray onto the Web was likely to bombard the explorer with trivia on esoteric topics. But finding the answer to a specific question could be hard indeed. One science writer had abandoned the Internet in frustration: "I have briefly signed up with a number of Internet providers, only to become exasperated by the maddening randomness of the Net." He complained that "the complicated searches . . . feel like a waste of time."[28]

The Web was an ever-growing library without reference librarians, quality control, or a card catalog. The most natural way to fix this was to produce a site full of links to other sites: a virtual library catalog complete with index and hierarchical subject headings. Yahoo (or as it branded itself, "Yahoo!"), the most successful of these, was founded in 1994 by Stanford University students Dave Filo and Jerry Yang as Jerry's Guide to the World Wide Web. They originally assembled its index themselves, hiring ever more people to keep up as the Web expanded. Access was free. When in August 1995 the site began running advertisements to generate revenue, a few complained but the service remained popular.[29] It got an early boost from Netscape, which put a link to it from its own site. In April 1996, Yahoo made an initial public offering of stock, providing a rapid payback for its initial investors and cementing its position as a leading Internet firm (figure 12.4). By then Yahoo employed fifty full-time staff to surf the Web and update its directory. It was hard for them to keep up with the proliferation of websites.[30]

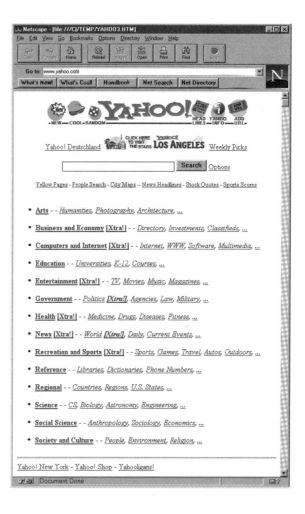

Figure 12.4

Yahoo's homepage in October 1996, when the site was still primarily a hierarchical catalog of the Web. Clicking one of the top-level headings on the front page brought up a page of subheadings, and so on. Shown running in Windows 95 on version 2.0 of the Netscape Navigator browser. The broken key in the bottom left reflects the addition of encryption to support credit card transactions.

As the Web grew, the limitations of the Yahoo approach soon became apparent. Users had to click through many levels of subject headings to find the site they were interested in. Yahoo cataloged entire sites rather than pages—fine if the user was looking for a website on a popular topic (such as the best Britney Spears fan site) but less useful if the search was for pages mentioning a particular book.

Directory sites like Yahoo faced growing competition from search engines, which used a software *bot* (short for *robot*) to *crawl* through Web sites and index the content on each page. To be useful, a search engine had to crawl as much of the Web as possible, following links from one site to another to track down obscure pages for indexing. It also had to deliver lightning-fast searches through that index, pulling all the pages holding a phrase and deciding which to show to the user first.

The early leader in search engines was Altavista, founded by the Silicon Valley lab of Digital Equipment Corporation. Altavista was good at finding pages holding search terms but terrible at deciding which pages it would show first to users. For example, it assumed that the more often the search term appeared in the code for a page the more relevant that page was. The result was search engine spam: unscrupulous Web publishers would draw traffic to their pages by repeating popular search terms hundreds of times in text visible only to search engines—for example, in tiny black text on a black background. Rather than improve its search results, Altavista, like Yahoo, focused on becoming a *Web portal* with a home page holding a grab bag of information and services from weather to horoscopes. This made it unsightly and slow to load.

It seemed that search engines had been beaten by spammers. In 1999, search engine analyst Danny Sullivan boldly announced that "This was the year that the humans won. [Earlier] you had one major search service, Yahoo, that used human beings to categorize sites while the others were trying to use technology to do the same thing. But now with six out of the top ten services, the main results you get will be by people."[31] Even as he wrote, search robots were mounting a decisive counterattack against the human indexers. Altavista's users were shifting to a new rival, Google, founded by Sergei Brin and Larry Page, who were then graduate students in Stanford's computer science program. Google's big advantage came in figuring out how to rank the pages that held a search term, which it did by favoring websites that had been linked to by large numbers of other sites. Spam pages were unlikely to be linked to and therefore fell to the bottom of the rankings.

This method was inspired by a system for the retrieval of scientific information developed by Eugene Garfield, called the *Science Citation Index*. It indexed scientific papers and ranked their impact by noting how many other papers referenced them. This, and thousands of other tweaks designed to improve the relevance of searches and defeat spam, meant that the first few search results provided by Google were likely to provide the information sought by a user. Google also enforced an aggressive simplicity on its home page which meant it was attractive and quick to load, consisting of little more than a box in which to enter search terms and two buttons, labeled Google Search and I'm Feeling Lucky.[32]

Web search and directory companies like Google and Yahoo, and a host of less successful rivals, were among the largest and most powerful Internet companies. This goes back to the simple, decentralized nature of Web publishing that let the Web leap to dominance in the first place. Web pages do not have to be tagged with metadata to identify, for example, authors or subjects. They show up anywhere on the Internet and do not have to be reported to any centralized catalog or database. There are millions of Web servers, many updating their content every few minutes. Everything that makes publishing to the Web easy makes indexing or cataloging it hard. Whether with humans, as Yahoo used to do, or with algorithms, as Google does, that is an enormous task requiring vast amounts of money and human talent.

Web Payments

The rise of the Web is inseparable from the efforts of companies, from Amazon to Zillow, to make money by computerizing every aspect of human interaction. We cannot tell all those stories here and so instead focus on the development of the Internet and the Web into universal platforms that could underpin diverse activities. The early Web was primarily an electronic publishing system, used to publish and view static pages. Minitel and AOL could bill the accounts of readers and share the proceeds with publishers. Because the Internet had not been designed for commercial use, neither it nor the Web provided any comparable mechanism for a website to charge its readers. The economics of conventional publishing were comparatively straightforward: publishers made money on each book or record sold. Selling more copies meant making more money, so that each hit underwrote the cost of many flops. In contrast, a popular website ran up huge bills for network bandwidth and servers without receiving any income from readers to cover this expense. Grabbing more readers meant bigger losses, not bigger profits.

Several companies were launched to create *electronic cash* systems, adapting Ted Nelson's micropayments scheme to the Web to provide a sound foundation for commercial Web publishing. These relied on encryption techniques to create secure and efficient billing systems to support very small transactions. The technology existed, but making it work as a business required simultaneous adaptations by browser makers, Internet users, and Internet publishers. Even in 1995 such a jump wasn't feasible. Encryption eventually gave Web browsers the ability to transmit credit card details securely—but even users willing to enter credit card details to buy a pair of shoes would balk at doing so to pay five cents to read a news story. In any event, transaction fees made credit cards unworkable for very small payments.

Websites could try to sell subscriptions rather than individual stories. That initially worked in two areas. Pornography drove much of the commercialization of the Internet, just as it did the early days of video recording and of motion pictures.[33] If Web surfers typed in www.whitehouse.com instead of www.whitehouse.gov, they were taken to a site that made millions of dollars by selling monthly subscriptions to pornographic materials. The other kind of Web content people were willing to pay for was financial information. The *Financial Times*, for example, never made full stories available to nonsubscribers.

The lack of a payment mechanism didn't slow the growth of Web publishing. Amateurs loved the idea of creating something that could draw thousands of visitors, existing publications wanted to be part of the Web revolution and reach new audiences, and deep-pocketed venture capitalists and technology companies decided to build audiences first and worry about the economics later. For example, *Slate* began in 1996 as a project of Microsoft to bolster MSN. It began by transplanting the practices of a politics and culture magazine online, under the direction of former *New Republic* editor Michael Kinsley. For a while it even created a print edition, to be sold in Starbucks stores.

Advertising

Instead of subscriptions or micropayments, the economics of Web publishing developed around advertising. The publisher would receive a tiny payment for each page displayed, but from an advertiser rather than the reader.[34] The shift to advertising favored large, commercial Web sites able to attract the interest of advertisers and deploy ad server technologies. In contrast, amateur Web publishers had no easy way to profit from their work, however popular their sites.

Slate, for example, had to support itself with advertising once Microsoft's patronage was withdrawn. An experimental effort to limit access to subscribers, in 1998, was abandoned when it attracted only twenty thousand readers. *Slate* remains popular through several reinventions, winning awards for its journalism, although its most successful writers tend to move on to other venues because online advertising provides nothing like the support once enjoyed by successful print magazines. (Slate's paywall returned in 2020, as falling advertising revenues forced many online publications to ask frequent readers to purchase subscriptions.)

Google and Search Advertising

Web directories and search engines were particularly well placed to attract advertisements. By the late 1990s, just three firms were estimated to be receiving 43 percent of all online advertising revenues: AOL, Yahoo, and Microsoft.[35] Internet navigation companies had a significant commercial advantage: they knew what their visitors were thinking about at any given moment. This information was harnessed most effectively by Google. In 2000, it began selling advertisements, with an approach it called AdWords.

Google realized that users disliked large, distracting advertisements and would not trust a search service in which the top results appeared only because users had paid for them. Google instead presented a single paid result, clearly labeled as a "sponsored link," and a handful of simple advertisements, each consisting of three short lines of text, grouped in a separate part of the screen.[36]

Advertisers entered bids online to display their ads when particular terms were searched for. Google copied this idea and its pay-per-click model, in which advertisers are paid only when a viewer clicks an advertisement, from an Internet advertising pioneer called Overture.com. In deciding which ads to display, Google's algorithms factored in how often advertisements were clicked on as well as the amount bid by the advertiser. Google's increasing popularity and ability to show advertisements directly related to whatever its users were researching made its advertising exceptionally profitable. Indeed, the very sparseness of its ads raised their value by decreasing the supply.

By 2006, Google was selling far more advertising than any other Internet firm and had overtaken traditional media powerhouses such as the broadcast television networks and Gannett (owner of more than a hundred daily newspapers including USA Today, twenty-three television stations, and more than a thousand other periodicals).[37] Much of the revenue came not from Google.com but from advertisements placed on other websites. Since the mid-1990s, specialist Web advertising firms had sold advertisement space in bulk. With no changes to browsers, Web publishers could deploy ad server programs on websites to incorporate banner and pop-up advertisements into each Web page delivered. The advertising company controlled which advertisements were shown and where. Google's AdSense system refined the concept, offering the same bid system and pay-per-click model used on its own website.[38] Revenue is shared between Google and the operators of the websites where advertisements are displayed. This helped to shift the economics of Web publishing back toward smaller amateur and semiprofessional ventures. (It also created a new and hugely profitable industry of sites holding nothing but Google advertising links, placed on attractive yet vacant domains such as clothes.com or on sites such as nytomes.com, yagoo.com, or ebey .com that were reached by mistyping more popular domain names.)

Dynamic Page Generation

Many early websites were structured as online magazines, newspapers, books, or exhibitions. These followed the original idea of the Web as a hypertext publishing system. Other sites were established to make money by selling goods. They were more like mail-order catalogs—users browsed merchandise pages, identified items of interest, and filled in a form to pay for them with a credit card. As we discussed previously, the Web lacked features like searching, micropayments, and permanent links that experts thought were essential for any public hypertext system. As a platform for online business, it was even less capable. Any website designed to support online sales had to have the capability to customize at least some of its pages, depending on what a shopper was looking for. Shoppers had to be able to choose their purchases, see a confirmation, enter shipping and payment details, and receive a receipt. Being able to search for products by name would also be helpful.

Nevertheless, *e-commerce* websites quickly emerged to push the development of the Web in more commercial directions. One of the first, and most popular, was Amazon.com, which began in 1994 as an online bookstore. Its website, like that of several other pioneering online stores, used code to track customers through a shopping session, accumulating their selections on a *shopping cart* page. That soon became an almost universal model for online sales.[39] Adding features of this kind required a lot of work to extend Web browsers and Web servers to support secure data entry and dynamic page generation. The one relevant capability the Web did have, right from the beginning, was forms for users to enter data. This had first been used to provide access to the CERN telephone directory. Users had to be able to submit to the Web server the name they were searching for. Instead of delivering a page stored on disk, the server passed the query to a script that interrogated a database, found the matching records, and formatted the results as HTML for the browser.[40] The Web browser was working somewhat like a terminal, sending inputs from a user to a remote computer and displaying its responses.

Over time, more and more of the Web was generated dynamically, to customize it for particular users. For example, Amazon began to fill its main page with items related to the interests and previous searches of each customer. This was made easier by the introduction of *cookies*—snippets of data stored in Web browsers so that a website could recognize a returning customer and easily trace her path from one screen to the next.[41]

Each interaction meant reloading the entire page. Web users filled values in predefined fields by typing in text boxes, clicking radio buttons, or choosing values from a list. As with a mainframe system, there was a noticeable lag between submitting a set of values and getting back the custom page holding the results. Clicking one button would add an item to a shopping cart. Then clicking another would start the checkout process, which would pass though several screens of data entry to specify address and credit card details.

Early Web servers relied on a simple mechanism to generate pages dynamically: the extension at the end of the URL let the server know whether to fetch a page from disk or to launch another program to generate the requested HTML data. In the early days, that extension was most often .pl, to mark it as a script interpreted by a program called Perl. Perl had been developed by Larry Wall, a systems administrator, back in 1987, and it spread rapidly as a convenient way of automating tasks on Unix systems.

As most early Web servers ran on Unix machines, Perl was a natural choice for tasks like generating Web pages based on the CERN telephone directory. Its programming language borrowed capabilities from many other programming languages, making it very flexible, but its particular strength is in grabbing information from input text and processing it. Perl was designed for quick programming rather than efficiency of execution or formal elegance. The book that popularized it, *Programming Perl* begins "Perl is a language for getting your job done," then promises to help readers "develop the three great virtues of a programmer: laziness, impatience, and hubris."[42] In 1998, journalist Andrew Leonard described Perl as "the indispensable duct tape, or glue, that holds the entire Web together."[43] Many large-scale websites, including Craigslist, Yahoo, and Priceline.com were coded largely in Perl.

Encryption

As well as dynamic page generation, online shipping sites such as Amazon depended on having a secure way to transmit data from browser to server. Web traffic, like everything else on the Internet, was broken up into TCP/IP data packets. Because of the Internet's noncommercial origins, these packets were not encrypted. They could be sifted for credit card numbers or passwords by people in any of the networks they passed through en route to the server.

It was too late to add encryption to the Internet's underlying protocols and infrastructure, but HTTP was a much less mature protocol. Because it ran over TCP/IP, it could be extended without having to modify networking equipment. HTTPS, first deployed by Netscape in 1994, added a new layer of code between applications and the Internet: the *secure sockets layer* (SSL).

SSL packaged for public use the results of a series of breakthroughs in *public key encryption* made during the 1970s and 1980s. The technique was introduced with the RSA algorithm (named after Ron Rivest, Adi Shamir, and Leonard Adelman). One early application was the *pretty good privacy* (PGP) system for email messages. Someone wishing to receive secure messages shares a public key used to encrypt the messages heading for them. These can be decrypted only with the corresponding private key, which the recipient keeps secret. The scheme for Web encryption was more complicated than PGP, as packets needed to be encrypted in both directions. This relied on another technique called Diffie-Hellman key exchange (named after Whitfield Diffie and Martin Hellman) to safely negotiate an encryption key for a session before encryption was turned on. Websites had to prove the authentic ownership of the keys they provided by sharing a *digital certificate*. The digital certificates were issued by Verisign, a company spun out from RSA Security for the purpose. Both the RSA and Diffie-Hellman teams won Turing awards for their achievements in securing electronic communication. The US government made ongoing efforts to restrict the availability and export of strong encryption, leading to repeated clashes with the cyberlibertarian community. One of those battles, over an effort to mandate use of a Clipper Chip whose keys were recoverable by law enforcement officials, peaked in 1994 just as the Web began to take off.[44] Through all this, most online users found encryption too complicated to bother protecting their email messages.

Making the Web secure enough to handle payments meant finding a way to build encryption seamlessly into browsers and servers. That got off to a rocky start. The US government initially forbade export of versions of Netscape able to use harder-to-crack 128-bit keys. The digital certificates were confusing to users. Serious security flaws were discovered in early versions, necessitating their rapid replacement. Website operators often let their certificates expire, causing confusing warnings. Over time, however, SSL and its successors became a seamless part of the Internet experience. They spread beyond banking and shopping applications, to help less sensitive websites like those of the *New York Times* and Wikipedia protect user privacy and guard against tampering. This became particularly important as users shifted from wired connections to wireless networks, upon which eavesdropping is absurdly easy.

THE BROWSER WARS

During the second half of the 1990s no battle in the computer industry received more press, or seemed more important, than the "browser wars" fought primarily between Netscape and Microsoft. That might seem odd to current Web users. Today there are several popular browsers to choose from, but they all do basically the same job and compete on efficiency, stability, and user interface details. The choice seems one of brand loyalty, like Coke versus Pepsi, rather than of competing values. And even though Microsoft beat Netscape, its Internet Explorer browser eventually lost its dominant position. So did the struggle matter at all? Our answer is that it did, although perhaps not as much as it seemed at the time and for different reasons.

Netscape's Early Lead
Any well-equipped personal computer already had a copy of Microsoft Windows, a Microsoft Office Suite, and an antivirus program. In 1995 it seemed that a Web browser was about to become the fourth essential software purchase for every computer user. Netscape adopted a version of the classic shareware software business model: Navigator could be downloaded without payment, but commercial users were required to register and pay for the software after a trial period. The basic price was ninety-nine dollars a copy, with discounts for corporate licensing.[45] Shrink-wrapped copies could be

purchased in computer stores.[46] It had locked up 80 percent of the market for browser software, having eclipsed Mosaic and dozens of smaller companies trying to sell browsers.

On August 9, Netscape went public, setting off a frenzy among reporters and investors as its stock soared. Netscape, investors believed, would be the Microsoft of the Internet era—earning huge profits by selling server software to Web publishers and browser software to companies and individuals. Its software existed at the intersection of two worlds: the Internet world of academic research in which code was usually given away, and the commercial world of personal computer software, in which most desktop or server software was purchased in shrink-wrapped boxes. Backers of Netscape had expected the commercial model to apply to Web browsers. They turned out to be wrong, but there is nothing inevitable about the fact that we pay for tax software but not for Web browsers.[47]

Microsoft's Response

Microsoft wanted to make sure that it was the Microsoft of the Internet era. It was a deeply competitive company, which had already vanquished once-dominant foes in the markets for word processing, spreadsheet, business presentation, compiler, and database software. It was promoting its own graphics, desktop publishing, multimedia, and personal finance packages to challenge the few remaining mass markets where independent PC software companies could earn healthy profits.

Microsoft had the Internet in mind when developing Windows 95, borrowing solid code for TCP/IP data packets from BSD Unix, and integrated the capabilities needed to run those packets over a dial-up modem link. It could smoothly multitask, running bundled software for email or terminal emulation along with whatever else a user was working on. In fact, the Windows 95 CD included every essential Internet tool except for a Web browser. Instead, Microsoft was prominently promoting its own MSN online service. Microsoft had licensed the Mosaic code to produce a simple browser, Internet Explorer (IE), but it was available only as part of Microsoft Plus, a bonus disk designed to extract an additional $49.95 from eager customers for a grab bag of fun extras like a pinball game and set of desktop themes.

Suddenly the Web seemed like the biggest opportunity of all. In May 1995, Bill Gates sent a now-famous memo entitled "The Internet Tidal Wave" to his fellow executives. It warned that "the Internet is critical to every part of our business . . . the most important single development . . . since the IBM PC" upon which Microsoft had built its business. As a first step, Gates ordered the shift of "all Internet value added from the Plus pack into Windows 95 as soon as we possibly can."[48] Improved Web capabilities were to be "the most important element" of all new releases of Microsoft application software, and to make "every product plan go overboard on Internet features." Gates even delayed publication of his book *The Road Ahead* to insert references to the Internet into its lengthy discussion of the "information highway."[49]

Internet Explorer jumped from the backwaters of Microsoft Plus to the heart of the firm's product plans. IE version 3.0, released just a year after Windows 95, gave Netscape its first real competition, thanks to the efforts of around a hundred programmers. It was free for personal or business use, as was the Web server that Microsoft began to distribute for Windows. This reinforced its dominant position in the operating system market, while threatening Netscape's most obvious sources of revenue and hence its plans to become profitable. Microsoft even started giving away Internet Explorer for Macintosh and Unix computers to make sure that commercial markets for browser software never emerged.[50]

New Browser Capabilities

The browser wars between Microsoft and Netscape were fought on a large scale. By 1999, Microsoft reportedly had more than one thousand people working on the development of Internet Explorer. When one army of programmers added a new feature, the other rushed to match it.[51] The HTML standard was maintained by the World Wide Web Consortium, founded by Berners-Lee in 1994 when he left CERN and moved to MIT. In the mid-1990s it was evolving rapidly, often working backward to standardize new features that had already been added by Microsoft or Netscape. Websites taking advantage of new features included badges identifying their sites as "Best viewed with Netscape Navigator" (or Internet Explorer), which could be clicked to download the latest version of the browser. The additions gave sites more control over the appearance of Web pages: fonts, text size, color, placement of graphics, and so on.

Browsers integrated new capabilities. Hyperlinks had always been able to link to other kinds of Web resource, like FTP sites, Usenet newsgroups, and email addresses. Early browsers were configured to open other software when these links were clicked, but Microsoft and Netscape did not want to provide openings for other software companies. Both added features to their browsers and programs to their bundles to handle these functions.

Both major browsers included *plug-in* capabilities so that other companies would produce software as extensions to the browser rather than free-standing alternatives. This was a period of widespread experimentation to add capabilities to the Web. Some produced brief excitement and failed. Silicon Graphics made a big push for adding explorable 3D worlds to Web pages. After a short period of enthusiasm for this technology as the future of Web interaction, it proved to be a clumsy novelty. Virtual reality modeling language (VRML) never became a standard feature of browsers. In contrast, streaming audio quickly went from a curiosity to a core use of the Internet. It was pioneered by the RealAudio Player package, launched in 1995 to stream audio from radio stations and sports games. RealNetworks gave away this client software, triggered when someone clicked a link to an audio stream. It made money selling the server software needed to encode audio streams. Real's technology dominated Internet streaming in the late 1990s, expanding to video as well as audio, but faded once Microsoft and Apple produced their own alternative encoders at no extra cost.

Not content just to build a better browser, Microsoft looked for ways to enlist its dominant Windows and Office products to beat Netscape. Microsoft preinstalled Internet Explorer on every computer, making it the default choice. It forced PC companies to stop distributing Netscape on their computers, and it made a deal with AOL to make Internet Explorer the default browser for its customers.[52] It integrated the software into Windows, so that Web channels could place content directly onto the computer desktop and users could browse files on their local drives with a Web-like interface. One technology reporter suggested that it "sticks its tendrils into every part of your operating systems and will wreak havoc on your computer" with a "nasty habit of crashing every thirty minutes."[53] Microsoft then insisted that Internet Explorer was so deeply integrated that it was technologically impossible to let users remove it or substitute another browser. Few outside the company found this claim credible.

Microsoft added features to its Office software that worked only with Internet Explorer, so that companies wishing to take advantage of them would have to switch browsers. It marketed the newfound ability of Word, Excel, and PowerPoint to output documents as Web pages as the key feature of the next release of Office. It also acquired a separate Web editor program, FrontPage. Many of the new features required the use of Microsoft servers to function properly, using the popularity of its desktop software to build a stronger position in the booming market for corporate intranets (internal networks based on Internet technologies). These tactics were part of its push to embrace and extend Internet technologies, gradually turning them from open standards into proprietary systems controlled by Microsoft.

Netscape Loses

Netscape wilted before the onslaught, making some significant mistakes. It produced browsers for many platforms, devoting effort to obscure versions of Unix that could have been better spent on improving core products. It packaged its browser into a suite with email and editing software, loading these with new features that slowed them down. New releases were rushed out without enough testing, leaving them bloated and buggy. By the time Netscape 4 went up against IE 4 in 1997, reviewers noted that Internet Explorer was faster and more stable than Netscape Navigator. Netscape was losing the browser wars and laying off employees, although that didn't stop America Online acquiring it for stock worth more than ten billion dollars in March 1999.[54]

That was part of an aggressive expansion by AOL, which culminated with its 2001 acquisition of Time Warner, the world's largest media company, controlling brands from HBO and CNN to Warner Brothers movies, *Time* magazine, and *Sports Illustrated*. This marked the high point of the dot-com (for URLs ending in .com) boom in Internet-related shares. The persistent appetite of investors for anything Internet related had pushed the stock prices of established companies, like AOL and Microsoft, up to stratospheric levels and created a ready market for initial public offerings by obscure start-up companies. Venture capitalists would fund a company with a plausible-seeming idea without worrying too much about whether it would ever make a profit. As long as it used their investment to expand rapidly, by giving away its services or selling products at a loss, the firm could make an initial public offering within a year and repay their investment handsomely. This was a classic speculative bubble, justified at the time through talk of a "new economy" in which traditional business practices and methods of stock valuation were hopelessly outmoded. The market began to teeter in mid-2000 and came crashing down the following year, just as the AOL Time Warner merger closed. Within the merged company, the broader collapse in the market for technology shares was mirrored by an implosion of AOL's management team, revenues, and reputation.[55] In 2002, the firm suffered what remains to this day the biggest corporate annual loss ever, a remarkable $98.7 billion, as it wrote off almost the entire value of the AOL assets left on its account books.

The chaos at AOL didn't help efforts to revamp the Netscape browser, which had already bogged down. When AOL finally disbanded Netscape in 2003, conceding defeat, Internet Explorer had more than 90 percent of the market. Microsoft took great risks to accomplish this. The US government and twenty states sued Microsoft in 1998 for illegally thwarting competition. Its actions against Netscape were a major part of the case. When a parade of industry executives from firms including Sun and Intel testified to its abuses, the firm's public image was badly shaken. Some of Microsoft's evidence turned out to have been tampered with, and several of its executives appeared arrogant or untrustworthy. Most damagingly, Bill Gates himself appeared evasive and dishonest when deposed, repeatedly claiming not to recall events described in his own emails.[56] The trial proved a turning point for Gates, who since 1995 had been the richest person in the world.[57] In its aftermath he shifted away from hands-on management of Microsoft and toward a second career running what eventually became the world's largest charitable foundation.

In 2000, Judge Thomas Penfeld Jackson ruled that Microsoft had used anticompetitive means to maintain a monopoly in the market for personal computer operating systems. It had illegally tied Internet Explorer to that Windows monopoly, destroying the browser market by adopting predatory pricing: "Microsoft paid vast sums of money, and renounced many millions more in lost revenue every year, in order to induce firms to take actions that would help enhance Internet Explorer's share of browser usage at Navigator's expense."[58] Given that Microsoft had promised never to charge for its browser, Jackson continued, this could "only represent a rational investment" if its purpose was to protect the Windows monopoly from the emergence of a rival applications platform.[59] He ordered that Microsoft be split into two separate firms, one with Windows and the other with Office and other applications, to prevent it from using its monopoly in the first to crush competitors in the second. Microsoft appealed. Before the process could be completed, the Clinton administration had given way to the more conservative administration of George W. Bush. Microsoft was allowed to settle the case by agreeing to share its application programming interfaces with other firms, so that Microsoft applications would not benefit from access to secret Windows features. The European Union was less sympathetic, levying record fines totaling several billion dollars against Microsoft in a series of judgements between 2006 and 2013.

Having won the browser wars, Microsoft announced an end to new browser releases except as a feature of new versions of Windows. It had released eight major versions of Internet Explorer during its six-year battle with Netscape, culminating in the arrival of version 6.0 in 2001.[60] The next major upgrade arrived a full five years later, prompted by the emergence of new competition from an unexpected source.

THE TRIUMPH OF OPEN SOURCE

Microsoft won the browser wars, but it never truly gained the power to redefine the Web as a closed system. That had a lot to do with what was happening at the other end of the HTTP connection: the Web server. Microsoft never came close to replicating the strong-to-monopolistic positions it held in personal computer, office server, and Web browser software in the market for Web servers. That had a lot to do with the popularity of a new Unix-like operating system, Linux, which incorporated a great deal of software produced by Richard Stallman's GNU project.

GNU Is Not Unix

Stallman's odyssey had begun at the MIT Artificial Intelligence Laboratory, home of the Incompatible Timesharing System and work on machine architectures optimized for Lisp. In the early 1980s he felt that the ethic of the lab, in which programmers freely shared their work, was being undermined by the founding of a company to commercialize its Lisp machines (figure 12.5).[61]

Stallman resolved to produce software to be freely given away. In a Usenet post in September 1983 Stallman wrote: "Starting this Thanksgiving I am going to write a

Figure 12.5
Richard Stallman, MIT hacker and originator of the GNU project. Taken from the cover of the O'Reilly book *Free as in Freedom: Richard Stallman's Crusade for Free Software* under the Creative Commons Attribution-Share Alike 3.0 Unported license.

complete Unix-compatible software system called GNU (for Gnu's Not Unix), and give it away free to everyone who can use it. Contributions of time, money, programs and equipment are greatly needed." Later in the same posting he stated his philosophy of free software: "I consider that the golden rule requires that if I like a program I must share it with other people who like it. I cannot in good conscience sign a nondisclosure agreement or a software license agreement. So that I can continue to use computers without violating my principles, I have decided to put together a sufficient body of free software so that I will be able to get along without any software that is not free."[62]

Interest in the project was boosted by increasing restrictions on the availability of actual Unix. In 1984, the Bell System was broken up by court order. AT&T exchanged its local phone companies and its monopoly for the ability to compete directly in the computer business. Its plans centered on making money from Unix-based systems. AT&T's demands for hefty license payments and habit of suing vendors suspected of using its code slowed and fragmented the adoption of Unix.

Stallman found that some components of a Unix system were already free—the X Window system for displaying graphics on a terminal and Donald Knuth's TeX typesetting program, among others. To those, Stallman added his own widely used Emacs text editor, which he rewrote for Unix.[63] With the help of a few colleagues, he began producing substitutes for one after another of the Unix tools, building up a body of free software that was comparable to, or even better than, what it took large teams of people to create in the commercial or academic world. He did much of this himself, in an office generously loaned to him by MIT. For a while that was also his sleeping quarters. Stallman's excellent Gnu C Compiler replaced the standard Unix version to establish a de facto standard for the C programming language.

Just as important was the legal agreement Stallman crafted to distribute this work under. With the help of an attorney he developed a GNU General Public License (GPL). In his words, "It is a legal instrument that requires those who pass on a program to include the rights to use, modify, and redistribute the code; the code and the freedoms become legally inseparable."[64] The license not only stipulated that GNU software was free to use and modify, but also required that those who improve it make *their* modifications available under the same terms. It is this last provision that was so radical. Giving away programs and source code was nothing new. In the 1950s, groups like SHARE and companies like IBM did that all the time, as discussed in chapter 2. For the GNU enthusiasts, giving away software was an ideological commitment, not just a pragmatic distribution choice. Stallman insists that the *free* in *free software* did not simply mean provided without payment, like "free beer." The software is "free as in freedom" because its users are empowered to study, modify, and repurpose the code.

Linux Is Almost Unix

Stallman ran into difficulties in developing an operating system kernel, the most vital component needed to turn GNU into a full replacement for Unix. It came eventually from an unexpected source. In 1991, Linus Torvalds, then 22 years old, bought (on an installment plan) a powerful 386-based PC. He was familiar with Unix from a course he had taken at the University of Helsinki. The closest thing he could obtain without spending a lot of money was Minix, developed by Andrew Tanenbaum of Vrije University in Amsterdam, as a simplified Unix-like system for instructional use.[65] Almost as soon as he had it running, Torvalds found that Minix fell short of his needs. Specifically, he needed a terminal-emulation program that allowed him to access the University's software resources and online discussion groups.[66] In a post to a Usenet discussion group in August 1991, he explained "I'm doing a (free) operating system (just a hobby, won't be big and professional like gnu) for 386 (486) AT clones. . . . I'd like any feedback on things people like/dislike in minix, as my OS resembles it somewhat."[67] The next month he posted his work online. At the suggestion of Ari Lemmke, who hosted his files, the program was called Linux. As work progressed, Torvalds began to think of Linux as an alternative to Unix itself, not just a derivative of Minix. That led to a break with Tanenbaum, who registered his disapproval of Torvalds's approach in a posting to the comp.os.minix newsgroup in early 1992. After a heated exchange with Tanenbaum, Linux discussion moved to its own newsgroup.

Linux grew quickly on the shoulders of a worldwide, voluntary group of enthusiasts. Torvalds exploited the Internet's ability to distribute his work cheaply and receive almost instant feedback from users. Torvalds gradually turned over portions of the project to those among that group whom he trusted. The success of Linux rested on all the open source tools already produced by the GNU team. (Stallman has for some decades now been reminding people that the operating system should really be called GNU/Linux rather than just Linux.) Because Linux was free, many different versions, known as *distributions* in the tradition of the Berkeley Standard Distribution of Unix, were created.

Eric Raymond was among those programmers who saw the merits of this model of software development. In an influential essay titled "The Cathedral and the Bazaar," he argued that by letting people look at and modify the source code, bugs are found and fixed faster than they possibly could be in a closed system. As Raymond said (paraphrasing Torvalds), "Given enough eyeballs, all bugs are shallow."[68] Raymond called the approach *open source* development, stressing its pragmatic benefits rather than Stallman's ideological struggle to free the world from software ownership. The distinction was passionately debated, even though, as anthropologist Chris Kelty has noted, the practices and licenses favored by the two sides were almost identical.[69]

The early volunteers of the free software movement were soon joined by paid employees of computer software and hardware companies. The most dramatic signal of this shift came when IBM embraced Linux as an alternative to its proprietary mainframe operating systems. That began as a "skunk works" project at IBM's German lab in Boeblingen, where in late 1999 a team of young programmers succeeded in porting Linux to an IBM 390 mainframe. IBM announced, "Linux is here and Linux is ready. Ready for business, Ready for e-business. Ready for enterprise." One ad showed a fuzzy black-and-white image of a penguin, the Linux mascot, walking through the towers of a mainframe installation.

Although companies distributing Linux were not allowed to charge more than a small fee to cover their costs, they could hope to make much more money selling support and service contracts. The most visible of the Linux companies, Red Hat, went public in 1999. Its stock tripled in price on their first day of trading, closing at a level that valued the fledgling company at three billion dollars. Unix companies had been trying for years to harmonize their systems so that programs written for one version would work on others, providing a unified platform to challenge Microsoft. That never happened, but Linux promised to provide a standardized core operating system. Costing nothing was another bonus. Corel, best known for its graphics software, saw Linux as a chance to challenge Windows as the standard desktop operating system. Unix was much harder to install, configure, and use, but Linux could be modified as needed. Corel invested millions of dollars to make Linux easier to use and install, releasing its first public version in 1999. Lindows, founded in 2001, had a similar goal, promising a version of Linux that made it easy to run programs written for Windows.

As a challenge to Windows on desktop computers, Linux flopped, although the threat did force Microsoft to price Windows more aggressively. Choosing Linux did not save a consumer much money, particularly because Microsoft forced PC makers to license Windows for every computer sold, whether customers wanted it or not. Most of the programs that ordinary users wanted to use were available for Windows but not for Linux. Linux remained daunting. For example, users had many choices for things like file management systems and desktop environments, which prevented software companies from producing applications that would work easily on all versions of Linux. Only software developers and open source enthusiasts were likely to run Linux on their personal computers, accounting for maybe one or two percent of the market in the early 2000s. That was less market share than Unix had had a decade earlier, back when expensive workstations were used for technical computing and financial modeling.

The LAMP Open Source Stack

For servers, the situation was quite different. In 1994, as the Web took off, Windows NT was beginning to build up momentum as a challenge to Novell servers for office use and to Unix servers for more demanding jobs. Although Windows NT could do a creditable job serving Web pages from cheap, standard PC hardware, it never dominated the market for servers the same way it did for desktop operating systems. Most early websites ran on Unix servers or on BSD, which had evolved from a package of Unix upgrades to a free-standing alternative with no AT&T code. Unix systems were expensive, which drove up the cost of operating Internet sites. Instead of shifting to Windows and PC hardware to reduce costs, Web companies saved even more by relying on the free Linux operating system.

By the early 2000s the other key software components of a Web application server were also increasingly likely to be free software. The first to gain dominance was Apache, which has been the most widely used Web server since 1996. The name was a play on words: it began as a collection of software patches to extend the NCSA Web server, itself based on code from CERN. Soon Apache became a robust and extensible piece of software in its own right. This killed the commercial market for Web servers before it had fully developed.[70]

Web application servers need a database management system to store information on users, products, and orders. Oracle was popular in the early days of the commercial Web, but like the proprietary Unix servers on which it ran, it was an expensive choice. MySQL, launched in 1995, was a free alternative. It was produced by a Swedish company that gave away the basic product and sold a version with extensions intended for larger companies. The first versions were suitable for simple applications only, making it a natural pairing with Web systems that often had been developed separately from a company's main databases. Over the next fifteen years, it gradually closed the edge with Oracle in features, adding capabilities such as triggers and stored procedures. Since 2010, MySQL has been owned by Oracle following that firm's purchase of Sun, which had previously acquired MySQL. This illustrates the extent to which free software has strengthened, rather than eliminated, powerful technology firms like Oracle and IBM.

Most Web applications run on the stack of software called LAMP, which stands for Linux, Apache, MySQL, and PHP, a system that gradually replaced Perl as the default choice for coding Web applications. PHP is the most popular programming language created specifically for the Web. Web application servers such as PHP processed code fragments interleaved with regular HTML content within Web pages. Then the server filled in the results of running this code before it sent the pages to the Web browser.

Application servers integrated support for SQL statements to be run on database management systems, giving an easy way to pull information from databases and format it within Web pages. That was a crucial requirement for most Web applications. Application servers also overcame challenges produced by the way in which Web servers receive each request for a page as an isolated event. For serving simple, unchanging pages that posed no problem, but to do something like tracking the products had added to an online shopping cart, a website had to track its customers from one page to another. Web application coders had to store separate sets of variables for each user *session* and track page requests against the users making them. Web application servers did both things automatically, simplifying the job of coders.

Companies found that licensing a commercial application server software such as ColdFusion, which could cost $5,000 for each large server, would quickly pay for itself with quicker application development and fewer bugs. Like MySQL, PHP began as a crude but free alternative. In 2001 *PC Magazine* rated PHP just 2 out of 5, behind commercial alternatives from companies like Microsoft, Borland, and IBM.[71] Despite being deemed "too immature for a high traffic business environment," PHP was nevertheless popular for small, low budget jobs—as reflected in its name, which originally stood for *personal home page*. Code libraries appeared to extend its capabilities by adding missing features. PHP was becoming a credible competitor, even though its core language and libraries remained frustratingly inconsistent because of its incremental development path. By the 2010s, PHP was a standard part of most Web installations, underpinning a huge number of popular websites and applications. For example, the popular Web publishing systems WordPress, Joomla, and Drupal are all written in PHP. Considered together, they power more than a third of the world's websites.

The bigger point here is that Microsoft was never able to turn the Web into a proprietary system because it couldn't match its domination of the browser side of the Web with similar control over the servers that generated Web pages. If Microsoft's Internet information server had also held a market share of over 90 percent, then Microsoft could have gradually shifted the Web from a system based on open standards to a system in which an all-Microsoft stack of software would suffice. As most websites used free software, even the success of Internet Explorer did not give Microsoft the power to unilaterally set Web standards for its own benefit.

The Firefox Browser

The Web's continuing openness made new challenges to Internet Explorer possible. It also reduced Microsoft's strategic interest in fighting vigorously to defend its position. For example, in 2003 the firm discontinued Internet Explorer for Macintosh, which it had previously paid Apple to make the default browser.

In 1998, as Netscape's browser business unraveled and its corporate strategy shifted toward the Web-portal market, its management had gambled open source development as a way to leverage its limited resources against Microsoft, releasing the full source code for its browser suite in the hope that this might produce a flood of improved variants. The resulting Mozilla browsers were clumsy to use and burdened with a mass of extraneous features, winning support only from the most dedicated open source supporters.

Then, seemingly suddenly in 2004, national publications such as the *Wall Street Journal* and *New York Times* took notice of a new browser, Firefox.[72] Version 1.0 was released in November, after the program had already won a large and loyal following. In 2002, a Netscape programmer and a teenager working as an intern had resolved to create a new version of Mozilla, pared down to its essentials.[73] The project gained steam rapidly, recruiting experienced developers, and eventually became the flagship product of the Mozilla project. Its appearance coincided with a rash of security attacks exploiting Internet Explorer's tight integration with Windows, giving users a convincing reason to shift. Firefox boasted other simple but useful features missing from Explorer, particularly the ability to open several pages as *tabs* within a single window and shift rapidly between them. Its capabilities were quickly extended by dozens of add-on programs, customizing Firefox to do things like block Internet advertisements.

Within a year of its official release, Firefox had more than one-tenth of the worldwide browser market. At its peak in 2009–10, Firefox had almost a third of the market. These gains eroded the position of Microsoft, whose share had already fallen to about half of all browser users. Firefox was the first open source desktop computer application widely used by Windows and Macintosh users. Its triumph signaled a shift in the computing landscape. Microsoft's hold on desktop operating systems and office applications remained secure, but the firm's attempts to dominate and enclose the Internet were visibly crumbling.

13 THE COMPUTER BECOMES A NETWORK

One of us has a twelve-year-old son. Perhaps unsurprisingly, the boy is at home with technology, easily navigating his Android phone, our iPads and Rokus, the school's Chromebooks, and several game consoles. He's also at home with Windows, using a dual-screen PC for video gaming, his writing assignments, and producing audio and video with Adobe's creative suite. Yet, unlike his father, he has never used Microsoft Word or any other desktop office software. He uses Google Docs to write and edit text and doesn't understand why the older generation sticks with Microsoft's applications. When it is time to write, he fires up the Chrome browser and opens his cloud documents store. He opens the same browser when he wants to watch a YouTube video, play offensive cartoon-like video games, or log into his school account.

When he began to use PCs during the late 2010s, the Web was already much more than the hypertext publishing system we described in the preceding chapter. Rather than retrieving and displaying pages stored as static files, browsers have become a universal interface for online applications running in the *cloud*—a distributed network of gigantic data centers each composed of thousands of computers. Even grownups began to find it normal to use one application, the Web browser, to read email, access company functions like work time reporting, find a date, buy a book, or keep in touch with friends.

This modern Web bridges the different modes of computing we have explored in previous chapters. Mainframes could handle high-capacity online applications with thousands of transactions a second, managing giant databases, but they were enormously expensive. Dealing with them meant using clunky text-based terminal interfaces. Personal computers could devote far more processing power to each user, and they had fast, colorful graphics and interactive applications. But through the 1980s, they were rarely networked. Users worked with their own private islands of data. Combining these models fulfilled a longstanding goal, once called *distributed computing*. Sun's bold slogan, "The Network is the Computer," emphasized that its workstations were designed to be networked to servers. Client server applications were increasingly common by the early 1990s but were typically used inside organizations, not for publicly accessible services.

The Web changed all that, gradually rolling back the personal computer model of using applications installed on the computer to work with data held on its hard drive. Having customers do their own data entry via Web browsers eliminated much of the work done by data entry clerks (the descendants of key punch operators) and by call-center employees. People could view and update their data with Web applications, with no humans standing between them and the corporate databases in which it resided. Until this point, the investments made by business in computers from the 1950s onward had not produced clear productivity benefits. In fact, productivity improvements across the economy slowed in the 1970s, just as information technology was becoming a significant fraction of overall business investment. The disconnection between the apparent benefits of computerization on individual firms and the plodding pace of economic growth was dubbed the *productivity paradox*.[1] From the late 1990s to 2005, however, the American economy experienced a jump in productivity that most economists attribute primarily to efficiency improvements caused by adoption of Internet technologies.

DATA CENTERS AND THE CLOUD

Early Internet companies relied on expensive Unix-based servers with multiple processors and custom, high-reliability disk storage and memory chips. These were operated in heavily air-conditioned data centers. To avoid service disruptions companies *mirrored* hard disks to maintain duplicate copies, and *clustered* servers so that a synchronized backup was ready when needed. By the 1990s, many mainframe and minicomputer firms like Unisys (the heir to Univac) and Data General had reoriented themselves to sell powerful servers based on standard processor chips. The spread of the Internet expanded this market. Major websites soon outstripped the capabilities of any single server, even a late-1990s flagship Unisys server with thirty-two Intel processors. Companies set up *farms* of servers running Web applications, with a *load balancing* system to route each new request to the least busy server. Storage area networks provided ultra-high-speed connections between servers and disk pools. The technological lines separating mainframes, minicomputers, and personal computers were starting to blur.

Itanium—The Future That Wasn't

PC-based servers could provide performance and reliability to rival traditional mainframes, but as with mainframes, this required expensive components. Most servers used special Xeon branded versions of Intel's Pentium chips. In 2001, after a joint development project that consumed billions of dollars and ran years behind schedule, Intel and Hewlett-Packard began to sell chips based on an entirely new architecture. Itanium was expected to first replace existing processors in higher-end servers, before reaching workstations, and finally ordinary PCs. Its 64-bit architecture supported very large amounts of memory and ran complex computations more effectively.

Itanium never caught on. Part of that was weak technology. Itanium chips had much less advantage than expected over Intel's existing processors. Only Hewlett-Packard itself used the chip extensively, in servers costing several hundred thousand dollars. More fundamentally, however, the high-performance server market that Itanium targeted was becoming less important. Internet companies obtained the daunting quantities of storage and processor power they needed by using massive numbers of ordinary computer boards joined together rather than spending huge amounts of money for smaller numbers of specialized processors with exotic storage and memory systems.

Google's Data Centers

The new approach was pioneered by Google. It became one of the world's most valuable companies by providing much better results than its competitors in two areas: Web search and Web advertising. Its success is usually attributed to superior algorithms, particularly the PageRank algorithm its founders created as graduate students. That gives only a part of the picture. Google's algorithms provided better search results to its users, and its advertising system made more money by selecting relevant ads that users might click. But running those clever algorithms consumed more processor cycles and RAM than the simpler approaches of its competitors.

In its earliest days, as a Stanford project, Google's search engine used an expensive dual processor Sun server. Google's business model hinged on giving access to services that consumed spectacularly large amounts of computer power. To have a chance of ever becoming profitable, the firm needed to dramatically lower the cost of computation. As Steven Levy described in his book *In the Plex*, when Google moved off campus it began to build its own servers. In 1999, it hired Urs Hölzle, who arranged for the purchase of two thousand cheap PC motherboards. Google staffers assembled these into cut-price servers (see figure 13.1). As Google grew it used hundreds of thousands, and eventually millions, of these computers running together, saving the cost of a graphics card and a case. The difference of scale became clear to outsiders in 2004 when Google launched its Gmail service with a full gigabyte of free storage, two hundred fifty times more than Yahoo's market-leading email service provided.

Conventional servers were expensive because they used more reliable, higher performance components. Google achieved reliability and performance with an extra layer of software. Motherboards and disk drives failed all the time, but users never even

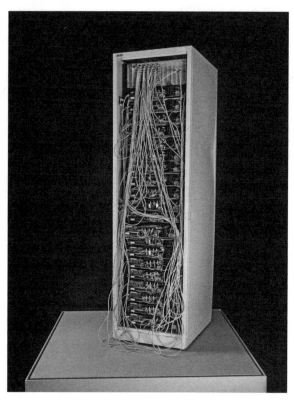

Figure 13.1

Google "corkboard" server, 1999, one of thirty servers built by Larry Page and Sergey Brin early in the company's history, establishing a tradition of cheap homebrew hardware. Each row in the rack held four motherboards and eight hard drives resting on a piece of cork. Courtesy of Google, Inc. Image provided by National Museum of American History, Smithsonian Institution.

noticed. A piece of software, the Google file system, provided reliable storage by keeping data redundantly. The core capability Google developed, the cryptically named MapReduce, let programmers specify huge tasks in a way that facilitated spreading them across many processors. It automatically split the work into small pieces and reassembled the results. If a processor suddenly failed, its share of the job would be repeated by a different unit. As Google developed other services, these were also coded to take advantage of MapReduce. The aim was to "treat the data center itself as one massive warehouse-scale computer."[2] As Google grew large enough to build its own data centers, its architects were similarly creative. For example, Google cut the cost of cooling substantially by letting data centers get hotter and using water rather than air conditioning to carry heat away. (Many companies also experimented with building data center equipment into standard shipping containers, so that data centers could easily be relocated or expanded.)

The MapReduce approach was taken up by software designer Doug Cutting, then working for Yahoo, and computer science graduate student Mike Cafarella for a new open source project, Apache Hadoop. Its first version, in 2006, was immediately applied by Yahoo, then trying to get ahead of Google in Web search. By early 2008, Yahoo was running a Hadoop cluster with more than ten thousand processor cores. Facebook, LinkedIn, and eBay quickly followed, rebuilding their services around Hadoop and contributing their own improvements to its code. Users, and even programmers, had no idea which of thousands of computers scattered across the world were storing their emails or hosting their word processor.

The new approach boosted the popularity of a catchy term: cloud computing (figure 13.2). The metaphor hid all the messiness, energy use, and human labor involved in data centers. Instead data seemed to float weightless in a realm of pure information.[3] Microsoft, IBM, and Amazon all developed large-scale cloud services so that companies could sign a contract to have their company's email systems or custom Web applications hosted. Companies spent a huge amount maintaining their own server farms, most of it not for hardware but for IT staff, backup and disaster recovery capabilities, and so on. Cloud services could spread those costs over many users, making computing more efficient and manageable. Amazon Web Services, the leading provider, brought in more than \$35 billion during 2019 by providing cloud infrastructure for firms like BMW, Korean Airlines, Samsung, GE, and Unilever.

The processor architecture that evolved from the humble 8088 chip used in the original IBM PC had become the engine for cloud computing clusters, high performance servers, and even mainframes, as well as laptop and desktop computers. Even Itanium's signature feature, 64-bit support, was matched by mainstream chips when Intel's rival, AMD, found a way to add it to Pentium-compatible chips without compromising their ability to run existing code. In 2004, Intel copied this system for its own processors, effectively setting the industry standard for 64-bit computing.

Multiple Core Processors

The idea of boosting performance by running processors in parallel applied within individual computers as well as data centers. From the 1970s to the early 2000s, almost all personal computers had a single processor chip. Following Moore's law, the power of these chips improved rapidly and consistently, driven by three main shifts.

First, processors were driven through more cycles each second. In 1984, IBM's fastest personal computer, the PC AT, ran its processor at 6 MHz. Twenty years later, Intel offered a version of its Pentium 4 that cycled more than six hundred times as fast.

Second, processors did more work in each cycle. For example, the shift from 8-bit to 64-bit architectures improved performance by letting the processor work on larger

Figure 13.2

Top: Data center, Ashburn, Virginia. Cloud servers are located around the world. The data centers in Ashburn, just north of Dulles Airport in Loudoun County, Virginia, may be the epicenter of cloud storage but their exteriors give little away. Photo by Paul Ceruzzi. Bottom: inside a T-Systems (Deutsche Telecom) data center in Biere, Germany, in 2014. A technician removes a standard-sized rack mount unit from a chassis into which servers, network switches, backup power supplies, and storage arrays all fit. These rack-mounted servers placed a PC motherboard, drives, and expansion cards into a compact, easily swappable case. Photo: Thomas Trutschel via Getty Images.

chunks of data. Pipelining used different units within the processor more efficiently by letting them work simultaneously on parts of different instructions.

Finally, system designers decoupled the core of the processor from the rest of the system. As RAM chip access times continued to lag behind processor speed, they relied on larger and more complex cache systems to keep the processor working productively.

All three of these changes were made possible by the ability to create smaller transistors and interconnections. Smaller transistors needed less power and could be driven faster without heating enough to melt the chip. More could be squeezed onto a single chip, providing room for logic to implement features such as pipelining and for cache memory to boost performance.

By the early 2000s, Intel was running up against the limitations of this approach. At high clock speeds, transistors were leaking charge through insulating materials, as the rules of quantum mechanics came into play. Processors were consuming more power, necessitating ever more elaborate cooling systems. The Pentium 4 was more complex than the Pentium 3 but got less done in each cycle because of trade-offs made to enable its very high clock rates. Some versions of the Pentium 4 had as many as thirty-one stages in their instruction pipelines. Working so far ahead hurt performance whenever an unexpected branch made it necessary to empty and refill the pipeline.

Rather than make the anticipated transition to 4 GHz and beyond, Intel shifted its strategy. Newer processors improved performance by taking fewer clock cycles to execute the average instruction, rather than by cycling faster. Power efficiency was now as important as raw performance, given the growing popularity of laptops and the threat posed by ARM's ultra-efficient architecture. In 2006 Intel introduced the Core brand, relegating its older Celeron and Pentium brands to budget-priced chips. The Core microarchitecture boosted overall performance while lowering clock speeds. Computer marketing began to de-emphasize clock speed, formerly the biggest selling point on most PC advertisements. That had always been misleading—like marketing a sports car according to the maximum RPM of its engine rather than the acceleration this accomplished.

The most important performance feature was the presence of two complete processor cores on each Core Duo chip, increasing to four cores with the Core 2 Quad models launched in 2008. Since the 1990s, Intel chips had often been used in dual processor workstations. But the cost of two high-performance processors, and the special motherboards able to hold them, meant that this approach never took off for normal computer users. Squeezing two or more cores onto one chip made it the norm.

Since 2010, Intel's branding has not revealed to consumers how many cores their processor has or what microarchitecture it uses, as tags like Pentium MMX and Core 2 Quad had previously done. Its product numbering was purely relative: Core i3 processors offered a good balance of power versus price, Core i5 processors gave higher performance in more expensive computers, and Core i7 models were for the old work-station market: users willing to pay a significant premium for faster and sometimes more power-hungry chips (see figure 13.3). New technology and features usually appeared at the high end and trickled down. The main designation of i3, i5, or i7 was followed by digits that gave experts a general sense of the relative performance of a specific chip. The scheme seems to have been borrowed from BMW, which used numbers to make sure that no special knowledge was needed to understand how impressive a neighbor's car was. A car with a 540 badge was clearly more powerful, expensive, and full of features than the 320 parked down the street.

Desktop and mobile versions of chips sold under the same brand could be quite different. The eighth generation Core chips introduced in 2018, for example, spanned the gamut from a mobile version of the i3–8130U, with just two cores, up to the new i9–9980XE desktop chips, with as many as 18 cores and a price tag of $2,000. The latter had only a 3 GHz base clock speed, cycling more slowly than Intel's flagship had a decade earlier, but offered greatly improved performance.

Intel's chips even found their way into mainframes. Designing and building custom processors for the mainframe market was no longer economically viable. Unisys undertook a lengthy project of migrating the distinct mainframe operating systems it inherited from its two constituent parts, Univac and Burroughs, onto a new range of Intel-powered hardware. This was a feat of emulation, making extensive use of Intel's virtualization features to provide reliable high-performance computing. According to the project's leader, "we didn't expect clients to ever recompile their program," meaning that executable programs compiled back in 1972 had to run without modification. The raw power of modern chips was enough to overcome the performance penalty imposed by emulating the old hardware. In 2013, a customer "that processes transactions totaling $2 trillion each day [probably the SWIFT banking hub] transitioned seamlessly, overnight, from a proprietary Libra system to one based on . . . Intel Xeon technology—and nobody noticed."[4]

A desktop computer with eighteen processor cores on a single giant chip brings an impressive degree of parallel processing. Compared to the latest graphics hardware, though, that is barely parallel at all. As of early 2019, Nvidia lists its flagship Titan V graphics card for a price that might give pause to even the most dedicated gamer: $3,000. Yet given its 5,750 processor cores, it could also be viewed as an extremely affordable alternative to a large building stuffed with mainframes. Nvidia markets the Titan for "deep learning" applications, claiming a throughput that would have made it the world's most powerful supercomputer just fifteen years earlier. Similar chips drive the world's fastest supercomputers, including the Titan system that Cray built for Oak Ridge National Laboratory in 2012 and its successor, the Summit computer delivered

Figure 13.3

A home-built gaming PC. The limited-edition case (2019) was designed by iBuyPower to celebrate the *Fallout* series of role-playing videogames. Its glass side shows off the components within, lit by two Corsair fans cycling through a range of colors. The compact Micro ATX scale motherboard, filling barely half the available area and decorated with racing-inspired red and black trim, is dominated by the lightly glowing Sapphire Nitro+ RX Vega 64 video card (2017) which blocks three of its four expansion slots and weighs three and a half pounds. Modern PCs are built for heat dissipation: a liquid cooling system pumps heat from the quad core, 4 GHz Intel processor (a sixth generation Core i7 variety from 2015), hidden under the Nuka Cola cap, to a radiator at the rear of the machine. The graphics card's three large fans (bottom) and inte-grated radiator draw away heat generated by thousands of graphics processing units. Other functions that would once have needed cards are built onto the motherboard, including Ethernet and sound. The case is close to the dimensions of the original IBM PC, but the space at the front where earlier PCs housed floppy, hard, or optical drives is filled with an ornamental bobble head. Instead a tiny Samsung Evo solid state drive screwed to the motherboard provides a terabyte of ultrafast storage. The power supply, much of the cabling, and a traditional hard drive are hidden behind partitions. Photo: Thomas Haigh.

by IBM in 2018. The latter holds 9,216 IBM Power processors and 27,648 Nvidia chips, each similar to the one on the Titan V card.

The corollary to desktop computers with supercomputer performance is users who continue to do fine with older and less powerful hardware. In the 1990s, Windows ran slowly at best, pushing users to upgrade to the latest hardware every year or two if possible. New releases slowed older hardware to the point of unusability. There was a saying, in quasi-biblical language, "What Intel giveth, Microsoft taketh away." Intel may have done its job too well for its own good. Today even a decade-old computer can run the latest versions of Office and Windows reasonably well. PC sales peaked in 2011, and without the need for frequent replacement fell in each of the next six years. There are still a lot of PCs in the world—around two billion, with about 250 million sold each year.

Virtualization

Demarcating what constitutes a computer got increasingly hard during the early 2000s. As we saw in the discussion of cloud computing, the platform targeted by programmers at a company like Google or Facebook is a software layer, not an individual server. Their code is run simultaneously across a pool of thousands of individual computers, to the extent that an entire data center, or even a global network of data centers, can be programmed like a giant computer.

The other side of this is virtualization: the ability to run multiple separate operating system instances simultaneously on a single piece of hardware. In combination, these two trends let IT staffs change the hardware assigned to a virtual computer by adding or removing the number of processors and amount of memory assigned to it without users, or even programmers, being aware of any change.

Virtualization began on IBM mainframes in the 1970s. We already talked about the importance of virtual memory to modern computing: giving each software process its own private memory space safe from interference from other programs. Chunks of memory were *paged* to disk storage as needed when programs needed more storage than was available internally. This helped the operating system to give users, and application programs, the illusion that they were running on their own private, *virtual* computers. Creating that illusion placed a heavy burden on the operating system.

The next step was to create virtual machines so robust, and so faithful to the underlying hardware, that they could run entire operating systems as well as application programs. This was first attempted in IBM's control program/Cambridge monitor system (CP/CMS), developed at its Cambridge Scientific Center between 1968 and 1972 as a response to IBM's difficulties in developing the TSS timesharing operating system for its System 360/67 mainframe. Supporting hundreds of simultaneous users on a single processor required a hugely complex operating system to juggle tasks and resources. A small research team led by Robert Creasy found a different approach: give each user a very simple operating system (CMS) running on a virtual processor. The control program's job was simulating multiple virtual processors using a single physical processor. CP/CMS could host more users than TSS, with better performance.

IBM seized on virtual machines, which became the hallmark of its System/370 range. Its virtual machine (VM) operating system centered on a control program to manage the virtual machines and assigned resources to them. Each virtual machine could run a regular OS/360-family batch-oriented operating system, the high-performance interactive CMS, or any other piece of software (including another copy of VM). This gave IBM mainframes of the late 1970s an unrivaled flexibility. A room full of minicomputers offered a lot more processing power for the money, but the mainframe could focus its entire power on a single huge job when needed. Operators might give more resources to virtual machines running interactive applications during the day and shift them to batch processing during the night shift. One mainframe could support development and production systems, isolating them completely from each other. Over time, virtual machines became valuable as a means of providing backward compatibility for older applications and operating systems.

Like many other architectural features, virtual machines eventually made their way from mainframes via minicomputers to desktop computers. The difference between virtualization and emulation is that emulation imposes a significant performance hit, whereas virtualization uses the native instruction set and hardware capabilities of the processor. Software to support virtual machines on Intel chips was launched in 1999 by VMWare, using ingenious techniques to compensate for the lack of hardware support. Intel and AMD began to add hardware instructions to support virtualization in 2005, improving the technology and rolling it out across their full range of processors over the next decade. Ordinary computer users may have experienced virtualization without realizing it: recent versions of Windows have relied on virtual machines to support older software that relied on insecure features eliminated from the main operating system.

Virtualization had its biggest impact on servers. Companies tended to accumulate large numbers of servers spread around different departments, some of them running software for specialized functions. Newer hardware could support enormous workloads on each server, but consolidating unrelated tasks onto the same server could introduce security risks and create unpredictable interactions between processes. Instead, system administrators converted obsolete servers to virtual machines. As workloads shifted, these virtual machines could be shuffled from one physical server to another without their users even noticing. The entire virtual server could be backed up as a single file and, in the event of a disaster, easily restored. Virtualization brought the old vision of a computer utility closer, as organizations could enjoy the flexibility that came with configuring their own (virtual) servers without needing to build and staff networks of data centers.

Streaming Video

More processor power and network bandwidth is now devoted to transmitting and decompressing streaming video than to any other task. We have already discussed the shift from analog television to high-definition digital television, and from analog video tapes to DVD and Blu-ray discs. Those televisions and disk players are special purpose computers.

In the early days of digital video, only optical discs or broadcast signals could channel enough data to these computers to produce good-quality images. Even with aggressive compression, video streaming over dial-up connections delivered poor results and was used mostly by pornographic sites. That began to change in the early 2000s, with the shift from dial-up modem connections to higher speed DSL and cable modems as the main form of Internet access. It also took the upgrading of Internet backbones around ever more effective forms of optical communication over fiber optic cables.

Less obviously though, the transition to streaming video relied on efficient cloud computing infrastructure. Most video is advertising-supported or offered for a low monthly subscription. The leading free video site, offering millions of videos uploaded by ordinary people and by companies, is YouTube. It was launched in 2005 and grew so rapidly that just eighteen months later Google agreed to acquire it for stock worth more than $1.6 billion. That seemed like a lot of money at the time, but it turned out to be a great investment as Google had the infrastructure to handle massive growth in traffic as YouTube's video quality and usage grew rapidly. YouTube became the most common way of viewing music videos, sharing home movies with friends, or streaming live events. After Google started sharing advertising revenues with the YouTubers who upload videos, some of its stars became wealthy celebrities thanks to their skills at chatting while playing video games, recording comedy sketches, or offering beauty tips. By 2013, YouTube was receiving more than a billion unique visitors each month. Its many subcultures developed their own jokes and practices, such as sharing *meme* videos recaptioning standard video sequences or images.

The most popular subscription service is Netflix, started as a mail-order DVD rental service before transitioning to streaming video from 2007 onward. Its selection of movies and TV shows for streaming was limited, particularly in comparison to its huge disk library, and the video quality was initially worse. But viewers appreciated the convenience and low price of streaming video. By 2011, the company was de-emphasizing its DVD rental business in order to market its streaming service separately. Over time, Netflix improved its video quality, adding high-definition video and then ultrahigh-definition content. Because television and movie studios were becoming less willing to license streaming rights to a competitor, Netflix shifted into the production of unique content, producing hundreds of its own TV shows and movies. By 2016, it was operating in every major country in the world except China, offering local material for most of them. At the end of 2020, it had around 200 million subscribers watching many billions of hours of video every year.

At peak times, these videos account for around a third of all the Internet data transmitted in the United States. That requires a spectacular amount of server capacity, but since 2016 Netflix hasn't run a single data center. Instead, its systems are hosted by Amazon Web Services, which made possible its rapid global expansion. Netflix also partnered with internet companies to place self-contained appliances in their data centers, so that videos are usually delivered without having to pass over Internet backbones between networks.

Netflix, together with competitors such as Hulu and Amazon Prime Video, have changed the way people watch television. Cable television subscriptions peaked in 2012 before beginning a gradual decline, and disc sales have dropped faster.[5] To younger people, the idea of having to be home at a certain time to watch a show seems strange. Streaming video originally took a Web browser running special plug-in software, running on a powerful PC. That equipment was unlikely to be found in a living room, where television has traditionally been watched, so Netflix worked with the producers of disc players to add Netflix capabilities and a special remote button to make it easier to watch on a television, and produced downloadable apps for tablets, phones, and game consoles so that pretty much anything with a screen or the capability to be plugged into a TV can stream Netflix. In the early 2010s, this brought a second life to many Nintendo Wii consoles. Eventually watching Netflix became even easier, as more expensive televisions were given the *smart* capability to download applications and stream video without extra hardware, and the popular Roku and Amazon Fire streaming boxes and sticks added the same capabilities to older and cheaper models.

Watching Netflix became an easy and constantly available activity, leading to the practice of "binge watching" entire seasons of television shows in a day or two. That could be undertaken alone or to pass time with friends, although by 2015, thanks to the somewhat mysterious evolution of language on the Internet, an invitation to "Netflix and chill" was generally understood as an invitation to sexual intimacy.

SOCIAL MEDIA

The term Web 2.0 was popularized by technical publisher Tim O'Reilly in the early 2000s. According to his definition, Web 2.0 sites would focus on user communities, deriving their value from content shared by members and interacted with by other members. For example, new *blogging* services let would-be journalists and publishers put material online quickly without having to mess around with HTML editing software, file uploads, and Unix command lines.[6] *Blogs* (a contraction of *Web log*) could function as public diaries, as venues for personal essays, or as newspapers or magazines devoted to particular aspects of politics, culture, or technology.[7]

Wikipedia was similarly showcasing the power of collaborative communities to produce compelling material, whereas Amazon's appeal came as much from its vast numbers of user-generated reviews as from its low prices and large selection of merchandise. On a technical level, Web 2.0 sites were dynamically generated, and they used techniques such as Ajax to provide a more responsive experience.

Early Social Media

The Web 2.0 concept was soon entangled with another new phrase: *social media*. A wave of companies providing Web-based tools for the editing of websites and blogs, such as Geocities and LiveJournal, was giving way to newer models for creating and sharing online content. This combined the content-posting features of blogging sites with a focus on establishing personal profiles and finding old friends borrowed from more tightly focused social networking sites such as Classmates.com. The model for these sites was Friendster.com, launched in 2002. It was an instant success, signing up three million users within a few months and receiving extensive coverage in newspaper and television profiles. According to the *New York Times*, its creator "conceived of Friendster as a dating site, but people's social curiosity turned it into a place where everyone becomes the center of an unfolding drama (or comedy) of connections." "Porn queens and venture capitalists," marveled the *Times*, "share the site with neo-Nazis and garden-variety hipsters."[8]

By 2005, Friendster had been overtaken by a rival, Myspace, which proved particularly popular with teenagers and musicians. Myspace profiles were heavily customizable, and tended to fill up with clashing colors, autoplaying music, and animated graphics. Myspace was the first social network to attract a hundred million users. For a few years it defined a new kind of fame for its popular successful users, whose personae chimed with its cheerfully trashy aesthetic. The most prominent of all, nude model and aspiring musician Tila Tequila, parlayed her Myspace success into a starring role in the MTV reality show *A Shot at Love with Tila Tequila*. Like Myspace itself, Tequila's success was short lived. In later years, media coverage dwindled and refocused on her professed admiration for Adolf Hitler and belief in a flat earth.

Social media services, podcasting, blogging sites, and platforms such as Tumblr, which let users share and annotate photographs, blurred the lines between the producers, consumers, and distributors of material. As with the influence of sampling on music, users remixed, ironically captioned, or reinterpreted material as they shared it.

Practices developed in closed subcultures, such as science fiction fandom, could now spread digitally on a scale previously possible only for the professional products of mass media. Media scholar Jean Burgess called this "vernacular creativity," a way for the voices of ordinary people to be heard.[9]

Facebook

Facebook, which in 2009 displaced Myspace as the leading social network, evolved from a Harvard University student's project and quickly expanded to other elite campuses. Starting with college students made sense, as they were open to new technology, had good Internet access, and were quick to learn what their classmates were using. Facebook became a company in the summer of 2004. The site opened membership to people without a school affiliation in 2006. Despite lagging the first commercial websites by a decade, its timing turned out to be perfect. The social network concept works only if a large cohort of friends and acquaintances are prepared to join the system. By 2006, the public was spending more time online with faster connections, making large, graphics-heavy pages quick to load.[10]

Facebook began as PHP code, and much of its core platform remained in PHP as the system grew. The firm produced many extensions for PHP, including a module to output C++ code to be compiled for improved performance. Its applications center on a gigantic database, which Facebook CEO Mark Zuckerberg called the *social graph*. The term *graph* comes from mathematics. It indicates that the company's focus is on the connections between things. Data on "likes," posts, pages, and messages all link to the platform's users—giving it the world's most extensive collection of data about the habits and tastes of individual people. Like Google's services, Facebook runs on thousands of personal computers rather than a giant mainframe.

Facebook quickly became the glue holding together the lives of its heaviest users and a prominent part of American culture. Just six years after its launch, the story of Facebook's founding and early growth was told in the Oscar-winning hit movie *The Social Network*. The movie cemented the position of Zuckerberg, still in his mid-20s and already a paper billionaire, as the new Bill Gates. Like Gates, who had also dropped out of Harvard, Zuckerberg was both mocked and celebrated as a ruthless nerd visionary, crushing enemies and former partners alike in his pursuit of power and barely imaginable wealth. In 2012, when Facebook made its initial public offering of stock, it commanded a record valuation of more than a hundred billion dollars, underpinned by its reported pool of more than eight hundred million active users and advertising-driven profits of more than a billion dollars a year. The company's further growth was driven by the popularity of mobile Internet platforms able to access Facebook to send messages, browse newsfeeds, and post pictures from anywhere.

Facebook's design and most widely used functions have changed fundamentally over time. In 2009, for example, it was promoting itself as a platform, on which other companies could build applications to exploit the connections established by its users. FarmVille, the most popular of these applications, was a game that centered on repeatedly clicking cows to milk them, clicking chickens to collect their eggs, and clicking fields to plow, plant, and harvest. Within a year more than eighty million people were playing it, inspiring game critic Ian Bogost to create a parody, Cow Clicker, which itself proved alarmingly popular.[11]

Facebook encouraged other websites to embed buttons to share content on Facebook and display the number of *likes* received. Facebook users accumulated so many friends and *liked* so many pages that new updates arrived every few seconds in its personalized News Feed. In 2008, Facebook hired Sheryl Sandberg, formerly the leader of Google's advertising team, as chief operating officer. A series of changes followed. To counter the sudden popularity of Twitter, a platform defined by the public posting of messages restricted to a terse 140 characters, Facebook reworked the algorithm used to prioritize updates by deciding which to highlight and which to bury. The new system de-emphasized updates from close friends in favor of paid advertising posts, posts from celebrities, and potentially viral items widely shared elsewhere on the network.[12] As mobile Internet devices became more popular, Facebook emphasized messaging and chat features.

Facebook's success undermined the appeal of blogging to more casual users—Facebook posts were more likely to be shared and commented on than blog entries. In contrast, the more successful blogs that attracted large numbers of visitors and comments evolved into media companies. They hired staff members, professionalized their editing and production, and supported themselves with user subscriptions. For example, the historian turned Web publisher Josh Marshall launched his political blog Talking Points Memo in November 2000, during heated discussion of the disputed presidential vote count in Florida. His personal skills as a blogger drew enough readers to underwrite hiring a growing team of reporters. Similar trajectories were followed by blogs in other areas such as technology law (Techdirt), celebrity gossip (PerezHilton.com), consumer technology reviews (Engadget), and personal productivity (LifeHacker).

Advertising Triumphant

Although Facebook has insisted that it is a communications platform and not a media company, Facebook's choices direct the attention of far more people than any single newspaper or television network. Partner firms complained that sudden tweaks to Facebook interfaces or its algorithms could destroy projects in which they had invested their future. In 2015, for example, it had begun to heavily promote itself as a platform for video publishing, encouraging media companies to "pivot to video" in the hope that Facebook would bring viewers and advertising revenue. Then Facebook abruptly de-emphasized shared videos amid accusations that it had faked the viewing figures that briefly made their creation seem like the best hope for making online publishing economically viable.

What held all these tactical shifts and feints together was a consistent strategic goal to encourage users to spend more and more time within Facebook and share more and more of their personal information. Instead of searching or browsing the Web, sending email, and using a variety of services to do things like share pictures or send invitation events, users should do everything within Facebook. That let Facebook sell more advertising and, potentially more importantly, gather more data on its users by cross referencing clues from all their online activities.

Facebook is an extreme example of a larger phenomenon: almost all the world's most visited websites are supported by advertising. The model spread from Web publishers to new kinds of Web services, including social media sites, video sharing, and eventually even pornography. By the mid-2010s, free sites Pornhub and xvideos.com had largely wiped out the once profitable businesses of subscription porn websites and DVD sales. They stream videos uploaded by users, many of them pirated with no royalties to the performers or copyright holders.[13] Today's performers rely on guest appearances in strip clubs and personalized videos for the bulk of their income.

The model is not quite universal. Craigslist and Wikipedia, two of the most popular Internet platforms, run as community services rather than profit-maximizing machines. Wikipedia is owned by a nonprofit foundation and, like National Public Radio in the US, supports itself largely by asking its users to make donations. Its development was shaped by the ideology of the free software movement, particularly the *wiki*, a tool for collaborative online editing.[14] Craigslist is an online site for classified advertisements, largely supplanting that formerly central source of revenue for local newspapers. Craigslist is a private business, but rather than go public, its founder Craig Newmark realized that he could bring in more than enough money to live well simply by charging a small fee to companies posting job openings. The site has retained an aggressively old-fashioned design, with few visible changes since the 1990s. The site displays no advertising (other than the classifieds that people visit to read) and most sellers pay nothing to list their goods and services.[15] The success of Craigslist and Wikipedia proves that the rampant commercialism of so much of the Internet is a choice rather than an inevitability.

APPLICATIONS HEAD FOR THE WEB

Client-server applications took a lot of maintenance work. On the PC side, IT staff had to install client programs, keep Windows up to date, and establish the necessary network and database connections. This meant supporting a variety of file, application, and database servers running in data rooms out of sight. Companies had to purchase, install, configure, and maintain applications. Large firms could easily spend hundreds of millions of dollars getting a major package like SAP in place.

Java and the Network Computer

The first big push to turn Web browsers into a platform for highly interactive online applications came with Java in the mid-1990s. Beginning in 1991, James Gosling, along with a small team of other programmers at Sun, developed a language to allow interactive applications to be downloaded to digital cable television boxes. Renamed *Java*, it was publicly announced in March 1995.[16]

Sun's pitch of "write it once, run it anywhere" promised that a Java program could run on any computer, large or small, without modification. This revived an old idea. For example, the IBM PC was announced in 1981 with three operating system choices: MS-DOS, CP/M 86 (released too late and priced too high to take off), or the *p-system* from the University of California, San Diego (UCSD). The p-system was derived from technology used to implement the Pascal language (the same technology was sold for the Apple II as Apple Pascal, which also replaced the usual operating system). Applications for the p-system were distributed not as executable machine language but as code for an imaginary *pseudomachine*. When the application program was run, the code was interpreted into instructions for the actual computer. The advantage was portability, because only the p-code interpreter needed to be rewritten to work on specific computers. One reason the system never caught on was that the IBM PC quickly became a standard, so it made more sense for applications to directly target its hardware, which yielded faster performance and access to its full capabilities.

A dozen years later, the Internet interconnected a diverse range of computers, and there were plans to extend access to televisions, hand-held organizers, and cell phones. Java's arrival coincided with the rise of Netscape, which seized on it as a way for Web page designers to give their pages animation, movement, and interactivity. HTML defined static pages: nothing changed until a user clicked a button or a link, which loaded another page. Java applets filled part of a Web page with interactive controls. When the user clicked something, code running on the PC reacted immediately, without waiting for a Web server to generate a new page and the browser to load it.

Companies rushed to add Java support to their devices and operating systems, programmers to learn Java, and software tools companies to produce Java development aids. Java was heralded as a way to break Microsoft's hold on personal computing. If people could access whatever software they needed via the Internet, who needed to buy Microsoft Office? If those programs were written in Java, who cared if the computer even ran Windows? Netscape's Marc Andreessen made the threat explicit when he promised that his firm could turn Windows into nothing more than a "poorly debugged set of device drivers" used to run its browser and Sun's Java engine.[17]

Microsoft's rivals had met little success competing with Windows, but Java raised the possibility of taking down the personal computer itself. Sun, Oracle, and IBM joined together to promulgate a new standard for the *network computer*. This was a hybrid between a personal computer and a terminal. Like a terminal, it worked only when connected to a network and had no disk drive of its own. Like a personal computer, it had a capable processor to run Java programs locally rather than rely entirely on the processor power of the server as terminals did. The network computer was promoted with the concept of *total cost of ownership* (TCO), an idea originated by the Gartner Group. Gartner analysts pointed out that the purchase price of a PC was a small part of its actual cost to a business. In 1996, Gartner estimated the total five-year cost of each PC to be around $44,000. That was more than double the cost of owning a DOS PC a decade earlier.[18] Windows was complicated. Most of the expense came from IT support, network infrastructure, software, upgrades, and work time lost as users tinkered with screen savers and tried to figure out why documents wouldn't print. That was particularly true in a corporate environment. Custom client applications had to be installed and kept current on each PC, together with the network configurations and database connections needed to communicate with their server components. The real savings from a network PC would come from slashing support costs by reducing the time users and support staff spent fiddling with each computer.

Corel, known primarily for its drawing software, had by 1996 snapped up WordPerfect and other application software to build a competitor to Microsoft's Office suite. Jumping on the Java bandwagon, it announced plans to rewrite its office suite in Java so that it could run on Windows, Macintosh, and Unix as well as network computers and other emerging platforms. That was an exciting strategy, but the software Corel produced underlined the limitations of Java: agonizingly slow, prone to hanging, and missing many of the features of the Windows versions. The Java virtual machine prevented printing, saving files locally, or cutting and pasting between applications.[19] After several preview versions, Corel abandoned its Java initiative without ever having released an actual product.

Network computers were marketed primarily by Oracle, which finally gave up in 2000. With few Java applications available, they were used primarily as graphical Unix terminals, with all the processing done on servers. Internet access remained slow, particularly at home, limiting the usefulness of a device that had to download all its programs and data. PC costs had fallen rapidly, closing the cost gap between a fully featured PC and a network computer. Microsoft reacted to the threat by building centralized administration capabilities into Windows, so that system administrators could roll out updates automatically and lock down systems to prevent users from tinkering with them.

Java applets also flopped. They sat awkwardly within Web pages, like a screen within a screen. Waiting for a Java-heavy page to load through a slow telephone connection could be a mind-numbing experience. The biggest problem was compatibility. Java was supposed to provide a standard virtual machine, but in practice, implementations differed and were constantly being updated. Programs might work in only one browser release with one version of Java and abruptly fail when either was updated. Interest in applets quickly faded, especially as hackers learned to exploit security holes in Sun's Java platform. By the early 2010s, experts were urging users to disable Java support in their browsers.

Java was somewhat more successful in environments where things were more controlled, such as the interactive cable boxes for which it had originally been designed. Cellphones of the early 2000s began to incorporate Java as a way of running application programs. But it enjoyed its biggest successes in two unanticipated areas. One was as a teaching language: computer science departments embraced Java as a substitute for C++ and Pascal. It was object oriented but easier to learn than C++ and, because of its virtual machine, able to catch and diagnose problems that would cause a C++ program to crash. The other was as a language for writing code snippets to run on servers, including code to run on Web servers (dubbed *servelets*) or inside other tools like database management systems. Java remains one of the world's most widely used programming languages.

Software as a Service

In the early 2000s, the computer industry was gripped by discussion of an exciting new idea: *Software as a Service* (SaaS). A company would purchase a subscription to an application accessed via Web browsers. No other software had to be installed, because the browser worked as a universal client. Data would be stored somewhere in the cloud, freeing companies from the responsibility for backing up or securing it. New features would appear in the application without any need to migrate servers to a new version or install updated client software.

This succeeded where the earlier push for Java and network computers had failed. The first major success for the new model was Salesforce.com, established in 1999. Its core product was a *customer relationship management* application, consolidating all the capabilities sales people needed to log information on actual and potential customers, track appointments, and log issues to be followed up by service personnel. There were already plenty of packages to handle these needs, but Salesforce was a perfect fit with the Web because salespeople spend most of their time out of the office. SaaS was also popular with smaller businesses. For example, most gyms do not have servers or IT staff onsite, so the industry switched rapidly to subscription-based services in which all the capabilities needed to log visitors, bill for membership renewals and small purchases, and so on were taken care of. By the 2010s, the new model was spreading to larger companies and more complex applications. Even SAP started promoting the cloud model for its enterprise applications.

Historians, having long memories, like to quibble about exactly how new the model really was.[20] Back in the 1970s, for example, timesharing companies were popular more for the access they offered to online applications rather than for simple access to an interactive computer. Terminals did not need to have any software installed on them. With a longer perspective, enthusiasm for freestanding personal computers in the 1980s and for client-server applications in the 1990s may look like an odd departure from the historical norm. But as the story of Java shows, it took considerable work to remake Web browsers into a smooth and capable interface for online applications, able to serve as a modern replacement for the text terminals of the 1970s.

Ajax—Interactive Applications in a Browser Window

The Web of the mid-1990s offered two rival approaches for the developers of Web-based applications. One approach was a fast-loading HTML page with no interactivity, so that responding to an option selected by a user involved generating and transmitting a whole new page. The other approach was an interactive Java applet or Shockwave animation that could respond immediately to inputs but would load slowly and work only if a user had previously installed the right plug-in to handle it.

A language called JavaScript, introduced by Netscape in late 1995, provided a third way. It had no real connection to Java—the name was a piece of marketing intended to exploit the Java hype. JavaScript was a clunky language, designed in ten days and initially very buggy. But what it did was very useful: developers could embed code snippets inside a Web page to be run when the value in an input box was updated or a user clicked on a button. These tiny programs could check that a date was in the correct format, hide controls that weren't needed with the option selected, or warn users that data had not been saved. Microsoft quickly copied JavaScript for Internet Explorer, and by 1997 it had been standardized.

By the mid-2000s, Web developers were beginning to discover creative ways to use existing technologies to create much more complex and efficient kinds of interactive pages. The key was to pull in new data from the server without reloading the whole page. Many people saw the capability for the first time in Google's Gmail service, which let users compose and filter messages within a dynamic page, and Google's Maps system, which let users scroll and zoom the map with the mouse, filling in blank parts of the map without a page reload. These capabilities worked on different browsers and operating systems without requiring any plugins or downloads. The technique combined JavaScript with XML, a data formatting method. In a 2005 article describing the new approach, Web designer Jesse James Garrett dubbed it Ajax, which stood for asynchronous JavaScript + XML.[21]

The name stuck and so did the approach. Ajax underpinned smooth and powerful Web applications without the need for Flash or Java. The techniques enabled the interactive Web interfaces of new based platforms like Facebook and Instagram. The capabilities of Ajax have continued to advance, along with improvements to HTML and allied technologies such as cascading style sheets.[22]

Online Office Applications

Since the 1970s, word processing and spreadsheet programs had been among the most widely used computer applications. Both put a premium on interactivity over processing power, making them well suited to PCs. But with the rise of Ajax techniques for Web applications, even these tasks became something that users could carry out using their Web browsers.

Google Docs, the first widely used online office suite, was based on products acquired by Google in 2006 and 2007. Features were limited, particularly in its early days. For example, the word processor did not initially support footnotes or mail merge—features common in high-end packages since the 1980s. Counterbalancing this, the cloud model had a big advantage in editing documents from public computers, such as those in campus computer laboratories or libraries, or when people in different locations needed to collaboratively edit a document. Team members have access to the latest version without having to exchange files, and revisions can easily be tracked. This delivered on the vision of collaborative editing showcased by Doug Engelbart's NLS system forty years earlier. Other systems followed, including online versions of Microsoft's own Office products. These systems are generally free for personal use, relying on revenue from advertisers and from subscription fees for premium accounts with more features or online storage space. This extended the advertising-supported model of Internet publishing into the traditional core of the personal computer software market.

Browsers gained the ability to install extensions, small packages of code able to perform specific tasks or accelerate the performance of particular cloud applications. Online office suites became more useful as technologies were developed to allow cloud applications to be partially downloaded so that users could continue to edit documents when they were temporarily offline. One popular extension allowed the Google Docs suite to run completely offline, opening and closing documents stored locally. In recognition of the shift to cloud applications, HTML itself was subsequently extended from a simple page description language to a platform optimized for interactive applications.

Chrome and Chromebooks

Firefox showed that Microsoft's monopolization of the browser market was reversible, but the knockout blow against Internet Explorer came from Google. Its stream of new Web-based applications like Gmail, Google Maps, and Google Docs relied on Ajax programming techniques. Existing browsers ran their complex JavaScript code slowly and unreliably. A small team of Firefox developers within Google began to put together a new browser, optimized for interactive applications. Google Chrome was launched in 2008, offering a characteristically minimalist visual design and unmatched stability and performance. By 2012, it had overtaken Internet Explorer as the most widely used browser, and at the time of writing it has about 70 percent of the browser market, whereas Microsoft and Firefox each has around 10 percent. Microsoft abandoned the fight in 2019, announcing that its browser would henceforth be based on Chrome technology rather than attempting to produce its own faster but potentially incompatible page-rendering engine.

Google used the success of Chrome, and the emergence of Web browsers as powerful clients for interactive applications, to relaunch the idea of the network computer. Its Chromebook design, announced in 2011, was taken up by companies including Lenovo, Samsung, and Hewlett-Packard as the basis of cheap, easy-to-manage laptops. They ran the Chrome operating system, a derivative of Linux streamlined and optimized for the Web. Users signed into their Google accounts and used Google applications to work with data held safely online in Google's servers. Applications like Gmail and Google's calendar could work with downloaded data even when users were offline, but Chromebooks were designed to be used almost exclusively when connected to the Internet. They were adopted enthusiastically by schools, for which their low cost of purchase and ownership, security, and control over the applications usable by students were powerful advantages over conventional laptops.

New Programming Systems

PHP, Perl, and Java are still widely used to code the server side of online applications, but they have been joined by the extremely popular languages Python and Ruby. Both are interpreted languages with object-oriented capabilities designed by individuals but (unlike Perl) standardized and implemented by many different groups. Python was created by Guido van Rossum, a Dutch systems programmer, and Ruby by Yukihiro Matsumoto, a Japanese computer scientist. They are general-purpose languages with particular strengths for the rapid development and modification of online systems. Python is seen as particularly welcoming to newcomers, so it is often used in programming courses for students with no computer science experience. Partisans spend hours arguing the merits of each language. Ruby, for example, is praised for the internal consistency of its design. This appeals to computer scientists repelled by the arbitrary and inconsistent syntax of PHP and Perl.

For most projects, however, the choice of language is a secondary matter, determined by the *application framework* chosen. Interactive Web applications involve a hugely complex collection of software technologies spread between browsers and servers. Developing all the necessary code from scratch would be prohibitively expensive. As with Hadoop for cloud computing, an application framework bundles together the code libraries and commonly used features, such as authentication, needed for online applications. Ruby's spread beyond the academic world took place largely because of the huge popularity of the Rails application framework, which underpins major sites such as Hulu, Airbnb, and Groupon. Other popular frameworks rely on Python, PHP, and Java.

The effect of all these new tools was to make the development of a major interactive website a much cheaper and faster proposition in the 2010s than it was in the 1990s, despite the enormously greater complexity of modern websites. This put a huge premium on programmers able to work with the exact blend of technologies needed for a project. Start-ups particularly prized the "full stack developer" whose skills extended from operating systems through databases, server code, and browser-based code, to interface design, requirements analysis, and project management. Such ideal programmers were rare—some said almost nonexistent—but the search for them signaled a shift from the old corporate model of systems development with large, specialized teams.

The shift was accompanied by a tendency of skilled programmers to identify as *coders*, and a new industry of *coding boot camps*. Back in the 1950s, coding had been identified as the most routine, and worst paid, aspect of programming. That work was soon automated by software tools, and the job title went out of use during the 1960s. Title inflation followed—programmers were called *analysts* or *software engineers*. The programming staff at firms like Google are usually called engineers, despite efforts by the traditional engineering professions to reserve the title for people achieving the status of *professional engineer* (a four-year accredited degree and professional examination, followed by a period of supervised work experience, culminating in a state licensing test).[23] The word *architect* was eventually applied to almost every part of the systems development process, perhaps spurred by design expert Don Norman's high-profile appointment as Apple's User Experience Architect in 1993. This similarly enraged real architects. According to a 2010 report, the "down economy has only sharpened their displeasure, as thousands of un- and underemployed architects sift through job listings for software architects, systems architects, data architects, and information architects: in short, every kind of 'architect' except their own kind."[24]

Coder, in contrast, was an aggressively unpretentious identity celebrating the practical business of sticking together computer instructions. Perhaps IT development work has reached a level of pay and respect such that efforts to appropriate job titles from better established professions are no longer necessary. A *New Yorker* article from 2014 claimed that elite developers were the new rock stars or professional athletes, able to earn huge sums negotiated by slick agents.[25]

Subscriptions and Cloud Storage

Ubiquitous high-speed Internet access also changed the way that traditional software packages were sold. In the 1990s, customers purchased a shrink-wrapped box holding a CD or a set of floppy disks. That gave them a permanent right to use the software, which is why companies liked to release major new versions of their programs every couple of years. Microsoft, in particular, advertised heavily to sell discounted copies of the new version to users of the now obsolete release. Microsoft occasionally made minor updates, usually comprising bug fixes and performance enhancements, available freely, but until the Internet came along most computer users did not bother to obtain and install these.

The first change was the easy distribution of new updates. Internet access made patches easier to obtain, but an Internet-connected computer's security vulnerabilities easy to exploit. With Windows 98, Microsoft began to automate the process, adding a Windows Update service to quickly distribute updates. Microsoft also started to deliver significant changes and improvements to Windows as free downloads rather than save them for the next major upgrade.

As Internet connections grew faster, more users shifted to purchasing software via downloads rather than on disc. Software companies eventually moved away entirely from the model of selling major releases every few years. Microsoft has stated that Windows 10, launched in 2015, was the last numbered Windows release and that it will receive two significant updates every year.

Although Microsoft has continued to number the releases of Office, most recently with Office 2019, its focus has been on selling annual subscription packages via its Office 365 program. Both Adobe and Microsoft push users to create accounts and log in to use their software, and to place files by default in cloud storage rather than on local hard drives. Users who stop paying lose access to the applications and cloud data entirely. Adobe has gone even further in this direction; in 2013, it halted sales of its popular Creative Suite bundle of applications. Anyone needing updated versions of the software would have to subscribe to its Creative Cloud service. This initially caused a lot of grumbling, especially after hackers stole and shared passwords for all Adobe accounts. Yet Adobe's revenues doubled over the following five years as customer payments became more predictable and frequent. Customers had few good alternatives to Adobe's software.

With the transition of Microsoft and Adobe, the two biggest PC software companies, to subscription plans and cloud-based storage, two different models of computing have almost completed their convergence. One model had its origin in the personal computer world, and the other in the timesharing industry. In the mid-1990s, when network computers and Java were first proposed, the idea of combining automatically downloaded applications with Internet file storage and centralized computer management seemed like a radical alternative to personal computing. But little by little, the models have grown closer together. Chromebooks and offline helpers let people use their cloud applications and files while they are temporarily offline. Windows automatically updates itself, and it includes management features to let IT staff exert remote control over every aspect of a PC. Users download applications, pay for them with subscriptions, and keep their files in far-off data centers. The PC has become a network computer and the network has finally become the computer—just as Sun promised back in the 1980s.

14 THE COMPUTER IS EVERYWHERE AND NOWHERE

In the 2010s personal computer sales began to drop, reversing a four-decade trend of constant growth. People were using the Internet more than ever, but they were accessing it less often from laptops and desktops and more often from smartphones and tablets. As their prices fell, these devices went from convenient adjuncts to conventional computers to—for lower-income consumers and countries, at least—an affordable and flexible alternative. Billions of people around the world who have never owned (and probably will never own) a personal computer have had their primary experience of computing with the new devices.

Was the entire previous history of computing made irrelevant by the launch of the iPhone in 2007? As with the Internet, situating the smartphone within the longer and broader story of computing lets us understand it more deeply. The smartphone was made possible by high-speed cellular data networks, but its rise to ubiquity was underpinned by the cloud computing and online applications that we discussed in the last chapter. Users comfortable keeping their documents and photographs in the cloud and using Web-based applications to handle items like email and customer data were increasingly able to rely on lightweight, portable devices. Yet, in the history of computing as in the history of almost everything else, new things sit alongside old things. Personal computers did not go away in 2007, any more than mainframes went away in the 1970s or minicomputers in the 1980s.

SPECIALIZED MOBILE DEVICES

A tech enthusiast going on a trip in the early 2000s was likely to pack a bag full of portable computers. As well as a laptop there would be a cellphone, an electronic personal organizer, a digital camera, a digital music player, and a GPS unit. Fortunately, carpenter jeans and cargo shorts were enjoying an otherwise nonutilitarian vogue, providing plenty of pockets to hold the gadgets. A decade later, the same enthusiast would have needed room for only one device: a smartphone. The smartphone integrates all the computing and media technologies discussed in this book, reabsorbing into a new and more universal computing platform the specialized devices developed to perform tasks for which PCs were impractical.

Handheld Computers

Handheld computers had been around since the arrival of microprocessors. There was always something magical about a computer that fit in a pocket and could be taken anywhere, but the constraints of tiny keys, limited screens, and very limited storage meant that programmable calculators for financial and engineering work were the only really successful early models.

By the late 1980s a second use for a small, constantly available computer was emerging: electronic address books and diaries. *Yuppies*, as ambitious young professionals had newly been dubbed, had already embraced the Filofax, a ring binder stuffed with pages for diaries, contact information, notes, and optional insets for items like subway maps, expense logs, and business cards. The Filofax users formed a large potential market for a computer of similar size that could handle the same functions. One of the first usable electronic alternatives, the Psion Organizer II, was launched in 1986. It looked a bit like a calculator, with a two-line LCD display above a small keyboard. It could be programmed by companies to give their employees access to custom mobile applications, but it also had built-in diary and address book features. Its RAM, which could be as small as 8 KB, was kept powered by the battery to preserve data. Psion followed in 1991 with its Series 3 range of organizers that folded like miniature laptops, allowing for larger screens and keyboards and including a much richer set of built-in programs. Psion sold more than a million organizers.

Pen Computers

The big problem with a tiny computer was the tiny keyboard. What if the keyboard were removed and users wrote directly onto the screen? The pen could replace a mouse, too. This was the natural extension of Alan Kay's dreams for Dynabook hardware, which drove his work on Smalltalk and informed Xerox's development of graphical user interfaces. His Dynabook idea was to package a computer as a slate-like screen with a thin keyboard attached. Dropping the keyboard shrank the package further and removed a source of potential intimidation, as more people were comfortable using an ordinary pen. By the early 1990s, the technology to package a computer and screen into a "slate" that could be held in one hand finally seemed to have arrived. The story was told by Jerry Kaplan in his book *Startup*.[1] In 1987, Kaplan co-founded GO Corporation, producers of the PenPoint operating system. This helped to send a wave of enthusiasm for *pen computing* through the computer industry. GO was a few years ahead of the capabilities of hardware, although Kaplan blamed its failure primarily on Microsoft's hype for a nebulous Windows-based alternative.

The most promising of the early systems to reach the market was Apple's Newton MessagePad series, launched in 1993. It was an attempt to deliver on the promises of John Sculley, head of Apple, who had been lured from Pepsi by Steve Jobs. Sculley had tried to establish himself as a technological visionary with a slick 1987 video promoting the Knowledge Navigator, a hypothetical device smart enough to work like an executive assistant, holding a conversation with its human boss and then looking up information, sending messages, or making reservations in response to those requests. The idea drove research into areas like intelligent software agents. In 1992, Apple chose the *personal digital assistant* tag for the Newton to suggest that it could fulfill this role (see figure 14.1).

Unlike GO's system, the Newton was marketed as a companion for a conventional computer and not a replacement for it. Its size and weight, about a pound, was comparable to a well-stuffed Filofax, although the price, $699, was quite a bit more.

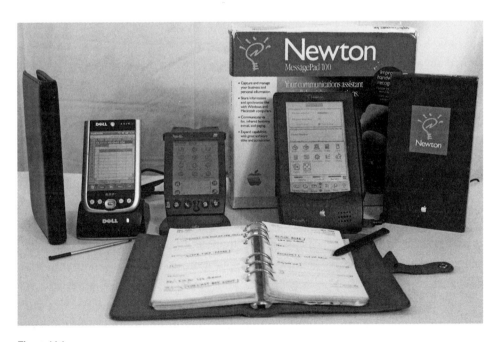

Figure 14.1

These three stylus-controlled personal digital assistants sought to replace the Filofax ring-binder organizer (front; far left, in slimline version). The 1995 Apple Newton MessagePad 100 (right, with tutorial video) was the largest and most ambitious. The pocket-sized Palm series (center, budget-priced Palm IIIe, 1999) was smaller, cheaper, and far more commercially successful. The Dell Axim 50v from 2004 (left) represents the last major generation of PDAs. It was far more powerful, with a bright color screen, Wi-Fi, and a 624 MHz ARM family processor (versus 20 MHz for the Message Pad). Its Microsoft operating system attempted to transplant aspects of Windows, including the start menu and Excel spreadsheet program, to the tiny screen.

A monochrome screen occupied most of the MessagePad's front, with a control strip at the bottom to launch frequently used functions like note taking, addresses, and appointments. With a modem card, it could send and receive faxes.

Sales of around 50,000 in four months fell well short of expectations, as did Newton's usefulness. Users entered text freehand, and the MessagePads used handwriting recognition software with a built-in dictionary to guess at the matching words. The technology wasn't ready for this ambitious goal. Initial performance was particularly bad, as it took a while for the machine and the user to learn each other's quirks. Newton's handwriting recognition was mocked by both *The Simpsons* (then at the height of its popularity) and the widely read *Doonesbury* comic strip.

Subsequent versions were more polished and much more capable. MessagePads integrated Internet email with simple word processing and spreadsheet capabilities better than any previous device had. Even the handwriting recognition improved. To the shock of its small but enthusiastic user community, Newton was abruptly canceled in 1998 when Steve Jobs returned to the top job at Apple.

By that point, a smaller, cheaper handheld computer had appeared. The Palm Pilot, released in 1996, was less technically ambitious than the MessagePad in every way. This was clearest in the handwriting recognition. Newton tried to recognize cursive text written anywhere on the screen. Palm required users to write characters one at a time in the input box, letters on the left, numbers on the right. They weren't even ordinary letters: each was replaced by a stylized representation in a new alphabet called Graffiti. Once users had adjusted to this system, text entry was reliable.

Stripping things down to the essentials left a compact six-ounce box. Under a small, low-resolution screen was a touch-sensitive area with printed icons for frequently used functions such as the calculator application. Below that were buttons for the built-in to-do list, notes, calendar, and address book. A Palm device had to be regularly synched with a Windows computer (or later, a Macintosh) by placing it in a cradle to synchronize contacts, appointments, and email with Windows applications such as Microsoft Outlook.

Synchronizing would also transfer new applications to the Palm. Thousands were produced, such as games, eBook readers, and news readers. The first Palm devices used low-power Motorola microcontrollers based on its classic 68000 design. But later models (from 2002 onward), and all MessagePads, used ARM technology. In fact the Newton range helped establish ARM as a viable competitor. In 1991, Apple had invested $2.5 million in exchange for 43 percent of the fledgling company.[2] When Apple liquidated this investment, it made a profit of almost $800 million, earning back all the money it had sunk into Newton over the years.

Palm made money more directly. By early 1999, Palm had sold more than three million devices.[3] As the century closed, in the tech world it seemed like the only choice was whether to carry a boxy Palm III or a Palm V with a curvaceous metal case and rechargeable battery. Geeks exchanged business cards over its infrared link and, admiring the efficiency of Graffiti, started writing it on whiteboards.

Cellphones, Pagers, and Beepers

Mobile two-way radio has been around for a long time, particularly for military use. By the 1950s, police cars and taxis were routinely fitted with two-way radio systems. The first radio telephones worked the same way, bridging the radio signal from a car into the telephone network at a base station. This severely limited the number of radio telephones that could be used in a city. New York, for example, had just twelve frequencies available and so allowed the installation of only 730 sets.[4]

Cellphones arrived in the 1980s, after a technological trial that began in 1978 in Chicago. Being *cellular* meant that the phone would automatically connect to a nearby ground station, shifting from one cell to another as it moved through the city. Using lower-powered short-range radios shrank phones, made battery power feasible, and opened up network capacity as signals no longer blanketed an entire city. Although the radio signals were analog, the phones relied on digital electronics, running programs on an embedded microprocessor to register the phone's location, make outgoing connections, and respond to incoming calls.

Smallness and portability were relative, of course (see figure 14.2). The Motorola DynaTAC 8000X, the first to go on sale, has become an iconic piece of obsolete technology, symbolizing the rapid advance of mobile electronics. It cost $4,000, weighed almost two pounds, and provided up to an hour of talk time. It was nicknamed The Brick and was featured in the 1987 movie "Wall Street," where it was always on the ear of junk bond magnate Gordon Gekko. By the end of the 1980s, phones had shrunk to fit comfortably on belt holsters. Soon they rested in pockets. Motorola's StarTAC, a distant descendant of the DynaTAC, was nine times lighter. Some people wore them around the neck like a piece of jewelry. Tens of millions were produced.

The same cellular technologies supported another kind of device: the pager. Pagers had small numerical displays, like calculators. After someone called the pager service to leave a number, it would appear almost instantly on the screen. This was often a telephone number to call, but some pager users maintained code books for more complex communication. The stereotypical pager users were doctors on call, whose employers appreciated the reliability of pager transmission; and drug dealers, who appreciated the anonymity of pagers.

Second generation (2G) networks and phones were based on a handful of standards, creating large international markets. They encoded voice calls digitally, which improved security, extended battery life, and used radio spectrum more efficiently. The

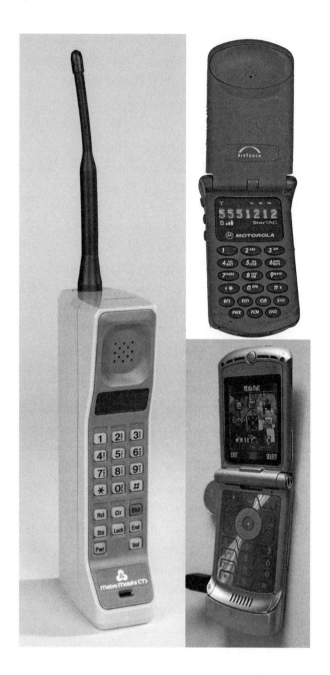

Figure 14.2

Three generations of Motorola Phones. The original 1983 DynaTAC "brick" phone (left) weighed 28 ounces. It was soon replaced by smaller alternatives, culminating in the 1996 StarTAC (top right) which weighed just three ounces. It was so compact that it had to be unfolded to reach both mouth and ear. The RAZR V3 (bottom right), a hugely popular second-generation phone launched in 2004, offered SMS messaging, mobile email, and even Web access, but the main draw was its slim metal case. DynaTAC courtesy Cooper Hewitt, Smithsonian Design Museum; StarTAC by Wikimedia user Nkp911m500, shared under Creative Commons Attribution-Share Alike 3.0 Unported license; RAZR by Thomas Haigh.

most widely used standard, GSM, began as a European effort to permit roaming across borders. Second generation phones subsumed and extended the functionality of pagers with the SMS text-based message service. Texting began to replace traditional calls among younger people, following a pattern established in Europe where users typically did not have prepaid call allowances.

Cellphone technology began to converge with computer communications in the early 2000s. Some analog phones had included modems for wireless Internet use, but their transmission speeds were very low. Second generation phones were already digital, and network enhancements to route Internet data packets to and from phones let them send data at speeds comparable with land-line modems, supporting Web browsing and email as standard features. The earliest efforts to provide mobile Web access had depended on website producers' support for special protocols, rather than on attempts to display standard HTML pages on tiny screens. These were most widely adopted in Japan, through the widely used iMode service launched in 1999.[5]

New phones spread fast because the industry had settled on a pricing mechanism that maximized sales but reduced the up-front cost to new customers. Every two years, the mobile phone company gave each customer a nominally free basic phone or a discounted phone with the latest features, locked to work only on the network of that phone operator. The hardware cost was hidden in inflated monthly bills. The model gave mobile phone operators an effective veto over the features included in new handsets.

The main factor limiting the miniaturization of phones was the need to reach both mouth and ear. The iconic phone of the mid-2000s, the Motorola RAZR V3, folded in half for storage. Motorola sold more than 130 million worldwide over four years, thanks to a shiny aluminum case, a slick glowing metal keypad, and a second screen on the outside so that it could display the time and call information even when folded. It was spectacularly thin and light, although not particularly powerful. Built-in applications included a crude Web browser, a calculator, and an address book, but there was no way to add more. Downloading a ringtone was the limit on customization.

Some second generation phones, bulkier and more powerful than the RAZR, were experienced by their users as hand-held computers. They held more standard applications and had the option to install new ones. Many could even run Java programs. The most widely used operating system for these *smartphones* was Symbian, a descendent of Psion's personal organizer software. Symbian phones could run thousands of applications, send and receive email, and browse the Web—although as most had tiny screens and numerical keyboards, this was a fiddly process. Data transmission rates rose in the mid-2000s as technologies such as enhanced data rates for GSM evolution (EDGE) brought mobile data closer to broadband, but the mobile Web experience remained clunky.

Mobile email users favored the rival BlackBerry system, introduced in 1999 by the Canadian firm Research in Motion (RIM), thanks to keyboards on which they could type at high speed using both thumbs. The experience of constant communication proved addictive for BlackBerry users, who nicknamed it the "CrackBerry." Barack Obama fought successfully to keep hold of his Blackberry after a recommendation by the Secret Service that he surrender it on becoming president. BlackBerries evolved from pagers, RIM having earlier developed the first model able to send as well as receive messages. By 2003, they could also browse the Web, send text messages, and make voice calls. RIM's core market was corporations, which could tie its devices into their email servers to provide secure mobile communication.

GPS Navigation Systems

By the late 1990s, computers able to direct motorists to their desired destination were appearing in the dashboards of luxury cars. This was made possible by the deployment of the global positioning system (GPS), a constellation of navigation satellites constantly transmitting time signals supplied by on-board atomic clocks. Receiving signals from four satellites allows the calculation of location to within a few meters. The timing information sent by the satellites is also used by cell phone providers to hand off phone calls from one cell to another, by financial traders to record the precise time of a trade, and by electric utilities to coordinate the transmission of power over their networks.

The system's architecture was set in the early 1970s, just as the ARPANET was being constructed. Both had their origins in military programs. GPS is operated and controlled by the US Air Force. The US military reserves some capabilities for its own use, but regular GPS service is free to all users, laying the foundation for commercial exploitation by companies like Apple, Google, and Uber. In this regard, GPS is, like the Internet itself, a government-sponsored technology that became the foundation for huge private wealth. The European Union, Russia, and China have each developed or planned to develop similar satellite-based systems (Galileo, GLONASS, and BeiDou, respectively).

GPS supplies accurate position and time, and nothing more. That was useful to military units, sailors, and hikers, but the development of a large commercial market for GPS units depended on the digitization of maps (a huge ongoing task), efficient route-finding algorithms, and the development of cheap, low-powered portable hardware. Navigation systems using other technologies were pioneered in Japan, long before GPS was available. The first Western car with a built-in navigation computer was the BMW 7 series in 1994. This set the template for later systems: a color screen in the dashboard, a map CD in the trunk, and voice instructions played through the car's stereo. In 1996, when local map data was ready, navigation reached the US as a $4,000 BMW option for a hugely expensive car already adorned with gadgets such as headlight washers.[6] Journalists could barely hide their schadenfreude when, on Christmas

Day 1998, a German motorist obediently following the computer's instructions drove the car *Motortrend* had called "a chariot for capitalist gods" straight into the Havel river. A ferry link had been miscoded as a bridge in its database. Similar stories have since become staples of slow news days, but worldwide reporting of the incident gave many readers their first awareness of the existence of GPS navigation.

Eventually, GPS navigation units made their way from gods to mortals, although cost-conscious customers generally preferred attaching a small box to the windshield to paying extra for an integrated GPS. One of the leading producers, the Dutch firm TomTom, began selling a GPS add-on and software for Windows PDAs in 2002. Within two years it was selling an all-in-one navigation system, with its own screen and CPU, for about $600. TomTom and its main competitor, the American GPS pioneer Garmin, sold tens of millions of navigation devices over the next few years. Later models used cellular data connections to fetch real-time traffic updates. By the time prices fell to little more than $100, the paper road atlas was as obsolete as the film camera.

ENTER THE IPHONE

The success of Palm was taken, like the later success of the iPod, as evidence that people wanted a pocket computer that did one useful thing well. As one analyst observed, its devices "hit a nice little balance point" between capabilities and simplicity. Palm's head of marketing insisted that customers did "not want the complexity of Windows on their handhelds, and they don't want a general-purpose device."[7] Yet this specialization meant that some customers were carrying around many little electronic boxes, creating a major opportunity to integrate their functions without compromising their usefulness. The miniature computers sold as GPS units, PDAs, cellphones, digital cameras, and MP3 players had a lot in common. They differed mostly in the software burned onto their ROM chips and their peripheral hardware. Combining the functions of two or more of these devices into a single box made sense.

Cell phones had begun incorporating cameras by the early 2000s, although the picture quality fell far below that of free-standing cameras. Some phones acquired memory card slots and the ability to play back MP3 files. PDAs also started to accumulate new features. Sony licensed Palm's operating system and used its expertise in consumer electronics to produce variants with higher resolution screens, built in cameras, and MP3 players. Handspring, another Palm-based company, included an expansion slot in its PDAs so that users could swap in extra modules as needed. The logical goal was a fully featured hybrid of cell phone and PDA, complete with touch screen, stylus, keyboard, cellular Internet access, and the ability to make voice calls. The first to reach the market were the Treo series, introduced by Handspring in 2002.

Microsoft, as usual, reacted to a successful new product by trying to crush it with a Windows-branded equivalent. Its Windows CE operating system, introduced in 1996, had gotten off to a slow start. Users preferred the simplicity of Palm's system to Microsoft's insistence on miniaturizing the Windows experience, complete with a Start button and "pocket" versions of Word, Excel, and the other Office applications. As hardware became more powerful, that began to change. Palm struggled to update its operating system for newer processors, while Hewlett-Packard, Compaq, and Dell produced capable Microsoft-powered PDAs. As the market shifted from traditional PDAs to smartphone hybrids, Microsoft established a dominant position. By 2006, even Palm had licensed Windows CE for one of its Treo models.

The First iPhone

It seemed that Microsoft had set the standard for smartphones as effectively as for desktop computers. Yet, today few people even remember that smartphones existed long before the launch of the Apple iPhone in 2007. Apple's attempt to enter a crowded market initially seemed rash, but its phone eventually wiped out every existing producer of smartphone hardware and software. As he prepared to unveil the iPhone, Jobs called it "a revolutionary product . . . that changes everything." For once this habitual hyperbole was merited. Jobs teased "three things: a widescreen iPod with touch controls, a revolutionary mobile phone, and a breakthrough Internet communications device" before confiding that they were all "one device, and we are calling it iPhone."[8]

Current smartphones, Jobs argued, "are not so smart and they're not so easy to use." The key was to "get rid of all those buttons—just make a giant screen." The iPhone was small—not much bigger than the iPod—but with no keyboard to take up space, its screen, measuring 3.5 inches diagonally, could fill almost the entire front. Like the original Macintosh, the iPhone was shaped by the uncompromising demands Jobs made on his design team. Jobs had the same vision for both: a small, beautifully designed box with no capability for internal expansion running an elegant operating system, which becomes the constant companion of its users. In 1984, Jobs's dictates almost sank the Macintosh. By 2007, the tradeoffs were different. Memory upgrades and expansion slots were no longer essential, whereas portability and elegance were more important than ever. Like a stopped clock, Jobs's priorities were fixed, but like the clock, the world eventually moved far enough to make him right. (Unlike the clock, Jobs stayed right for the rest of his short life.)

The iPhone had up to 8 GB of built-in storage, an ARM processor running at 400 MHz and an operating system, later called iOS, that was controlled with swipes and pinches rather than key presses and mouse clicks. Traditional PDAs could sense only one contact point at a time, using a stylus to point precisely. Apple designed its operating system around a new touchscreen technology, able to sense multiple contacts. As Jobs put it, "We're gonna use a pointing device that we're all born with—we're born with ten of them." The precision sacrificed by abandoning the stylus was more than made up for with intuitive gestures. Move two fingers closer together to zoom in; further apart to move out. Underneath this unorthodox interface lay a reworking of Apple's Macintosh operating system, with a common heritage reaching back to BSD Unix.

The first iPhone was developed with AT&T Wireless, which had an exclusive deal with Apple. Prices started at $500, exceptionally high for a phone, but the real commitment was a two-year contract to pay AT&T $60 a month for service. The total cost was well over $2,000, once taxes and fees were added. What customers received in return had not previously been available at any price: a portable Web browsing device that could view real Web pages, was fun to use, and fit easily in a pocket. The iPhone became so popular that major websites were redesigned around its capabilities, dropping Adobe Flash animations (which Jobs refused to support) and automatically reformatting pages based on the width of the screen viewing them.

A year after the original launch, Jobs returned to the stage in San Francisco's Moscone Center to announce a replacement that fixed the iPhone's most serious flaws. Its cellular data transfers were much faster, thanks to the new 3G networks built with mobile data applications in mind, and Apple had squeezed in a set of new sensors including GPS. Apple sold more than a million on the first weekend. The second iPhone replaced more devices, more completely, than any previous hybrid. As a PDA, it had a more elegant user interface and excelled at Web browsing, email, and messaging despite the lack of a keyboard. It had an unusually good camera and an excellent music player based on the capabilities of the iPod, and it made an outstanding portable games machine. It was smaller and more convenient than any dedicated navigation unit.

Since then a new iPhone has arrived every year. Processors got faster, storage capacities larger, and cameras and screens crisper (figure 14.3). In 2010 the iPhone got a second camera, mounted on the front for video cameras and the taking of "selfies." Older phones received new versions of the operating system for a few years but tended to slow down under the burden. For Apple's most devoted users, the purchase of a new phone became a seasonal event, like the falling of leaves or the swarming of ants.

The App Store

The iPhone came with a good selection of applications, including Google Maps, weather information, stock prices, and a notepad, but had no way to download or install more. Other applications would have to run inside its Web browser. As usual, Jobs wanted to control what users could and couldn't do with his device. Apple relented within a year, producing development tools and an App Store (modeled on the iTunes store) to distribute the results. Some apps repackaged popular websites and services for the iPhone's

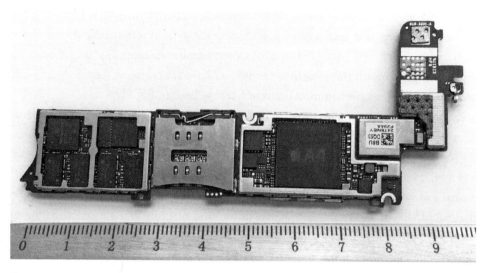

Figure 14.3

An iPhone 4 circuit board from 2010. The large A4 system on chip (SOC), also used in the first iPad, integrates an ARM-based microprocessor with a graphics processor and, sandwiched into a second layer above the processor, 512 MB of RAM. Other chips integrate up to 32 GB of flash memory, radio receivers and transmitters, a GPS receiver, an accelerometer, and a magnetic compass. The entire board is less than 10 cm long. Photo: Paul Ceruzzi.

small screen. Others offered social media and communication services already popular with personal computer users, such as Facebook, YouTube, Netflix, and Skype.

Apple's affluent customer base and thriving App Store made it the leading target of smartphone developers even after rival platforms with the same capabilities became available. The growing complexity of professional applications, plus the investment needed to get a new product distributed and advertised, had consolidated the PC software industry around a handful of large publishers. In the 1990s, entrepreneurs and investors had shifted their focus from software packages to websites. In contrast, iPhone apps were relatively simple programs, and modern development tools made it easier for small groups to produce polished software.

The App Store gave those groups a shot at distributing that software to a large market. By 2016, the App Store had more than two million apps available for download, with about 100,000 new or updated titles arriving every week.[9] Using the App Store was safer than downloading a conventional program from the Internet and offered something the Web lacked: a built-in payment system. The vast majority of apps sank without trace, but there were enough stories of authors earning huge payouts from self-published programs to inspire many others to try their luck.

Many of the most popular apps were accessible titles that exploited the iPhone's novel interface capabilities, like tilting the phone or flicking a finger. The *Angry Birds* series, whose players rescue eggs from pigs by throwing birds at them with a slingshot, received more than a billion downloads in its first three years. Hits like that, created by small teams, underwrote the development of major development studios employing hundreds or thousands of people.

Making money got harder over time. Early games had sold for a few dollars. Then a Maltese developer, King, scored the biggest iPhone game hit of all with *Candy Crush*, released in 2012. It popularized the *fremium* model for games–the download is free, but players make repeated purchases to unlock advantages or advance to the next level without a long wait. This model eroded the market for games that were purchased up front, favoring bigger companies able to wait for revenues and keep adding new items to their games. King itself was eventually acquired by Activision Blizzard, the biggest video game company of all.

Apple took a 30 percent cut of the purchase price and, more controversially, of any purchases made inside the applications.[10] It vetted their contents, banning those with sexual content or drug themes. Apple also prevented apps from challenging its own control of the platform. Until 2010, for example, it banned apps from making telephone calls using cellular data, to protect AT&T's ability to charge iPhone users for call time.

The Mobile Cloud

iPhone apps had almost constant access to cloud data centers, via fast Wi-Fi connections in homes, offices, or hotels and over slower cellular data links during travel. It was years before the phone could be purchased without an unlimited data plan, although early users sometimes received bills for thousands of dollars after short trips abroad. The success of the smartphone as a universal computing platform owes much to this flexibility: some applications ran entirely on the phone, but others were more like timesharing applications because most of the work was done in the cloud by server farms processing gigantic databases. Only the user interface had to fit into the phone.

The combination of these capabilities with built-in GPS equipment laid the foundation for entirely new businesses and created new opportunities for services that had been viable on PCs. Facebook, for example, now uploaded photographs directly from iPhones and tagged them with the location where they were taken. OpenTable, a restaurant booking site, had been useful with a Web browser running on a PC. As a mobile app, it could instantly list nearby restaurants with immediately available seats, confirming a reservation with a tap of the finger.

The most important apps used the platform to create opportunities that had not previously existed. Instagram, a social media network focused on photo and video sharing, was created in 2010 as an iPhone application to take pictures, apply effects, and immediately upload them. Grindr, launched in 2009 for men seeking sexual encounters with other men, was another iPhone original. It displayed pictures and profiles of nearby users, closest first, and allowed the exchange of messages. Together with the earlier rise of Web-based online dating sites, Grindr undermined gay bars, for decades a fixture of urban landscapes, by enabling connections in places like small towns and airports.[11] The model needed some tweaking for other sexual communities, but Tinder, launched 2012, became one of the most popular smartphone apps. The photos and profiles were still there, but users had to indicate mutual interest before messages could be exchanged. This was signaled by swiping right on a photo, creating a new figure of speech. By 2018 a reported 26 million matches were made on Tinder every day. Unlike traditional dating services, which matched profiles based on self-reported interests, Tinder and Grindr used the capabilities of the iPhone to mimic the offline search for a potential mate: snap judgments based on appearance and proximity.

Thanks to GPS, iPhone apps could locate rides, as well as lovers and food. The Uber *ride sharing* service launched in 2009, followed by Lyft in 2012. They provided convenient alternatives to taxi service. Opening the app shows a list of nearby drivers. Searching for a destination brings up a price. Agree to the price, and the app displays the license plate and live position of a car, which arrives a few minutes later. After arriving, one jumps out, and the trip is automatically charged to a credit card. The companies work as what economists call *two sided* platforms: the drivers are also users of the app. They log in and click to accept rides. Some work full time, others take occasional shifts to help cover the cost of car ownership. Uber and Lyft describe themselves as online marketplaces bringing together supply and demand for rides (in part to sidestep the obligations of categorizing their drivers as employees). When demand spikes around major events or bad weather, the algorithm raises prices to rebalance. Uber expanded rapidly, without concern for local laws regulating taxi services. Its low prices were subsidized by the unprecedented $24 billion it received from venture capitalists and private equity investors. Between 2014 and 2019, Uber lost more than $16 billion. Like Lyft, its long-term plan involves replacing human drivers entirely to cut costs.

Hundreds of start-ups announced their intention to become the Uber for *X* where *X* was something like walking dogs, assembling IKEA furniture, or delivering pot. The common factor was using a smartphone to order a service at a predefined price. A pool of willing workers had to log into the other side of the service to take the jobs. Surveying these firms a few years later, Alexis Madrigal concluded that about half remained in business, although only four, all offering delivery services, had been highly successful. As he summarized the "consumer Internet's" development during the 2010s: "Venture capitalists have subsidized the creation of platforms for low-paying work that deliver on-demand servant services to rich people, while subjecting all parties to increased surveillance."[12] Other researchers concluded that reports that these services had increased the number of people carrying out "gig" work were much exaggerated.

In 2018, *Bloomberg News* ran the baffling headline "Bird Races to Become the First Scooter Unicorn." This reflected renewed enthusiasm for Internet-based services among global investment funds. Business writers had started to call firms like Uber *unicorns* because they were valued at more than a billion dollars but had not made an initial public offering of stock. Bird was one of many services blanketing cities with *dockless* bicycles and electric scooters (see figure 14.4). Vehicles unlocked themselves after being rented using an app. Users abandoned Bird's scooters at their destinations, but at the end of each day underemployed people could earn bounties of a few dollars by recharging a scooter and returning it to its nest. The dockless vehicles were Chinese, and so were services such as Ofo and Mobike, which quickly expanded into European cities. It was far from clear that these businesses were sustainable, given the tendency of locals to vandalize the vehicles, but the millions of bicycles and scooters suddenly blocking sidewalks and piling up in canals around the developed world provided material evidence of vast sums of invisible capital mobilized around smartphone apps, wireless data, GPS services, and cloud infrastructure.

Voice Controlled Assistants

Although early devices didn't really live up to the "personal digital assistant" branding, one necessary technology for a true electronic assistant, speech recognition, advanced

Figure 14.4
Dockless scooters owned by the *unicorn* start-up Bird and local rival Skip obstruct a Washington, DC, sidewalk in 2018. A computer and cellular modem in the box of electronics mounted to the handlebar reported the position of the scooter to cloud servers and unlocked it when rented via a smartphone app. Photograph: Paul Ceruzzi.

rapidly during the 1990s. Better algorithms and faster chips made voice control and dictation workable on ordinary PCs by the end of the decade. But the same tricks were out of reach of pocket computers, with their slower processors and limited battery life.

The 2011 annual iPhone launch focused on a new software feature: Siri, the "intelligent voice assistant." Siri's name (a traditional Norwegian one) reflected its developer SRI International, a research and development organization funded by DARPA to develop the technology as an assistant to military commanders. Apple had purchased the technology the previous year. When users asked questions, Siri searched information from their calendars, email, and the Web to come up with an answer. Siri could be asked to carry out actions, such as making an appointment, playing a song, or starting navigation. Siri ran on the cloud systems in Apple's data centers, reducing the load on the phone's own processor and giving the algorithm access to huge volumes of data. Early reviewers found the experience inconsistent: Siri sometimes responded with a startling illusion of intelligence, but on other occasions was baffled by the simplest request. Other tech companies soon introduced their own helpers: Google Assistant, Microsoft Cortana, and Amazon Alexa. These matched and, according to many, outstripped Siri's capabilities within a few years.

Voice assistants started showing up in other places. Amazon put Alexa into its popular Fire TV streaming devices and began to sell a line of Echo battery-powered speakers. The speakers were Linux computers without screens or keyboards, connected to the Internet over home networks. Amazon priced them aggressively, creating a large user base that in turn inspired the developers of home automation devices such as the Nest thermostat to add support for Alexa. Google countered with its Home line of smart speakers, sold for as little as $30. This made home audio the latest front in the ongoing battle of cloud services between the tech giants.

The success of Siri and the other assistants was part of a huge boom in so-called *artificial intelligence* (AI). Despite being a major research area of computer science since the 1950s, AI had struggled to deliver on its grand promise to replicate human thought by building programs able to reason logically. After several cycles of hype and disappointment, researchers largely abandoned formal logic, focusing instead on statistical approaches. Large volumes of data were used to train systems to select the most appropriate responses. Initial guesses were scored and refined, to incrementally improve performance. That approach had proven successful with natural language parsing in the 1970s, and it underlay the success of Google in determining which results were most relevant to a query (and what the user meant to search for). Its proponents originally called it *machine learning* to avoid the stigma associated with AI, but by the mid-2010s they had grown confident enough to embrace and redefine artificial intelligence on their own terms.

Apple Wins

The iPhone's striking, and initially unique, combination of a large screen with finger control introduced a new mode of personal computing. By 2011, four years after entering the business, Apple had displaced Nokia as the firm selling the most smartphones. Measured by profit, it wasn't even close: Apple was already making more money than all other cell phone producers put together.[13] In 2012, Apple became the first company with a market valuation of half a trillion dollars ($500,000,000,000) and in 2018 the first to a trillion.

It helped that Apple itself didn't make any phones. Although iPhones displayed the "Designed by Apple in California" tag that adorned all Apple products, they were manufactured under contract by other firms, mostly in the gigantic Chinese factories of Foxconn, a Taiwanese multinational corporation. With no factories or production workers, most of Apple's employees worked in its minimalist boutiques. These enjoyed the highest sales per square foot of any chain in the country.[14] Apple outsourced even the assembly and testing processes, with container loads of fully packaged phones shipped across the Pacific. This transition completed the process begun with the shift of PC component production to Asia in the 1980s.

Apple's competitors took years to reorient themselves from keyboard or stylus control. Nokia had been the world's biggest phone maker, but its Symbian operating system stood no chance against the iPhone. In 2011, Nokia's CEO compared its position to a North Sea oil worker whose platform had caught fire, leaving only the choice between being burned to death or jumping into freezing water. Nokia chose a desperate leap into business with Microsoft, which was reworking its suddenly obsolete Windows CE into a new, iPhone-like product called Windows Phone. The result was praised by many reviewers as a cleanly designed and powerful mobile operating system, but duplicating the vast array of applications available for iPhones was harder than matching Apple's operating system. Microsoft later committed billions to purchase Nokia's struggling mobile phone business outright, but it still couldn't make headway. In 2016, with less than 1 percent of the market, Microsoft began to abandon its phones.

BlackBerry devices peaked with more than a third of the US smartphone market but also declined quickly after 2011. Research in Motion slashed jobs and announced a switch to an iPhone-like touch-controlled operating system. Its arrival in 2013 failed to stem the defection of its users. By 2016, the firm had abandoned the phone design business. Palm vanished more quickly. It had produced a promising new Linux-based operating system but was acquired by Hewlett-Packard in 2010 and dismantled the next year after a change of corporate leadership.

The iPhone was wiping out whole product categories, as well as competing products. The market for GPS navigation boxes began to collapse. The success of the iPhone also shrank the market for hand-held games consoles from Nintendo and Sony. Games for these were expensive; games for smartphones were cheap or free. Music players, whose sale Apple already dominated with the iPod, also began to vanish. Apple's later iPods were Touch variants—essentially iPhones with the cellular connection stripped out. Digital camera sales peaked in 2010, falling by about 80 percent over the next six years. Traditional cameras had better lenses than smartphones and bigger image sensors, but Apple had a lot of processor power and a huge development budget to throw at the problem. Soon independent movies shot on iPhones were being released in theaters.

Tablets

Apple followed up the iPhone with a new, and initially even more successful, application of the same technologies. The iPad tablet, launched in 2010, was essentially an iPhone with a much larger and higher resolution screen. It had the same operating system and ran the same apps. Although it didn't have a stylus, the iPad finally fulfilled the hopes expressed twenty years earlier for "slate computers." It could run for about ten hours on a single charge, and it made a formidable book or document reader, Web browser, or video player. Because Apple had removed telephone capabilities (leaving cellular data as an option) the iPad did not lock its users into a long-term contract. That made it more affordable. The second version, announced a year later, was faster, lighter, and held front and back cameras for video chat. The third revision quadrupled the screen resolution and introduced a quad core processor.

Since peaking in 2014, iPad sales have fallen rapidly. Part of that was competition from cheaper tablets. Another factor was the increasingly large-screen size of smartphones, reducing the need to carry an additional device. Mostly, however, the iPad was a victim of its own adequacy. Older models continued to work well for many years.

Apple attempted to revive demand with its iPad Pro range, designed for use with precise electronic pencils and detachable keyboards (figure 14.5). Their screens and processors and the storage capacity of the most capable models exceeded those of most laptops, but so did their prices. Reviewers agreed that the wonderful hardware was let down by the limitations of Apple's iOS for demanding work.[15] In time the descendants of this machine—or of Microsoft's rival Surface tablets, which have the advantage of running standard PC applications, will surely replace the traditional laptop. Apple took another step toward convergence in 2020 when it announced it would be replacing the Intel processors used in its Macintosh computers with faster and more power efficient ARM units derived from those used in its iPhones and iPads.

Figure 14.5

2020 editions of the Apple iPad Pro 12.9-inch (left, with optional Apple Pencil and Logitech Slim Folio Pro detachable keyboard sleeve) and the Lenovo X1 Carbon (right), a premium business-oriented laptop. With keyboard, the iPad (3.0 pounds) was thicker and heavier than the X1 (2.6 pounds). In an example of convergent evolution, the former grew from a smartphone and the latter shrank from a desktop PC. They met in the middle with comparable pricing and hardware: large vivid touch screens, powerful processors (a 6-core Intel i7 versus an 8-core custom Apple ARM architecture chip) and flash memory storage. With five cameras and four microphones the iPad was ideal for video conferencing, and its graphics chip gave it an edge for video games. Although both could be connected to full-size peripherals and used to run office applications, Lenovo's Windows 10 operating system retained a significant edge for most tasks. For example, the iPad could not display more than two applications at once. Photograph: Thomas Haigh.

SMARTPHONES EVOLVE

The only effective competition to the Apple iPhone came from a new operating system, Google's Android. Android began in 2003 as a start-up company developing operating systems for mobile devices. Purchased by Google in 2005, it became a team building a Linux-based platform for mobile phones. The arrival of the iPhone forced a course change midstream, away from a conventional smartphone with keys and toward hardware designs and a user interface similar to Apple's.

Android

Google gave away its software and sample ARM-based hardware designs, letting companies produce competitive phones without starting from scratch. The core of Android is open source, which phone manufactures combine with proprietary software including drivers to work with their specific hardware. From Google's viewpoint, the money it spent producing Android was an investment against the long-term threat of Apple using a monopoly on the smartphone market to limit mobile access to Google's hugely profitable online services (as when Apple abruptly dropped Google Maps and replaced it with a homegrown rival). Google required phone producers to feature its full suite of applications and services on Android devices, which eventually led to a large anti-trust fine from the European Union. In effect, Google had extended the Internet's advertising-supported model to operating system software and hardware design.

Some Android phones were premium products from well-known brands. The early showcase was Motorola's Droid phone, launched in 2009. It had a slide-out keyboard as well as a large touch screen. In 2011, Google purchased Motorola for its expertise in mobile electronics and its patent portfolio. The most successful Android range came from Samsung, whose beautiful hardware often sported features not yet matched by Apple. Its Note range, for example, had an integrated stylus as an alternative to finger control. Samsung built the first phones whose curved screens wrapped to their edges, and it pioneered oversized "phablets" to fill the gap between tablets and conventionally sized phones. They were particularly attractive to users who, unlike Apple's affluent fans, could not afford both a premium phone and a tablet. After a few years Apple responded with its own oversized models.

Apple's phones were marketed and priced as luxury goods without significant discounts from retail price, like Chanel perfumes. Prices went up over time, not down. In 2017, when the iPhone celebrated its tenth anniversary, prices for its exotically shaped iPhone X crossed the $1,000 mark. Most Android phones, adorned with unfamiliar Asian brands, resembled the "smells like" discount imitations sold by street vendors. By 2015, a usable smartphone could be purchased for less than $100. They were slower and less capable than Apple's models, with washed out screens and fuzzy cameras, but did the job for a fraction of the price.

Phone producers had considerable latitude to tweak Android. That proved a mixed blessing. As delivered by Google, Android was an elegant system. When loaded down with visual widgets, personalized assistants, and duplicate applications for features like photography, it became a confusing mess. This customization also compromised security. Apple and Google both produced regular updates and security patches for their operating systems. Apple's went directly to its users, whereas Google's went to phone manufacturers who at best took months to test, customize, and distribute them. Cheaper phones might never receive a single update. To showcase the benefits of an uncluttered, up-to-date Android experience, Google started to brand and directly sell phones produced to its specifications: the Nexus range, introduced in 2010, and its successor, the Pixel series, from 2016.

Smartphones in the Developing World

Android devices have been outselling iPhones since 2011, particularly in markets outside the US. For much of the world's population, an Android phone was a first telephone as well as a first computer. In countries with unreliable electricity and little landline telephone infrastructure, this had a transformative impact. By the end of 2018, an estimated 3.3 billion people, well over half the global population aged fifteen or older, had a smartphone.

About 800 million of those people were Chinese. When the iPhone took off, Western activists protested Apple's reliance on Asian contract manufacturing. Their campaigns contrasted Apple's huge profits and the comfortable lives of iPhone users with the low wages and unpleasant working conditions of the people making them. This narrative, previously applied to other expensive products such as sneakers, did not fully capture the changing relationship of China with the developed world. By 2015, Apple was selling more iPhones to China's fast-growing middle class than to Americans.[16]

China built a "great firewall" around its Internet users, enforcing censorship and blocking access to many Western sites. Together with limitations on Western access to Chinese markets this ensured the success of local companies. China has Weibo instead of Facebook, Baidu instead of Google, and Alibaba instead of Amazon. Those are now among the world's most successful businesses. In 2005, pioneering American Internet firm Yahoo made a major investment in Alibaba. Thanks to their divergent fortunes, by 2017 that stake was worth more than the whole of Yahoo, which was thus harming its shareholders by remaining in business. Admitting defeat, Yahoo's management sold its struggling core businesses to Verizon, leaving happy shareholders with unadulterated Alibaba. The biggest Chinese success of the smartphone era was WeChat, released in 2011. Analysts call it the "everything app," noting that Chinese users rely on it so heavily that the choice between Android and iOS barely matters. It provides text messaging, voice and video chat, Facebook-like social features, and bill payment services.

The use of phones to make small payments took off first in lower-income countries, such as Kenya where millions of people were making payments made via the text messaging infrastructure in 2010, even before the arrival of smartphones there. WhatsApp was another big success driven by the needs of the developing world. As sending international texts could be expensive, its free text messaging service was popular for people with friends and relatives abroad. WhatsApp was acquired by Facebook in 2014 for more than $19 billion. As of 2018, it claimed more than one and a half billion users, sending sixty billion messages every day. For much of the world, such as the vast Indian population, it had become a key vehicle for communicating and sharing news.[17]

After the iPhone?

At the end of 2018, Apple announced that it would no longer be disclosing the number of iPhones sold in its financial results. Fears that this policy reflected the beginning of a significant slowdown in sales were borne out when Apple issued a profit warning to investors. iPhone sales had been declining in 2015, and the rising price of each phone could no longer compensate. Analysts began to talk of "peak iPhone," a reference to the old concern with "peak oil."

Apple had faced particular challenges that year, including weak sales in China as a trade war pushed customers away from American brands. More fundamentally, though, the news reflected the maturing of the smartphone. The new iPhones were more stunning than ever: screens with pixels too small to see, six core main processors paired with four core graphics processors, cameras good enough for professional use, and up to 512 GB of storage. The problem was that its phones from a year or two earlier were almost as wonderful, as were competitors selling for less than half the price. As with PCs and tablets before them, smartphone sales declined once there was no need for most users to upgrade frequently.

Smartphones with more power than the supercomputers of the 1980s are small enough to carry in a purse or a pocket. The same small, light, and more energy-efficient technologies opened up other product opportunities. Design teams asked which of the almost infinite possible forms for a tiny computer might complement, or eventually replace, the shiny slab of the smartphone. The first big hit was the Fitbit, a step counter that synchronized with a smartphone application. Cell phones could track steps themselves, but not everyone carries a phone when exercising. Fitbit brought a social element to fitness, letting users view the step counts of their friends and send taunting or encouraging messages. The original Fitbit cost $100 and appeared in 2009.[18] Within a few years the most popular Fitbit models had grown small screens and morphed into wristwatches. That helped to justify their increasing price tags. $200 is a lot for a pedometer, even with the addition of heart rate monitoring, but not much in comparison with a luxury watch.

Meanwhile, in an example of convergent evolution, *smart watches* added fitness tracking capabilities. Smart watches were little watch-shaped computers. Their screens displayed the time, and other things like text messages. In the 1990s Microsoft had promoted Timex watches able to download data from Windows PCs by reading flashing patterns. It tried again in 2004 with wristwatches using built in FM radios to receive data. Like Fossil's efforts to miniaturize a Palm Pilot to create a smartwatch controlled with a tiny stylus, these miniature marvels turned out to be solutions in search of a problem.[19]

Smart watches were revived as peripherals for phones, which would usually travel with the watch, rather than for computers. This began in 2013 with the Pebble, a hugely successful Kickstarter project that provided a cautionary tale on the difficulty of building a successful company around a hot product. Apple's Watch arrived with a big splash in

2015. Prices ran from $350 for a utilitarian model up to around $1,000 for an elegant metal version intended to compete with luxury watches. The watch wasn't obviously useful except to display phone alerts: the tiny screen was so awkward that users still found it quicker to pull phones from their pocket when performing other tasks. Later versions have improved and offer their own cellular connections, letting users stay online without carrying phones. In 2017 Apple sold about $6 billion of watches and accessories, a disappointment compared to the iPhone itself but enough to displace Rolex as the watch industry's top brand. Because people are used to wearing watches, it seems possible that smart watches will eventually be very widely used.

The other device that many of us are used to wearing is eyeglasses. Embedding a tiny display gives the illusion of a huge screen that can overlay information onto the real world, a technique dubbed augmented reality. The tech world got excited in 2012 when word spread that Google was developing a pair of glasses with a built-in screen and camera, controlled with voice input and a small touchpad. Google Glass was supposed to have applications in medicine, journalism, and film making. Preview users called "Glass Explorers" paid $1,500 for a development version. Google lined up an impressive range of partners promising apps, services, and designer frames. Somehow it all fell apart before Glass even went on sale to the public. As privacy concerns mounted over what were essentially spy cameras, organizations such as cinemas, hospitals, strip clubs, and casinos started to ban people from wearing them. Users discovered that wearing what had recently been the height of geek chic now got them branded as "glassholes."[20] Google didn't officially kill Glass, but in 2015 it conceded that the push to turn it into a consumer product had been premature and reoriented its development work toward finding business applications for the technology. The basic idea seems to us like something that will eventually work in one form or another, just as pen computing failed repeatedly before succeeding.

Glasses and watches aren't the only things that sprouted computers and hooked up to the Internet. People started to talk about the *Internet of Things*, a strange phrase as the Internet has always interconnected things. In practice, the term means *Internet of things that aren't recognizable as computers*. Businesses hooked up things like electricity meters, environmental monitors, and vending machines. Domestic users built Internet-connected home automation systems, revitalizing an idea popular back in the 1980s. Products based on the increasingly popular Z-Wave wireless protocols allowed a programmable home controller to react to inputs from control panels, smartphones, or light switches by triggering actions such as unlocking a door, dimming a light, or sounding an alarm. This uses a mesh network, so each device acts as a booster to spread control signals throughout the house. Configuring and programming these systems took a lot of fiddling.

By the mid-2010s, Wi-Fi and Bluetooth chips were small and cheap enough that companies were building them into things like lightbulbs for the benefit of more casual users. Philips launched the popular Hue line of color-shifting bulbs that could be controlled with a smartphone, Nest (acquired by Google) created a modish computerized thermostat that promised to automatically program itself by observing usage patterns, and Ring sold doorbells with video cameras to record and screen visitors. One idea that was repeatedly tried and failed was adding an Internet connection to refrigerators—to watch movies or access family calendars on the fridge door, or to scan food items and track freshness. Few potential customers found these features compelling, but smart fridges did win the approval of spam emailers, who hacked them and used them to distribute messages.[21]

In contrast, smartphones are the dominant computing platform of the modern world. Their processor architecture stretches back to an obscure British personal computer and, more generally, to John von Neumann's 1945 *First Draft of a Report on the EDVAC*. To the computer scientist they are computers, but most of their users do not think of them as computers or of what they do with them as computing. The vast majority of computers, like those that replaced the guts of telephones, televisions, and music players, are hidden from their owners. Normal people think of the PC they use at work or the old laptop on the shelf as computers, but not their tablet, Fitbit, or television. At the end of our story, the computer is everywhere and nowhere.

15 EPILOGUE: A TESLA IN THE VALLEY

We started this book with the grand promises made at the dawn of the computer era in sources such as the ENIAC press releases, Edmund Berkeley's *Giant Brains: Or Machines That Think*, and Roddy Osborne's *Harvard Business Review* article touting the accomplishments of General Electric's first Univac computer. They promised new worlds in which computers automated an ever-growing range of human activities. Since then, waves of hype have risen and fallen, cresting in the mid-1960s with business enthusiasm for totally integrated management information systems, the late 1970s when Christopher Evans's *Micro Millennium* promised the imminent arrival of ultra-intelligent machines and a leisure-filled retirement for humanity, and again in the late 1990s when financial markets were gripped with irrational enthusiasm for Internet stocks. The broad case for the revolutionary importance of computers was always made by sketching out utopian futures that remained just out of reach.

The idea is captured wonderfully in a *Byte* cover from January 1977, aimed at the magazine's rapidly growing base of personal computing hobbyists (figure 15.1). Its background is a grim and polluted cityscape visible through a large window. Railroad tracks, factories, and power lines partially block the view of a small clump of downtown towers. The gloomy picture of urban decay captured the mood of the times—the Environmental Protection Agency and Earth Day had been established earlier in the decade, and cities were depopulating rapidly as affluent white residents fled for the suburbs. In the foreground lies the dream system of many *Byte* readers: an Altair microcomputer equipped with a floppy disk drive and a video terminal. Its screen displays a gleaming image of clear skies and verdant rolling hills, atop which gleam white, geometric structures recalling the classic futurism of early science fiction movies such as 1936's *Things to Come*. *Byte* readers probably interpreted the image as a suggestion that their machines had the capability to bring about the transition to the *postindustrial* society then being touted by intellectuals and policymakers. Doing what

Figure 15.1
Robert Tinney's cover for *Byte*'s January 1977 issue positioned an Altair 8800 computer, symbol of the emerging personal computer industry, with floppy disks and paper tape in front of a depressing and polluted cityscape. Its video terminal seems to promise a computer utopia. Courtesy Robert Tinney.

they loved would make the world a better place—a smug assumption deeply engrained in Silicon Valley culture, which has woven threads from 1960s utopian thinking together with an older faith in technological progress that promotes technology as a force outside human control that can fix social problems.

We now live in a future in which almost every technological prediction made for the advancement of computer technology has come true (humanlike intelligence is the one glaring failure). Has the computer utopia arrived? Looking at the *Byte* cover today suggests a darker reading. We spend more time than ever looking at imaginary worlds on computer screens, whether to play video games or to immerse ourselves in social media worlds expertly tailored to shield us from unwelcome information. The gleaming world on the screen might be a distraction from the actual social, environmental, and political disasters unfolding outside our windows, rather than a promise that computer technology will somehow fix them. So let us finish by pulling back our focus from the evolution of computing to survey a world remade around computer technology.

In these last few chapters, we have focused on the stealthy proliferation of computers inside other consumer devices, from televisions to wristwatches, and the rise of smartphones to replace personal computers as the most widely used computing platform. Yet, most of the computing power owned by a typical middle-class American family of the late 2010s was not in the laptop in the den, or the tablets in its children's bedrooms, or in the big screen televisions hung on its walls, or in the smartphones its members carried everywhere. Although each of these devices comfortably outstrips the power of a Cray 1, their biggest cluster of supercomputers was parked in the garage.

Computers started showing up in cars in the 1970s, to control antilock braking systems and manage the injection of fuel into engines to boost their power and efficiency. They control airbags, play music, give driving directions, and monitor tire pressure. In the mid-2010s the convergence of cars and computers was most apparent in the Tesla Model S, an expensive electric car manufactured in Silicon Valley and sold in mall boutiques like a piece of consumer electronics rather than through the traditional network of independent dealers. Elon Musk, its CEO, said the firm had "designed the Model S to be a very sophisticated computer on wheels." Musk boasted that Tesla was a Silicon Valley "software company as much as it is a hardware company."[1]

INTO THE VALLEY

By that point, Silicon Valley seemed to be the symbolic heart of a new global order as much as it was a place. It stood for a world transformed by *disruptive innovation* through the Internet, smartphones, and cloud computing. Back in the 1940s, electronic computers were used by a few hundred people around the world to carry out calculations. By the end of our story, they were used by almost everyone, in the developed and middle-income world, to do almost everything. In 1990, of all the hardware and software companies only IBM ranked among the world's most valuable corporations, up with General Electric and Exxon Mobil. By 2020, as globalization and tax cuts boosted corporate profits, the first four companies to achieve market valuations of a trillion dollars were Apple, Amazon, Alphabet (Google), and Microsoft. Even the word *technology* had been redefined to mean computers and the Internet so that, for example, eBay was considered a "tech company," but not Ford or Boeing.

The groups developing new computer technologies and applications in most of this book had no single center. Work initially centered on the mid-Atlantic area and in the UK. IBM blossomed in upstate New York, with labs around the world; minicomputers and the earliest workstations were produced primarily in the Boston area. The IBM PC was developed in Florida. Compaq cloned it in Texas. Britain briefly led the world in cheap personal computers; France got more people online than the entire rest of the world; and the Web was developed in Switzerland. Nintendo of Japan dominated home video game design.

In the last few chapters of the book, that has changed as Silicon Valley's initial focus on microchips and the closely related development of personal computers has built up networks of venture capitalists, entrepreneurs, and technical talent that makes it the center of wave after wave of new businesses, from biotechnology and green energy, to personal computers, smartphone software, and social networks. Most of the companies that we have discussed in the last few chapters are clustered on a strip of land only a few miles wide, between the San Francisco Bay and the largely undeveloped hills that separate the valley from the ocean. The silicon chips plants themselves are long gone, as are the factories assembling electronic devices. By the 2010s, Tesla was unusual in building things there, which it did in a defunct GM/Toyota factory.

In February, one of those cars, a high-performance Model S costing more than $100,000, was collected from a delivery hub close to the factory in Fremont, on the less desirable side of the Bay, and set out for its first drive home. The car signaled its allegiance to Silicon Valley, rather than Detroit or Bavaria, with a 17-inch touch screen. This worked like a giant tablet computer, controlling navigation and entertainment options with a responsive touch-based interface supplemented by voice control. When the Model S was introduced in 2012, reviewers reached for phrases like "rolling tablet" to describe it. Its connection to the trends discussed in the last few chapters was unmistakable.

Yet that visible computer, powered by a powerful triple core processor, was just one of more than sixty installed throughout the Tesla. Most were dedicated to individual functions such as air suspension, the parking brake, the rear-view camera, and the power seats. One computer controlled the wipers and looked for signs of rain. Four computers, one in each door, listened for radio signals from a smartphone, ready to pop out touch-sensitive door handles if their owner came near.

Silicon Valley is not very big, even though definitions of its scope have broadened greatly since the 1970s to encompass most of the Bay Area. According to the car's navigation system, without traffic, even the longer route around the bottom of the bay would get it from Fremont to the garage of a luxury apartment in San Francisco near the Salesforce tower in just an hour and a quarter. The same tsunami of investment capital from around the world that funded the car's development and purchase meant that accomplishing this was a virtual impossibility—rapid growth in highly paid jobs had combined with restrictive zoning, intended to maintain the suburban feel of the area and maximize the wealth of existing homeowners, to create a permanent housing shortage. That, in turn, had institutionalized two-hour commutes from California's interior and chronic traffic jams up and down the Valley.

That route passed close to Adobe, PayPal and Cisco in San Jose before reaching Netflix in Los Gatos, and then swung north toward Interstate 101, passing highway exits for Apple and Seagate in Cupertino; Intel, Nvidia, and AMD in Santa Clara; Google and Symantec in Mountain View; Hewlett-Packard in Palo Alto; Facebook in Menlo Park; Oracle in Redwood City; Genentech in South San Francisco; and Uber and Lyft in San Francisco. Most of the start-ups whose assimilation had ensured their continued dominance, such as Android, WhatsApp, Instagram, and YouTube had also resided in nondescript low-rise buildings spread up and down the Valley. So had the venture capitalists, huddled just outside Stanford's campus, and the incubators hatching dozens of fledgling firms in the hope that one of them might, in a mixed metaphor, grow into a unicorn.

Visible signs of the region's roots in government and aerospace contacts had faded with, for example, the demolition a few years earlier of the "Blue Cube" that once towered over Sunnyvale: a computer-filled Air Force control center for military satellites. NASA's sprawling Ames Research Center remained hard to miss, although its bulky wind tunnel, the world's largest, had been largely supplanted by computer modelling and Google was leasing its airfield and hangars.

Throughout this book, the computer has become many new things. If we follow the Tesla through the Valley, we can glimpse all these modes of computing working together. As the car passed silently past a toll collection station, in an express lane of SR 237 that offered a brief escape from congestion, the transponder mounted to its dashboard provided a code to identify the owner. Billing and payment were handled by the FastTrak tolling system, a credit card network, and the driver's bank, using batch processing systems as discussed in chapter 3, in which the computer became a data processing tool. Most of these systems still ran on mainframes, using 1970s-style CODASYL database management systems.

In chapter 4, the computer became a real-time control system. NASA used an embedded computer to control the engines of its Saturn V rocket, solving the problems that led its Soviet rival to explode on each of its four test flights. Musk's other company, SpaceX, took advantage of huge increases in computer power over the subsequent forty years to do something even more difficult: handle the minute adjustments in thrust needed to safely land a rocket on a designated spot, which lowered the cost of spaceflight by making launch vehicles reusable. The same control and stabilization technologies may soon make the long-promised "flying car" a reality as a battery-powered robot taxi built by scaling up drone technology, rather than a personally owned car with wings.

Most of the processors in the Tesla performed similar real-time control functions. Its battery pack alone held sixteen computers to manage the charging and discharging of a battery pack of lithium-ion cells weighing around 1,200 pounds. This was the same power technology used in laptops, but the Tesla needed about a thousand times the capacity to provide its then-unique driving range of several hundred miles. Lithium-ion batteries hold a lot of power, but drawing power from them too quickly or overcharging them will shorten their lifespan and can start a fire. It took the availability of cheap embedded computers to make electric cars viable, or to equip the performance model of the Model S, a family car, with the "ludicrous mode" power burst needed to accelerate to 60 miles per hour in 2.4 seconds. Videos of Teslas beating out exotic sports cars at the drag strip were a popular subgenre among the firm's large fan base.

In chapter 5, the development of timesharing turned the computer into an interactive tool. That narrative culminated in the development of Unix, which, reimplemented as Linux, powers the main touch-screen system controlling the Tesla and most other modern cars. Its user interface is the distant descendant of the 1980s innovations covered in chapter 9, when the computer became a graphical tool, and the cousin of the smartphone systems discussed more recently.

In chapter 6, the computer became a communications platform with the rise of networking. All modern cars include a local area network for their many computers. The standard interconnection is the CAN (controller area network) bus, developed during the 1980s. Unlike Ethernet, it is optimized for guaranteed response time rather than throughput, because control signals from pedals and steering wheels must be delivered quickly and reliably. The spread of higher bandwidth applications, such as cameras, forced companies to run other kinds of local connections through their vehicles, such as optical fibers. Other networks connected the car to the outside world. Like many other cars of the era, the Tesla had a built-in cellular data connection, used to download operating system updates and traffic information for navigation. The computer behind the touch-screen ran apps including a Web browser, calendar, and text messaging program—the technologies that made the computer into a publishing platform and dissolved it into the network in our final chapters. Adding Internet capabilities to cars opened up their internal networks to attacks by hackers. In 2014, researchers demonstrated vulnerabilities in several popular models that would let attackers take full control, even turning off engines during driving.[2]

In chapter 7, the computer became a personal plaything. Tesla even built a selection of classic Atari video games into a 2018 update for the Model S firmware. Cars, for obvious safety reasons, do not permit the same level of individual tweaking as early personal computers, but Tesla has been successful in building a large community of enthusiastic users who identify emotionally with its products, obsess over each change, and support the company against detractors. Their evangelic fervor recalls the culture of early personal computer users.

In chapter 11, the computer became a media device. Digital media saturated the Tesla. It was too aggressively modern to include a CD or DVD player, still found on most cars of the era, but could replay audio files from USB memory sticks. The main controller could stream Internet radio stations using built-in apps such as Spotify. Bluetooth hooked smartphones into the car's local network, to play voice calls or audio books through its speakers or pop up text messages on the screen.

The car relied on digital video as well as digital audio. By the mid-2010s, rearview digital cameras had already migrated from luxury vehicles down to economy cars, in preparation for a law mandating their fitment as standard features starting in 2018. More expensive cars placed cameras around the vehicle, using software able to knit their perspectives together to fake an overhead view of the car and its surroundings. Some had front-mounted camera to read speed limits from road signs. Coupled with radar, these let a vehicle automatically maintain a safe distance from a car in front, braking automatically if a collision seemed likely.

Tesla made aggressive use of cameras and computers in its Autopilot feature, initially developed for the Model S in collaboration with the Israeli company Mobileye. That version relied on a single monochrome camera, a radar system, and sonar sensors for short-range obstacles during parking. Software released to customers in 2015 used these sensors to automate many driving functions. The car could park itself and drive along the highway, speeding up and slowing down and even changing lanes, as needed.

Work toward this self-driving car had been funded by DARPA for decades, beginning with its Strategic Computing initiative (figure 15.2).[3] Its competitions made self-driving cars a focus for university-based robotics research teams, whose members provided the core for various start-ups attempting to commercialize the technology. In the second chapter, we explored the development of the computer into a scientific supertool for nuclear weapons labs and weather simulation. Video processing takes a huge amount of computer power—in the 1980s, it had taken a Cray supercomputer just to render high-resolution clips for movies. The personal transportation of the upper middle class posed the much more demanding challenge of interpreting images and radar data in real time to identify objects and their trajectories. Test cars had been roaming Mountain View for

Figure 15.2
Stanley, a self-driving Volkswagen, won the 2005 DARPA Grand Challenge and its $2 million prize for a Stanford University team. GPS receivers, roof-mounted LIDAR units, and video cameras helped it navigate a 132 mile desert tract without human intervention. Photo by Mark Avino, Smithsonian National Air and Space Museum (NASM 2012–01952).

years thanks to Waymo, which began as a Google project to automate street photography for its Google Maps service.

Users loved the Autopilot system. They posted videos to YouTube showing themselves watching movies or reading as their cars drove themselves for long distances. One of those video makers was Navy veteran Joshua Brown.[4] In May 2016, his Tesla continued straight as a large truck made a turn in front of it, shearing off its roof and killing Brown as it passed underneath. The investigation showed that he had kept the Autopilot system engaged for almost his entire trip, touching the wheel for only about thirty seconds during a period of thirty-eight minutes and ignoring multiple reminders to return his hands to the wheel. He had set the car to 74 miles an hour, well over the speed limit, on a road with cross traffic (unlike the highways for which Autopilot was designed).[5] Tesla responded by putting more limits on the system, although it remained more aggressive than traditional automakers that employed similar technology but limited its use to low speeds or short periods, branding their features more cautiously as "traffic jam assist," "driving assistant plus," or "Super Cruise."[6]

By February 2020, the newly produced Model S was equipped with more sensors, giving its system access to signals from eight cameras. It ran internally developed software on a custom AI chip with hugely increased processing power. Since 2016, Tesla had been taking full prepayment from its more credulous customers for "full self-driving" software to be delivered later, on a schedule that kept slipping. Musk promised that they would turn a profit on car ownership by letting their vehicles serve as robot taxis while they worked or slept. Uber also was keen to deploy self-driving cars, to do away with its reliance on human drivers and finally make the service profitable. It had rushed its own fleet of test cars into use in Arizona in 2018 with promises that they would be carrying paying passengers within months. Then one of them struck and killed a pedestrian, forcing a halt to the public test program.

Whether or not Tesla will ever deliver real self-driving capabilities for its current cars, by February 2020, the industry-wide changes set in motion by its Model S were already irreversible. Automakers now competed as much on the strength of their software engineering and interface design as on their mechanical engineering. Cars had been fundamentally reshaped around the possibilities opened up by cheap, tiny, and reliable computers. More and more of the tasks formerly undertaken by humans and mechanical control systems were being given over to software and digital electronics. These systems accounted for an ever-increasing share of the cost of building a car and the human effort needed to design one.

The convergence of computing and communications, predicted since the 1970s, changed the way we live and work through technologies such as the Internet, smartphone, and streaming video. Today, an equally important convergence of computing and transportation appears to be underway. As well as driverless cars, drones, and dockless scooters this underpinned dramatic reductions in the cost of space flight, which enthusiasts predict will create a new industry of space tourism.

A ONE-WAY TRIP TO AN UNCERTAIN FUTURE

The country the Tesla drove through was prosperous but anxious and increasingly divided. In the seven decades since ENIAC flickered to life, computer technology had made remarkable progress. Enshrined in Moore's Law, its exponential growth in performance and reduction in size and cost came to seem like a bountiful feature of nature. This in turn created the illusion of an era of unprecedented innovation and ever-accelerating technological change. Other sectors, such as education, health care, and journalism, were compared to the progress made by computer technology and found lacking. Technological fixes, such as the adoption of automated online classes, were prescribed. A book by the editor of *Wired*, called *What Technology Wants*, captured the quasireligious belief that technology had its own evolutionary agenda that humanity must strive to understand.[7]

One of the hottest Silicon Valley start-ups of the 2010s, Theranos, had taken on the inefficiencies of traditional lab work. It developed a refrigerator-sized automated lab able to run hundreds of tests on a single thumbprick of blood, at a fraction of the existing cost. The firm lined up well over a billion dollars in funding and was valued at around ten billion dollars, without ever disclosing how its breakthrough *lab on a chip* technology worked. Elizabeth Holmes, its young founder, turned herself into a kind of Steve Jobs tribute act—clad constantly in black turtlenecks, dropping out of college, and ignoring the advice of engineers who suggested that her magic box might not be feasible. In 2015, Theranos started to collapse as journalists revealed that its breakthrough technology had never worked. Holmes had simply ignored the constraints of physics and chemistry, hiding the failure of Theranos's development efforts by filling the basement with conventional testing machines hidden from visitors and investors.[8] Her attempt to emulate Jobs by bending the universe to her will ended in a federal fraud indictment and the suicide of the firm's chief scientist.

Theranos provides a dramatic example of a broader problem: rapid improvements in computer technology had not led to proportionally great social advances or economic developments. The diverting capabilities of our smartphones, a device at last commensurate with our capacity for wonder, have failed to bring about an economic revolution. In fact, the introduction of the iPhone in 2007 coincided with the end of the short burst of high labor productivity growth caused largely by business adoption of Internet technologies. From 2008 to 2018, productivity growth in the US averaged just 1.3 percent a year, well below the 2.7 percent achieved from 2000 to 2007. It is also well below the average achieved by the American economy from the Industrial Revolution to the 1970s. Ridesharing, video streaming, and online dating have altered our everyday lives, but their impact pales in comparison to that of earlier changes such as the adoption of cars and resulting suburbanization, antibiotics, electric lighting and power, air conditioning, or refrigeration. For societies that were already rich in 1980, the last four decades have been marked by economic stagnation and technological stasis.

Computerization has been more successful in redistributing wealth than making societies richer. The typical American worker of 2018 was no better paid, after adjusting for inflation, than the typical American worker of the 1970s. Economic inequality, which had previously been declining, increased substantially over the period. Well-off Americans, in the top 10 percent of households by income, were better off than their predecessors, having seen their share of national income rise significantly. Political resentment has focused increasingly on the top 0.1 percent of earners, whose income after tax quintupled from 1980 to 2018.[9] Not all this can be explained by technology— political shifts such as tax cuts on investors and high earners and the decline of labor unions played important roles. But economists Erik Brynjolfsson and Andrew McAfee argued persuasively that a shift to automation and online marketplaces had produced a *winner take all* economy in which economic returns were concentrated in the hands of a small number of very successful people.[10]

The promise of self-driving cars and delivery trucks pointed to an automated future in which specialists in robotics and AI grew even wealthier but ordinary Americans struggled to find employment of any kind. Computer technology now threatened to dissolve occupations and industries as well as other machines. Around the world, journalists and politicians were growing increasingly alarmed about the growing concentration of power in the hands of companies such as Google and Facebook and concentration of wealth in the tiny club of their early investors and founders.

Meanwhile, the promise that online communication via Facebook and Twitter would lead to a future of universal democracy, freedom, and tolerance was not holding up well. Twitter took credit for helping activists topple repressive governments during the Arab Spring of 2011. But it, and other online platforms, proved just as useful to neo-Nazis and white supremacists. The speed with which tweets could *go viral* created a mob culture of perennial outrage, whether persecuting women who dared to write about video games or destroying the careers of people who sent messages judged to be racially insensitive. Donald Trump, the most successful Twitter user of all time, began most of his days as US president with a few hours spent watching Fox News while tweeting out incendiary messages, sudden policy changes, and government firings.

Technological utopianism had been well and truly discredited. Across the developed world, long dominant political parties were being eclipsed by antiestablishment movements centered on populism, anti-immigrant sentiment, or environmental protest. Discussion of smartphones focused on the harm they were doing to personal relationships, to democratic institutions, and to civil society. The shift to online discourse and erosion of traditional media practices seemed to be a crucial ingredient in the development of what was called a *post truth* society.

Much of this can be traced back to the Internet's lack of a payment mechanism, forcing Web publishing to rely on advertising revenues. The most successful companies have been those best able to monopolize the attention of their users and to collect and mine user data. Facebook came under fire by privacy campaigners for the lack of protection it gave users to control the use and sharing of their data. Criticism grew louder and far more widespread after the 2016 elections in the US and the Brexit referendum in the UK. It became apparent that Facebook's algorithms had favored inflammatory fake news, inherently likely to be clicked and shared, to such an extent that Russian propaganda and far-right conspiracy theorists were able to exert a potentially decisive influence on election results.[11] Others complained that tacit racism, class prejudice, and sexism were built into search results, speech recognition systems, and other algorithms whose importance increases constantly.[12]

Facebook's data gathering had gone far beyond its own social network, extending even to people who had never created Facebook accounts. Facebook was buying user data from many popular smartphone apps, including the ovulation cycles of millions of women.[13] That horde of data lets Facebook sell the attention of its users to advertisers with great precision, choosing the education level, income, employment, relationship status, race, age, political beliefs, hobbies, and shopping patterns of its users. Facebook could even sell advertisers an audience experiencing a specified life event such as divorce, new motherhood, or a wedding anniversary. Such practices inspired a "techlash" against what the veteran scholar of information technology and organizations Shoshana Zuboff called *The Age of Surveillance Capitalism*.[14] Companies were obtaining information cheaply and processing it into enormously valuable databases. There was a boom in dystopian science fiction showcasing the looming perils of a society saturated with electronic media and corporate surveillance.

TECHNOLOGY AND PANDEMIC

By April 2020, traffic on US 101 was moving freely but the new Tesla had been parked for a while: amid the COVID-19 pandemic, the Valley's offices, schools, and restaurants were locked down, along with most of its shops. Even the Tesla plant had been forced to close by local authorities. Across Europe and North America unemployment spiked to record levels and stock markets cratered. Some local businesses were in trouble. The scooter unicorns and pizza robots were already looking sickly. Now giants like Uber and Airbnb were plunged into crisis as investors questioned the viability of firms designed to bring strangers into close physical contact with each other.

Overall, however, the pandemic was deepening the world's reliance on computer technology. Confined to home, people streamed more video than ever from Netflix and its competitors. Video conferencing replaced in-person discussions as office meetings, college classes, and grade school education shifted to online platforms. Because companies had already rebuilt their business processes around Internet technology, workers could continue to answer customer calls or process mortgage applications from home. The closure of the New York Stock Exchange's trading floor made barely a ripple on Wall Street. Even television news and entertainment shows were filmed from living rooms and basements while studios sat empty. Dates and birthday parties were streamed over video chat, and thousands of families made their final goodbyes over Apple's FaceTime as loved ones neared death in hospitals. Network infrastructure met the challenge well, although in Europe, Netflix and Amazon were forced to lower video quality to preserve bandwidth for other needs.

Technology had made it possible to do some kinds of work online, but not others. This deepened economic and racial inequalities further. White collar professionals with large, comfortable houses and well-equipped home offices might enjoy being freed from lengthy commutes. Silicon Valley companies suggested that many of their employees might work from home permanently. During 2020 rents in San Francisco fell more than 25 percent as tech workers poured out of the city, though they remained the nation's highest. Firms such as Tesla, Palantir, and Oracle announced plans to relocate facilities to states with lower costs. Oracle chairman and founder Larry Ellison relocated to Hawaii, where he had purchased almost all of the island of Lanai.

In contrast, restaurant workers, hotel staff, hairdressers, and store clerks across the country lost their jobs by the millions and were confined with their children in less pleasant neighborhoods. The surge in online shopping and grocery delivery lowered the exposure of customers to the deadly virus and boosted Amazon's revenues while putting the lives of lower-paid warehouse and delivery workers at risk.

The power and longevity of computer code was highlighted as Congress readied legislation to supplement the often meager unemployment benefits offered by states to the tens of millions of newly unemployed. Most states relied on ancient and inflexible COBOL programs, so that a fixed weekly supplement, regardless of previous income level, was the only change that could plausibly be mandated for rapid adoption. Conservative politicians complained that this would discourage poorly paid workers from finding jobs. As a flood of news reports reminded people than many billions of lines of COBOL were still powering mainframes throughout the world, IBM announced free online courses to train the next generation of COBOL programmers.

By June, anger over the slow asphyxiation of George Floyd by Minneapolis police, one of a series of killings of unarmed or already restrained African Americans, had led to a few nights of rioting followed by weeks of nationwide racially integrated mass protests unique in American history. Such police killings were nothing new, but the ubiquity of smartphones meant that they were captured on video and watched by millions on social media, which was also used to organize the protests.

Meanwhile the technology-heavy Nasdaq stock index was setting new highs as investors bet that the new world created by the virus would be more reliant than ever on companies like Zoom and Microsoft. Record profit reports from the tech giants soon followed, as Apple became the first company valued by investors at two trillion dollars. By the end of the year Tesla's valuation had exceeded that of the world's nine highest-volume car makers combined, and shortly thereafter Elon Musk displaced Amazon's Jeff Bezos as the world's richest person. By then broader market indices were also at all-time highs, even as the virus raged out of control across the US killing thousands of people each day, unemployment stabilized at very high levels, and economists predicted long-term reductions in growth rates.

Analysts attributed these striking juxtapositions in part to an increasing disconnection between the fortunes of large companies, buoyed by tax cuts and stimulus measures, and those of ordinary people. Some experts also highlighted the role of smartphone technology in the hands of amateur investors, specifically Robinhood: an investment app with a game-like interface and social media elements that turned stock speculation into a fun diversion. With sports betting impossible and casinos closed, gamblers chased Reddit tips to inflate bubbles in the stocks of troubled, or even bankrupt, companies. The stock of Fangdd, a Chinese real estate company, shot up thirteenfold in two hours as frenzied investors mistook it for the popular FANG group (consisting of Facebook, Amazon, Netflix, and Google).[15]

In the early years of the twenty first century, politicians and tech industry leaders espoused a shared faith that Internet technology, free trade, and human rights would reinforce each other to strengthen liberal democracy around the world. This proposition, already shaken by events of the 2010s, crumbled during the pandemic. The United States had identified its own political culture with the Internet: free speech, limited regulation, and structural decentralization of power. Yet these features, combined with a self-absorbed president, incompetent administration, and badly eroded institutions contributed during the summer of 2020 to the least effective virus-containment efforts and highest death tolls of any major industrialized nation. China, its leading geopolitical rival, had initially got off to a poor start because its political apparatus tried to suppress evidence that the virus was spreading between humans. Yet as the pandemic unfolded, China became a showpiece for the effectiveness of computer technology used without respect for human rights.

Communist rule had collapsed in states such as East Germany in part because of the huge expense and attendant economic inefficiency of running a totalitarian state that depended on a domestic security operation employing vast numbers of spies and informers. Smartphones, data mining algorithms running in cloud data centers, and ubiquitous video cameras can do the job much more cheaply. China's Internet is heavily censored and policed. The country's rapid embrace of face recognition, smartphone applications and electronic payments gives the state access to the activities of its citizens. The same capabilities let China set up a platform to monitor the health status and the movements of all residents. Anyone who had recently been close to a confirmed case would be flagged for self-quarantine. A mandatory smartphone app displayed a code to be shown when entering a public space, to make sure only those with algorithmic authorization could enter. Thanks to such measures, China claimed to have almost eliminated internal transmission of the virus within two months of its first major outbreak after suffering a relatively small number of deaths.

Westerners tend to treat the Internet as experienced in the US as the true Internet, but simple numbers suggest this assumption is no longer viable. China has far more smartphone users than any other country and its internal ecommerce market is twice the combined size of those of the US and Europe. Despite China's increasingly overt regulation of personal freedom and economic activity it appears poised to emerge from the pandemic on course to overtake the US as the world's largest economy more rapidly than previously expected. We can, at the very least, no longer assert that Internet technology has an inherent tendency to boost freedom and undercut autocracy. Nor is it clear that democratic systems of government will shape its continued development.

And so on, and so on. By the time you read this, many more things will have happened in the world and a good proportion of them will have been made possible by computer technology. Yet, this may be a good place to end our story. At the start of the 2020s, the story of modern computing has reached the beginning of an end, as the last great story of modernist technological progress fractures belatedly into postmodern chaos. Computing is deeply intertwined with vital structural developments in global relations, economics, society, and culture. Barring apocalypse, those connections will only deepen. As the computer has become a truly universal machine, the history of computing has become a part of the history of everything. Computer technology does not dictate the direction of history, but its affordances do create new possibilities, advantage some choices over others, and rearrange economic and political incentives. We began the story of the modern computer with the debut of a single machine on the front page of the *New York Times* in 1946 and have followed its legacy to the point of asking whether liberal democracy can survive the Internet. There is little prospect of squeezing an answer into this book. Once computers became part of every infrastructure, the idea of *the computer* as a machine in the tradition of ENIAC, a self-contained device whose users tackled different jobs by creating new programs, has become less relevant. The conceptual problem with the idea of a universal solvent was always that, if any such substance was ever concocted, no flask could contain it. Our protagonist, which dissolved so much in the world that once seemed permanent, has finally dissolved itself.

NOTES

BECOMING UNIVERSAL

1. Brian McCullough, *How the Internet Happened: From Netscape to the iPhone* (New York: Liveright, 2018).

2. Thomas J. Misa, "Understanding 'How Computing Has Changed the World'," *IEEE Annals of the History of Computing* 29, no. 4 (October–December 2007): 52–63.

3. The historical literature on the interaction of users with the Model T is much richer and deeper than that for the personal computer. For example, Kathleen Franz, *Tinkering: Consumers Reinvent the Early Automobile* (Philadelphia: University of Pennsylvania Press, 2005), and Ronald Kline and Trevor Pinch, "Users as Agents of Technological Change: The Social Construction of the Automobile in the Rural United States," *Technology and Culture* 37, no. 4 (October 1996): 763–795. It plays a central role in studies of industrial production, such as David Hounshell, *From the American System to Mass Production, 1800–1932: The Development of Manufacturing Technology in the United States* (Baltimore: Johns Hopkins University Press, 1984). There have been many broad studies of the role of the automobile in American life, such as James J. Flink, *The Automobile Age* (Cambridge, MA: MIT Press, 1988), and Clay McShane, *Down the Asphalt Path* (New York: Columbia University Press, 1994).

4. Liesbeth De Mol, "Turing Machines," Stanford Encyclopedia of Philosophy, September 24, 2018, https://plato.stanford.edu/entries/turing-machine/.

5. Part way through the production of this book, one of us outlined the new structure and its motivation in more detail in Thomas Haigh, *Finding a Story for the History of Computing* (Siegen, Germany: Media of Cooperation Working Paper Series, Siegen University, 2018).

6. Michael S. Mahoney and Thomas Haigh (ed.), *Histories of Computing* (Cambridge, MA: Harvard University Press, 2011), 64.

7. Herman H. Goldstine, *The Computer from Pascal to von Neumann* (Princeton, NJ: Princeton University Press, 1972); Michael R. Williams, *A History of Computing Technology* (Englewood Cliffs, NJ: Prentice Hall, 1985); and Martin Campbell-Kelly and William Aspray, *Computer: A History of the Information Machine* (New York: Basic Books, 1996).

8. Thomas Haigh, "The Tears of Donald Knuth," *Communications of the ACM* 58, no. 1 (Jan 2015): 40–44. Although not a conventional history, you can get a good sense of the development of computer science in Matti Tedre, *The Science of Computing: Shaping a Discipline* (New York: CRC Press, 2015).

9. The role of these adjectives inserted between "first" and "computer" is discussed in Michael R. Williams, "A Preview of Things to Come: Some Remarks on the First Generation of Computers," in *The First Computers: History and Architectures*, ed. Raúl Rojas and Ulf Hashagen (Cambridge, MA: MIT Press, 2000), 1–16.

10. Alice R. Burks and Arthur W. Burks, *The First Electronic Computer: The Atanasoff Story* (Ann Arbor, MI: University of Michigan Press, 1989).

11. Thomas Haigh and Mark Priestley, "Colossus and Programmability," *IEEE Annals of the History of Computing*.

CHAPTER 1

1. T. R. Kennedy Jr., "Electronic Computer Flashes Answers, May Speed Engineering," *New York Times*, February 15, 1946.

2. Press releases are in the binder "Sperry Rand vs. Bell Telephone labs . . . Goldstine Exhibits, 1943–46," Herman Heine Goldstine papers, American Philosophical Society, Philadelphia, PA.

3. Thomas Haigh and Mark Priestley, "Where Code Comes From: Architectures of Automatic Control from Babbage to Algol," *Communications of the ACM* 59, 1 (January 2016): 39–44.

4. *Computers and Their Future: Speeches Given at the World Computer Pioneer Conference* (Llandudno, Wales: Richard Williams and Partners, 1970), 7–3, 7–4.

5. Details on ENIAC here and in the rest of this section, are taken from Thomas Haigh, Mark Priestley, and Crispin Rope, *ENIAC In Action: Making and Remaking the Modern Computer* (Cambridge, MA: MIT Press, 2016).

6. John W. Mauchly, "Preparation of Problems for EDVAC-Type Machines," in *Proceedings of a Symposium on Large-Scale Digital Calculating Machinery, 7–10 January 1947*, ed. William Aspray (Cambridge, MA: MIT Press, 1985), 203–207.

7. Haigh, Priestley, and Rope, *ENIAC in Action*, chap. 2.

8. Haigh, Priestley, and Rope, *ENIAC in Action*, chap. 2.

9. The delay line was initially created for radar and held pulses for a single rotation of a radar antenna. Eckert was the first to realize that data could be stored indefinitely in a delay line by feeding pulses read from its output back into its input.

10. The *First Draft* has been republished in several places, most accessibly as John von Neumann, *First Draft of a Report on the EDVAC*, *IEEE Annals of the History of Computing* 15, no. 4 (October 1993): 27–75 or online as https://library.si.edu/digital-library/book/firstdraftofrepo00vonn. Its genesis and reception are discussed in Haigh, Priestley, and Rope, *ENIAC in Action*, 129–151. The approaches taken to memory in computers of the 1940s are analyzed in Mark Priestley and Thomas Haigh, "The Media of Programming," in *Exploring the Early Digital*, ed. Thomas Haigh (Cham, Switzerland: Springer, 2019): 135–158.

11. The three paradigms are taken from Haigh, Priestley, and Rope, *ENIAC in Action*, 142–149.

12. On cybernetics, see Ronald Kline, *The Cybernetics Moment, Or Why We Call Our Age the Information Age* (Baltimore, MD: Johns Hopkins University Press, 2015).

13. Alan Perlis, "Epigrams on Programming," *ACM SIGPLAN Notices* 17, no. 9 (September 1982).

14. Martin Campbell-Kelly and Michael R. Williams, eds., *The Moore School Lectures: Theory and Techniques for Design of Electronic Digital Computers* (Cambridge, MA: MIT Press, 1985).

15. The series of "Planning and Coding of Problems for an Electronic Computer" reports issued by the IAS computer group in between 1946 and 1948 are collected in John von Neumann, *Papers of John von Neumann on Computing and Computer Theory* (eds. William Aspray and Arthur Burks) (Cambridge, MA: MIT Press, 1987).

16. Haigh, Priestley, and Rope, *ENIAC in Action*. Klara von Neumann is vividly profiled in George Dyson, *Turing's Cathedral: The Origins of the Digital Universe* (New York: Pantheon Books, 2012). The phrase "Women of ENIAC" was introduced in W. Barkley Fritz, "The Women of ENIAC," *IEEE Annals of the History of Computing* 18, no. 3 (Fall 1996): 13–28, and the then-obscure position of the first six operators was challenged in the classic Jennifer S. Light, "When Computers Were Women," *Technology and Culture* 40, no. 3 (July 1999): 455–483.

17. For an overview of tube memory projects, see B. Jack Copeland et al., "Screen History: The Haeff Memory and Graphics Tube," *IEEE Annals of the History of Computing* 39, no. 1 (Jan–Mar 2017): 9–28.

18. Martin Campbell Kelly, "The Evolution of Digital Computing Practice on the Cambridge University EDSAC, 1949–1951," in *Exploring the Early Digital*, ed. Thomas Haigh (Cham, Switzerland: Springer, 2019), 117–134.

19. Simon Lavington, *Early British Computers* (Bedford, MA: Digital Press, 1980).

20. F. C. Williams and T. Kilburn, "A Storage System for Use with Binary-Digital Computing Machines," *Institution of Electrical Engineers, Proc. Part III* 96 (March 1949): 81–100.

21. Aiken's ignorance of the substance of Babbage's work at the time he designed the Mark I was demonstrated in I. Bernard Cohen, *Howard Aiken: Portrait of a Computer Pioneer* (Cambridge, MA: MIT Press, 1999), 61–72.

22. See Thomas Haigh and Mark Priestley, "Von Neumann Thought Turing's Universal Machine Was 'Simple and Neat.': But That Didn't Tell Him How to Design a Computer," *Communications of the ACM* 63, no. 1 (January 2020): 26–32. Some authors with backgrounds in logic or philosophy have assumed that the theoretical work of Alan Turing must have had a major influence on the invention of actual computers. For example, Martin Davis, *Engines of Logic: Mathematicians and the Origin of the Computer* (New York: Norton, 2001). In contrast, historians working from primary sources have seen little evidence of such influence. See Simon Lavington, ed., *Alan Turing and His Contemporaries* (Swindon, UK: British Informatics Society, 2012), and Thomas Haigh, "Actually, Turing Did Not Invent the Computer," *Communications of the ACM* 57, no. 1 (January 2014): 36–41.

23. Priestley and Haigh, "The Media of Programming." ACE and its relatives are explored in B. Jack Copeland, ed., *Alan Turing's Automatic Computing Engine: The Master Codebreaker's Struggle to Build the Modern Computer* (New York: Oxford University Press, 2005).

24. Konrad Zuse, "Planfertigungsgeräte," 1944; Zuse Collection, Deutsches Museum Archives, Munich.

25. Maurice Wilkes, *Memoirs of a Computer Pioneer* (Cambridge, MA: MIT Press, 1985).

26. Martin Campbell-Kelly, "Programming the EDSAC: Early Programming Activity at the University of Cambridge," *Annals of the History of Computing* 2, no. 1 (October 1980): 7–36; also Maurice V. Wilkes, David J. Wheeler, and Stanley Gill, *The Preparation of Programs for an Electronic Digital Computer* (Cambridge, MA: Addison-Wesley, 1951).

27. Haigh, Priestley, and Rope, *ENIAC in Action*, 238–243.

28. Franz L. Alt, "Fifteen Years ACM," *Communications of the ACM* 5, no. 6 (1962): 300–307. There is no comprehensive history of the ACM, but facets of its story are explored in the papers collected as Thomas J. Misa, ed., *Communities of Computing: Computer Science and Society in the ACM* (San Rafael, CA: Morgan & Claypool [ACM Books], 2017). Berkeley's story is told in Bernadette Longo, *Edmund Berkeley and the Social Responsibility of Computer Professionals* (San Rafael, CA: Morgan & Claypool [ACM Books], 2015).

29. Testimony by Cannon, Hagley Museum, Honeywell v. Sperry Rand papers, Series III, box 140, p. 17,680.

30. Cohen, *Howard Aiken: Portrait of a Computer Pioneer*, 283–293.

31. For a discussion of the fate of the EDVAC, see Michael R. Williams, "The Origins, Uses, and Fate of the EDVAC," *IEEE Annals of the History of Computing* 15, no. 1 (Jan–Mar 1993): 22–38.

32. Mauchly to J. P. Eckert Jr. et al., 1/12/1948; Hagley Museum, Sperry Univac Company Records, Series I, box 3.

33. Herman Luckoff, *From Dits to Bits: A Personal History of the Electronic Computer;* (Portland, OR: Robotics Press, 1979), chap. 9.

34. Luther A. Harr, *The Univac System, a 1954 Progress Report* (Remington Rand Corporation, 1954), 6.

35. James C. McPherson, "Census Experience Operating a UNIVAC System," in *Symposium on Managerial Aspects of Digital Computer Installations* (Washington, DC: US Office of Naval Research, 1953), 30–36.

36. Nancy Beth Stern, *From ENIAC to UNIVAC: An Appraisal of the Eckert-Mauchly Computers* (Bedford, MA: Digital Press, 1981), 148–151.

37. J. Presper Eckert, "Thoughts on the History of Computing," *IEEE Computer* 9, no. 12 (December 1976): 58–65.

38. I. Bernard Cohen and Gregory W. Welch, eds., *Makin' Numbers: Howard Aiken and the Computer* (Cambridge, MA: MIT Press, 1999), 247.

39. The development of the Univac tape drives was a major engineering challenge, described in Arthur L. Norberg, *Computers and Commerce: A Study of Technology and Management at Eckert-Mauchly Computer Company, Engineering Research Associates, and Remington Rand, 1946–1957* (Cambridge, MA: MIT Press, 2005), 186–192.

40. For a broader account of use of technology in business administration, see JoAnne Yates, *Control Through Communication: The Rise of System in American Management* (Baltimore, MD: Johns Hopkins University Press, 1989). Punched card technology and its use in business have been explored in JoAnne Yates, *Structuring the Information Age* (Baltimore: Johns Hopkins University Press, 2005); Lars Heide, *Punched-Card Systems and the Early Information Explosion, 1880–1945* (Baltimore: Johns Hopkins University Press, 2009); and Martin Campbell-Kelly, *ICL: A Technical and Business History* (New York: Oxford University Press, 1989).

41. Roddy F. Osborn, "GE and UNIVAC: Harnessing the High-Speed Computer," *Harvard Business Review* 32, no. 4 (July–August 1954): 99–107.

42. McPherson, "Census Experience Operating a UNIVAC System."

43. Remington Rand, *UNIVAC Fac-Tronic System* (brochure), archive.org, c. 1951, https://archive.org/details/UNIVACFacTronicSystemBrochure/mode/2up.

44. R. Dorfman, "The Discovery of Linear Programming," *Annals of the History of Computing* 6, no. 3 (July 1984): 283–295.

45. L. R. Johnson, "Installation of a Large Electronic Computer," *Proc. ACM Meeting, Toronto* (1952): 77–80.

46. Harr, *The Univac System*.

47. Lawrence Livermore Laboratory, *Computing at Lawrence Livermore Laboratory*, UCID Report 20079, 1984.

48. Osborn, "GE and UNIVAC."

49. Harr, *The Univac System*.

50. Univac Conference, OH 200. Oral history on 17–18 May 1990, Washington, DC. Charles Babbage Institute, University of Minnesota, Minneapolis, http://purl.umn.edu/104288.

51. John Diebold, *Automation, the Advent of the Automatic Factory* (New York: Van Nostrand, 1952).

52. John Diebold, "Factories Without Men: New Industrial Revolution," *The Nation,* September 19, 1953. See also David F. Noble, *Forces of Production: A Social History of Industrial Automation* (New York: Alfred A. Knopf, 1984), chapter 4.

53. Osborn, "GE and UNIVAC," quotation on p. 103.

CHAPTER 2

1. Emerson W. Pugh, *Memories That Shaped an Industry: Decision Leading to IBM System/360* (Cambridge, MA: MIT Press, 1984), 30.

2. Cuthbert C. Hurd, "Early IBM Computers: Edited Testimony," *Annals of the History of Computing* 3, no. 2 (April–June 1981): 163–182.

3. Katherine Davis Fishman, *The Computer Establishment* (New York: Harper & Row, 1981), 44.

4. Pugh, *Memories That Shaped an Industry*, chap. 2.

5. Kent C. Redmond and Thomas M. Smith, *Project Whirlwind: The History of a Pioneering Computer* (Bedford, MA: Digital Press, 1980), 206. The ENIAC core memory is discussed in Thomas Haigh, Mark Priestley, and Crispin Rope, *ENIAC In Action: Making and Remaking the Modern Computer* (Cambridge, MA: MIT Press, 2016), 223–224.

6. Arthur W. Burks, Herman Heine Goldstine, and John von Neumann, *Preliminary Discussion of the Logical Design of an Electronic Computing Instrument* (Princeton, NJ: Institute for Advanced Studies, 1946), 14.

7. Herman H. Goldstine and John von Neumann, *Planning and Coding Problems for an Electronic Computing Instrument. Part II, Volume 1* (Princeton, NJ: Institute for Advanced Study, 1947), 33.

8. C. Gordon Bell, J. Craig Mudge, and John E. McNamara, *Computer Engineering: A DEC View of Hardware Systems Design* (Bedford, MA: Digital Press, 1978), 256–257.

9. Alice R. Burks and Arthur W. Burks, *The First Electronic Computer: The Atanasoff Story* (Ann Arbor, MI: University of Michigan Press, 1989), chap. 1; also J. Presper Eckert, "A Survey of Digital Computer Memory Systems," *Proceedings of the IRE* 41, no. 10 (October 1953): 1393–1406.

10. Erwin Tomash and Arnold A. Cohen, "The Birth of an ERA: Engineering Research Associates, Inc., 1946–1955," *IEEE Annals of the History of Computing* 1, no. 2 (April–June 1979): 83–97.

11. The history of ERA has been told in detail in Arthur L. Norberg, *Computers and Commerce: A Study of Technology and Management at Eckert-Mauchly Computer Company, Engineering Research Associates, and Remington Rand, 1946–1957* (Cambridge, MA: MIT Press, 2005), chaps. 1 and 3. The development of the drum is described on pages 120–142.

12. Charles J. Murray, *The Supermen: The Story of Seymour Cray and the Technical Wizards Behind the Supercomputer* (New York: Wiley, 1997), 44–45.

13. Tomash and Cohen, "The Birth of an ERA," quotation p. 90. The story of the first ERA computers is told in Norberg, *Computers and Commerce*, 142–150 on Atlas and 154–159 on the ERA 101.

14. Norberg, *Computers and Commerce*, 159–166.

15. Seymour R. Cray, "Computer-Programmed Preventative Maintenance for Internal Memory Sections of the ERA 1103 Computer System," in *Proceedings of the WESCON Computer Sessions* (New York: The Institute of Radio Engineers, 1954), 62–66.

16. Engineering Research Associates, *High-Speed Computing Devices* (New York: McGraw-Hill, 1950), 322–339.

17. C. Gordon Bell and Allen Newell, *Computer Structures: Readings and Examples* (New York: McGraw-Hill, 1971).

18. Martin Campbell-Kelly, "Programming the Pilot Ace: Early Programming Activity at the National Physical Laboratory," *Annals of the History of Computing* 3, no. 2 (April 1981): 133–162.

19. R. Hunt Brown, "Computer Comparison and Census Chart," *Management and Business Automation* 4, no. 1 (August 1960): 34.

20. Early programming work done under Hopper's direction at the Eckert & Mauchly Computer Corporation is described in Norberg, *Computers and Commerce*, 192–205.

21. Grace Hopper, "Compiling Routines," internal memorandum, Eckert-Mauchly Computer Corporation, Philadelphia, Dec. 31, 1953; box 6, folder 9, Grace Murray Hopper Collection, Archives Center, National Museum of American History.

22. Grace M. Hopper, "Compiling Routines," *Computers and Automation* 2 (May 1953): 1–5.

23. J. H. Laning and N. Zierler, "A Program for Translation of Mathematical Equations for Whirlwind I," January 1954. Charles Babbage Institute, NBS Collection, box 39, folder 8.

24. Donald E. Knuth and Luis Trabb Pardo, "The Early Development of Programming Languages," in *A History of Computing in the Twentieth Century*, ed. N. Metropolis, J. Howlett, and Gian-Carlo Rota (New York: Academic Press, 1980), 197–273.

25. John Backus, "Programming in America in the 1950s—Some Personal Impressions," in *A History of Computing in the Twentieth Century*, ed. N. Metropolis, J. Howlett, and Gian-Carlo Rota (New York: Academic Press, 1980), 125–135.

26. Backus, "Programming in America," 130–131.

27. David Nofre, Mark Priestley, and Gerard Alberts, "When Technology Became Language: The Origins of the Linguistic Conception of Computer Programming, 1950–1960," *Technology and Culture* 55, no. 1 (January 2014): 40–75.

28. Grace Hopper, "Keynote Address," in *History of Programming Languages*, ed. Richard L. Wexelblat (New York: Academic Press, 1981), 7–20.

29. Paul Armer, "SHARE—A Eulogy to Cooperative Effort," *Annals of the History of Computing* 2, no. 2 (April 1980): 122–129. Armer says that SHARE did not stand for anything; others say it stood for Society to Help Avoid Redundant Effort.

30. Roy A. Larner, "FMS: The IBM FORTRAN Monitor System," in *Proceedings of the National Computer Conference* (AFIPS, 1987): 815–820.

31. G. F. Ryckman, "The Computer Operation Language," in *Proceedings of the Western Joint Computer Conference* (1960), 341–343.

32. SOS is described in K. V. Hanford, "The SHARE Operating System for the IBM 709," *Annual Review in Automatic Programming* 1 (1960):169–177. The early development of operating systems is explored in Maarten Bullynck, "What Is an Operating System? A Historical Investigation (1954–1964)," in *Reflections on Programming Systems: Historical and Philosophical Aspects*, ed. Liesbeth de Mol and Giuseppe Primiero (Cham, Switzerland: Springer, 2019), 49–79.

33. Atsushi Akera, "Voluntarism and the Fruits of Collaboration," *Technology and Culture* 42, no. 4 (October 2001): 710–736.

34. Ray Argyle, "25th Anniversary Issue," *University of Waterloo, Department of Computing Services Newsletter*, 1982. See also Scott Campbell, "'Wat For Ever:' Student-Oriented Computing at the University of Waterloo," *IEEE Annals of the History of Computing* 35, no. 1 (Jan–Mar 2013): 11–22.

35. "Industry Profile . . . Wes Graham of Waterloo U," *Computer Data: The Canadian Computer Magazine,* May 1976.

36. Paul Cress, Paul Dirksen, and J. Wesley Graham, *Fortran IV with WATFOR and WATFIV* (Englewood Cliffs, NJ: Prentice Hall, 1970).

37. Thomas Haigh, "Jack Dongarra: Supercomputing Expert and Mathematical Software Specialist," *IEEE Annals of the History of Computing* 30, no. 2 (April–June 2008): 74–81.

38. BNF is sometimes called Backus-Naur Form, as Naur edited the report in which it was first used, and some object on technical grounds to calling it a "normal form."

39. Gerald W. Brock, *The Telecommunications Industry: The Dynamics of Market Structure* (Cambridge, MA: Harvard University Press, 1981), 187–194.

40. Luckoff, *From Dits to Bits: A Personal History of the Electronic Computer*, J. L. Maddox, J. B. O'Toole, and S. Y. Wong, "The Transac S-1000 Computer," in *Proceedings of the 1956 Eastern Joint Computer Conference* (New York: Association for Computing Machinery, 1956), 13–16.

41. Simon Lavington, *A History of Manchester Computers* (Swindon, UK: The British Computer Society, 1998).

42. NASA, Ames Research Center, "A Justification of the Need to Replace the IBM 7040/7094 Direct Couple System," March 31, 1967, p. 8. NASA Ames History Archives.

43. NASA-Ames Research Center, "ADPE Acquisition Plan—Category A," memorandum Sept. 25, 1967.

44. Michael R. Williams, *A History of Computing Technology* (Englewood Cliffs, NJ: Prentice-Hall, 1985), 393–394.

45. Werner Buchholz, ed., *Planning a Computer System: Project Stretch* (New York: McGraw-Hill, 1962), 40, and Werner Buchholz, "Anecdote: Origin of the Word Byte," *Annals of the History of Computing* 3, no. 1 (January 1981): 72.

46. Thomas Watson Jr. and Peter Petre, *Father, Son & Co: My Life at IBM and Beyond* (New York: Bantam, 1990), 282–283.

47. Simon Lavington, "The Manchester Mark I and Atlas: A Historical Perspective," *Communications of the ACM* 21, no. 1 (January 1978): 4–12.

48. Bill Buzbee, "Oral History Interview by Thomas Haigh, 8 and 9 April, Westminster, CO," 2005, in Society for Industrial and Applied Mathematics.

49. James C. Worthy, "Control Data Corporation: The Norris Era," *IEEE Annals of the History of Computing* 17, no. 1 (January–March 1995): 47–53.

50. Jack Dongarra, "Oral History Interview by Thomas Haigh, 26 April, University of Tennessee, Knoxville, TN," 2005, in Society for Industrial and Applied Mathematics.

CHAPTER 3

1. LEO is discussed in G. Ferry, *A Computer Called LEO: Lyons Tea Shops and the World's First Office Computer.* (London: Fourth Estate, 2003), and David Caminer et al., eds., *User Driven Innovation: The World's First Business Computer* (London: McGraw-Hill, 1996). In the UK, operations and methods work was pioneered within the civil service: Jon Agar, *The Government Machine: A Revolutionary History of the Computer* (Cambridge, MA: MIT Press, 2003).

2. Edmund C. Berkeley, *Giant Brains or Machines That Think* (New York: John Wiley & Sons, 1949), vii.

3.   John M. Thesis, "Practical Application of Electronic Equipment," *Journal of Machine Accounting* 6, no. 3 (March 1955): 5, 7–8, 16–17.

4.   Caminer et al., *User Driven Innovation*, 52.

5.   John Aris, "The LEO Approach—An Evaluation," in *User Driven Innovation: The World's First Business Computer*, ed. David Caminer, et al. (London: McGraw-Hill Book Company, 1996), 320–326.

6.   Charles J. Bashe et al., *IBM's Early Computers* (Cambridge, MA: MIT Press, 1986), 129, appendix B.

7.   Bashe et al., *IBM's Early Computers,* 173–178.

8.   Bashe et al., *IBM's Early Computers*, 168.

9.   Erwin Tomash and Arnold A. Cohen, "The Birth of an ERA: Engineering Research Associates, Inc., 1946–1955," *IEEE Annals of the History of Computing* 1, no. 2 (April–June 1979).

10. Bashe et al., *IBM's Early Computers*, chaps. 3, 5.

11. Thomas Watson Jr. and Peter Petre, *Father, Son & Co: My Life at IBM and Beyond* (New York: Bantam, 1990), 224.

12. Knuth provided a detailed description of his pleasant evenings with the 650 in Donald Knuth, "The IBM 650: An Appreciation from the Field," *Annals of the History of Computing* 8, no. 1 (January–March 1986): 50–55.

13. Bell and Newell, *Computer Structures: Readings and Examples*, chap. 18; also Franklin M. Fisher, James W. McKie, and Richard B. Mancke, *IBM and the US Data Processing Industry: An Economic History* (New York: Praeger, 1983), 53.

14. "Six-Month Shipments Top 4,000 Mark," *Business Automation,* August 1966. Excluding "small" computer systems costing less than $100,000, such as those made by Friden, the 1400 family made up almost half of the world's installed base of US-built computers.

15. The first suggestion for using a magnetic disk in a computer seems to have come from Presper Eckert in 1944, who suggested using it as part of the control system for an electronic calculator. Eckert hoped that keeping code for the steps needed to carry out a complex operation such as multiplication as information on the disk would be cheaper than building complex hardware, lowering the cost enough to compete with mechanical desk calculators. Thomas Haigh, Mark Priestley, and Crispin Rope, *ENIAC In Action: Making and Remaking the Modern Computer* (Cambridge, MA: MIT Press, 2016), 232–238

16. Christopher Johnson, "Charlie Brown—Peanuts—Charles M. Schulz—Snoopy—Woodstock," ASCII Art, n.d., accessed June 2, 2020, https://asciiart.website/index.php?art=comics/peanuts.

17. Mitchell E Morris, "Professor RAMAC's Tenure," *Datamation,* April 1981.

18. The feasibility studies are discussed in Thomas Haigh, "The Chromium-Plated Tabulator: Institutionalizing an Electronic Revolution, 1954–1958," *IEEE Annals of the History of Computing* 23, no. 4 (October–December 2001): 75–104.

19. Thomas Haigh, "Charles W. Bachman: Database Software Pioneer," *IEEE Annals of the History of Computing* 33, no. 4 (October–December 2011): 70–80.

20. Thomas J. Watson, Jr, "Address by Thomas J. Watson, Jr., President, International Business Machines Corp.," in *Data Processing (1): 1958 Conference Proceedings*, ed. Charles H. Johnson (Chicago: National Machine Accountants Association, 1958), 15–19.

21. Haigh, "Chromium-Plated Tabulator."

22. Thomas J. Misa, "Gender Bias in Computing," in *Historical Studies in Computing, Information, and Society*, ed. William Aspray (Cham, Switzerland: Springer Nature, 2019), 113–133.

23. Thomas Haigh, "Masculinity and the Machine Man," in *Gender Codes: Why Women Are Leaving Computing*, ed. Thomas J. Misa (Hoboken, NJ: IEEE Computer Society Press, 2010): 51–71. In a case study of the British civil service, Marie Hicks argues that during the 1960s its leaders created deliberate barriers to promotion from "machine grade" operator jobs, largely filled by women, to programming work to avoid an influx of women into well-paid and high-status positions. Marie Hicks, *Programmed Inequality: How Britain Discarded Women Technologists and Lost Its Edge in Computing* (Cambridge, MA: MIT Press, 2017), 152–167 and 170–176.

24. John von Neumann, Letter to Herman Goldstine, May 8, 1945 in box 21 of the Herman H. Goldstine Papers, American Philosophical Society, Philadelphia. Von Neumann's work during this period is discussed in Mark Priestley, *Routines of Substitution: John von Neumann's Work on Software Development, 1945–1948* (Cham, Switzerland: Springer, 2018).

25. Donald Knuth, *The Art of Computer Programming Volume 3: Sorting and Searching* (Reading, MA: Addison-Wesley, 1973), 386.

26. Generalized file management software, including 9PAC, is discussed in Thomas Haigh, "How Data Got its Base: Information Storage Software in the 1950s and 1960s," *IEEE Annals of the History of Computing* 31, no. 4 (October-December 2009): 6–25. The Gruenberger quote is from the "Verbatim Transcript of the 9th Meeting of SHARE, Oct. 3, 1957" in the SHARE archives at the Charles Babbage Institute.

27. C. A. R. Hoare, "Quicksort," *Computer Journal* 5, no. 1 (1962): 10–15.

28. Juris Hartmanis and Richard E. Stearns, "On the Computational Complexity of Algorithms," *Transactions of the American Mathematical Society* 117 (1965): 285–306.

29. Work on COBOL was carried out by two committees in CODASYL: the COnference On DAta SYstems Languages. See Jean E. Sammet, *Programming Languages: History and Fundamentals* (Eaglewood Cliffs, NJ: Prentice Hall, 1969), section V.3.

30. Jean E. Sammet, "The Early History of Cobol," in *History of Programming Languages*, ed. Richard L. Wexelblat (New York: Academic Press, 1981), 199–242.

31. The code fragment is taken from Sammet, *Programming Languages*, 337.

32. Bob O. Evans, "System/360: A Retrospective View," *Annals of the History of Computing* 8, no. 2 (April–June 1986): 155–179.

33. John W. Haanstra and Bob O. Evans, "Processor Products—Final Report of SPREAD Task Group, December 28, 1961," *Annals of the History of Computing* 5, no. 1 (January 1983): 6–26.

34. Various, "Discussion of the SPREAD Report, June 23, 1982," *Annals of the History of Computing* 5, no. 1 (January 1983): 27–44; see p. 31.

35. A. Padegs, "System/360 and Beyond," *IBM Journal of Research and Development* 25, no. 5 (September 1981): 377–390.

36. Reprinted in Maurice Wilkes, "The Best Way to Design an Automatic Calculating Machine," *Annals of the History of Computing* 8, no. 2 (April–June 1986): 118–121.

37. Various, "Discussion of the SPREAD Report, June 23, 1982"; see p. 33.

38. Wilkes, "The Best Way to Design an Automatic Calculating Machine." Although Wilkes was the first to articulate this as a design method, the same approach was used in configuring ENIAC to run modern code. Haigh, Priestley, and Rope, *ENIAC in Action*.

39. Emerson W. Pugh, Lyle R. Johnson, and John H. Palmer, *IBM's 360 and Early 370 Systems* (Cambridge, MA: MIT Press, 1991), 163, 214–217.

40. The history of PL/I has been rather neglected in contrast to the attention given by historians to other early languages, but it is described by a participant in George Radin, "The Early History and Characteristics of PL/1," in *History of Programming Languages*, ed. Richard L. Wexelblat (New York: Academic Press, 1981), 551–574.

41. Bob Evans, quoted in T. A. Wise, "IBM's $5,000,000,000 Gamble," *Fortune,* September 1966.

42. Pugh, Johnson, and Palmer, *IBM's 360 and Early 370 Systems*, 169.

43. The interim versions of OS/360 were known informally by the acronyms BOS, TOS, PCP, and MFT.

44. Pugh, Johnson, and Palmer, *IBM's 360 and Early 370 Systems*, 321–331.

45. William Kahan, "Oral History Interview by Thomas Haigh, 5–8 August, 2005, Berkeley, California," 2005, in Society for Industrial and Applied Mathematics. The changes made by IBM are described in Harding, L. J., Jr. (1966), "Modifications of System/360 Floating Point," SHARE Secretary Distribution, pp. 11–27, SSD 157, C4470.

46. Pugh, Johnson, and Palmer, *IBM's 360 and Early 370 Systems*, chap. 7.

47. Pugh, Johnson, and Palmer, *IBM's 360 and Early 370 Systems*, 169–174.

48. Richard G. Canning and Roger L. Sisson, *The Management of Data Processing* (New York: John Wiley & Sons, 1967). This quotation gave its name to Haigh, "Chromium-Plated Tabulator," which explores the initial adoption of computers by American business.

49. Harold J. Leavitt and Thomas L. Whisler, "Management in the 1980s," *Harvard Business Review* 36, no. 6 (November–December 1958): 41–48.

50. W. Robert Widener, "New Concepts of Running a Business," *Business Automation* 13, no. 4 (April 1966): 38–43, 63

51. *Datamation*, March 1967, 2–3.

52. McKinsey and Company, *Unlocking the Computer's Profit Potential* (New York: McKinsey, 1968).

53. Richard L. Nolan, ed., *Managing the Data Resource Function* (New York: West, 1974), 27.

54. Charles W. Bachman, "Integrated Data Store—The Information Processing Machine That We Need!" Charles W. Bachman Papers (CBI 123), box 1, folder 11, Charles Babbage Institute, University of Minnesota. Quotation from page 6. Bachman recalled the origins of IDS in Charles W. Bachman, "The Origin of the Integrated Data Store (IDS): The First Direct-Access DBMS," *IEEE Annals of the History of Computing* 31, no. 4 (October–December 2009): 42–54.

55. The work of the CODASYL groups is summarized in T. William Olle, "Recent CODASYL Reports on Data Base Management," in *Data Base Systems*, ed. Randall Rustin (Englewood Cliffs, NJ: Prentice Hall, 1972), 175–184.

56. Charles W. Bachman, "The Programmer as Navigator," *Communications of the ACM* 16, no. 11 (November 1973): 653–658.

57. The DBA position and the development of the DBMS industry are discussed in Thomas J. Bergin and Thomas Haigh, "The Commercialization of Database Management Systems, 1969–1983," *IEEE Annals of the History of Computing* 31, no. 4 (October–December 2009): 26–41.

58. Material for the IRS case study comes from unpublished materials supplied to the authors by the IRS, including the film "History of ADP in IRS Service Centers," IRS Austin, Texas Service Center, Dec. 11, 1991. We also rely on Daniel Capozzoli, "The Early Years of Data Processing," *Computer Services Newsletter (Internal Revenue Service),* July 1987, and "ADP-History," undated typescript (c. 1970), IRS, Washington, DC.

59. Comptroller General of the United States, *Safeguarding Taxpayer Information—An Evaluation of the Proposed Computerized Tax Administration System (LCD-76–115)* (Washington, DC: Department of the Treasury, 1977), 4.

60. "Proposed IRS System May Pose Threat to Privacy," *Computerworld*, February 21, 1977, 1–6.

61. The question was posed by Rep. Clarence Miller of Ohio of the House subcommittee on the Treasury, Postal Service, and General Government Appropriations.

62. US House of Representatives, Hearings before a Subcommittee of the Committee on Appropriations, 95th Congress, March 12, 1978, p. 438.

63. The whole story is told in David Burnham, *The Rise of the Computer State* (New York: Random House, 1983).

64. Burton Grad, "A Personal Recollection: IBM's Unbundling of Software and Services," *IEEE Annals of the History of Computing* 24, no. 1 (January–March 2002): 64–71. Martin Campbell-Kelly, *From Airline Reservations to Sonic the Hedgehog: A History of the Software Industry* (Cambridge, MA: MIT Press, 2003), 109–115.

65. Thomas Haigh, "Larry A. Welke—Biography," *IEEE Annals of the History of Computing* 26, no. 4 (October-December 2004): 85–91.

66. Campbell-Kelly, *From Airline Reservations*, 165, 167.

67. Haigh, "Charles W. Bachman: Database Software Pioneer."

68. Timo Leimbach, "The SAP Story: Evolution of SAP within the German Software Industry," *IEEE Annals of the History of Computing* 30, no. 4 (October–December 2008): 60–76.

CHAPTER 4

1. For the history of analog computing, see Charles Care, *Technology for Modelling: Electrical Analogies, Engineering Practice, and the Development of Analogue Computing* (Cham, Switzerland: Springer, 2010); and Aristotle Tympas, *Calculation and Computation in the Pre-electronic Era* (Cham, Switzerland: Springer, 2017).

2. Nicholas Barr, "The History of the Phillips Machine," in *A.W.H. Phillips: Collected Works in Contemporary Perspective* (New York: Cambridge University Press, 2000), 89–114.

3. L. Searle, "The Bombsight War: Norden vs. Sperry," *IEEE Spectrum* 26, no. 9 (1989): 60–64.

4. Frederik Nebeker, *Signal Processing: The Emergence of a Discipline, 1948 to 1998* (New Brunswick, NJ: IEEE History Center, 1998); and James S. Small, *The Analogue Alternative: The Electronic Analogue Computer in Britain and the USA, 1930–1975* (New York: Routledge, 2001).

5. Peter Galison, "The Ontology of the Enemy: Norbert Wiener and the Cybernetic Vision," *Critical Inquiry* 21, no. 1 (Autumn 1994): 228–266.

6. Norbert Wiener, *Cybernetics, or Control and Communication in the Animal and the Machine* (Cambridge, MA: Technology Press, 1948).

7. Paul Ceruzzi, *Beyond the Limits: Flight Enters the Computer Age* (Cambridge, MA: MIT Press, 1989), chaps. 1 and 8.

8. Perry Orson Crawford Jr, "Automatic Control by Arithmetical Operations" (Massachusetts Institute of Technology, 1942).

9. Kent C. Redmond and Thomas M. Smith, *Project Whirlwind: The History of a Pioneer Computer* (Bedford, MA: Digital Press, 1980).

10. Atsushi Akera, *Calculating a Natural World: Scientists, Engineers, and Computers During the Rise of U.S. Cold War Research* (Cambridge, MA: MIT Press, 2007), chap. 5.

11. R. E. Everett and F. E. Swain, "Whirlwind I Computer Block Diagrams," Project Whirlwind Report mechanisms Laboratory, Massachusetts Institute of Technology, September 4, 1947.

12. Saul Rosen, "Programming Systems and Languages: A Historical Survey," in *Proceedings of the April 21–23, 1964 Spring Joint Computer Conference* (AFIPS, 1964), 1–15.

13. Paul N. Edwards, *The Closed World: Computers and the Politics of Discourse in Cold War America* (Cambridge, MA: MIT Press, 1996).

14. Edwards, *The Closed World*, 110.

15. IBM Corporation, "On Guard! The Story of SAGE," 1957.

16. Emerson W. Pugh, *Memories That Shaped an Industry: Decision Leading to IBM System/360* (Cambridge, MA: MIT Press, 1984), 102–117.

17. Edmund Van Deusen, "Electronics Goes Modern," *Fortune,* June 1955.

18. For example, Nathan Ensmenger, *The Computer Boys Take Over: Computers, Programmers, and the Politics of Technical Expertise.* (Cambridge, MA: MIT Press, 2010), 60, suggests that SDC's 700 programmers constituted 60 percent of the US workforce in 1956 and that by training 7,000 more in the following five years, SDC doubled the national supply of programmers.

19. The IBM training program is described in "Business Week Reports to Readers On: Computers," *Business Week,* 21 June 1958. The figure on SDC comes from Claude Baum, *The System Builders: The Story of SDC* (Santa Monica: System Development Corporation, 1981), 47.

20. Duncan G. Copeland, Richard O. Mason, and James L. McKenney, "SABRE: The Development of Information-Based Competence and Execution of Information-Based Competition," *IEEE Annals of the History of Computing* 17, no. 3 (Fall 1995): 30–57.

21. Ceruzzi, *Beyond the Limits*, chap. 9.

22. S. E. James, "Evolution of Real Time Computer Systems for Manned Spaceflight," *IBM Journal of Research and Development* 25, no. 5 (September 1981): 417–428, quotation p. 245; and Saul I. Gass, "The Role of Digital Computers in Project Mercury," in *Proceedings of the Eastern Joint Computer Conference* (AFIPS, 1961), 33–46.

23. Marilyn Scott and Robert Hoffman, "The Mercury Programming System," in *Proceedings of the Eastern Joint Computer Conference* (AFIPS, 1961), 47–53.

24. The IBM installations were highlighted in the best-selling book, Margot Lee Shetterly, *Hidden Figures* (New York: Harper Collins, 2016). The film adaptation takes some artistic liberties with the narrative but is correct in showcasing the importance of the 7090 mainframe in the early phases of the manned space program.

25. Jamie Parker Pearson, *Digital at Work: Snapshots from the First Fifty Years* (Burlington, MA: Digital Press, 1992), 6–9.

26. C. Gordon Bell, J. Craig Mudge, and John E. McNamara, *Computer Engineering: A DEC View of Hardware Systems Design* (Bedford, MA: Digital Press, 1978), 125–127.

27. Bell, Mudge, and McNamara, *Computer Engineering*, 129.

28. Bell, Mudge, and McNamara, *Computer Engineering*, 136–139; and Pearson, *Digital at Work*, 16–21.

29. Glenn Rifkin and George Harrar, *The Ultimate Entrepreneur: The Story of Ken Olsen and Digital Equipment Corporation* (Chicago: Contemporary Books, 1988).

30. Pearson, *Digital at Work*, 143.

31. C. Gordon Bell and Allen Newell, *Computer Structures: Readings and Examples* (New York: McGraw-Hill, 1971); and Charles J. Bashe et al., *IBM's Early Computers* (Cambridge, MA: MIT Press, 1986), 448–449.

32. Bell, Mudge, and McNamara, *Computer Engineering*, 64, 180, and 198–199.

33. Bell, Mudge, and McNamara, *Computer Engineering*, 181.

34. "A Chorus Line: Computerized Lighting Control Comes to Broadway," *Theatre Crafts,* November/December 1975.

35. Emerson W. Pugh, Lyle R. Johnson, and John H. Palmer, *IBM's 360 and Early 370 Systems* (Cambridge, MA: MIT Press, 1991), 448–451, quote p. 451.

36. Thomas J. Misa, "Military Needs, Commercial Realities, and the Development of the Transistor, 1948–1958," in *Military Enterprise and Technological Change*, ed. Merritt Roe Smith (Cambridge, MA: MIT Press, 1985), 253–287; and David C. Brock and David A. Laws, "The Early History of Microcircuitry: An Overview," *IEEE Annals of the History of Computing* 34, no. 1 (January–March 2012).

37. Mara Mills, "Hearing Aids and the History of Electronics Miniaturization," *IEEE Annals of the History of Computing* 33, no. 2 (April–June 2011): 24–44.

38. Jack S. Kilby, "Invention of the Integrated Circuit," *IEEE Transactions on Electron Devices* 23, no. 7 (July 1976): 648–654.

39. Charles Phipps, "The Early History of ICs at Texas Instruments: A Personal View," *IEEE Annals of the History of Computing* 34, no. 1 (2012): 37–47.

40. Arnold Thackray, David Brock, and Rachel Jones, *Moore's Law: The Life of Gordon Moore, Silicon Valley's Quiet Revolutionary* (New York: Basic Books, 2015), chap. 5.

41. Martha Smith Parks, *Microelectronics in the 1970s* (Anaheim, CA: Rockwell International, 1974), 64.

42. "Minuteman Is Top Semiconductor User," *Aviation Week and Space Technology*, July 26, 1965.

43. Jack Kilby, letter to Gwen Bell, June 26, 1984, formerly in the Computer Museum Archives, Boston, MA.

44. "Nineteen Sixty-Four: The Year Microcircuits Grew Up," *Electronics,* March 13, 1964.

45. Robert Noyce, "Integrated Circuits in Military Equipment," *IEEE Spectrum* 1, no. 6 (June 1964): 71–72.

46. David Mindell, *Digital Apollo: Human and Machine in Spaceflight* (Cambridge, MA: MIT Press, 2008), 125–127.

47. A. Michal McMahon, "The Computer and the Complex: A Study of Technical Innovation in Postwar America," unpublished paper, October 1986, NASA History Office. For the broader context of Fairchild's aggressive price reductions, see Christophe Lécuyer, *Making Silicon Valley: Innovation and the Growth of High Tech, 1930–70* (Cambridge, MA: MIT Press, 2006), 235–248.

48. Mindell, *Digital Apollo: Human and Machine in Spaceflight*, 286, fn. 288.

49. John Haanstra, "Monolithics and IBM," report of September 1964, unpaginated; IBM Archives. On SLT technology and IBM's initial work on chips, see Ross Knox Bassett, *To the Digital Age: Research Labs, Start-Up Companies, and the Rise of MOS Technology* (Baltimore: Johns Hopkins University Press, 2002), 66–78.

50. Don Lancaster, *TTL Cookbook* (Indianapolis, IN: Howard Sams, 1974).

51. "The Minuteman High Reliability Component Parts Program: A History and Legacy," Rockwell International, Autonetics Strategic Systems Division, Anaheim, CA, Report C81–451/201, July 31, 1981; National Air and Space Museum Archives.

52. Larry Waller, "Clean-Room Inventor Whitfield Leaves a Spotless Legacy," *Electronics,* February 4, 1985.

53. Philip J. Klass, "Reliability Is Essential Minuteman Goal," *Aviation Week,* October 19, 1959.

54. Although the problem was widely reported to be a missing hyphen in the code itself, a claim that made its way into the official investigation of the crash, the problem was actually a missing overbar in a mathematical equation included in the specifications from which the code was written. Ceruzzi, *Beyond the Limits,* 202–204.

55. Margaret Hamilton, interview with Paul Ceruzzi, unpublished.

56. The MIT engineers were not fluent in German. The German word for *eyeball* is *Augapfel,* literally translated "eye apple."

57. Don Eyles, *Sunburst and Luminary: An Apollo Memoir* (Boston, MA: Fort Point Press, 2018).

58. Margaret Hamilton, interview with Paul Ceruzzi, unpublished. Samantha Shorey and Daniela K. Rosner, "A Voice of Process: Re-presencing the Gendered Labor of Apollo Innovation," *communication +1* 7, no. 2 (2018): article 4.

59. Stephen Manes and Paul Andrews, *Gates: How Microsoft's Mogul Reinvented an Industry—And Made Himself the Richest Man in America* (New York: Touchstone, 2002), 52–55.

60. The engineering challenges of the shuttle are explored in Wayne Hale et al., eds., *Wings in Orbit: Scientific and Engineering Legacies of the Space Shuttle, 1971–2010* (Washington, DC: NASA, 2011), 226–241.

61. Bashe et al., *IBM's Early Computers,* 306.

62. "Evolution of the Space Shuttle General Purpose Computer," n.d. document, IBM Archives.

CHAPTER 5

1. Maarten Bullynck and Liesbeth De Mol, "Setting-up Early Computer Programs: D. H. Lehmer's ENIAC Computation," *Archive of Mathematical Logic* 49 (2010): 123–146.

2. Steven Levy, *Hackers: Heroes of the Computer Revolution* (Garden City, NY: Anchor Press/Doubleday, 1984), 14.

3. Gordon Bell, "Towards a History of (Personal) Workstations," in *A History of Personal Workstations,* ed. Adele Goldberg (New York: ACM Press, 1988), 4–36.

4. Levy, *Hackers.*

5. Sherry Turkle, *The Second Self: Computers and the Human Spirit* (New York: Simon and Schuster, 1984), chap. 6.

6. Levy, *Hackers.*

7. Joseph November, *Biomedical Computing: Digitizing Life in the United States* (Baltimore, MD: Johns Hopkins, 2012); chap. 3 gives a comprehensive history of LINC. Most sales were of the LINC-8/1 model, rebranded as the PDP-12. Clark told the story himself in Wesley Clark, "The LINC was Early and Small," in *A History of Personal Workstations,* ed. Adele Goldberg (New York: ACM Press, 1988), 345–400.

8. Jamie Parker Pearson, *Digital at Work: Snapshots from the First Thirty-Five Years* (Burlington, MA: Digital Press, 1992), 52.

9. John McCarthy, "John McCarthy's 1959 Memorandum," *IEEE Annals of the History of Computing* 14, no. 1 (January–March 1992): 19–23.

10. John McCarthy, "Time-Sharing Computer Systems," in *Computers and the World of Tomorrow,* ed. Martin Greenberger (Cambridge, MA: MIT Press, 1962), 221–248.

11. David Walden and Tom Van Vleck, eds., *The Compatible Time Sharing System (1961–1973)* (Washington, DC: IEEE Computer Society, 2011).

12. Arthur L. Norberg and Judy E. O'Neill, *Transforming Computer Technology: Information Processing for the Pentagon, 1962–1986* (Baltimore: Johns Hopkins University Press, 1996), 33–60.

13. November, *Biomedical Computing: Digitizing Life in the United States*, 136–137.

14. Fernando Corbató, Marjorie Merwin-Daggett, and Robert C. Caley, "An Experimental Timesharing System," in *Proceedings of the Spring Joint Computer Conference, Volume 21* (AFIPS, 1962).

15. Teletype Corporation, "Teletype, Model 33 Equipment for Fast, Economical 8-level Data Communications," product literature, c. 1966, AT&T archives.

16. John G. Kemeny, *Man and the Computer* (New York: Scribner, 1972), vii.

17. Thomas E. Kurtz, "BASIC Session," in *History of Programming Languages*, ed. Richard L Wexelblat (New York: Academic Press, 1981), 515–549.

18. Joy Lisi Rankin, *A People's History of Computing in the United States* (Cambridge, MA: Harvard University Press, 2018), 94. This book devotes three chapters to a thorough history of BASIC at Dartmouth.

19. Thomas Haigh, "Cleve Moler: Mathematical Software Pioneer and Creator of Matlab," *IEEE Annals of the History of Computing* 30, no. 1 (January–March 2008): 87–91.

20. H. R. J. Grosch, "High Speed Arithmetic: The Digital Computer as a Research Tool," *Journal of the Optical Society of America* 43, no. 4 (1953): 306–310.

21. Emerson W. Pugh, Lyle R. Johnson, and John H. Palmer, *IBM's 360 and Early 370 Systems* (Cambridge, MA: MIT Press, 1991), 360–363.

22. Pugh, Johnson, and Palmer, *IBM's 360 and Early 370 Systems*, 362–365.

23. Akera, *Calculating a Natural World*, 320–335 tells the story of early work at Michigan out of which MTS grew.

24. *The Michigan Terminal System: Volume 1, Reference R1001* (Ann Arbor, MI: University of Michigan, Information Technology Division, 1991), 13.

25. Tom Van Vleck, ed., "PL/1," Multicians.org, n.d., accessed June 12, 2019, https://multicians.org/pl1.html #EPL.

26. Peter J. Denning, "The Working Set Model for Program Behavior," *Communications of the ACM* 11, no. 5 (May 1968): 323–333.

27. Peter J. Denning and Edward Grady Coffman, *Operating Systems Theory* (Englewood Cliffs, NJ: Prentice-Hall, 1973).

28. Frederick P. Brooks, Jr, *The Mythical Man Month: Essays on Software Engineering* (Reading, MA: Addison-Wesley, 1975).

29. Max Palevsky, interview with R. Mapstone, February 15, 1973, Computer Oral History Collection, Archives Center, National Museum of American History, Smithsonian Institution, pp. 12–13.

30. Franklin M Fisher, James W McKie, and Richard B Mancke, *IBM and the US Data Processing Industry: An Economic History* (New York: Praeger, 1983), 267.

31. Tymshare is the best documented of the timesharing firms. Its story is told in Jeffrey R. Yost, *Making IT Work: A History of the Computer Services Industry* (Cambridge, MA: MIT Press, 2017), 159–172; and by a participant in its late-1970s growth, in Nathan Gregory, *The Tym Before: The Untold Origins of Cloud Computing* (self-pub., 2018).

32. Richard L. Crandall, "Oral History Interview by Paul Ceruzzi, 3 May 2002, Washington, D.C.," 2002, in Charles Babbage Institute, University of Minnesota.

33. Ann Hardy, "OH 458: Oral History by Jeffrey Yost, 2 April, Palo Alto, California," 2012, in Charles Babbage Institute, University of Minnesota, 15. Hardy gives more detail on this work in Ann Hardy, Oral History by David C. Brock, Marc Weber, and Hansen Hsu, Mountain View, July 11, 2016, Computer History Museum.

34. Dennis Ritchie, "An Incomplete History of the QED Text Editor," Bell Labs, n.d., accessed December 20, 2018, https://archive.org/stream/incomplete-history-qed/Image071517121025_djvu.txt.

35. Levy, *Hackers*, 121.

36. Specifically, the DECsystem-2020, a miniaturized version of the PDP-10 which fit into a single cabinet. "Digital Computing Timeline: 1978," VT100.net, n.d., https://vt100.net/timeline/1978.html.

37. Michael A. Banks, *On the Way to the Web* (Berkeley, CA: Apress, 2008), chap. 3.

38. Yost, *Making IT Work: A History of the Computer Services Industry*, 158.

39. Association of Data Processing Service Organizations, "Second Operating Ratios Survey," 1967, in ADAPSO Records (CBI 172), Charles Babbage Institute, University of Minnesota.

40. John K. Jerrehian, "The Computer Time Sharing Business," *ADAPSO News (CBI 172)* 8, no. 2 (April 1968): 3–5, quotation p. 5.

41. Stephen Manes and Paul Andrews, *Gates: How Microsoft's Mogul Reinvented an Industry—And Made Himself the Richest Man in America* (New York: Touchstone, 2002), 28–36.

42. Douglas K. Smith and Robert C. Alexander, *Fumbling the Future: How Xerox Invented, Then Ignored, the First Personal Computer* (New York: HarperCollins, 1989), 122. An insider account of Xerox's doomed acquisition of SDS is given in Paul A. Strassmann, *The Computers Nobody Wanted: My Years at Xerox* (New Canaan, CT: Information Economics Press, 2008), 21–47.

43. C. Gordon Bell et al., "The Evolution of the DECsystem 10," *Communications of the ACM* 21, no. 1 (January 1978): 44–63.

44. Digital Equipment Corporation, *PDP-10 Timesharing Handbook* (Maynard, MA: Digital Equipment Corporation, 1970).

45. Nick Montfort, *Twisty Little Passages: An Approach to Interactive Fiction* (Cambridge, MA: MIT Press, 2003), 9–10 and 85–93.

46. Tracy Kidder, *The Soul of a New Machine* (Boston, MA: Little Brown, 1981), 86–90 and 260–264.

47. Thomas Haigh, "Dijkstra's Crisis: The End of Algol and the Beginning of Software Engineering: 1968–72," draft discussed at the workshop History of Software, European Styles, at the Lorentz Center of the University of Leiden, Netherlands, 2010, http://www.tomandmaria.com/Tom/Writing/DijkstrasCrisis_LeidenDRAFT.pdf. Janet Abbate, *Recoding Gender: Women's Changing Participation in Computing* (Cambridge, MA: MIT Press, 2012), 97–105.

48. The history of Wagner's curve is discussed in Martin Campbell-Kelly, *From Airline Reservations to Sonic the Hedgehog: A History of the Software Industry* (Cambridge, MA: MIT Press, 2003), 91–94. The numbers were impressionistic rather than measured; later writers who produced their own versions in the 1970s and 1980s pushed back the timeline and neatened the percentages to fit the 80:20 ratio of Pareto's law.

49. William F. Atchison et al., "Curriculum 68: Recommendations for Academic Programs in Computer Science: A Report of the ACM Curriculum Committee on Computer Science," *Communications of the ACM* 11, no. 3 (March 1968): 151–197.

50. Edsger W. Dijkstra, "EWD 563: Formal Techniques and Sizeable Programs," in *Selected Writings on Computing: A Personal Perspective*, ed. Edsger W. Dijkstra (New York: Springer, 1982), 205–214.

51. Edsger Wybe Dijkstra, "The Structure of the 'THE'-Multiprogramming System," *Communications of the ACM* 11, no. 5 (May 1968): 341–346; and Edsger Wybe Dijkstra, "EWD 1303: My Recollections of Operating System Design," E. W. Dijkstra Archive, 2001, https://www.cs.utexas.edu/users/EWD/transcriptions/EWD13xx/EWD1303.html.

52. Peter Naur and Brian Randell, eds., *Software Engineering: Report on a Conference Sponsored by the NATO Science Committee, Garmisch, Germany, 7th to 11th October 1968* (Brussels: Scientific Affairs Division, NATO, 1969).

53. Edsger Wybe Dijkstra, "EWD 1175: The Strengths of the Academic Enterprise," E. W. Dijkstra Archive, 1994, https://www.cs.utexas.edu/users/EWD/transcriptions/EWD11xx/EWD1175.html.

54. Edsger Wybe Dijkstra, "The Humble Programmer," *Communications of the ACM* 15, no. 10 (October 1972): 859–866.

55. Edsger Wybe Dijkstra, "EWD1110: To the Members of the Budget Council (Confidential)," E. W. Dijkstra Archive, 1991, https://www.cs.utexas.edu/users/EWD/transcriptions/EWD11xx/EWD1110.html.

56. Brian Randell, "The 1968/69 NATO Software Engineering Reports, from unpublished proceedings of Dagstuhl-Seminar 9635: 'History of Software Engineering,' August 26–30 1996," 1996, archived at https://web.archive.org/web/20160303171551/http://homepages.cs.ncl.ac.uk/brian.randell/NATO/NATOReports/index.html.

57. Watts S. Humphrey, *Managing the Software Process* (Reading, MA: Addison-Wesley, 1989).

58. Dinesh C. Sharma, *The Outsourcer: The Story of India's IT Revolution* (Cambridge, MA: MIT Press, 2015), 177–178.

59. Unix has not yet had a comprehensive scholarly history, but its early story is told in Peter Salus, *A Quarter Century of Unix* (New York: Addison-Wesley, 1994), and by one of its creators in Dennis M. Ritchie, "Unix Time-Sharing System: A Retrospective," *Bell System Technical Journal* 57 (1978): 1947–1969.

60. Tom Van Vleck, "Myths," Multicians.org, n.d., accessed June 12, 2019, https://multicians.org/myths.html#water. Don B. Wagner, who helped to document McIlroy's EPL compiler, tells us that this is exaggerated and that he doubts there were more than five hundred pages in the systems programmer's manual even by 1967.

61. Brian Kernighan, *UNIX: A History and a Memoir* (Middletown, DE: Kindle Direct, 2020), 68.

62. M. D. McIlroy, "Oral History with Michael S. Mahoney," UNIX Oral History Project, 1989, https://www.princeton.edu/~hos/mike/transcripts/mcilroy.htm.

63. M. D. McIlroy, E. N. Pinson, and B. A. Tague, "Foreword (to Special Issue on UNIX Time-Sharing System)," *The Bell System Technical Journal* 57, no. 6, part 2 (July–August):1899–1904.

64. "Operating systems can be written in secure high-level languages. Most operating systems designers have abandoned secure languages in favor of the low-level language C." Per Brinch Hansen, ed., *Classic Operating Systems* (New York: Springer, 2001), 24.

65. Pearson, *Digital at Work*, 47, 59, and 67.

66. Digital Equipment Corporation, *PDP-11 Processor Handbook* (Maynard, MA: 1981), v.

67. Salus, *A Quarter Century of Unix*, 137–145 and 153–172.

68. Donald A. Norman, "The Trouble with UNIX," *Datamation,* November 1981.

69. Christopher M. Kelty, *Two Bits: The Cultural Significance of Free Software* (Durham, NC: Duke University Press, 2008), 132–135. It was finally published as John Lions, *Lions' Commentary on UNIX* (San Jose, CA: Peer-to-Peer Communications, 1996).

70. C. Gordon Bell, J. Craig Mudge, and John E. McNamara, *Computer Engineering: A DEC View of Hardware Systems Design* (Bedford, MA: Digital Press, 1978), 405–428.

71. Kidder, *The Soul of a New Machine.*

72. As vector graphics systems, the Tektronix terminals did not have pixels, but their controllers did use a coordinate system to situate lines on the screen. The early models used 5 bits to encode these coordinates, effectively imposing a 1,024 resolution grid onto the screen.

CHAPTER 6

1. Joseph Carl Robnet Licklider and Robert W. Taylor, "The Computer as a Communications Device," *Science and Technology:For the Technical Men in Management,* 1968. Thierry Bardini, *Bootstrapping: Douglas Engelbart, Coevolution, and the Origins of Personal Computing* (Stanford, CA: Stanford University Press, 2000).

2. Joseph Carl Robnet Licklider, "Memorandum for Members and Affiliates of the Intergalactic Computer Network," Kurzweil, December 11, 2001, 1963, https://www.kurzweilai.net/memorandum-for-members-and-affiliates-of-the-intergalactic-computer-network. The "network" at this point consisted of people and not, as is sometimes claimed, computers.

3. David Hemmendinger, "Messaging in the Early SDC Time-Sharing System," *IEEE Annals of the History of Computing*, 36 no. 1 (January–March 2014): 52–57.

4. Fletcher Knebel, "Potomac Fever," *Appleton Post-Crescent,* November 2, 1959.

5. Tom Van Vleck, "Electronic Mail and Text Messaging in CTSS, 1965–1973," *IEEE Annals of the History of Computing* 34, no. 1 (2012): 4–6.

6. Starr Roxanne Hiltz and Murray Turoff, *The Network Nation: Human Communication via Computer* (Reading, MA: Addison-Wesley, 1978).

7. Douglas C. Engelbart, "Presentation to 1968 Fall Joint Computer Conference (hosted on The Mouse Site as The Demo)," 1968, http://sloan.stanford.edu/MouseSite/1968Demo.html. Engelbart's group is described in detail in Bardini, *Bootstrapping.* Engelbart himself told its story in Douglas C. Engelbart, "The Augmented Knowledge Workshop," in *A History of Personal Workstations,* ed. Adele Goldberg (New York: ACM Press, 1988), 187–232.

8. Brian Dear, *The Friendly Orange Glow: The Untold Story of the PLATO System and the Dawn of Cyberculture* (New York: Pantheon, 2017), chap. 7.

9. Joy Lisi Rankin, *A People's History of Computing in the United States* (Cambridge, MA: Harvard University Press, 2018), 178–186.

10. Rankin, *A People's History,* 194.

11. Donald Bitzer, "The Million Terminal System of 1985," in *Computers and Communications: Implications for Education,* ed. Robert J. Seidel and Martin Rubin (New York: Academic Press, 1977), 59–70.

12. The use of Plato for communication is the main theme of chapter 7 of Rankin, *A People's History.*

13. Dear, *The Friendly Orange Glow,* 242–257.

14. That story was told from the viewpoint of a former leader of CDC in Robert M. Price, *The Eye for Innovation* (New Haven, CT: Yale University Press, 2005), 53–59, 115.

15. Thomas C. Hayes, "Logic Says that Plato's About to Pay Off," *New York Times,* April 26, 1981.

16. Arsenio Orloroso, Jr., "PLATO Buyer Must Seek Markets," *Crain's Chicago Business,* July 31, 1989.

17. ARPA's role in creating the ARPANET is placed in the broader context of its support for computer projects in Arthur L. Norberg and Judy E. O'Neill, *Transforming Computer Technology: Information Processing for the Pentagon, 1962–1986* (Baltimore: Johns Hopkins University Press, 1996). The standard history of the ARPANET and early Internet, on which we draw heavily, is Janet Abbate, *Inventing the Internet* (Cambridge, MA: MIT Press, 1999).

18. Abbate, *Inventing the Internet,* chap. 1. The contributions of Baran and Davies are parsed in Morten Bay, "Hot Potatoes and Postmen: How Packet Switching Became ARPANET's Greatest Legacy," *Internet Histories* 3, no. 1 (2019): 15–30.

19. Adele Goldberg, ed., *A History of Personal Workstations* (New York: Addison-Wesley/ACM, 1988), 151–152.

20. Norberg and O'Neill, *Transforming Computer Technology.*

21. Andrew L. Russell and Valerie Schafer, "In the Shadow of ARPANET and Internet: Louis Pouzin and the Cyclades Network in the 1970s," *Technology and Culture* 55, no. 4 (October 2014): 880–907.

22. Abbate, *Inventing the Internet,* 106–112.

23. Andrew L. Russell, "'Rough Consensus and Running Code' and the Internet-OSI Standards War," *IEEE Annals of the History of Computing* 28, no. 3 (July–September 2006): 48–61. He attributes the coinage to David Clark in 1992, though the approach it captured was older.

24. It is often said that mail accounted for three quarters of ARPANET data, according either to an "ARPANET study" of 1973 or a "MITRE report" of 1974. We've been unable to track down a source for this claim, but in email discussion, ARPANET pioneers Dave Crocker, Vint Cerf, and Dan Lynch agreed that mail spread fast enough to be the "dominant application" of the network by the end of 1973 and account for a majority of the bytes transmitted.

25. Bradley Fidler and Andrew L. Russell, "Financial and Administrative Infrastructure for the Early Internet: Network Maintenance at the Defense Information Systems Agency," *Technology and Culture* 59, no. 4 (October 2018): 899–924.

26. Craig Partridge, "The Technical Development of Internet Email," *IEEE Annals of the History of Computing* 30, no. 2 (April–June 2008): 3–29.

27. Peter T. Kirstein, "The Early Days of the Arpanet," *IEEE Annals of the History of Computing* 31, no. 3 (July–September 2009).

28. Brad Templeton, "Reaction to the DEC Spam of 1978," Templetons.com, n.d., accessed June 21, 2019, https://www.templetons.com/brad/spamreact.html#msg. The broader story of spam is told in Finn Brunton, *Spam: A Shadow History of the Internet* (Cambridge, MA: MIT Press, 2013).

29. Abbate, *Inventing the Internet,* 113–122.

30. Abbate, *Inventing the Internet,* 122–133.

31. Abbate, *Inventing the Internet,* 133–140. Fidler and Russell, "Financial and Administrative Infrastructure."

32. Abbate, *Inventing the Internet,* 140–143.

33. Paul Dourish, "The Once and Future Internet: Infrastructural Tragedy and Ambiguity in the Case of IPv6," *Internet Histories* 2, no. 1–2 (2018): 55–74.

34. The origins of netnews are recalled in Gregory G. Woodbury, "Net Cultural Assumptions," *Amateur Computerist* 6, no. 2–3 (1994): 7–9.

35. Abbate, *Inventing the Internet*, 183–186; and Peter A. Freeman, W. Richards Adrion, and William Aspray, *Computing and the National Science Foundation, 1950–2016* (New York: Association for Computing Machinery, 2019), 55–60.

36. Nathan Gregory, *The Tym Before: The Untold Origins of Cloud Computing* (self-pub., 2018).

37. Hiltz and Turoff, *The Network Nation*; Christopher Evans, *The Micro Millennium* (New York: Viking, 1979); and Alvin Toffler, *The Third Wave* (New York: William Morrow, 1980).

38. Thomas Lean, *Electronic Dreams: How 1980s Britain Learned to Love the Computer* (London: Bloomsbury, 2016), 149.

39. The failure of Viewtron and similar services made some experts doubt that ordinary people would ever use computers to work and shop from home. Tom Forester, "The Myth of the Electronic Cottage," in *Computers in the Human Context: Information Technology, Productivity, and People*, ed. Tom Forester (Cambridge, MA: MIT Press, 1989), 213–227.

40. Lean, *Electronic Dreams*.

41. Minitel is the subject of Julien Mailland and Kevin Driscoll, *Minitel: Welcome to the Internet* (Cambridge, MA: MIT Press, 2017), and Amy L. Fletcher, "France Enters the Information Age: A Political History of Minitel," *History and Technology* 18, no. 2 (2002): 103–117.

42. X.25 and its relationship to the Internet is discussed in Abbate, *Inventing the Internet*, chap. 5, and Russell, "'Rough Consensus and Running Code' and the Internet-OSI Standards War."

43. Jeffrey R. Yost, *Making IT Work: A History of the Computer Services Industry* (Cambridge, MA: MIT Press, 2017), 166.

44. Mailland and Driscoll, *Minitel: Welcome to the Internet*, 78.

45. Richard L. Crandall, "Oral History Interview by Paul Ceruzzi, 3 May 2002, Washington, D.C.," 2002, in Charles Babbage Institute, University of Minnesota.

46. Numbers were taken from the widely quoted Robert H'obbes' Zakon, "Hobbes' Internet Timeline 25," accessed July 20, 2020, https://www.zakon.org/robert/internet/timeline/#Growth, although their provenance is unclear. Historical host tables are available at https://github.com/ttkzw/hosts.txt, although the existence of classified networks, the shift to decentralized network directories with the domain name system, and the increasing importance of local area networks make direct comparisons over time very difficult.

47. Peter Salus, *A Quarter Century of Unix* (New York: Addison-Wesley, 1994).

48. Peter Denning, "The Science of Computing: The Internet Worm," *American Scientist* 77, no. 2 (March–April 1989): 126–128.

49. Abbate, *Inventing the Internet*, 191–194. On the founding of NSFNET, see Freeman, Adrion, and Aspray, *Computing and the National Science Foundation, 1950–2016*, 80–88.

50. Shane Greenstein, *How the Internet Became Commercial: Innovation, Privatization, and the Birth of a New Network* (Princeton, NJ: Princeton University Press, 2015), 76–84.

51. Ed Krol, *The Whole Internet User's Guide and Catalog* (Sebastopol, CA: O'Reilly, 1992), 353–354.

52. Greenstein, *How the Internet Became Commercial*, 84–87.

53. Janet Abbate, "Privatizing the Internet: Competing Visions and Chaotic Events, 1987–1995," *IEEE Annals of the History of Computing* 32, no. 1 (January 2010): 10–22.

54. Alexander R. Galloway, *Protocol: How Control Exists after Decentralization* (Cambridge, MA: MIT Press, 2004).

55. For a snapshot of the Internet experience in 1992, see Krol, *The Whole Internet User's Guide*.

CHAPTER 7

1. Fred Gruenberger, "RAND Symposium 14," 1973, in RAND Symposia Collection (CBI 78), Charles Babbage Institute, University of Minnesota. The first few meetings had been high-profile events sponsored by the RAND Corporation, hence the name, but they continued as more private affairs.

2. An Wang and Eugene Linden, *Lessons: An Autobiography* (Reading, MA: Addison-Wesley, 1986), 126–159.

3. "The Digital Age," *Electronics,* April 17, 1980. Quotation from pp. 397–398.

4. Chuck House, "Hewlett-Packard and Personal Computing Systems," in *A History of Personal Workstations*, ed. Adele Goldberg (New York: ACM Press, 1988), 401–432, quotation pp. 413–414.

5. Joseph Weizenbaum, *Computer Power and Human Reason: From Judgment to Calculation* (San Francisco, CA: W.H. Freeman, 1976), 116.

6. "65-Notes" *(Newsletter of the HP-65 Users' Club)*, 2, no. 1 (January 1975): 7. HP-65 customers were overwhelmingly male; the newsletter made a special note of the first female member to join the Users' Club, a year after its founding.

7. *Science* advertisement for the "new Hewlett-Packard 9100A personal computer" in October 1968.

8. Gordon E. Moore, "Moore's Law at 40," in *Understanding Moore's Law*, ed. David C Brock (Philadelphia, PA: Chemical Heritage Foundation, 2006), 67–84. The same volume includes a reprint of his original 1965 paper.

9. The development of this observation into a "law" is explored in Ethan R. Mollick, "Establishing Moore's Law," *IEEE Annals of the History of Computing* 28, no. 3 (July-September 2006): 62–75.

10. Ross Knox Bassett, *To the Digital Age: Research Labs, Start-Up Companies, and the Rise of MOS Technology* (Baltimore: Johns Hopkins University Press, 2002), 174–198.

11. Bassett, *To the Digital Age*, 277–279.

12. C. Gordon Bell, interview with Paul Ceruzzi, June 16, 1992.

13. Our description of the 4004's development is based on William Aspray, "The Intel 4004 Microprocessor: What Constituted Invention?," *IEEE Annals of the History of Computing* 19, no. 3 (July-September 1997): 4–15. Intel's early processor efforts are described in Bassett, *To the Digital Age*, 262–271.

14. *Electronic News*, November 15, 1971.

15. Dov Frohman and Robert Howard, *Leadership the Hard Way* (San Francisco, CA: Jossey-Bass, 2008), 37–41.

16. Arnold Thackray, David Brock, and Rachel Jones, *Moore's Law: The Life of Gordon Moore, Silicon Valley's Quiet Revolutionary* (New York: Basic Books, 2015), 313.

17. Robert Noyce and Marcian Hoff, "A History of Microprocessor Development at Intel," *IEEE Micro* 1, no. 1 (February 1981): 8–21.

18. Early Intel development boards are described in Zbigniew Stachniak, "This Is Not a Computer: Negotiating the Microprocessor," *IEEE Annals of the History of Computing* 35, no. 4 (October–December 2013): 48–54.

19. Robert Slater, *Portraits in Silicon* (Cambridge, MA: MIT Press, 1987), 251–261.

20. Thackray, Brock, and Jones, *Moore's Law*, 351–352.

21. Oliver Strimpel, "The Early Model Personal Computer Contest," *The Computer Museum Report*, Fall 1986.

22. Zbrigniew Stachniak, *Inventing the PC: The MCM/70 Story* (Montreal: McGill-Queen's University Press, 2011).

23. Susan Douglas, *Inventing American Broadcasting, 1899–1922* (Baltimore, MD: Johns Hopkins University Press, 1987).

24. Don Lancaster, "TV-Typewriter," *Radio-Electronics,* September 1973.

25. *QST*, March 1974, p. 154.

26. Jonathan Titus, "Build the Mark-8 Minicomputer," *Radio-Electronics,* July 1974.

27. H. Edward Roberts and William Yates, "Exclusive! Altair 8800: the Most Powerful Minicomputer Project Ever Presented—Can be Built for Under $400," *Popular Electronics,* January 1975.

28. Stanley Mazor, "Intel 8080 CPU Chip Development," *IEEE Annals of the History of Computing* 29, no. 2 (April 2007): 70–73.

29. Stan Veit, *Stan Veit's History of the Personal Computer* (Worldcomm Press, 1993), 43.

30. Kevin Gotkin, "When Computers Were Amateur." *IEEE Annals of the History of Computing*, 36, no. 2 (April–June 2014): 4–14.

31. Paul Freiberger and Michael Swaine, *Fire in the Valley: The Making of the Personal Computer* (Berkeley, CA: Osborne/McGraw-Hill, 1984).

32. Theodor H. Nelson, *Computer Lib/Dream Machines* (self-pub., 1974).

33. "Everyone Who Bought One of Those 30,000 Copies Started a Band," Quote Investigator, March 1, 2016, https://quoteinvestigator.com/2016/03/01/velvet/.

34. A network analysis of reported influences on 53,630 artists identified The Velvet Underground was ranked the fifth most influential act. Dan Kopf and Amy X Wang, "A Definitive List of the Musicians Who Influenced Our Lives Most," *Quartz*, October 7, 2017, https://qz.com/1094962/a-definitive-list-of-the -musicians-who-influenced-our-lives-most/.

35. Forrest Mims III, "The Tenth Anniversary of the Altair 8800," *Computers and Electronics,* January 1985.

36. Stephen Manes and Paul Andrews, *Gates: How Microsoft's Mogul Reinvented an Industry—And Made Himself the Richest Man in America* (New York: Touchstone, 2002), 63.

37. C. Gordon Bell, J. Craig Mudge, and John E. McNamara, *Computer Engineering: A DEC View of Hardware Systems Design* (Bedford, MA: Digital Press, 1978), 383.

38. Kevin Driscoll, "Professional Work for Nothing: Software Commercialization and 'An Open Letter to Hobbyists,'" *Information & Culture* 50, no. 2 (2015): 257–283.

39. Information on people in the picture from Matt Weinberger, "Where Are They Now?," Business Insider, January 26, 2019, https://www.businessinsider.com/microsoft-1978-photo-2016-10.

40. Emerson W. Pugh, Lyle R. Johnson, and John H. Palmer, *IBM's 360 and Early 370 Systems* (Cambridge, MA: MIT Press, 1991), 510–521.

41. Clifford Barney, "Award for Achievement [Alan F. Shugart]," *Electronics Week,* January 14, 1985.

42. Gary Kildall, "Microcomputer Software Design—a Checkpoint," in *AFIPS '75: Proceedings of the May 19–22, 1975, National Computer Conference and Exposition* (New York: ACM, 1975), 99–106.

43. Jim C. Warren, "First Word on a Floppy-Disc Operating System," *Dr. Dobb's Journal,* April 1976. Warren noticed that both systems specified the disk drive in use with a letter, used file names with a period and three-character extension, and had commands called DIR (directory), PIP, and DDT.

44. Slater, *Portraits in Silicon*, chap. 23.

45. Bell, Mudge, and McNamara, *Computer Engineering*, graph on p. 195.

46. C. Gordon Bell, interview with Paul Ceruzzi, June 16, 1992, Los Altos, California.

47. The TRS-80 tends to be marginalized in histories of computing despite its status as the best-selling early personal computer. Its story, which stretched well into the 1980s with follow-up models, is told with personal detail in David Welsh and Theresa Welsh, *Priming the Pump: How TRS-80 Enthusiasts Helped Spark the PC Revolution* (Ferndale, MI: Seeker Books, 2011).

48. Steve Wozniak and Gina Smith, *iWoz: Computer Geek to Cult Icon: How I Invented the Personal Computer, Co-Founded Apple, and Had Fun Doing It* (New York: W. W. Norton, 2006), 54–55, 71, and 150–172. The best overview of Apple's early history is Michael Moritz, *The Little Kingdom: The Private Story of Apple Computer* (New York: William Morrow, 1984).

49. Manes and Andrews, *Gates*, 111.

50. Stephen Wozniak, "The Apple II," *Byte* 2, no. 5 (May 1977): 34–43; and Wozniak and Smith, *iWoz*, 192–193.

51. Wozniak and Smith, *iWoz*, 211–219.

52. Advertisement for Apple, *Byte*, July 1978, pp. 14–15.

53. David L. Craddock, *Break Out: How the Apple II Launched the PC Gaming Revolution* (Atglen, PA: Schiffer, 2017), chap. 2.

54. Ward Christensen and Randy Suess, "Hobbyist Computerized Bulletin Board," *Byte* 3, no. 11 (November 1978): 150–157.

55. The most influential nonfiction portraits of early hackers were Bill Landreth and Howard Rheingold, *Out of the Inner Circle: A Hacker's Guide to Computer Security* (Bellevue, WA: Microsoft Press, 1985), discussed in Michael J. Halvorson, *Code Nation: Personal Computing and the Learn to Program Movement in America* (New York: ACM Books, 2020), 206–211; and Clifford Stoll, *The Cuckoo's Egg: Tracking a Spy Through the Maze of Computer Espionage* (New York: Doubleday, 1989).

56. Kevin Driscoll, "Demography and Decentralization: Measuring the Bulletin Board Systems of North America," *WiderScreen* 23, no. 2–3 (2020).

57. On the pre-electronic arcade era, see Michael Z. Newman, *Atari Age: The Emergence of Video Games in America* (Cambridge, MA: MIT Press, 2017), 22–36.

58. Henry Lowood, "Videogames in Computer Space: The Complex History of Pong," *IEEE Annals of the History of Computing* 31, no. 3 (July–September 2009): 5–19.

59. Walter Isaacson, *Steve Jobs* (New York: Simon & Schuster, 2011).

60. Wozniak and Smith, *iWoz*, 190–192.

61. The gendered dimensions of Pac Man are explored in Newman, *Atari Age*, chap. 6.

62. Newman, *Atari Age*, chap. 2.

63. Nick Montfort and Ian Bogost, *Racing the Beam: The Atari Video Computer System* (Cambridge, MA: MIT Press, 2009), chap. 2.

64. Montfort and Bogost, *Racing the Beam*, 73.

65. Montfort and Bogost, *Racing the Beam*, chap. 4 (*Pac-Man*) and chap. 6 (*Pitfall*).

66. On Logo, see Halvorson, *Code Nation*, 87–92. Soviet computer literacy campaigns focused on programmable calculators, which were produced in large numbers. Ksenia Tatarchenko, "The Man with a Micro-calculator," in *Exploring the Early Digital*, ed. Thomas Haigh (Cham, Switzerland: Springer, 2019),

179–200; and Ksenia Tatarchenko, "The Great Soviet Calculator Hack," *IEEE Spectrum* 55, no. 10 (October 2018): 42–47.

67. The BBC's "Computer Program" and "Making the Most of the Micro," both part of the Computer Literacy Project, loom large in studies of the era. See Alison Gazzard, *Now the Chips Are Down* (Cambridge, MA: MIT Press, 2016), 4–9; and Thomas Lean, *Electronic Dreams: How 1980s Britain Learned to Love the Computer* (London: Bloomsbury, 2016), 89–114.

68. Newman, *Atari Age*, 138–152 discusses this, with a focus on the advertising messages used to sell home computers. The program names come from Kenniston W. Lord, *Using the Radio Shack TRS-80 in Your Home* (New York: Van Nostrand Reinhold, 1981).

69. Jesse Adams Stein, "Domesticity, Gender and the 1977 Apple II Personal Computer," *Design and Culture* 3, no. 2 (2011): 193–216.

70. Theodore Jerome Cohen and Jacqueline H. Bray, *Melissa and John and the Magic Machine* (Peterborough, NH: BYTE/McGraw Hill, 1979).

71. Ian Adamson and Richard Kennedy, *Sinclair And the "Sunrise" Technology: The Deconstruction of a Myth* (Penguin, 1986), chaps. 5–6; and Lean, *Electronic Dreams*, 61–70.

72. Adamson and Kennedy, *Sinclair and the "Sunrise" Technology*, chap. 7; and Lean, *Electronic Dreams*, 115–124.

73. The early British videogame industry is described in Lean, *Electronic Dreams*, 173–209.

74. Gazzard, *Now the Chips Are Down*.

75. The story of the VIC-20 is told in Brian Bagnall, *On the Edge: The Spectacular Rise and Fall of Commodore* (Winnipeg, MB: Variant Press, 2005), 157–224.

76. Commodore International, "Commodore VIC-20 Job Interview Advertisement," YouTube, 1984, https://www.youtube.com/watch?v=c5tqmyl3XQk.

77. "The Price TI Is Paying for Misreading a Market," *Business Week,* September 19. 1983.

78. Andrew Pollack, "Retreat Set by Texas Instruments," *New York Times,* October 29, 1983.

79. Dennis Kneale, "Commodore Hits Production Snags in Its Hot-Selling Home Computer," *Wall Street Journal,* October 28, 1983. A contemporary summary of Commodore's quality issues was given in John J. Anderson, "Commodore," *Creative Computing* 10, no. 3 (March 1984): 56, 60. According to Bagnall, *On the Edge*, 262, defective power supplies burned down several houses.

80. Bagnall, *On the Edge*, 265–269.

81. Bagnall, *On the Edge*, 304–393.

82. Gazzard, *Now the Chips Are Down*, 91–108.

83. Lean, *Electronic Dreams*, 197–199.

84. Marc S. Blank and S. W. Galley, "How to Fit a Large Program into a Small Machine," *Creative Computing* 6, no. 7 (July 1980): 80–87. The *Zork* story is told in Montfort, *Twisty Little Passages*, 97–112 and 125–135; and Craddock, *Break Out*, chap. 3.

85. Craddock, *Break Out*, chaps. 4, 6–7, and 13.

86. Jane Margolis and Allan Fisher, *Unlocking the Clubhouse: Women in Computing* (Cambridge, MA: MIT Press, 2001); and J. McGrath Cohoon and William Aspray, eds., *Women and Information Technology: Research on Underrepresentation* (Cambridge, MA: MIT Press, 2006), 145–151.

87. Elaine Lally, *At Home with Computers* (New York: Berg, 2002).

88. Lean, *Electronic Dreams*, 153–157; and Chris Bourne, "Going On-line," *Sinclair User,* January 1986: 126–127, 132.

89. Prodigy is discussed in Michael A. Banks, *On the Way to the Web* (Berkeley, CA: Apress, 2008), 139–156.

90. Katie Hafner, *The Well: A Story of Love, Death & Real Life in the Seminal Online Community* (New York: Carroll & Graf, 2001). Fred Turner, "Where the Counterculture Met the New Economy: The WELL and the Origins of Virtual Community," *Technology and Culture* 46, no. 3 (July 2005): 485–512. Howard Rheingold, *The Virtual Community: Homesteading on the Electronic Frontier* (Reading, MA: Addison-Wesley, 1993).

91. For a breezy history of Nintendo's rise see Jeff Ryan, *Super Mario: How Nintendo Conquered America* (New York: Portfolio/Penguin, 2011). Nathan Altice, *I Am Error: The Nintendo Family Computer / Entertainment System Platform* (Cambridge, MA: MIT Press, 2015) gives an in-depth look at the programming techniques behind these games.

92. The rivalry between Sega and Nintendo is treated as an epic battle in the rather overheated Blake J. Harris, *Console Wars: Sega, Nintendo, and the Battle That Defined a Generation* (New York: HarperCollins, 2014).

CHAPTER 8

1. David Bradley, "The Creation of the IBM PC," *Byte* 15, no. 9 (September 1990): 414–420, quotation on 420.

2. The discussion of word processing here is based on Thomas Haigh, "Remembering the Office of the Future: The Origins of Word Processing and Office Automation," *IEEE Annals of the History of Computing* 28, no. 4 (October-December 2006): 6–31, and Matthew G. Kirschenbaum, *Track Changes: A Literary History of Word Processing* (Cambridge, MA: Harvard University Press, 2016).

3. Ulrich Steinhilper, *Don't Talk—Do It! From Flying to Word Processing* (Bromley, UK: Independent Books, 2006).

4. C. E. Mackenzie, *Coded Character Sets: History & Development* (Reading, MA: Addison-Wesley, 1980).

5. An Wang and Eugene Linden, *Lessons: An Autobiography* (Reading, MA: Addison-Wesley, 1986).

6. Edwin McDowell, "'No Problem' Machine Poses a Presidential Problem," *New York Times,* March 24, 1981.

7. Charles Kenney, *Riding the Runaway Horse: The Rise and Decline of Wang Laboratories* (New York: Little Brown, 1992).

8. Kenney, *Riding the Runaway Horse*, 68–73.

9. Stephen T. McClellan, *The Coming Computer Industry Shakeout: Winners, Losers, and Survivors* (New York: Wiley, 1984), 299–303.

10. Seymour Rubinstein, "Recollections: The Rise and Fall of WordStar," *IEEE Annals of the History of Computing* 28, no. 4 (October–December 2006): 64–72.

11. Kirschenbaum, *Track Changes*, chap. 5.

12. Jerry Pournelle, "Ulterior Motives, Lobo, Buying Your First Computer, JRT Update," *Byte* 8, no. 5 (May 1983): 298–324, quotation on 306.

13. Lily Hay Newman, "George R. R. Martin Writes on a DOS-Based Word Processor from the 1980s," Slate.com, May 14, 2014, http://www.slate.com/blogs/future_tense/2014/05/14/george_r_r_martin_writes _on_dos_based_wordstar_4_0_software_from_the_1980s.html. Martin's computer is a little more modern than Zeke: he uses WordStar 4.0 from 1987, running on an MS-DOS computer.

14. Mark Dahmke, "The Osborne 1," *Byte* 7, no. 6 (June 1982): 348–363.

15. David Thomas, *Alan Sugar: The Amstrad Story* (London: Century, 1990), 160–186 on the PCW and quotation on 247.

16. John Donaldson, "Benchtest: Amstrad PCW9512," *Personal Computer World* 9, no. 10 (October 1987): 98–102.

17. Carver Mead and Lynn Conway, *Introduction to VLSI Systems* (Reading, MA: Addison-Wesley, 1980). Their ideas had already circulated widely from the notes Conway prepared for a 1978 course at MIT. Lynn Conway, "Reminiscences of the VLSI Revolution," *IEEE Solid State Circuits* 4, no. 4 (Fall 2012): 8–31.

18. Peter A. McWilliams, *The Word Processing Book: A Short Course in Computer Literacy*, 5th ed. (Los Angeles: Prelude Press, 1983), 211.

19. Jerry Mar, "Word Processing on the Apple with WordStar and Diablo," *Creative Computing* 9, no. 3 (March 1983): 81.

20. Burton Grad, "The Creation and the Demise of VisiCalc," *IEEE Annals of the History of Computing* 29, no. 3 (July–September 2007): 20–31.

21. Steven Levy, "A Spreadsheet Way of Knowledge," *Harpers Magazine,* November 1984.

22. Dan Fylstra, Oral History Interview by Thomas Haigh, May 7, 2004, in Needham, MA (to be released by the Computer History Museum after Fylstra's death).

23. Peter Passell, "Economic Scene; Michael Milken's Other Accusers," *New York Times,* April 12 1989.

24. Fylstra, Oral History Interview.

25. Robert X. Cringely, *Accidental Empires: How the Boys of Silicon Valley Make their Millions, Battle Foreign Competition, and Still Can't Get a Date* (Reading, MA: Addison-Wesley, 1992), chap. 8.

26. Bradley, "The Creation of the IBM PC," James Chposky and Ted Leonsis, *Blue Magic: The People, Power, and Politics Behind the IBM Personal Computer* (New York: Facts on File, 1988).

27. Greg Williams, "A Closer Look at the IBM Personal Computer," *Byte* 7, no. 1 (January 1982): 36–68.

28. Ibid, 60.

29. Stephen Manes and Paul Andrews, *Gates: How Microsoft's Mogul Reinvented an Industry—And Made Himself the Richest Man in America* (New York: Doubleday, 1993).

30. Tim Paterson, telephone interview with Paul Ceruzzi, July 24, 1996.

31. Williams, "A Closer Look," 42.

32. Larry Augustin, "The Mainframe Connection: IBM's 3270 PC," *Byte* 9, no. 9 (Fall 1984): 231–237.

33. "Machine of the Year: The Computer Moves In," *Time Magazine,* January 3, 1983.

34. Michael Moritz, *The Little Kingdom: The Private Story of Apple Computer* (New York: William Morrow, 1984), 293–297.

35. Steve Wozniak and Gina Smith, *iWoz: Computer Geek to Cult Icon: How I Invented the Personal Computer, Co-Founded Apple, and Had Fun Doing It* (New York: W. W. Norton, 2006), 229.

36. Personal communication, 2019, in comments on draft manuscript.

37. Stephen S. Fried, "Evaluating 8087 Performance on the IBM PC," *Byte* 9, no. 9 (Fall 1984): 197–208.

38. That company was MicroWay. Its products were listed in an advertisement on page 101 of the April 30, 1985, issue of *PC Magazine*.

39. Tracy Kidder, *The Soul of a New Machine* (Boston, MA: Little Brown, 1981), 31.

40. Erik Sandberg-Diment, "The Little IBM Finally Arrives for a Test," *New York Times*, December 27, 1983.

41. Sierra's introduction of the graphical adventure, with an earlier game for the Apple II, is discussed in Laine Nooney, "Let's Begin Again: Sierra On-Line and the 'Origins' of the Graphical Adventure Game," *American Journal of Play* 10, no. 1 (2017): 71–98.

42. Gregg Williams, "Lotus Development Corporation's 1-2-3," *Byte* 7, no. 12 (December 1982): 182–198. The development of spreadsheet packages is explored in Martin Campbell-Kelly, "Number Crunching without Programming: The Evolution of Spreadsheet Usability," *IEEE Annals of the History of Computing* 29, no. 3 (July–September 2007): 6–19.

43. Mitch Kapor, "Oral History by William Aspray, November 19, Mountain View, California," 2004, 15.

44. Gerardo Con Díaz, *Software Rights: How Patent Law Transformed Software Development in America* (New Haven, CT: Yale University Press, 2019), 224–229.

45. Arthur Naiman, *Word Processing Buyer's Guide* (New York: BYTE/McGraw-Hill, 1983), 177.

46. Thomas J. Bergin, "The Origins of Word Processing Software for Personal Computers: 1976–1985," *IEEE Annals of the History of Computing* 28, no. 4 (October–December 2006): 32–47.

47. L. L. Beavers, "WordPerfect: Not Quite Perfect, But Certainly Superb," *Creative Computing* 9, no. 11 (November 1983): 74.

48. Lindsy Van Gelder, "WordPerfect Reaches for the Star," *PC Magazine* 1, no. 10 (March 1983): 431–437.

49. W. E. Peterson, *Almost Perfect: How a Bunch of Regular Guys Built WordPerfect Corporation* (Rocklin, CA: Prima, 1994). On market share, see Stan J. Liebowitz and Stephen E. Margolis, *Winners, Losers & Microsoft: Competition and Antitrust in High Technology* (Oakland, CA: The Independent Institute, 2001), 181.

50. dBase has not received a thorough historical examination, but its story is outlined in Martin Campbell-Kelly, *From Airline Reservations to Sonic the Hedgehog: A History of the Software Industry* (Cambridge, MA: MIT Press, 2003), 220–221 and 256–227.

51. Campbell-Kelly, *From Airline Reservations*, 257.

52. Donna K. H. Walters, "Lotus to Drop Copy Protection for Some: May Be Extended to All Customers Later," *Los Angeles Times,* August 14, 1986.

53. Sheldon Leemon, "PC-Write Word Processor For PC & PCjr," *Compute!,* no. 57 (February 1982): 82–86.

54. Mark J. Welch, "Expanding on the PC," *Byte* 8, no. 11 (November 1983): 168–184.

55. Phil Lemmons, "Victor Victorious," *Byte* 7, no. 11 (November 1982): 216–254, quotation on 254.

56. Rifkin and Harrar, *The Ultimate Entrepreneur: The Story of Ken Olsen and Digital Equipment Corporation*, chaps. 25, 29, and 30.

57. Pamela Archbold and John Verity, "The Datamation 100: Company Profiles," *Datamation* 31, no. 11 (June 1 1985): 58–182, quotation on 140.

58. Winn L. Rosch, "Playing Hardball Against the XT," *PC Magazine* 3, no. 6 (April 3 1984): 115–122.

59. Stan Miaskowski, "Software Review: Microsoft Flight Simulator," *Byte* 9, no. 3 (1984): 224–232. On Isaacson, see Stewart Alsop, "A Public Windows Pane to Make Compatibility Clearer," *InfoWorld,* January 31, 1994. For an example of the tests in use, see Peter Bright, "Benchtest: Epson PC," *Personal Computer World* 9, no. 2 (February 1986): 104–110.

60. Thomas, *Alan Sugar*, 224.

61. Thomas, *Alan Sugar*, 220–226.

62. Guy Kewney, "Benchtest: Amstrad PC1512," *Personal Computer World* 9, no. 10 (October 1986): 126–136.

63. Keith Ferrell, "IBM Compatibles: The Universe Expands," *Compute!*, no. 86 (July 1987): 14–24.

64. Paul Somerson, "IBM Brings Out the Big Guns," *PC Magazine*, November 13, 1984, quotation on p. 133.

65. *Byte*, December 1984, p. 148.

66. Robert M. Metcalfe, "How Ethernet Was Invented," *IEEE Annals of the History of Computing* 16, no. 4 (October–December 1994): 81–88.

67. Metcalfe, "How Ethernet Was Invented," 83.

68. R. Binder et al., "ALOHA Packet Broadcasting: a Retrospect," in *1975 National Computer Conference* (Montvale, NJ: AFIPS Press, 1975), 203–215. Robert Metcalfe and David R. Boggs, "Ethernet: Distributed Packet Switching For Local Computer Networks," *Communications of the ACM* 19, no. 7 (July 1976).

69. Metcalfe, "How Ethernet Was Invented," p. 85.

70. Kerry Elizabeth Knobelsdorff, "IBM's Four-Month-Old PS/2 Has Put Computer World on Hold," *Christian Science Monitor*, August 19, 1987.

71. Cringely, *Accidental Empires: How the Boys of Silicon Valley Make their Millions, Battle Foreign Competition, and Still Can't Get a Date*, 285–286.

72. David E. Sanger, "IBM Offers a Blitz of New PC's," *New York Times*, April 3, 1987.

73. Rod Canion, *Open: How Compaq Ended IBM's PC Domination and Helped Invent Modern Computing* (Dallas, TX: BenBella, 2013), 91–106.

74. Peter Jackson, "Benchtest: Compaq DeskPro 386," *Personal Computer World* 8, no. 11 (November 1986): 138–144.

75. William D. Marbach and Karen Springen, "Compaq Chips Away at IBM's Strength," *Newsweek*, September 22, 1986.

76. Canion, *Open*, 35–41.

77. Michael Dell and Catherine Fredman, *Direct from Dell* (New York: Harper Business, 1999).

78. Derek Cohen, "Cover Benchtest: Atari Portfolio," *Personal Computer World* 11, no. 8 (August 1989): 130–136.

79. Specifically, the Norton SI 4.0 benchmark, although that typically overestimated the performance gains provided by cache memory.

80. Alfred Poor, "25-MHz Computers: Dell System 325," *PC Magazine*, February 24, 1989.

81. Bureau of Labor Statistics, "Issues in Labor Statistics: Computer Ownership Up Sharply in the 1990s," March, 1999, https://www.bls.gov/opub/btn/archive/computer-ownership-up-sharply-in-the-1990s.pdf.

CHAPTER 9

1. Victor K. McElheny, "Xerox Fights to Stay Ahead in the Copier Field," *New York Times*, February 21, 1977.

2. Michael Hiltzik, *Dealers of Lightning: Xerox PARC and the Dawn of the Computer Age* (New York: Harper-Business, 1999).

3. Quoted in David Dickson, *The New Politics of Science* (New York: Pantheon Books, 1984), 122.

4. Arthur L. Norberg and Judy E. O'Neill, *Transforming Computer Technology: Information Processing for the Pentagon, 1962–1986* (Baltimore: Johns Hopkins University Press, 1996).

5. Hiltzik, *Dealers of Lightning*.

6. Charles P. Thacker, "Personal Distributed Computing: The Alto and Ethernet Hardware," in *A History of Personal Workstations*, ed. Adele Goldberg (New York: ACM Press, 1988), 267–289.

7. Alan Kay and Adele Goldberg, "Personal Dynamic Media," *Computer* 10, no. 3 (March 1977): 31–41.

8. Kristen Nygaard and Ole-Johan Dahl, "The Development of the Simula Languages," in *History of Programming Languages*, ed. Richard L Wexelblat (New York: Academic Press, 1981), 439–480.

9. Alan C. Kay, "The Early History of Smalltalk," in *History of Programming Languages, II,* ed. Thomas J. Bergin and Rick G. Gibson (New York: ACM Press, 1996), 511–598.

10. *Byte* devoted most of the nonadvertising space in the issue, thirteen articles, to detailed examinations of specific aspects of Smalltalk. The introduction was Adele Goldberg, "Introducing the Smalltalk-80 System," *Byte* 6, no. 8 (1981): 14–26.

11. Bjarne Stroustrup, *The C++ Programming Language* (Reading, MA: Addison-Wesley, 1985).

12. Andrew Binstock, "Interview with Alan Kay," Dr. Dobb's, July 10, 2012, http://www.drdobbs.com /architecture-and-design/interview-with-alan-kay/240003442#.

13. Lawrence G. Tesler, "How Modeless Editing Came to Be," *IEEE Annals of the History of Computing* 40, no. 3 (July–September 2018): 55–67.

14. Hiltzik, *Dealers of Lightning*.

15. Phillip Ein-Dor, "Grosch's Law Re-revisited," *Communications of the ACM* 28, no. 2 (February 1985): 142–151.

16. Andrew D. Birrell and Bruce Jay Nelson, "Implementing Remote Procedure Calls," *ACM Transactions on Computer Systems* 2, no. 1 (February 1984): 39–59.

17. Hiltzik, *Dealers of Lightning*.

18. David Canfield Smith, "Designing the Star User Interface," *Byte* 7, no. 4 (April 1982): 242–282.

19. David Canfield Smith et al., "The Star User Interface: An Overview," in *Proceedings of the AFIPS National Computer Conference* (AFIPS: 1982), 515–528

20. Douglas K. Smith and Robert C. Alexander, *Fumbling the Future: How Xerox Invented, Then Ignored, the First Personal Computer* (New York: HarperCollins, 1989).

21. Thomas W. Starnes, "Design Philosophy behind Motorola's MC68000: Part 1," *Byte* 8, no. 4 (April 1983): 70–92, quotation from p. 70.

22. Mark Hall and John Barry, *Sunburst: The Ascent of Sun Microsystems* (Chicago: Contemporary Books, 1990), chap. 1.

23. David F. Hinnant, "Benchmarking UNIX Systems," *Byte* 9, no. 8 (August 1984): 132–135, 400–409.

24. Gregg Williams, "The Lisa Computer System," *Byte* 8, no. 2 (February 1983): 33–50.

25. Apollo reported $18 million in sales for calendar year 1982. If the average sales price was $40,000 that suggests unit sales of 450 during its second year selling workstations. For 1983 it reported $80 million in sales, suggesting several thousand workstations shipped.

26. The efforts of engineers to sneak a slot (unsuccessfully) and extra memory lines (successfully) into the original Macintosh are discussed in Andy Hertzfeld, *Revolution in The Valley: The Insanely Great Story of How the Mac Was Made* (Sebastopol, CA: O'Reilly, 2004), 60–61.

27. Steven Levy, *Insanely Great: The Life and Times of Macintosh, the Computer that Changed Everything* (New York: Viking, 1994), 187.

28. Levy, *Insanely Great: The Life and Times of Macintosh*, 186.

29. On early languages and resources for Mac programming, see Michael J. Halvorson, *Code Nation: Personal Computing and the Learn to Program Movement in America* (New York: ACM Books, 2020), 188–200.

30. Donna Osgood, "The Difference in Higher Education," *Byte* 12, no. 2 (February 1987): 165–178.

31. Suzanne Crocker, "Paul Brainerd, Aldus Corporation, and the Desktop Publishing Revolution," *IEEE Annals of the History of Computing* 41, no. 3 (July–September 2019): 35–41.

32. The development of the language is described in John E Warnock, "The Origins of PostScript," *IEEE Annals of the History of Computing* 40, no. 3 (July–September 2018): 68–76. PostScript pages were expressed as a computer program that, when executed, would produce the desired output.

33. Peter Bright, "Checkout: PageMaker & LaserWriter," *Personal Computer World* 8, no. 10 (October 1985): 166–171.

34. John Scull and Hansen Hsu, "The Killer App That Saved the Macintosh," *IEEE Annals of the History of Computing* 41, no. 3 (July–September 2019): 42–52.

35. Phillip Robinson and Jon R. Edwards, "The Atari 1040ST," *Byte* 11, no. 3 (March 1986): 84–93.

36. Guy Swarbrick, "Cover Benchtest: Macintosh Portable," *Personal Computer World* 11, no. 10 (October 1989): 130–136.

37. "Mac Portable's Pluses Outweigh the Negatives," *Computerworld*, December 11, 1989.

38. Guy Swarbrick, "Cover Benchtest: Atari Stacey," *Personal Computer World* 11, no. 12 (December 1989):130–136.

39. The Amiga retains a loyal fanbase decades later, who celebrated its technical accomplishments in Jimmy Maher, *The Future Was Here: The Commodore Amiga* (Cambridge, MA: MIT Press, 2012), and bemoaned the failings of Commodore's managers in Brian Bagnall, *Commodore: The Amiga Years* (Winnipeg, MB: Variant Press, 2017).

40. Randall E Stross, *Steve Jobs and the NeXT Big Thing* (New York: Scribner, 1993).

CHAPTER 10

1. On the "Dummies" books, see Michael J. Halvorson, *Code Nation: Personal Computing and the Learn to Program Movement in America* (New York: ACM Books, 2020), 183–187.

2. Chris Larson, "MS DOS 2.0: An Enhanced 16-Bit Operating System," *Byte* 8, no. 11 (November 1983): 285–290.

3. John Markoff, "Five Window Managers for the IBM PC," *Byte* 9, no. 9 (1984): 65–87.

4. Lawsuits between Apple, Microsoft and Xerox are described in Gerardo Con Díaz, *Software Rights: How Patent Law Transformed Software Development in America* (New Haven, CT: Yale University Press, 2019), 214–217. Apple did manage to scare Digital Research into stripping GEM of the ability to move or resize desktop windows.

5. Microsoft Annual Report, 1992.

6. Frank Rose, *West of Eden: the End of Innocence at Apple Computer* (New York: Penguin, 1989), chap.11.

7. The dynamics of the office software industry in the 1980s and 1990s are discussed in Martin Campbell-Kelly, *From Airline Reservations to Sonic the Hedgehog: A History of the Software Industry* (Cambridge, MA: MIT Press, 2003), 251–264.

8. Stephen Manes and Paul Andrews, *Gates: How Microsoft's Mogul Reinvented an Industry—And Made Himself the Richest Man in America* (New York: Doubleday, 1993), 423.

9. Peter Norvig, "PowerPoint: Shot With Its Own Bullets," *The Lancet* 362, no. 9381 (2003): 343–344.

10. Douglas Coupland, "Microserfs," *Wired*, January 1994.

11. Ellen Ullman, *Close to the Machine: Technophilia and its Discontents* (San Francisco: City Lights Books, 1997).

12. Lohr, *Go To*, 93–98. Halvorson, *Code Nation*, 156–165.

13. Ullman, *Close to the Machine*, 103.

14. E. F. Codd, "A Relational Model of Data for Large Shared Databanks," *Communications of the ACM* 13, no. 6 (June 1970): 377–390.

15. Robert Preger, "The Oracle Story, Part 1: 1977–1986," *IEEE Annals of the History of Computing* 34, no. 4 (October–December 2012): 51–57.

16. Donald J. Haderle and Cynthia M. Saracco, "The History and Growth of IBM's DB2," *IEEE Annals of the History of Computing* 35, no. 2 (April–June 2013): 54–66.

17. Bob Epstein, "History of Sybase," *IEEE Annals of the History of Computing* 35, no. 2 (April–June 2013): 31–41.

18. Lawrence Rowe, "History of the Ingres Corporation," *IEEE Annals of the History of Computing* 34, no. 4 (October–December 2012): 58–70.

19. Preger, "The Oracle Story, Part 1: 1977–1986." The relational database industry was focused largely in Silicon Valley—perhaps the first major software field to center there. Martin Campbell-Kelly, "The RDBMS Industry: A Northern California Perspective," *IEEE Annals of the History of Computing* 34, no. 4 (October–December 2012): 18–29.

20. Oracle's boom years are documented in Andrew Mendelsohn, "The Oracle Story: 1984–2001," *IEEE Annals of the History of Computing* 35, no. 2 (April–June 2013): 10–23.

21. Ullman, *Close to the Machine*, 108.

22. Christopher Koch, "The Integration Nightmare: Sounding the Alarm," *CIO Magazine*, November 15, 1996.

23. Rod Canion, *Open: How Compaq Ended IBM's PC Domination and Helped Invent Modern Computing* (Dallas, TX: BenBella, 2013), 123–179.

24. For an exhaustive roundup of the graphics card market in late 1992, by which point Windows accelerators were being widely produced and the first VESA local bus cards were appearing, see Alfred Poor, "Video Technology: Making a Choice in an Era of Change," *PC Magazine*, January 12, 1993.

25. C. Gordon Bell, J. Craig Mudge, and John E. McNamara, *Computer Engineering: A DEC View of Hardware Systems Design* (Bedford, MA: Digital Press, 1978), chap. 17.

26. William Strecker, "VAX-11/780—A Virtual Address Extension to the PDP-11 Family," in *Proceedings of the National Computer Conference* (New York: AFIPS, 1978), 967–980.

27. The "wild duck" memo is described in Herbert R. J. Grosch, *Computer: Bit Slices from a Life* (Novato, California: Third Millennium, 1991), 258.

28. George Radin, "The 801 Minicomputer," *IBM Journal of Research and Development* 27 (May 1983): 237–246.

29. David A. Patterson, "Reduced Instruction Set Computers," *Communications of the ACM* 28, no. 1 (January 1985): 8–21; and David Patterson and John L. Hennessy, *Computer Architecture: A Quantitative Approach* (San Mateo, CA: Morgan Kaufmann, 1990).

30. Patterson and Hennessy, *Computer Architecture*, 190; and Mark Hall and John Barry, *SunBurst: The Ascent of Sun Microsystems* (Chicago: Contemporary Books, 1990), 163.

31. Patterson and Hennessy, *Computer Architecture*, 190.

32. Nick Baran, "Two Powerful Systems from SUN," *Byte* 14, no. 5 (May 1989): 108–112.

33. Tom Yager and Ben Smith, "Son of SPARCstation," *Byte* 15, no. 13 (December 1990): 140–146; and Roger C. Alford, "NCR's S486/MC33 Has Unique Approach to Reliability," *Byte* 15, no. 13 (December 1990): 191–193.

34. Michael A. Cusumano and Richard W. Selby, *Microsoft Secrets: How the World's Most Powerful Software Company Creates Technology, Shapes Markets, and Manages People* (New York: Free Press, 1995), 36, 269–270.

35. G. Pascal Zachary, *Show Stopper! The Breakneck Race to Create Windows NT and the Next Generation at Microsoft* (New York: Free Press, 1994), chap. 1.

36. Raymond Ga Côté and Barry Nance, "Pentium PCs: Power to Burn," *Byte* 18, no. 8 (July 1993): 94–102, quotation on p. 101.

37. Steve Apiki and Rick Grehan, "Fastest NT Workstations," *Byte* 20, no. 3 (March 1995): 115–122.

38. Michelle Campanale, "Eight Heavy-Hitting NT Workstations," *Byte* 23, no. 1 (January 1998): 98.

39. Associated Press, "Windows 95 Sales Plunge from Peak," *New York Times*, September 8, 1995.

40. IBM's crisis of the early 1990s, followed by a streamlined revival under new leadership, is a discussed in detail in James W. Cortada, *IBM: The Rise and Fall and Reinvention of a Global Icon* (Cambridge, MA: MIT Press, 2019).

41. Michael J. Miller, "Computers: More Than One Billion Sold," *PC Magazine*, September 3, 2002. The one billion figure reportedly applied to all personal computers, going back to the 1970s, not just to IBM-compatible PCs. Even so, a majority would have been IBM-compatible machines sold from mid-1995 to mid-2001 and hence almost certainly pre-installed with a variant of Windows 95.

42. Gil Amelio and William L. Simon, *On the Firing Line: My 500 Days at Apple* (New York: Harperbusiness, 1998).

43. US Census Bureau, "Home Computers and Internet Use in the United States: August 2000," US Department of Commerce, September, 2001, https://www.census.gov/prod/2001pubs/p23-207.pdf.

CHAPTER 11

1. Christopher Evans, *The Micro Millennium* (New York: Viking, 1979), 79.

2. Mark Weiser, Rich Gold, and John Seely Brown, "The Origins of Ubiquitous Computing Research at PARC in the Late 1980s," *IBM Systems Journal* 38, no. 4 (1999): 693–696

3. James W. Cooley and John W. Tukey, "An Algorithm for the Machine Calculation of Complex Fourier Series," *Mathematics of Computation* 19 (1965): 297–301.

4. Allen Newell, *Intellectual Issues in the History of Artificial Intelligence (Report CMU-CS-142)* (Carnegie-Mellon University, Department of Computer Science, 1982), 9.

5. Tom Sito, *Moving Innovation: A History of Computer Animation* (Cambridge, MA: MIT Press, 2013), 182–183.

6. Robert Lucky, quoted in Frederik Nebeker, *Signal Processing: The Emergence of a Discipline, 1948 to 1998* (New Brunswick, NJ: IEEE History Center, 1998), 88.

7. Trevor Pinch and Frank Trocco, *Analog Days: The Invention and Impact of the Moog Synthesizer* (Cambridge, MA: Harvard University Press, 2002).

8. Simon Reynolds, "Song from the Future: The Story of Donna Summer and Giorgio Moroder's 'I Feel Love,'" *Pitchfork*, June 29, 2017, https://pitchfork.com/features/article/song-from-the-future-the-story-of-donna-summer-and-giorgio-moroders-i-feel-love/.

9. Frederic D. Schwarz, "The Casio Effect," *Innovation & Technology* 18, no. 1 (2002).

10. Specifically, the version of the song from the concert film *Stop Making Sense*, in which the entire musical production of the band appears to come from a "boombox" cassette player placed on stage.

11. Jack Hamilton, "808s and Heart Eyes," Slate, December 16, 2016, http://www.slate.com/articles/arts/music_box/2016/12/_808_the_movie_is_a_must_watch_doc_for_music_nerds.html.

12. T. G. Stockham, T. M. Cannon, and R. B. Ingebretsen, "Blind Deconvolution Through Digital Signal Processing," *Proceedings of the IEEE* 63, no. 4 (April 1975): 678–692.

13. J. Gordon Holt, "Sony CDP-101 Compact Disc Player," *Stereophile,* January 23, 1983.

14. Changing practices in digital mixing and attitudes toward the ostensibly perfect sound of CDs are discussed in Kieran Downes, "'Perfect Sound Forever': Innovation, Aesthetics, and the Re-Making of Compact Disc Playback," *Technology and Culture* 51, no. 2 (April 2010): 305–331.

15. Jimmy McDonough, *Shakey: Neil Young's Biography* (New York: Random House, 2002), 568.

16. The functioning of the CD-ROM is described, and early dreams for its potential documented, in the evocative Steve Lambert and Suzanne Ropiequet, eds., *CD ROM the New Papyrus: The Current and Future State of the Art* (Redmond, WA: Microsoft Press, 1986).

17. Bob Strauss, "Wing Commander III: Heart of the Tiger," *Entertainment Weekly,* February 10, 1995.

18. Jonathan Coopersmith, *Faxed: The Rise and Fall of the Fax Machine* (Baltimore, MD: Johns Hopkins University Press, 2015).

19. Coopersmith, *Faxed,* 146.

20. Coopersmith, *Faxed,* 156.

21. "NHS Told to Ditch 'Absurd' Fax Machines," BBC News, December 9, 2018, https://www.bbc.com/news/uk-46497526.

22. Elizabeth R. Petrick, *Making Computers Accessible: Disability Rights and Digital Technology* (Baltimore, MD: Johns Hopkins University Press, 2015), chap. 2.

23. John Markoff, "Now, PC's That Read A Page and Store It," *New York Times,* August 17, 1988.

24. Commodore's Amiga was well suited to video production, thanks to high resolution video modes that functioned well with inexpensive genlock and frame grabber hardware. Jimmy Maher, *The Future Was Here: The Commodore Amiga* (Cambridge, MA: MIT Press, 2012), chap. 5.

25. On the history of spy satellites, see William E. Burrows, *Deep Black: Space Espionage and National Security* (New York: Random House, 1986).

26. R. W. Smith and J. N. Tatarewicz, "Replacing a Technology: The Large Space Telescope and CCDs," *Proceedings of the IEEE* 73, no. 7 (July 1985): 1221–1235.

27. Chris O'Falt, "Pixelvision: How a Failed '80s Fisher-Price Toy Camera Became One of Auteurs' Favorite '90s Tools," IndieWire, August 2018, https://www.indiewire.com/2018/08/pixelvision-pxl-2000-fisher-price-toy-experimental-film-camera-lincoln-center-series-1201991348/.

28. Martin Hand, *Ubiquitous Photography* (Malden, MA: Polity Press, 2012).

29. Jonah Engel Bromwich, "Once $50,000. Now, VCRs Collect Dust," *New York Times,* July 21 2016.

30. Joel Brinkley, "HDTV: High Definition, High in Price," *New York Times,* August 26, 1998.

31. Evans, *The Micro Millennium*, 219.

32. Michael Lewis, "Boom Box," *New York Times,* August 13, 2000.

33. The development of the MP3 format is described in Jonathan Sterne, *MP3: The Meaning of a Format* (Durham, NC: Duke University Press, 2012), chaps. 4 and 5. In preceding chapters, Sterne emphasizes its deeper historical roots in telecommunications and biological research.

34. Sterne, *MP3*, chap. 6. Stephen Witt, *How Music Got Free* (New York: Viking Press, 2015), 53–98, tells the parallel stories of the developers of MP3 and the pirates who copied music from inside CD distribution plants.

35. Jennifer Sullivan, "Napster: Music Is for Sharing," *Wired,* November 1999. The era is well captured in John Alderman, *Sonic Boom: Napster, MP3, and the New Pioneers of Music* (New York: Basic Books, 2001).

36. Nate Anderson, "Thomas Verdict: Willful Infringement, $1.92 Million Penalty," Ars Technica, June 18, 2009, https://arstechnica.com/tech-policy/2009/06/jammie-thomas-retrial-verdict/. The amount of damages awarded increased and decreased through three trials and an appeal.

37. Walter S. Mossberg, "Apple Brings Its Flair for Smart Designs to Digital Music Player," *Wall Street Journal,* November 1, 2001.

38. Steven Levy, *The Perfect Thing: How the iPod Shuffles Commerce, Culture, and Coolness* (New York: Simon & Schuster, 2006), 1.

39. Stephen Silver, "The iPod Touch Is a Worthy End to the Iconic Music Line," AppleInsider, September 5, 2018, https://appleinsider.com/articles/18/09/05/the-ipod-touch-is-a-worthy-end-to-the-iconic-music-line. Apple had taken out a full-page advertisement in the *Wall Street Journal* to celebrate the 100,000,000th iPod sold.

40. "Apple Introduces Revolutionary New Laptop with No Keyboard," The Onion, January 5, 2009, https://www.theonion.com/apple-introduces-revolutionary-new-laptop-with-no-keybo-1819594761.

41. Dick Pountain, "Benchtest: Acorn Archimedes," *Personal Computer World* 9, no. 8 (August 1987): 98–104.

42. Jacob Gaboury, *Image Objects: An Archaeology of Computer Graphics* (Cambridge, MA: MIT Press, 2021), 147.

43. The development of the first commercial system for bitmapped 3D graphics is described in Nick England, "The Graphics System for the 80's," *IEEE Computer Graphics and Applications* 40, no. 3 (May/June 2020): 112–119.

44. Early excitement around virtual reality was stoked by Howard Rheingold, *Virtual Reality* (New York: Summit Books, 1991).

45. Gibson first used the term *cyberspace* in a short story, "Burning Chrome" published in July 1982, but his vision of an immersive electronically generated virtual reality was popularized by the novel William Gibson, *Neuromancer* (New York: Ace, 1984). Gibson knew nothing about computers at the time, which made his fiction more exciting but hard for user interface designers to try to turn into a reality.

46. The rise and fall of VR is remembered by participants in Adi Robertson and Michael Zelenko, "Voices from a Virtual Past," The Verge, 2014, https://www.theverge.com/a/virtual-reality/oral_history. Virtual reality hype returned in the 2010s, with much better technology, but it has yet to establish mass appeal beyond dedicated videogame players.

47. David Kushner, *Masters of Doom: How Two Guys Created An Empire and Transformed Pop Culture* (New York: Random House, 2003).

48. Henry Lowood, "Game Engine," in *Debugging Game History: A Critical Lexicon*, ed. Henry Lowood and Raiford Guins (Cambridge, MA: MIT Press, 2016), 202–209.

49. Trent Ward, "Quake Review," June 22, 1996, https://www.gamespot.com/reviews/quake-review/1900 -2532549/.

50. Thomas Pabst, "3D Accelerator Card Reviews: Diamond Monster 3D," Tom's Hardware, November 9, 1997, https://www.tomshardware.com/reviews/3d-accelerator-card-reviews,42-7.html.

51. Rich Brown, "GeForce 8800 GTX Review," CNET, November 8, 2006, https://www.cnet.com/reviews /geforce-8800-gtx-review/.

52. Similar efforts continued with later generations of Xbox hardware. Brendan I. Koerner, "The Young and the Reckless," *Wired*, May 2018.

53. George Kuriakose Thiruvathukal and Steven E. Jones, *Codename Revolution: The Nintendo Wii Platform* (Cambridge, MA: MIT Press, 2012).

CHAPTER 12

1. US Census Bureau, "Computer and Internet Use in the United States: 2003," US Department of Commerce, October 2005, https://www.census.gov/prod/2005pubs/p23-208.pdf.

2. Michael A. Banks, *On the Way to the Web* (Berkeley, CA: Apress, 2008), 95–101 and 115–121.

3. Wendy Grossman, *net.wars* (New York: New York University Press), 1997, chap. 1.

4. Kara Swisher, *aol.com: How Steve Case Beat Bill Gates, Nailed the Netheads, and Made Millions in the War for the Web* (New York: Random House, 1998), 103. AOL's growth in the first half of the 1990s is also discussed in Banks, *On the Way to the Web*, 127–137.

5. Vannevar Bush, "As We May Think," *The Atlantic Monthly* 176, no. 1 (July 1945): 101–108. For an earlier and broader history of efforts to organize knowledge with new technology, see W. Boyd Rayward, ed., *Information beyond Borders: International Cultural and Intellectual Exchange in the Belle Époque* (Burlington, VT: Ashgate, 2014), and W. Boyd Rayward, "Visions of Xanadu: Paul Otlet (1868–1944) and Hypertext," *Journal of the American Society for Information Science* 45, no. 4 (May 1994): 235–250.

6. Theodor H. Nelson, *Computer Lib/Dream Machines* (self-published, 1974), DM 44–45.

7. Nelson, *Computer Lib/Dream Machines*, DM 19.

8. Belinda Barnet, "Hypertext before the Web—or, What the Web Could Have Been," in *The SAGE Handbook of Web History*, ed. Niels Brügger and Ian Milligan (Thousand Oaks, CA: Sage, 2019), 215–226.

9. Academic work on hypertext was summarized in a special issue of the *Communications of the ACM*, 31 (July 1988).

10. Philip L. Frana, "Before the Web There Was Gopher," *IEEE Annals of the History of Computing* 26, no. 1 (January–March 2004): 20–41.

11. Tim Berners-Lee, "WWW: Past, Present, and Future," *IEEE Computer* 29, no. 10 (October 1996): 69–77, p. 70.

12. Berners-Lee, "WWW," 71.

13. The pre-Mosaic browsers are discussed most thoroughly in James Gillies and Robert Cailliau, *How the Web Was Born: The Story of the World Wide Web* (Oxford, UK: Oxford University Press, 2000).

14. On the incorporation of graphics into early browsers, see Marc Weber, "Browsers and Browser Wars," in *The SAGE Handbook of Web History*, ed. Niels Brügger and Ian Milligan (Thousand Oaks, CA: Sage, 2019), 270–296.

15. Matthew Gray, "Measuring the Growth of the Web: June 1993 to June 1995," MIT personal site, 1996, http://www.mit.edu/people/mkgray/growth/.

16. Gary Wolfe, "The (Second Phase of the) Revolution Has Begun," *Wired Magazine*, October 1994.

17. This story is told from the inside in Jim Clark and Owen Edwards, *Netscape Time: The Making of the Billion-Dollar Start-Up That Took on Microsoft* (New York: St. Martin's Press, 1999), and with more zest in Michael Lewis, *The New New Thing* (New York: W. W. Norton, 2000).

18. Web server numbers through 1997 measure the number of distinct Web server host names. The date for 10 million servers is for "active websites," a substantially lower number because of domain squatting, search engine spam sites, and similar practices that gained popularity. All are taken from Netcraft, "June 2020 Web Server Survey," June 25, 2020, https://news.netcraft.com/archives/2020/06/25/june-2020-web-server-survey.html.

19. Swisher, *aol.com*. Shane Greenstein, *How the Internet Became Commercial: Innovation, Privatization, and the Birth of a New Network* (Princeton, NJ: Princeton University Press, 2015), 225–229 on America Online and WorldNet, and 268 on the later dominance of AOL as an Internet service provider.

20. Nina Munk, *Fools Rush In: Steve Case, Jerry Levin, and the Unmaking of AOL Time Warner* (New York: HarperCollins, 2004), 118.

21. Greenstein, *How the Internet Became Commercial*, 252.

22. Julian Dibbell, "The Unreal Estate Boom," *Wired* 11, no. 1 (January 2003).

23. John Perry Barlow, "A Declaration of the Independence of Cyberspace," Electronic Frontier Foundation, February 8, 1996, https://www.eff.org/cyberspace-independence.

24. Fred Turner, *From Counterculture to Cyberculture: Stewart Brand, the Whole Earth Network, and the Rise of Digital Utopianism* (Chicago: University of Chicago Press, 2006).

25. Martin Dodge and Rob Kitchin, *Atlas of Cyberspace* (New York: Addison-Wesley, 2001).

26. Benjamin M. Compaine, *The Digital Divide: Facing a Crisis or Creating a Myth?* (Cambridge, MA: MIT Press, 2001).

27. Tim Berners-Lee and Mark Fischetti, *Weaving the Web: The Original Design and Ultimate Destiny of the World Wide Web by Its Inventor* (San Francisco: Harper, 1999).

28. Charles C. Mann, "Is the Internet Doomed?," *Inc* 17, no. 9 (June 13 1995): 47–50, 52, 54.

29. Robert H. Reid, *Architects of the Web: 1,000 Days that Built the Future of Business* (New York: John Wiley & Sons, 1997), chap. 6.

30. Karen Angel, *Inside Yahoo! Reinvention and the Road Ahead* (New York: John Wiley & Sons, 2002).

31. Paul Festa, "Web Search Results Still Have Human Touch," News.com, December 27, 1999, http://news.com.com/2100-1023-234893.html.

32. The early history of Google is best told in Steven Levy, *In the Plex: How Google Thinks, Works, and Shapes Our Lives* (New York: Simon & Schuster, 2011), chap. 1.

33. Jonathan Coopersmith, "Pornography, Technology, and Progress," *Icon* 4 (1998): 94–125.

34. Reid, *Architects of the Web*, 280–320.

35. Angel, *Inside Yahoo!*, 140.

36. Google's adoption of advertising is discussed in Levy, *In the Plex*, 83–99.

37. Google reported revenue of $2.69 billion for the third quarter of 2006, up 70% from the previous year. Sara Kehaulani Goo, "Surge in Profit Reflects Google's Widening Lead," *Washington Post,* October 20 20006. Gannett reported operating revenues of $1.9 billion over the same period.

38. Levy, *In the Plex*, 99–120.

39. Brad Stone, *The Everything Store: Jeff Bezos and the Age of Amazon* (New York: Little, Brown, 2013).

40. The CERN phone book was the original target for browser development, according to Berners-Lee and Fischetti, *Weaving the Web*, 32–33.

41. Meg Leta Jones, "Cookies: A Legacy of Controversy," *Internet Histories* 4, no. 1 (2020): 87–104.

42. Larry Wall and Randal L. Schwartz, *Programming Perl* (Sebastopol, CA: O'Reilly, 1991), xiv. The history of PERL is discussed in Michael Stevenson, "Having It Both Ways: Larry Wall, Perl and the Technology and Culture of the Early Web," *Internet Histories* 2, no. 3–4 (2018): 264–280.

43. Andrew Leonard, "The Joy of Perl," Salon, October 13, 1998, https://www.salon.com/1998/10/13/feature_269/.

44. Steven Levy, "Battle of the Clipper Chip," *New York Times Magazine*, June 12, 1994.

45. Netscape Communications, "Netscape Communications Offers New Network Navigator Free on the Internet," October 13, 1994, accessed May 30, 2006.

46. Richard Karpinski, "Netscape Sets Retail Rollout," *Interactive Age* 2, no. 16 (June 5 1995): 1; and Netscape Communications, "Netscape Communications Offers New Network Navigator."

47. Greenstein, *How the Internet Became Commercial.*

48. Bill Gates, "The Internet Tidal Wave," US Department of Justice, May 26, 1995, accessed May 20, 2006, http://www.usdoj.gov/atr/cases/exhibits/20.pdf.

49. Bill Gates, Nathan Myhrvold, and Peter Rinearson, *The Road Ahead* (New York: Viking, 1995). The book appeared at the very end of 1995, more than a year behind schedule.

50. These choices are reviewed from the viewpoint of economics in Greenstein, *How the Internet Became Commercial*, 303–314.

51. The tactics used by Netscape during this period are explored at length in Michael A. Cusumano and David B. Yoffie, *Competing on Internet Time* (New York: Free Press, 1998).

52. Microsoft's anticompetitive efforts to hobble Netscape are analyzed in Greenstein, *How the Internet Became Commercial*, 314–320, who concludes that "Gates authorized such an approach because he wanted users and developers not to work with anyone else . . . regardless of what users or developers wanted." (320).

53. Myles White, "'Explorer' Closes Gap," *Toronto Star,* September 4 1997.

54. AOL is often said to have paid $4.2 billion for Netscape, reflecting the value of the AOL stock offered when the deal was announced. By March 1999 when the deal closed, however, that stock was worth around $10 billion.

55. Munk, *Fools Rush In*.

56. Ken Auletta, *World War 3.0: Microsoft and Its Enemies* (New York: Random House, 2001)

57. David Bank, *Breaking Windows: How Bill Gates Fumbled the Future of Microsoft* (New York: The Free Press, 2001).

58. Joel Brinkley, "U.S. Judge Says Microsoft Violated Antitrust Laws with Predatory Behavior," *New York Times*, April 4, 2000.

59. Thomas Penfield Jackson, "Excerpts From the Ruling That Microsoft Violated Antitrust Law," *New York Times*, April 4, 2000.

60. As well as the integer releases of Internet Explorer, this count includes 4.5 and 5.5 as versions with significant new features.

61. Stallman's story is told in Sam Williams, *Free as in Freedom: Richard Stallman's Crusade for Free Software* (Sebastopol, CA: O'Reilly, 2002), and more analytically, in Christopher M. Kelty, *Two Bits: The Cultural Significance of Free Software* (Durham, NC: Duke University Press, 2008), 182–209.

62. Glyn Moody, *The Rebel Code: The Inside Story of Linux and the Open Source Revolution* (Cambridge, MA: Perseus, 2001), 14–19.

63. Richard M. Stallman, "What Is a GNU/Linux System?," *GNU's Bulletin* 1, no. 23 (1997):4–5.

64. "What Is Copyleft?," *GNU's Bulletin* 1, no. 23 (July 1997).

65. Andrew S. Tanenbaum, *Operating Systems: Design and Implementation* (Englewood Cliffs, NJ: Prentice Hall, 1987).

66. Linus Torvalds and David Diamond, *Just for Fun: The Story of an Accidental Revolutionary*, 1st ed. (New York: HarperBusiness, 2001), 61–62.

67. Torvalds and Diamond, *Just for Fun*, 85.

68. Eric S. Raymond, *The Cathedral and the Bazaar* (Sebastopol, CA: O'Reilly, 2001).

69. The origins and significance of the free/open source distinction are discussed in Kelty, *Two Bits*, 98–117.

70. Kelty, *Two Bits*, 223–229.

71. Timothy Dyck, "Web Server Brains & Brawn," *PC Magazine*, May 22, 2001.

72. Byron Acohido, "Firefox Ignites Demand for Alternative Browser," *New York Times*, November 10, 2004; and Walter S. Mossberg, "Security, Cool Features of Firefox Web Browser Beat Microsoft's IE," *Wall Street Journal*, December 30, 2004.

73. Josh McHugh, "The Firefox Explosion," *Wired* 13, no. 2 (February 2005).

CHAPTER 13

1. Erik Brynjolfsson, "The Productivity Paradox of Information Technology," *Communications of the ACM* 36, no. 12 (December 1993): 66–77; and Paul Strassmann, *The Squandered Computer* (New Canaan, CT: Information Economics Press, 1997).

2. Steven Levy, "Google Throws Open Doors to Its Top-Secret Data Center," *Wired*, November 2012.

3. Although historians of computing have not yet grappled with cloud computing, the cloud metaphor has drawn considerable attention from media theorists, including John Durham Peters, *The Marvelous Clouds: Towards a Philosophy of Elemental Media* (Chicago, IL: University of Chicago Press, 2015), and Tung-Hui Hu, *A Prehistory of the Cloud* (Cambridge, MA: MIT Press, 2015).

4. Jim Thompson, "How Unisys Transitioned from Proprietary to Open Architecture," Enterprise Tech, July 28, 2015, https://www.enterpriseai.news/2015/07/28/how-unisys-transitioned-from-proprietary-to-open-architecture/.

5. Gerry Smith, "Who Killed the Great American Cable-TV Bundle?," Bloomberg, 2018, accessed August 8, https://www.bloomberg.com/news/features/2018-08-08/who-killed-the-great-american-cable-tv-bundle.

6. Tim O'Reilly, "What Is Web 2.0," O'Reilly, September 30, 2005, accessed October 4, 2006, http://www.oreillynet.com/pub/a/oreilly/tim/news/2005/09/30/what-is-web-20.html.

7. Ignacio Siles, "Blogs," in *The SAGE Handbook of Web History*, ed. Niels Brügger and Ian Milligan (Thousand Oaks, CA: Sage, 2019), 359–371.

8. Michael Erard, "Decoding the New Cues in Online Society," *New York Times,* November 27, 2003.

9. Jean Burgess, "Hearing Ordinary Voices: Cultural Studies, Vernacular Creativity and Digital Storytelling," *Continuum: Journal of Media and Cultural Studies* 20, no. 2 (June 2006): 201–214.

10. The most comprehensive history of Facebook is Steven Levy, *Facebook: The Inside Story* (New York: Blue Rider, 2020).

11. Ian Bogost, "The Aesthetics of Philosophical Carpentry," in *The Nonhuman Turn*, ed. Richard Grusin (Minneapolis, MN: University of Minnesota Press, 2015), 112–131. Facebook's period as a platform for third-party applications is described in Levy, *Facebook: The Inside Story*, chap. 7.

12. Levy, *Facebook: The Inside Story*, 128–131 and 137–144 on the origins of News Feed, 260–263 on changes to mimic Twitter and promote viral content, and 295–297 on the integration of advertising into News Feed.

13. Susanna Paasonen, "Online Pornography," in *The SAGE Handbook of Web History*, ed. Niels Brügger and Ian Milligan (Thousand Oaks, CA: Sage, 2019), 551–563.

14. Andy Famiglietti, "Wikipedia," in *The SAGE Handbook of Web History*, ed. Niels Brügger and Ian Milligan (Thousand Oaks, CA: Sage, 2019), 315–329. Andrew Lih, *The Wikipedia Revolution* (New York: Hyperion, 2009).

15. Jessa Lingel, "Socio-technical Transformations in Secondary Markets: a Comparison of Craigslist and VarageSale," *Internet Histories* 3, no. 2 (2019).

16. Lohr, *Go To*, chap. 10.

17. Andreessen himself says he borrowed the famous quote from Bob Metcalf, of Ethernet fame. Cliver Anderson, "The Man Who Makes the Future: Wired Icon Marc Andreessen," *Wired*, May 2012.

18. Carol Hildebrand, "The PC Price Tag," *CIO Enterprise* 11, no. 2 (1997): 42–46.

19. Dana Cline, "Corel Office for Java," Dr. Dobb's, October 1, 1997, archived at https://web.archive.org/web/20130924045624/http://www.drdobbs.com/corel-office-for-java/184415588.

20. Martin Campbell-Kelly, "The Rise, Fall, and Resurrection of Software as a Service," *Communications of the ACM* 52, no. 5 (May 2009): 28–30.

21. Jesse James Garrett, "Ajax: A New Approach to Web Applications," Adaptive Path, February 18, 2005, accessed October 10, 2018, archived at https://immagic.com/eLibrary/ARCHIVES/GENERAL/ADTVPATH/A050218G.pdf.

22. In fact, the name Ajax has become a misnomer, as modern browsers rely on JSON rather than XML for the background data transfer.

23. For example, software engineers were warned to seek professional engineer status or risk prosecution under state licensing laws in John R. Speed, "What Do You Mean I Can't Call Myself a Software Engineer." *IEEE Software* 16, no. 6 (November–December 1999): 45–50.

24. Amanda Kolson Hurley, "Everyone's an Architect," July 8, 2010, https://www.architectmagazine.com/design/everyones-an-architect_o.

25. Lizzie Widdicombe, "The Programmer's Price," *New Yorker*, November 24, 2014.

CHAPTER 14

1. Jerry Kaplan, *Startup: A Silicon Valley Adventure*. Boston: Houghton Mifflin, 1995.

2. Owen W. Linzmayer, *Apple Confidential 2.0: The Definitive History of the World's Most Colorful Company* (San Francisco, CA: No Starch Press, 2004), 188.

3. Andrea Butler and David Pogue, *Piloting Palm: The Inside Story of Palm, Handspring and the Birth of the Billion Dollar Handheld Industry* (New York: John Wiley, 2002), 250.

4. Jon Agar, *Constant Touch: A Global History of the Mobile Phone* (Cambridge: Icon Books, 2004), 36.

5. Gerard Goggin, "Emergence of the Mobile Web," in *The SAGE Handbook of Web History*, ed. Niels Brügger and Ian Milligan (Thousand Oaks, CA: Sage, 2019), 297–311.

6. Wesley Allison, "1995 BMW 740iL—Long-Term Wrapup," November 1, 1996, https://www.motortrend.com/cars/bmw/7-series/1995/1995-bmw-740-il/.

7. Robert Lemos, "New Pilot Adds Technology to Simplicity," ZDNet, March 9, 1998, https://www.zdnet.com/article/new-pilot-adds-technology-to-simplicity/.

8. Apple Inc., "Steve Jobs iPhone 2007 Presentation," YouTube, 2013, https://www.youtube.com/watch?v=vN4U5FqrOdQ.

9. Sarah Perez, "Apple's Big App Store Purge Is Now Underway," TechCrunch, November 15, 2016, https://techcrunch.com/2016/11/15/apples-big-app-store-purge-is-now-underway/.

10. Brian X. Chen, "Apple Change Quietly Makes iPhone, iPad Into Web Phones," *Wired*, January 28, 2010, https://www.wired.com/2010/01/iphone-voip/.

11. Hugo Greenhalgh, "Grindr and Tinder: The Disruptive Influence of Apps on Gay Bars," *Financial Times*, December 11 2017; Matt Kapp, "Grindr: Welcome to the World's Biggest, Scariest Gay Bar," May 27, 2011, https://www.vanityfair.com/news/2011/05/grindr-201105.

12. Alexis C. Madrigal, "The Servant Economy," *The Atlantic*, March 6, 2019, https://www.theatlantic.com/technology/archive/2019/03/what-happened-uber-x-companies/584236/.

13. This remained true even as Apple's market share fell. In late 2017, for example, Apple was estimated to produce only 19 percent of the smartphones sold worldwide but to generate 87 percent of the industry's profits. Chuck Jones, "Apple Continues to Dominate the Smartphone Profit Pool," *Forbes*, March 2, 2018, https://www.forbes.com/sites/chuckjones/2018/03/02/apple-continues-to-dominate-the-smartphone-profit-pool/.

14. Connie Guglielmo, "Apple Touts Itself as Big Job Creator in the US," *Forbes*, March 2, 2012, https://www.forbes.com/sites/connieguglielmo/2012/03/02/apple-touts-itself-as-big-job-creator-in-the-u-s/#568d37b1a606.

15. For example, the iPad Pro was called a "story of impressive hardware and untapped potential" in Scott Stein, "iPad Pro (2018) Review: A Powerful, Beautiful Tablet That Needs a Software Overhaul," *CNet*, December 14, 2018, https://www.cnet.com/reviews/apple-ipad-pro-2018-review/. Addressing such criticism, in mid-2019, Apple announced a new "iPadOS" that would increasingly diverge from iOS.

16. Brian X. Chen, "China Becomes Apple's Hottest iPhone Market," *New York Times*, April 27, 2015. Apple's "greater China" sales included those in Taiwan and Hong Kong.

17. Farhad Manjoo, "Psst. WhatsApp Needs Fixing. Pass It On," *New York Times*, October 24, 2018.

18. Tim Stevens, "Fitbit Review," Engadget, September 15, 2009, https://www.engadget.com/2009/10/15 /fitbit-review.

19. Joe Thompson, "A Concise History of the Smartwatch," Bloomberg, January 8, 2018, https://www .bloomberg.com/news/articles/2018-01-08/a-concise-history-of-the-smartwatch.

20. Mat Honan, "I Glasshole: My Year With Google Glass," *Wired* (December 30, 2013).

21. "Fridge Sends Spam Emails as Attack Hits Smart Gadgets," BBC News, January 17, 2014, https://www .bbc.com/news/technology-25780908.

CHAPTER 15

1. Jerry Hirsch, "Elon Musk: Model S Not a Car but a 'Sophisticated Computer on Wheels,'" Los Angeles Times, March 19, 2015, https://www.latimes.com/business/autos/la-fi-hy-musk-computer-on-wheels -20150319-story.html.

2. "Deus ex Vehiculum," *The Economist,* June 23, 2015.

3. Alex Roland and Philip Shiman, *Strategic Computing: DARPA and the Quest for Machine Intelligence* (Cambridge, MA: MIT Press, 2002).

4. Rachel Abrams and Annalyn Kurtz, "A Driver's Zeal, an Engineer's Worry," *New York Times*, July 1, 2016.

5. Johana Bhuiyan, "A Federal Agency Says an Overreliance on Tesla's Autopilot Contributed to a Fatal Crash," Recode, September 12, 2017, https://www.vox.com/2017/9/12/16294510/fatal-tesla-crash-self -driving-elon-musk-autopilot.

6. A study by the Insurance Institute for Highway Safety found that the "Autopilot" branding made motorists more likely to assume it was safe to text or to take their hand off the wheel for a long period. Kyle LaHuick, "Tesla's Autopilot Found Most Likely to Confuse Drivers on Safety," Bloomberg, June 20, 2019, https://www.bloomberg.com/news/articles/2019-06-20/tesla-s-autopilot-found-most-likely-to-confuse -drivers-on-safety.

7. Kevin Kelly, *What Technology Wants* (New York: Viking, 2010).

8. John Carreyrou, *Bad Blood: Secrets and Lies in a Silicon Valley Startup* (New York: Alfred A. Knopf, 2018).

9. David Leonhardt, "How the Upper Middle Class Is Really Doing," *New York Times*, February 24, 2019.

10. Andrew McAfee and Erik Brynjolfsson, *The Second Machine Age: Work, Progress, and Prosperity in a Time of Brilliant Technologies* (New York: W. W. Norton, 2014).

11. One of us attempted a taxonomy of conceptions of fake news in Maria Haigh and Thomas Haigh, "Fighting and Framing Fake News," in *The Sage Handbook of Propaganda*, ed. Paul Baines, Nicholas O'Shaughnessy, and Nancy Snow (Thousand Oaks, CA: Sage Publishing, 2020).

12. To choose just three of many recent books on the abuse of algorithms: Cathy O'Neil, *Weapons of Math Destruction: How Big Data Increases Inequality and Threatens Democracy* (New York: Crown Books, 2016); Virginia Eubanks, *Automating Inequality* (New York: Picador, 2019); and Frank Pasquale, *The Black Box Society: The Secret Algorithms That Control Money and Information* (Cambridge, MA: Harvard University Press, 2015).

13. Sam Schechner and Mark Secada, "Apps Send User Secrets to Facebook," *Wall Street Journal,* February 23, 2019.

14. Shoshana Zuboff, *The Age of Surveillance Capitalism* (New York: Public Affairs, 2019).

15. Sarah Ponczek and Vildana Hajric, "Robinhood Market Made Bursting Bubbles Wall Street's Obsession," Bloomberg, June 13, 2020, https://www.bloomberg.com/news/articles/2020-06-13/robinhood-market -made-bursting-bubbles-wall-street-s-obsession.

BIBLIOGRAPHY

Abbate, Janet. *Inventing the Internet*. Cambridge, MA: MIT Press, 1999.

Abbate, Janet. "Privatizing the Internet: Competing Visions and Chaotic Events, 1987–1995." *IEEE Annals of the History of Computing* 32, no. 1 (January 2010): 10–22.

Abbate, Janet. *Recoding Gender: Women's Changing Participation in Computing*. Cambridge, MA: MIT Press, 2012.

Abrams, Rachel, and Annalyn Kurtz. "A Driver's Zeal, an Engineer's Worry." *New York Times*, July 1, 2016, B1.

Acohido, Byron. "Firefox Ignites Demand for Alternative Browser." *USA Today*, November 10, 2004, B1.

Adamson, Ian, and Richard Kennedy. *Sinclair and the "Sunrise" Technology: The Deconstruction of a Myth*. Harmondsworth, UK: Penguin, 1986.

Agar, Jon. *Constant Touch: A Global History of the Mobile Phone*. Cambridge: Icon Books, 2004.

Agar, Jon. *The Government Machine: A Revolutionary History of the Computer*. Cambridge, MA: MIT Press, 2003.

Akera, Atsushi. *Calculating a Natural World: Scientists, Engineers, and Computers during the Rise of U.S. Cold War Research*. Cambridge, MA: MIT Press, 2007.

Akera, Atsushi. "Voluntarism and the Fruits of Collaboration." *Technology and Culture* 42, no. 4 (October 2001): 710–736.

Alderman, John. *Sonic Boom: Napster, MP3, and the New Pioneers of Music*. New York: Basic Books, 2001.

Alford, Roger C. "NCR's S486/MC33 Has Unique Approach to Reliability." *Byte* 15, no. 13 (December 1990): 191–193.

Allison, Wesley. "1995 BMW 740iL—Long-Term Wrapup." November 1, 1996. https://www.motortrend.com/cars/bmw/7-series/1995/1995-bmw-740-il/.

Alsop, Stewart. "A Public Windows Pane to Make Compatibility Clearer." *InfoWorld*, January 31, 1994, 102.

Alt, Franz L. "Fifteen Years ACM." *Communications of the ACM* 5, no. 6 (1962): 300–307.

Altice, Nathan. *I Am Error: The Nintendo Family Computer / Entertainment System Platform*. Cambridge, MA: MIT Press, 2015.

Amelio, Gil, and William L. Simon. *On the Firing Line: My 500 Days at Apple*. New York: Harperbusiness, 1998.

Anderson, Chris. "The Man Who Makes the Future: Wired Icon Marc Andreessen." *Wired*, May 2012.

Anderson, John J. "Commodore." *Creative Computing* 10, no. 3 (March 1984): 56, 60.

Anderson, Nate. "Thomas Verdict: Willful Infringement, $1.92 Million Penalty." Ars Technica, June 18, 2009. https://arstechnica.com/tech-policy/2009/06/jammie-thomas-retrial-verdict/.

Angel, Karen. *Inside Yahoo! Reinvention and the Road Ahead.* New York: John Wiley & Sons, 2002.

Apiki, Steve, and Rick Grehan. "Fastest NT Workstations." *Byte* 20, no. 3 (March 1995): 115–122.

Apple Inc. "Steve Jobs iPhone 2007 Presentation," YouTube, 2013. https://www.youtube.com/watch?v=vN4U5 FqrOdQ.

Archbold, Pamela, and John Verity. "The Datamation 100: Company Profiles." *Datamation* 31, no. 11 (June 1, 1985): 58–182.

Argyle, Ray. "25th Anniversary Issue." *University of Waterloo, Department of Computing Services Newsletter* 1982, 2.

Aris, John. "The LEO Approach—An Evaluation." In *User Driven Innovation: The World's First Business Computer,* edited by David Caminer, John Aris, Peter Hermon, and Frank Land, 320–326. London: McGraw-Hill, 1996.

Armer, Paul. "SHARE—A Eulogy to Cooperative Effort." *Annals of the History of Computing* 2, no. 2 (April 1980): 122–129.

Aspray, William. "The Intel 4004 Microprocessor: What Constituted Invention?" *IEEE Annals of the History of Computing* 19, no. 3 (July–September 1997): 4–15.

Associated Press. "Windows 95 Sales Plunge from Peak." *New York Times* 1995, D6.

Association of Data Processing Service Organizations. "Second Operating Ratios Survey," 1967, ADAPSO Records (CBI 172), Charles Babbage Institute, University of Minnesota, Minneapolis.

Atchison, William F., Samuel D. Conte, John W. Hamblen, Thomas E. Hull, Thomas A. Keenan, William B. Kehl et al. "Curriculum 68: Recommendations for Academic Programs in Computer Science: A Report of the ACM Curriculum Committee on Computer Science." *Communications of the ACM* 11, no. 3 (March 1968): 151–197.

Augustin, Larry. "The Mainframe Connection: IBM's 3270 PC." *Byte* 9, no. 9 (1984): 231–237.

Auletta, Ken. *World War 3.0: Microsoft and its Enemies.* New York: Random House, 2001.

Bachman, Charles W. "The Origin of the Integrated Data Store (IDS): The First Direct-Access DBMS." *IEEE Annals of the History of Computing* 31, no. 4 (October–December 2009): 42–54.

Bachman, Charles W. "The Programmer as Navigator." *Communications of the ACM* 16, no. 11 (November 1973): 653–658.

Backus, John. "Programming in America in the 1950s—Some Personal Impressions." In *A History of Computing in the Twentieth Century,* edited by N. Metropolis, J. Howlett, and Gian-Carlo Rota, 125–135. New York: Academic Press, 1980.

Bagnall, Brian. *Commodore: The Amiga Years.* Winnipeg, MB: Variant Press, 2017.

Bagnall, Brian. *On the Edge: The Spectacular Rise and Fall of Commodore.* Winnipeg, MB, Canada: Variant Press, 2005.

Bank, David. *Breaking Windows: How Bill Gates Fumbled the Future of Microsoft.* New York: The Free Press, 2001.

Banks, Michael A. *On the Way to the Web.* Berkeley, CA: Apress, 2008.

Baran, Nick. "Two Powerful Systems from SUN." *Byte* 14, no. 5 (May 1989): 108–112.

Bardini, Thierry. *Bootstrapping: Douglas Engelbart, Coevolution, and the Origins of Personal Computing.* Stanford, CA: Stanford University Press, 2000.

Barlow, John Perry. "A Declaration of the Independence of Cyberspace." Electronic Frontier Foundation, February 8, 1996. https://www.eff.org/cyberspace-independence.

Barnet, Belinda. "Hypertext before the Web—or, What the Web Could Have Been." In *The SAGE Handbook of Web History*, edited by Niels Brügger and Ian Milligan, 215–226. Thousand Oaks, CA: Sage, 2019.

Barney, Clifford. "Award for Achievement [Alan F. Shugart]." *Electronics Week*, January 14, 1985, 40–44.

Barr, Nicholas. "The History of the Phillips Machine." In *A.W.H. Phillips: Collected Works in Contemporary Perspective*, 89–114. New York: Cambridge University Press, 2000.

Bashe, Charles J., Lyle R. Johnson, John H. Palmer, and Emerson W. Pugh. *IBM's Early Computers*. Cambridge, MA: MIT Press, 1986.

Bassett, Ross Knox. *To the Digital Age: Research Labs, Start-Up Companies, and the Rise of MOS Technology*. Baltimore: Johns Hopkins University Press, 2002.

Baum, Claude. *The System Builders: The Story of SDC*. Santa Monica, CA: System Development Corporation, 1981.

Bay, Morten. "Hot Potatoes and Postmen: How Packet Switching Became ARPANET's Greatest Legacy." *Internet Histories* 3, no. 1 (2019): 15–30.

Beavers, L. L. "WordPerfect: Not Quite Perfect, But Certainly Superb." *Creative Computing* 9, no. 11 (November 1983): 74.

Bell, C. Gordon. "Towards a History of (Personal) Workstations." In *A History of Personal Workstations*, edited by Adele Goldberg, 4–36. New York: ACM Press, 1988.

Bell, C. Gordon, A. Kotok, T. N. Hastings, and R. Hill. "The Evolution of the DECsystem 10." *Communications of the ACM* 21, no. 1 (January 1978): 44–63.

Bell, C. Gordon, J. Craig Mudge, and John E McNamara. *Computer Engineering: a DEC View of Hardware Systems Design*. Bedford, MA: Digital Press, 1978.

Bell, C. Gordon, and Allen Newell. *Computer Structures: Readings and Examples*. New York: McGraw-Hill, 1971.

Bergin, Thomas J. "The Origins of Word Processing Software for Personal Computers: 1976–1985." *IEEE Annals of the History of Computing* 28, no. 4 (October–December 2006): 32–47.

Bergin, Thomas J., and Thomas Haigh. "The Commercialization of Database Management Systems, 1969–1983." *IEEE Annals of the History of Computing* 31, no. 4 (October–December 2009): 26–41.

Berkeley, Edmund C. *Giant Brains or Machines That Think*. New York: John Wiley & Sons, 1949.

Berners-Lee, Tim. "WWW: Past, Present, and Future." *IEEE Computer* 29, no. 10 (October 1996): 69–77.

Berners-Lee, Tim, and Mark Fischetti. *Weaving the Web: The Original Design and Ultimate Destiny of the World Wide Web by Its Inventor*. San Francisco: Harper, 1999.

Bhuiyan, Johana. "A Federal Agency Says an Overreliance on Tesla's Autopilot Contributed to a Fatal Crash." Recode, September 12, 2017. https://www.vox.com/2017/9/12/16294510/fatal-tesla-crash-self-driving-elon-musk -autopilot.

Binder, R., N. Abramson, F. Kuo, A. Okinaka, and D. Wax. "ALOHA Packet Broadcasting: A Retrospect." In *1975 National Computer Conference*, 203–215. Montvale, NJ: AFIPS Press, 1975.

Binstock, Andrew. "Interview with Alan Kay." Dr. Dobb's, July 10, 2012. http://www.drdobbs.com/architecture -and-design/interview-with-alan-kay/240003442#.

Birrell, Andrew D., and Bruce Jay Nelson. "Implementing Remote Procedure Calls." *ACM Transactions on Computer Systems* 2, no. 1 (February 1984): 39–59.

Bitzer, Donald. "The Million Terminal System of 1985." In *Computers and Communications: Implications for Education*, edited by Robert J. Seidel and Martin Rubin, 59–70. New York: Academic Press, 1977.

Blank, Marc S., and S. W. Galley. "How to Fit a Large Program into a Small Machine." *Creative Computing* 6, no. 7 (July 1980): 80–87.

Bogost, Ian. "The Aesthetics of Philosophical Carpentry." In *The Nonhuman Turn*, edited by Richard Grusin, 112–131. Minneapolis, MN: University of Minnesota Press, 2015.

Bourne, Chris. "Going On-line." *Sinclair User*, January 1986, 126–127, 132.

Bradley, David. "The Creation of the IBM PC." *Byte* 15, no. 9 (September 1990): 414–420.

Bright, Peter. "Benchtest: Epson PC." *Personal Computer World* 9, no. 2 (February 1986): 104–110.

Bright, Peter. "Checkout: PageMaker & LaserWriter." *Personal Computer World* 8, no. 10 (October 1985): 166–171.

Brinch Hansen, Per, ed. *Classic Operating Systems*. New York: Springer, 2001.

Brinkley, Joel. "HDTV: High Definition, High in Price." *New York Times*, August 26, 1998, G1.

Brinkley, Joel. "U.S. Judge Says Microsoft Violated Antitrust Laws with Predatory Behavior." *New York Times*, April 4, 2000, A1.

Brock, David C., and David A. Laws. "The Early History of Microcircuitry: An Overview." *IEEE Annals of the History of Computing* 34, no. 1 (January–March 2012).

Brock, Gerald W. *The Telecommunications Industry: The Dynamics of Market Structure* Cambridge, MA: Harvard University Press, 1981.

Bromwich, Jonah Engel. "Once $50,000. Now, VCRs Collect Dust." *New York Times*, July 21, 2016, B1.

Brooks, Frederick P, Jr. *The Mythical Man Month: Essays on Software Engineering*. Reading, MA: Addison-Wesley, 1975.

Brown, R. Hunt. "Computer Comparison and Census Chart." *Management and Business Automation* 4, no. 1 (August 1960): 34.

Brown, Rich. "GeForce 8800 GTX Review." CNET, November 8, 2006. https://www.cnet.com/reviews/geforce-8800-gtx-review/.

Brunton, Finn. *Spam: A Shadow History of the Internet*. Cambridge, MA: MIT Press, 2013.

Brynjolfsson, Erik. "The Productivity Paradox of Information Technology." *Communications of the ACM* 36, no. 12 (December 1993): 66–77.

Bucholz, Werner. "Anecdote: Origin of the Word Byte." *Annals of the History of Computing* 3, no. 1 (January 1981): 72.

Bucholz, Werner, ed. *Planning a Computer System: Project Stretch*. New York: McGraw-Hill, 1962.

Bullynck, Maarten. "What Is an Operating System? A Historical Investigation (1954–1964)." In *Reflections on Programming Systems: Historical and Philosophical Aspects*, edited by Liesbeth de Mol and Giuseppe Primiero, 49–79. Cham, Switzerland: Springer, 2019.

Bullynck, Maarten, and Liesbeth De Mol. "Setting-up early computer programs: D. H. Lehmer's ENIAC computation." *Archive of Mathematical Logic* 49 (2010): 123–146.

Bureau of Labor Statistics. "Issues in Labor Statistics: Computer Ownership Up Sharply in the 1990s." March 1999. https://www.bls.gov/opub/btn/archive/computer-ownership-up-sharply-in-the-1990s.pdf.

Burgess, Jean. "Hearing Ordinary Voices: Cultural Studies, Vernacular Creativity and Digital Storytelling." *Continuum: Journal of Media and Cultural Studies* 20, no. 2 (June 2006): 201–214.

Burks, Alice R., and Arthur W. Burks. *The First Electronic Computer: The Atanasoff Story*. Ann Arbor: University of Michigan Press, 1989.

Burks, Arthur W., Herman Heine Goldstine, and John von Neumann. *Preliminary Discussion of the Logical Design of an Electronic Computing Instrument*. Princeton, NJ: Institute for Advanced Studies, 1946.

Burnham, David. *The Rise of the Computer State*. New York: Random House, 1983.

Burrows, William E. *Deep Black: Space Espionage and National Security*. New York: Random House, 1986.

Bush, Vannevar. "As We May Think." *The Atlantic Monthly* 176, no. 1 (July 1945): 101–108.

"Business Week Reports to Readers On: Computers." *Business Week*, June 21, 1958, 68–92.

Butler, Andrea, and David Pogue. *Piloting Palm: The Inside Story of Palm, Handspring and the Birth of the Billion Dollar Handheld Industry*. New York: John Wiley, 2002.

Buzbee, Bill. "Oral History Interview by Thomas Haigh, April, Westminster CO," 2005, Society for Industrial and Applied Mathematics, Philadelphia, PA.

Caminer, David, John Aris, Peter Hermon, and Frank Land, eds. *User Driven Innovation: The World's First Business Computer*. London: McGraw-Hill, 1996.

Campanale, Michelle. "Eight Heavy-Hitting NT Workstations." *Byte* 23, no. 1 (January 1998): 98.

Campbell, Scott. "'Wat For Ever:' Student-Oriented Computing at the University of Waterloo." *IEEE Annals of the History of Computing* 35, no. 1 (Jan–Mar 2013): 11–22.

Campbell-Kelly, Martin. "The Evolution of Digital Computing Practice on the Cambridge University EDSAC, 1949–1951." In *Exploring the Early Digital*, edited by Thomas Haigh, 117–134. Cham, Switzerland: Springer, 2019.

Campbell-Kelly, Martin. *From Airline Reservations to Sonic the Hedgehog: A History of the Software Industry*. Cambridge, MA: MIT Press, 2003.

Campbell-Kelly, Martin. *ICL: A Technical and Business History*. New York: Oxford University Press, 1989.

Campbell-Kelly, Martin. "Number Crunching without Programming: The Evolution of Spreadsheet Usability." *IEEE Annals of the History of Computing* 29, no. 3 (July–September 2007): 6–19.

Campbell-Kelly, Martin. "Programming the EDSAC: Early Programming Activity at the University of Cambridge." *IEEE Annals of the History of Computing* 2, no. 1 (October 1980): 7–36.

Campbell-Kelly, Martin. "Programming the Pilot Ace: Early Programming Activity at the National Physical Laboratory." *Annals of the History of Computing* 3, no. 2 (April 1981): 133–162.

Campbell-Kelly, Martin. "The RDBMS Industry: A Northern California Perspective." *IEEE Annals of the History of Computing* 34, no. 4 (October–December 2012): 18–29.

Campbell-Kelly, Martin. "The Rise, Fall, and Resurrection of Software as a Service." *Communications of the ACM* 52, no. 5 (May 2009): 28–30.

Campbell-Kelly, Martin, and William Aspray. *Computer: A History of the Information Machine*. New York: Basic Books, 1996.

Campbell-Kelly, Martin, and Michael R. Williams, eds. *The Moore School Lectures: Theory and Techniques for Design of Electronic Digital Computers*. Cambridge, MA: MIT Press, 1985.

Canion, Rod. *Open: How Compaq Ended IBM's PC Domination and Helped Invented Modern Computing*. Dallas, TX: BenBella, 2013.

Canning, Richard G., and Roger L. Sisson. *The Management of Data Processing*. New York: John Wiley & Sons, 1967.

Capozzoli, Daniel. "The Early Years of Data Processing." *Computer Services Newsletter* (Internal Revenue Service), July 1987.

Care, Charles. *Technology for Modelling: Electrical Analogies, Engineering Practice, and the Development of Analogue Computing.* Cham, Switzerland: Springer, 2010.

Carreyrou, John. *Bad Blood: Secrets and Lies in a Silicon Valley Startup* New York: Alfred A. Knopf, 2018.

Ceruzzi, Paul. *Beyond the Limits: Flight Enters the Computer Age.* Cambridge, MA: MIT Press, 1989.

Chen, Brian X. "Apple Change Quietly Makes iPhone, iPad Into Web Phones." *Wired*, January 28, 2010. https://www.wired.com/2010/01/iphone-voip/.

Chen, Brian X. "China Becomes Apple's Hottest iPhone Market." *New York Times*, April 27, 2015, B1.

"A Chorus Line: Computerized Lighting Control Comes to Broadway." *Theatre Crafts*, November/December 1975, 6–11, 26–29.

Chposky, James, and Ted Leonsis. *Blue Magic: The People, Power, and Politics behind the IBM Personal Computer.* New York: Facts on File, 1988.

Christensen, Ward, and Randy Suess. "Hobbyist Computerized Bulletin Board." *Byte* 3, no. 11 (November 1978): 150–157.

Clark, Jim, and Owen Edwards. *Netscape Time: The Making of the Billion-Dollar Start-Up That Took on Microsoft.* New York: St. Martin's Press, 1999.

Clark, Wesley. "The LINC Was Early and Small." In *A History of Personal Workstations*, edited by Adele Goldberg, 345–400. New York: ACM Press, 1988.

Cline, Dana. "Corel Office for Java." Dr. Dobb's, October 1, 1997. http://www.drdobbs.com/corel-office-for-java/184415588.

Codd, E. F. "A Relational Model of Data for Large Shared Databanks." *Communications of the ACM* 13, no. 6 (June 1970): 377–390.

Cohen, Derek. "Cover Benchtest: Atari Portfolio." *Personal Computer World* 11, no. 8 (August 1989): 130–136.

Cohen, I. Bernard. *Howard Aiken: Portrait of a Computer Pioneer.* Cambridge, MA: MIT Press, 1999.

Cohen, I. Bernard, and Gregory W. Welch, eds. *Makin' Numbers: Howard Aiken and the Computer.* Cambridge, MA: MIT Press, 1999.

Cohen, Theodore Jerome, and Jacquelin H. Bray. *Melissa and John and the Magic Machine.* Peterborough, NH: BYTE/McGraw Hill, 1979.

Cohoon, J. McGrath, and William Aspray, eds. *Women and Information Technology: Research on Underrepresentation.* Cambridge, MA: MIT Press, 2006.

Commodore International. "Commodore VIC-20 Job Interview Advertisement." YouTube, 1984. https://www.youtube.com/watch?v=c5tqmyl3XQk.

Compaine, Benjamin M. *The Digital Divide: Facing a Crisis or Creating a Myth?* Cambridge, MA: MIT Press, 2001.

Comptroller General of the United States. *Safeguarding Taxpayer Information—An Evaluation of the Proposed Computerized Tax Administration System (LCD-76-115).* Washington, DC: Department of the Treasury, 1977.

Computers and Their Future: Speeches Given at the World Computer Pioneer Conference. Llandudno, Wales: Richard Williams, 1970.

Conway, Lynn. "Reminiscences of the VLSI Revolution." *IEEE Solid State Circuits* 4, no. 4 (Fall 2012): 8–31.

Cooley, James W., and John W. Tukey. "An Algorithm for the Machine Calculation of Complex Fourier Series." *Mathematics of Computation* 19 (1965): 297–301.

Coopersmith, Jonathan. *Faxed: The Rise and Fall of the Fax Machine*. Baltimore, MD: Johns Hopkins University Press, 2015.

Coopersmith, Jonathan. "Pornography, Technology, and Progress." *Icon* 4 (1998): 94–125.

Copeland, B. Jack, ed. *Alan Turing's Automatic Computing Engine: The Master Codebreaker's Struggle to Build the Modern Computer*. New York: Oxford University Press, 2005.

Copeland, B. Jack, Andre A. Haeff, Peter Gough, and Cameron Wright. "Screen History: The Haeff Memory and Graphics Tube." *IEEE Annals of the History of Computing* 39, no. 1 (January–March 2017): 9–28.

Copeland, Duncan G., Richard O. Mason, and James L. McKenney. "SABRE: The Development of Information-Based Competence and Execution of Information-Based Competition." *IEEE Annals of the History of Computing* 17, no. 3 (Fall 1995): 30–57.

Corbató, Fernando, Marjorie Merwin-Daggett, and Robert C. Caley. "An Experimental Timesharing System." In *Proceedings of the Spring Joint Computer Conference, Volume 21*. N.p.: AFIPS, 1962.

Cortada, James W. *IBM: The Rise and Fall and Reinvention of a Global Icon*. Cambridge, MA: MIT Press, 2019.

Côté, Raymond Ga, and Barry Nance. "Pentium PCs: Power to Burn." *Byte* 18, no. 8 (July 1993): 94–102.

Coupland, Douglas. "Microserfs." *Wired*, January 1994.

Craddock, David L. *Break Out: How the Apple II Launched the PC Gaming Revolution*. Atglen, PA: Schiffer, 2017.

Crandall, Richard L. "Oral History Interview by Paul Ceruzzi, 3 May 2002, Washington, D.C.," 2002, Charles Babbage Institute, University of Minnesota, Minneapolis.

Crawford, Perry Orson, Jr. "Automatic Control by Arithmetical Operations." Master's thesis, Massachusetts Institute of Technology, 1942.

Cray, Seymour R. "Computer-Programmed Preventative Maintenance for Internal Memory Sections of the ERA 1103 Computer System." In *Proceedings of the WESCON Computer Sessions*, 62–66. New York: Institute of Radio Engineers, 1954.

Cress, Paul, Paul Dirksen, and J. Wesley Graham. *Fortran IV with WATFOR and WATFIV*. Edgewood Cliffs, NJ: Prentice Hall, 1970.

Cringely, Robert X. *Accidental Empires: How the Boys of Silicon Valley Make their Millions, Battle Foreign Competition, and Still Can't Get a Date*. Reading, MA: Addison-Wesley, 1992.

Crocker, Suzanne. "Paul Brainerd, Aldus Corporation, and the Desktop Publishing Revolution." *IEEE Annals of the History of Computing* 41, no. 3 (July–September 2019): 35–41.

Cusumano, Michael A., and Richard W. Selby. *Microsoft Secrets: How the World's Most Powerful Software Company Creates Technology, Shapes Markets, and Manages People* New York: Free Press, 1995.

Cusumano, Michael A., and David B. Yoffie. *Competing on Internet Time*. New York: Free Press, 1998.

Dahmke, Mark. "The Osborne 1." *Byte* 7, no. 6 (June 1982): 348–363.

Davis, Martin. *Engines of Logic: Mathematicians and the Origin of the Computer*. New York: Norton, 2001.

Dear, Brian. *The Friendly Orange Glow: The Untold Story of the PLATO System and the Dawn of Cyberculture*. New York: Pantheon, 2017.

Dell, Michael, and Catherine Fredman. *Direct from Dell*. New York: Harper Business, 1999.

De Mol, Liesbeth. "Turing Machines." Stanford Encyclopedia of Philosophy, September 24, 2018. https://plato.stanford.edu/entries/turing-machine.

Denning, Peter. "The Science of Computing: The Internet Worm." *American Scientist* 77, no. 2 (March–April 1989): 126–128.

Denning, Peter J. "The Working Set Model for Program Behavior." *Communications of the ACM* 11, no. 5 (May 1968): 323–333.

Denning, Peter J., and Edward Grady Coffman. *Operating Systems Theory*. Prentice Hall, 1973.

"Deus ex Vehiculum." *The Economist*, June 23, 2015.

Díaz, Gerardo Con. *Software Rights: How Patent Law Transformed Software Development in America*. New Haven, CT: Yale University Press, 2019.

Dibbell, Julian. "The Unreal Estate Boom." *Wired* 11, no. 1 (January 2003).

Dickson, David. *The New Politics of Science*. New York: Pantheon Books, 1984.

Diebold, John. *Automation, the Advent of the Automatic Factory*. New York: Van Nostrand, 1952.

Diebold, John. "Factories without Men: New Industrial Revolution." *The Nation*, September 19, 1953, 227–228, 250–251, 271–222.

"The Digital Age." *Electronics*, April 17, 1980, 373–414.

"Digital Computing Timeline: 1978." VT100.net, n.d. https://vt100.net/timeline/1978-2.html.

Digital Equipment Corporation. *PDP-10 Timesharing Handbook*. Maynard, MA: Digital Equipment Corporation, 1970.

Digital Equipment Corporation. *PDP-11 Processor Handbook*. Maynard, MA: Digital Equipment Corporation, 1981.

Dijkstra, Edsger W. "EWD 563: Formal Techniques and Sizeable Programs." In *Selected Writings on Computing: A Personal Perspective*, edited by Edsger W. Dijkstra, 205–214. New York: Springer, 1982.

Dijkstra, Edsger Wybe. "EWD1110: To the Members of the Budget Council (Confidential)." E. W. Dijkstra Archive, 1991. https://www.cs.utexas.edu/users/EWD/transcriptions/EWD11xx/EWD1110.html.

Dijkstra, Edsger Wybe. "EWD 1175: The Strengths of the Academic Enterprise." E. W. Dijkstra Archive, 1994. https://www.cs.utexas.edu/users/EWD/transcriptions/EWD11xx/EWD1175.html.

Dijkstra, Edsger Wybe. "EWD 1303: My Recollections of Operating System Design." E. W. Dijkstra Archive, 2001. http://userweb.cs.utexas.edu/users/EWD/transcriptions/EWD13xx/EWD1303.html.

Dijkstra, Edsger Wybe. "The Humble Programmer." *Communications of the ACM* 15, no. 10 (October 1972): 859–866.

Dijkstra, Edsger Wybe. "The Structure of the 'THE'-Multiprogramming System." *Communications of the ACM* 11, no. 5 (May 1968): 341–346.

Dodge, Martin, and Rob Kitchin. *Atlas of Cyberspace*. New York: Addison-Wesley, 2001.

Donaldson, John. "Benchtest: Amstrad PCW9512." *Personal Computer World* 9, no. 10 (October 1987): 98–102.

Dongarra, Jack. "Oral History Interview by Thomas Haigh, April 26, University of Tennessee, Knoxville TN." 2005, Society for Industrial and Applied Mathematics, Philadelphia, PA.

Dorfman, R. "The Discovery of Linear Programming." *Annals of the History of Computing* 6, no. 3 (July 1984): 283–295.

Douglas, Susan. *Inventing American Broadcasting, 1899–1922*. Baltimore: Johns Hopkins University Press, 1987.

Dourish, Paul. "The Once and Future Internet: Infrastructural Tragedy and Ambiguity in the Case of IPv6." *Internet Histories* 2, no. 1–2 (2018): 55–74.

Downes, Kieran. "'Perfect Sound Forever': Innovation, Aesthetics, and the Re-Making of Compact Disc Playback." *Technology and Culture* 51, no. 2 (April 2010): 305–331.

Driscoll, Kevin. "Demography and Decentralization: Measuring the Bulletin Board Systems of North America." *WiderScreen* 23, no. 2–3 (2020).

Driscoll, Kevin. "Professional Work for Nothing: Software Commercialization and 'An Open Letter to Hobbyists.'" *Information & Culture* 50, no. 2 (2015): 257–283.

Dyck, Timothy. "Web Server Brains & Brawn." *PC Magazine*, May 22, 2001, 124–142.

Dyson, George. *Turing's Cathedral: The Origins of the Digital Universe*. New York: Pantheon Books, 2012.

Eckert, J. Presper. "A Survey of Digital Computer Memory Systems." *Proceedings of the IRE* 41, no. 10 (October 1953): 1393–1406.

Eckert, J. Presper. "Thoughts on the History of Computing." *IEEE Computer* 9, no. 12 (December 1976): 58–65.

Edwards, Paul N. *The Closed World: Computers and the Politics of Discourse in Cold War America*. Cambridge, MA: MIT Press, 1996.

Ein-Dor, Phillip. "Grosch's Law Re-revisited." *Communications of the ACM* 28, no. 2 (February 1985): 142–151.

Engelbart, Douglas C. "The Augmented Knowledge Workshop." In *A History of Personal Workstations*, edited by Adele Goldberg, 187–232. New York: ACM Press, 1988.

Engelbart, Douglas C. "Presentation to 1968 Fall Joint Computer Conference (hosted on The Mouse Site as The Demo)." 1968. http://sloan.stanford.edu/MouseSite/1968Demo.html.

Engineering Research Associates. *High-Speed Computing Devices*. New York: McGraw-Hill, 1950.

England, Nick. "The Graphics System for the 80's." *IEEE Computer Graphics and Applications* 40, no. 3 (May/June 2020): 112–119.

Ensmenger, Nathan. *The Computer Boys Take Over: Computers, Programmers, and the Politics of Technical Expertise*. Cambridge, MA: MIT Press, 2010.

Epstein, Bob. "History of Sybase." *IEEE Annals of the History of Computing* 35, no. 2 (April–June 2013): 31–41.

Erard, Michael. "Decoding the New Cues in Online Society." *New York Times*, November 27 2003, G1.

Eubanks, Virginia. *Automating Inequality*. New York: Picador, 2019.

Evans, Bob O. "System/360: A Retrospective View." *Annals of the History of Computing* 8, no. 2 (April–June 1986): 155–179.

Evans, Christopher. *The Micro Millennium*. New York: Viking, 1979.

"Everyone Who Bought One of Those 30,000 Copies Started a Band." Quote Investigator, March 1, 2016. https://quoteinvestigator.com/2016/03/01/velvet/.

Eyles, Don. *Sunburst and Luminary: an Apollo Memoir*. Boston: Fort Point Books, 2018.

Famiglietti, Andy. "Wikipedia." In *The SAGE Handbook of Web History*, edited by Niels Brügger and Ian Milligan, 315–329. Thousand Oaks, CA: Sage, 2019.

Ferrell, Keith. "IBM Compatibles: The Universe Expands." *Compute!*, no. 86 (July 1987): 14–24.

Ferry, G. *A Computer Called LEO: Lyons Tea Shops and the World's First Office Computer*. London: Fourth Estate, 2003.

Festa, Paul. "Web Search Results Still Have Human Touch." News.com, December 27, 1999. http://news.com .com/2100-1023-234893.html.

Fidler, Bradley, and Andrew L. Russell. "Financial and Administrative Infrastructure for the Early Internet: Network Maintenance at the Defense Information Systems Agency." *Technology and Culture* 59, no. 4 (October 2018): 899–924.

Fisher, Franklin M., James W. McKie, and Richard B. Mancke. *IBM and the US Data Processing Industry: An Economic History*. New York: Praeger, 1983.

Fishman, Katharine Davis. *The Computer Establishment*. New York: Harper & Row, 1981.

Fletcher, Amy L. "France Enters the Information Age: A Political History of Minitel." *History and Technology* 18, no. 2 (2002): 103–117.

Flink, James J. *The Automobile Age*. Cambridge, MA: MIT Press, 1988.

Forester, Tom. "The Myth of the Electronic Cottage." In *Computers in The Human Context: Information Technology, Productivity, and People*, edited by Tom Forester, 213–227. Cambridge, MA: MIT Press, 1989.

Frana, Philip L. "Before the Web There Was Gopher." *IEEE Annals of the History of Computing* 26, no. 1 (January–March 2004): 20–41.

Franz, Kathleen. *Tinkering: Consumers Reinvent the Early Automobile*. Philadelphia: University of Pennsylvania Press, 2005.

Freeman, Peter A., W. Richards Adrion, and William Aspray. *Computing and the National Science Foundation, 1950–2016*. New York: Association for Computing Machinery, 2019.

Freiberger, Paul, and Michael Swaine. *Fire in the Valley: The Making of the Personal Computer*. Berkeley, CA: Osborne/McGraw-Hill, 1984.

"Fridge Sends Spam Emails as Attack Hits Smart Gadgets." BBC News, January 17, 2014. https://www.bbc.com/ news/technology-25780908.

Fried, Stephen S. "Evaluating 8087 Performance on the IBM PC." *Byte* 9, no. 9 (Fall 1984): 197–208.

Fritz, W Barkley. "The Women of ENIAC." *IEEE Annals of the History of Computing* 18, no. 3 (Fall 1996): 13–28.

Frohman, Dov, and Robert Howard. *Leadership the Hard Way*. San Francisco, CA: Jossey-Bass, 2008.

Gaboury, Jacob. *Image Objects: An Archaeology of Computer Graphics*. Cambridge, MA: MIT Press, 2021.

Galison, Peter. "The Ontology of the Enemy: Norbert Wiener and the Cybernetic Vision." *Critical Inquiry* 21, no. 1 (Autumn 1994): 228–266.

Galloway, Alexander R. *Protocol: How Control Exists after Decentralization*. Cambridge, MA: MIT Press, 2004.

Garrett, Jesse James. "Ajax: A New Approach to Web Applications," Adaptive Path, February 18, 2005. Accessed October 10, 2018. http://adaptivepath.org/ideas/ajax-new-approach-web-applications/.

Gass, Saul I. "The Role of Digital Computers in Project Mercury." In *Proceedings of the Eastern Joint Computer Conference*, 33–46, 1961.

Gates, Bill. "The Internet Tidal Wave," US Department of Justice, May 26, 1995. Accessed May 20, 2006. http://www.usdoj.gov/atr/cases/exhibits/20.pdf.

Gates, Bill, Nathan Myhrvold, and Peter Rinearson. *The Road Ahead*. New York: Viking, 1995.

Gazzard, Alison. *Now the Chips Are Down*. Cambridge, MA: MIT Press, 2016.

Gibson, William. *Neuromancer*. New York: Ace Books, 1984.

Gillies, James, and Robert Cailliau. *How the Web Was Born: The Story of the World Wide Web*. Oxford: Oxford University Press, 2000.

Goggin, Gerard. "Emergence of the Mobile Web." In *The SAGE Handbook of Web History*, edited by Niels Brügger and Ian Milligan, 297–311. Thousand Oaks, CA: Sage, 2019.

Goldberg, Adele, ed. *A History of Personal Workstations*. New York: Addison-Wesley/ACM, 1988.

Goldberg, Adele. "Introducing the Smalltalk-80 System." *Byte* 6, no. 8 (1981): 14–26.

Goldstine, Herman H. *The Computer from Pascal to von Neumann*. Princeton, NJ: Princeton University Press, 1972.

Goldstine, Herman H., and John von Neumann. *Planning and Coding Problems for an Electronic Computing Instrument. Part II, Volume 1*. Princeton, NJ: Institute for Advanced Studies, 1947.

Goo, Sara Kehaulani. "Surge in Profit Reflects Google's Widening Lead." *Washington Post*, October 20, 2006, D.01.

Gotkin, Kevin. "When Computers Were Amateur." *IEEE Annals of the History of Computing* 36, no. 2 (April–June 2014): 4–14.

Grad, Burton. "The Creation and the Demise of VisiCalc." *IEEE Annals of the History of Computing* 29, no. 3 (July–September 2007): 20–31.

Grad, Burton. "A Personal Recollection: IBM's Unbundling of Software and Services." *IEEE Annals of the History of Computing* 24, no. 1 (January–March 2002): 64–71.

Gray, Matthew. "Measuring the Growth of the Web: June 1993 to June 1995." MIT personal site, 1996. http://www.mit.edu/people/mkgray/growth/.

Greenhalgh, Hugo. "Grindr and Tinder: The Disruptive Influence of Apps on Gay Bars." *Financial Times*, December 11, 2017.

Greenstein, Shane. *How the Internet Became Commercial: Innovation, Privatization, and the Birth of a New Network*. Princeton, NJ: Princeton University Press, 2015.

Gregory, Nathan. *The Tym Before: The Untold Origins of Cloud Computing*: Self-published, 2018.

Grosch, H. R. J. "High Speed Arithmetic: The Digital Computer as a Research Tool." *Journal of the Optical Society of America* 43, no. 4 (1953): 306–310.

Grosch, Herbert R. J. *Computer: Bit Slices from a Life*. Novato, California: Third Millennium, 1991.

Grossman, Wendy. *net.wars*. New York: New York University Press, 1997.

Gruenberger, Fred. "RAND Symposium 14," 1973, RAND Symposia Collection (CBI 78), Charles Babbage Institute, University of Minnesota, Minneapolis.

Guglielmo, Connie. "Apple Touts Itself as Big Job Creator in the US." Forbes, March 2, 2012. https://www.forbes.com/sites/connieguglielmo/2012/03/02/apple-touts-itself-as-big-job-creator-in-the-u-s/#568d37b1a606.

Haanstra, John W, and Bob Evans. "Processor Products—Final Report of SPREAD Task Group, December 28, 1961." *Annals of the History of Computing* 5, no. 1 (January 1983): 6–26.

Haderle, Donald J., and Cynthia M. Saracco. "The History and Growth of IBM's DB2." *IEEE Annals of the History of Computing* 35, no. 2 (April–June 2013): 54–66.

Hafner, Katie. *The Well: A Story of Love, Death & Real Life in the Seminal Online Community*. New York: Carroll & Graf, 2001.

Haigh, Maria, and Thomas Haigh. "Fighting and Framing Fake News." In *The Sage Handbook of Propaganda*, edited by Paul Baines, Nicholas O'Shaughnessy, and Nancy Snow. Thousand Oaks, CA: Sage, 2020.

Haigh, T. "Larry A. Welke—Biography." *IEEE Annals of the History of Computing* 26, no. 4 (October–December 2004): 85–91.

Haigh, Thomas. "Actually, Turing Did Not Invent the Computer." *Communications of the ACM* 57, no. 1 (January 2014): 36–41.

Haigh, Thomas. "Cleve Moler: Mathematical Software Pioneer and Creator of Matlab." *IEEE Annals of the History of Computing* 30, no. 1 (January–March 2008): 87–91

Haigh, Thomas. "Charles W. Bachman: Database Software Pioneer." *IEEE Annals of the History of Computing* 33, no. 4 (October–December 2011): 70–80.

Haigh, Thomas. "The Chromium-Plated Tabulator: Institutionalizing an Electronic Revolution, 1954–1958." *IEEE Annals of the History of Computing* 23, no. 4 (October–December 2001): 75–104.

Haigh, Thomas. "Dijkstra's Crisis: The End of Algol and the Beginning of Software Engineering: 1968–72." Draft discussed at the workshop History of Software, European Styles, at the Lorentz Center of the University of Leiden, Netherlands, 2010, 2010. http://www.tomandmaria.com/Tom/Writing/DijkstrasCrisis Leiden DRAFT.pdf.

Haigh, Thomas. *Finding a Story for the History of Computing*. Siegen, Germany: Media of Cooperation Working Paper Series, Siegen University, 2018.

Haigh, Thomas. "How Data Got Its Base: Information Storage Software in the 1950s and 1960s." *IEEE Annals of the History of Computing* 31, no. 4 (October–December 2009): 6–25.

Haigh, Thomas. "Jack Dongarra: Supercomputing Expert and Mathematical Software Specialist." *IEEE Annals of the History of Computing* 30, no. 2 (April–June 2008): 74–81.

Haigh, Thomas. "Masculinity and the Machine Man." In *Gender Codes: Why Women Are Leaving Computing*, edited by Thomas J Misa, 51–71. Hoboken, NJ: IEEE Computer Society Press, 2010.

Haigh, Thomas. "Remembering the Office of the Future: The Origins of Word Processing and Office Automation." *IEEE Annals of the History of Computing* 28, no. 4 (October–December 2006): 6–31.

Haigh, Thomas. "The Tears of Donald Knuth." *Communications of the ACM* 58, no. 1 (January 2015): 40–44.

Haigh, Thomas. "Von Neumann Thought Turing's Universal Machine Was 'Simple and Neat.' But That Didn't Tell Him How to Design a Computer." *Communications of the ACM* 63, no. 1 (January 2020): 26–32.

Haigh, Thomas. "Where Code Comes From: Architectures of Automatic Control from Babbage to Algol." *Communications of the ACM* 59, no. 1 (January 2016): 39–44.

Haigh, Thomas, and Mark Priestley. "Colossus and Programmability." *IEEE Annals of the History of Computing* 40, no. 4 (October–December 2018): 5–17.

Haigh, Thomas, Mark Priestley, and Crispin Rope. *ENIAC in Action: Making and Remaking the Modern Computer*. Cambridge, MA: MIT Press, 2016.

Hale, Wayne, Helen Lane, Gail Chapline, and Kamlesh Lulla, eds. *Wings in Orbit: Scientific and Engineering Legacies of the Space Shuttle, 1971–2010*. Washington, DC: NASA, 2011.

Hall, Mark, and John Barry. *SunBurst: The Ascent of Sun Microsystems*. Chicago: Contemporary Books, 1990.

Halvorson, Michael J. *Code Nation: Personal Computing and the Learn to Program Movement in America*. New York: ACM Books, 2020.

Hamilton, Jack. "808s and Heart Eyes." Slate, December 16, 2016. http://www.slate.com/articles/arts/music_box/2016/12/_808_the_movie_is_a_must_watch_doc_for_music_nerds.html.

Hand, Martin. *Ubiquitous Photography*. Malden, MA: Polity Press, 2012.

Hanford, K V. "The SHARE Operating System for the IBM 709." *Annual Review in Automatic Programming* 1 (1960): 169–177.

Hardy, Ann. "OH 458: Oral History by Jeffrey Yost, 2 April, Palo Alto, California" 2012, Charles Babbage Institute, University of Minnesota, Minneapolis.

Harr, Luther A. *The Univac System: A 1954 Progress Report*. N.p.: Remington Rand Corporation, 1954.

Harris, Blake J. *Console Wars: Sega, Nintendo, and the Battle That Defined a Generation*. New York: HarperCollins, 2014.

Hartmanis, Juris, and Richard E. Stearns. "On the Computational Complexity of Algorithms." *Transactions of the American Mathematical Society* 117 (1965): 285–306.

Hayes, Thomas C. "Logic Says that Plato's About to Pay Off." *New York Times*, April 26, 1981.

Heide, Lars. *Punched-Card Systems and the Early Information Explosion, 1880–1945*. Baltimore: Johns Hopkins University Press, 2009.

Hemmendinger, David. "Messaging in the Early SDC Time-Sharing System." *IEEE Annals of the History of Computing* 36, no. 1 (January–March 2014): 52–57.

Hertzfeld, Andy. *Revolution in The Valley: The Insanely Great Story of How the Mac Was Made*. Sebastopol, CA: O'Reilly, 2004.

Hicks, Marie. *Programmed Inequality: How Britain Discarded Women Technologists and Lost Its Edge in Computing*. Cambridge, MA: MIT Press, 2017.

Hildebrand, Carol. "The PC Price Tag." *CIO Enterprise* 11, no. 2 (1997): 42–46.

Hiltz, Starr Roxanne, and Murray Turoff. *The Network Nation: Human Communication via Computer*. Reading, MA: Addison-Wesley, 1978.

Hiltzik, Michael. *Dealers of Lightning: Xerox PARC and the Dawn of the Computer Age*. New York: HarperBusiness, 1999.

Hinnant, David F. "Benchmarking UNIX Systems." *Byte* 9, no. 8 (August 1984): 132–135, 400–409.

Hirsch, Jerry. "Elon Musk: Model S Not a Car but a 'Sophisticated Computer on Wheels.'" *Los Angeles Times*, March 19, 2015. https://www.latimes.com/business/autos/la-fi-hy-musk-computer-on-wheels-20150319-story.html.

Hoare, C. A. R. "Quicksort." *Computer Journal* 5, no. 1 (1962): 10–15.

Holt, J. Gordon. "Sony CDP-101 Compact Disc Player." *Stereophile*, January 23 1983.

Honan, Mat. "I Glasshole: My Year with Google Glass." *Wired* (December 30, 2013).

Hopper, Grace. "Keynote Address." In *History of Programming Languages*, edited by Richard L. Wexelblat, 7–20. New York: Academic Press, 1981.

Hopper, Grace M. "Compiling Routines." *Computers and Automation* 2 (May 1953): 1–5.

Hounshell, David. *From the American System to Mass Production, 1800–1932: The Development of Manufacturing Technology in the United States*. Baltimore: Johns Hopkins University Press, 1984.

House, Chuck. "Hewlett-Packard and Personal Computing Systems." In *A History of Personal Workstations*, edited by Adele Goldberg, 401–432. New York: ACM Press, 1988.

Hu, Tung-Hui. *A Prehistory of the Cloud*. Cambridge, MA: MIT Press, 2015.

Humphrey, Watts S. *Managing the Software Process*. Reading, MA: Addison-Wesley, 1989.

Hurd, Cuthbert C. "Early IBM Computers: Edited Testimony." *Annals of the History of Computing* 3, no. 2 (April–June 1981): 163–182.

Hurley, Amanda Kolson. "Everyone's an Architect." July 8, 2010. https://www.architectmagazine.com/design /everyones-an-architect_o.

"Industry Profile . . . Wes Graham of Waterloo U." *Computer Data: The Canadian Computer Magazine*, May 1976, 29–30.

Isaacson, Walter. *Steve Jobs*. New York: Simon & Schuster, 2011.

Jackson, Peter. "Benchtest: Compaq DeskPro 386." *Personal Computer World* 8, no. 11 (November 1986): 138–144.

Jackson, Thomas Penfield. "Excerpts from the Ruling That Microsoft Violated Antitrust Law." *New York Times*, April 4, 2000, C14.

James, S. E. "Evolution of Real Time Computer Systems for Manned Spaceflight." *IBM Journal of Research and Development* 25, no. 5 (September 1981): 417–428.

Jerrehian, John K. "The Computer Time Sharing Business." *ADAPSO News (CBI 172)* 8, no. 2 (April 1968): 3–5.

Johnson, Christopher. "Charlie Brown—Peanuts—Charles M. Schulz—Snoopy—Woodstock." ASCII Art, n.d. accessed June 2, 2020. https://asciiart.website/index.php?art=comics/peanuts.

Johnson, L. R. "Installation of a Large Electronic Computer." *Proc. ACM Meeting, Toronto* (1952): 77–80.

Jones, Chuck. "Apple Continues to Dominate the Smartphone Profit Pool." Forbes, March 2, 2018. https:// www.forbes.com/sites/chuckjones/2018/03/02/apple-continues-to-dominate-the-smartphone-profit-pool/.

Jones, Meg Leta. "Cookies: A Legacy of Controversy." *Internet Histories* 4, no. 1 (2020): 87–104.

Kahan, William. "Oral History Interview by Thomas Haigh, 5–8 August 2005, Berkeley, California," 2005, Society for Industrial and Applied Mathematics, Philadelphia, PA.

Kaplan, Jerry. *Startup: A Silicon Valley Adventure*. Boston: Houghton Mifflin, 1995.

Kapor, Mitch. "Oral History by William Aspray, November 19, Mountain View, California," 2004.

Kapp, Matt. "Grindr: Welcome to the World's Biggest, Scariest Gay Bar." May 27, 2011. https://www.vanityfair .com/news/2011/05/grindr-201105.

Karpinski, Richard. "Netscape Sets Retail Rollout." *Interactive Age* 2, no. 16 (June 5, 1995): 1.

Kay, Alan C. "The Early History of Smalltalk," edited by Thomas J Bergin and Rick G Gibson, 511–598. New York, NY: ACM Press, 1996.

Kay, Alan, and Adele Goldberg. "Personal Dynamic Media." *Computer* 10, no. 3 (March 1977): 31–41.

Kelly, Kevin. *What Technology Wants*. New York: Viking, 2010.

Kelty, Christopher M. *Two Bits: The Cultural Significance of Free Software*. Durham, NC: Duke University Press, 2008.

Kemeny, John G. *Man and the Computer*. New York: Scribner, 1972.

Kennedy, T. R. Jr. "Electronic Computer Flashes Answers, May Speed Engineering." *New York Times*, February 15, 1946, 1, 16.

Kenney, Charles. *Riding the Runaway Horse: The Rise and Decline of Wang Laboratories*. New York: Little Brown & Company, 1992.

Kernighan, Brian. *UNIX: A History and a Memoir*. Middletown, DE: Kindle Direct Publishing, 2020.

Kewney, Guy. "Benchtest: Amstrad PC1512." *Personal Computer World* 9, no. 10 (October 1986): 126–136.

Kidder, Tracy. *The Soul of a New Machine*. Boston: Little Brown, 1981.

Kilby, Jack S. "Invention of the Integrated Circuit." *IEEE Transactions on Electron Devices* 23, no. 7 (July 1976): 648–654.

Kildall, Gary. "Microcomputer Software Design—A Checkpoint." In *AFIPS '75: Proceedings of the May 19–22, 1975, National Computer Conference and Exposition*, 99–106. New York: ACM, 1975.

Kirschenbaum, Matthew G. *Track Changes: A Literary History of Word Processing*. Cambridge, MA: Harvard University Press, 2016.

Kirstein, Peter T. "The Early Days of the Arpanet." *IEEE Annals of the History of Computing* 31, no. 3 (July–September 2009).

Klass, Philip J. "Reliability Is Essential Minuteman Goal." *Aviation Week*, October 19, 1959, 13F.

Kline, Ronald. *The Cybernetics Moment, or Why We Call Our Age the Information Age*. Baltimore, MD: Johns Hopkins University Press, 2015.

Kline, Ronald, and Trevor Pinch. "Users as Agents of Technological Change: The Social Construction of the Automobile in the Rural United States." *Technology and Culture* 37, no. 4 (October 1996): 763–795.

Kneale, Dennis. "Commodore Hits Production Snags In Its Hot-Selling Home Computer." *Wall Street Journal*, October 28, 1983, 33.

Knebel, Fletcher. "Potomac Fever." *Appleton Post-Crescent*, November 2, 1959, 6.

Knobelsdorff, Kerry Elizabeth. "IBM's Four-Month-Old PS/2 Has Put Computer World on Hold." *Christian Science Monitor*, August 19, 1987, 19.

Knuth, Donald. *The Art of Computer Programming Volume 3: Sorting and Searching*. Reading, MA: Addison-Wesley, 1973.

Knuth, Donald. "The IBM 650: An Appreciation from the Field." *Annals of the History of Computing* 8, no. 1 (January–March 1986): 50–55.

Knuth, Donald E., and Luis Trabb Pardo. "The Early Development of Programming Languages." In *A History of Computing in the Twentieth Century*, edited by N. Metropolis, J. Howlett and Gian-Carlo Rota, 197–273. New York: Academic Press, 1980.

Koch, Christopher. "The Integration Nightmare: Sounding the Alarm." *CIO Magazine*, 15 November 1996.

Koerner, Brendan I. "The Young and the Reckless." *Wired*, May 2018.

Krol, Ed. *The Whole Internet User's Guide and Catalog*. Sebastopol, CA: O'Reilly, 1992.

Kurtz, Thomas E. "BASIC Session." In *History of Programming Languages*, edited by Richard L. Wexelblat, 515–549. New York: Academic Press, 1981.

Kushner, David. *Masters of Doom: How Two Guys Created an Empire and Transformed Pop Culture*. New York: Random House, 2003.

LaHuick, Kyle. "Tesla's Autopilot Found Most Likely to Confuse Drivers on Safety." Bloomberg, June 20, 2019. https://www.bloomberg.com/news/articles/2019-06-20/tesla-s-autopilot-found-most-likely-to-confuse-drivers-on-safety.

Lally, Elaine. *At Home with Computers*. New York: Berg, 2002.

Lambert, Steve, and Suzanne Ropiequet, eds. *CD ROM the New Papyrus: The Current and Future State of the Art.* Redmond, WA: Microsoft Press, 1986.

Lancaster, Don. *TTL Cookbook.* Indianapolis, IN: Howard Sams, 1974.

Lancaster, Don. "TV-Typewriter." *Radio-Electronics,* September 1973, 43–52.

Landreth, Bill, and Howard Rheingold. *Out of the Inner Circle: A Hacker's Guide to Computer Security.* Bellevue, WA: Microsoft Press, 1985.

Larner, Roy A. "FMS: The IBM FORTRAN Monitor System." In *Proceedings of the National Computer Conference,* 815–820. N.p.: AFIPS, 1987.

Larson, Chris. "MS DOS 2.0: An Enhanced 16-Bit Operating System." *Byte* 8, no. 11 (November 1983): 285–290.

Lavington, Simon, ed. *Alan Turing and His Contemporaries.* Swindon, UK: British Informatics Society, 2012.

Lavington, Simon. *Early British Computers.* Bedford, MA: Digital Press, 1980.

Lavington, Simon. *A History of Manchester Computers.* Swindon, UK: The British Computer Society, 1998.

Lavington, Simon. "The Manchester Mark I and Atlas: A Historical Perspective." *Communications of the ACM* 21, no. 1 (January 1978): 4–12.

Lean, Thomas. *Electronic Dreams: How 1980s Britain Learned to Love the Computer.* London: Bloomsbury, 2016.

Leavitt, Harold J., and Thomas L. Whisler. "Management in the 1980s." *Harvard Business Review* 36, no. 6 (November–December 1958): 41–48.

Lécuyer, Christophe. *Making Silicon Valley: Innovation and the Growth of High Tech, 1930–70.* Cambridge, MA: MIT Press, 2006.

Leemon, Sheldon. "PC-Write Word Processor For PC & PCjr." *Compute!,* no. 87 (February 1985), 82–86.

Leimbach, Timo. "The SAP Story: Evolution of SAP within the German Software Industry." *IEEE Annals of the History of Computing* 30, no. 4 (October 2008): 60–76.

Lemmons, Phil. "Victor Victorious." *Byte* 7, no. 11 (November 1982): 216–254.

Lemos, Robert. "New Pilot Adds Technology to Simplicity." ZDNet, March 9, 1998. https://www.zdnet.com /article/new-pilot-adds-technology-to-simplicity/.

Leonard, Andrew. "The Joy of Perl," Salon, October 13, 1998. https://www.salon.com/1998/10/13/feature_269/.

Leonhardt, David. "How the Upper Middle Class Is Really Doing." *New York Times,* February 24, 2019, 23.

Levy, Steven. "Battle of the Clipper Chip." *New York Times Magazine,* June 12, 1994, 44–51, 60, 70.

Levy, Steven. *Facebook: The Inside Story.* New York: Blue Rider Press, 2020.

Levy, Steven. "Google Throws Open Doors to Its Top-Secret Data Center." *Wired,* November 2012.

Levy, Steven. *Hackers: Heroes of the Computer Revolution.* Garden City, NY: Anchor Press/Doubleday, 1984.

Levy, Steven. *In the Plex: How Google Thinks, Works, and Shapes Our Lives.* New York: Simon & Schuster, 2011.

Levy, Steven. *Insanely Great: The Life and Times of Macintosh, the Computer that Changed Everything.* New York: Viking, 1994.

Levy, Steven. *The Perfect Thing: How the iPod Shuffles Commerce, Culture, and Coolness.* New York: Simon & Schuster, 2006.

Levy, Steven. "A Spreadsheet Way of Knowledge." *Harper's Magazine,* November 1984, 58–64.

Lewis, Michael. "Boom Box." *New York Times,* August 13, 2000, 36, 41, 51, 65–67.

Lewis, Michael. *The New New Thing*. New York: W. W. Norton, 2000.

Licklider, Joseph Carl Robnet. "Memorandum for Members and Affiliates of the Intergalactic Computer Network." Kurzweil, December 11, 2001, 1963. https://www.kurzweilai.net/memorandum-for-members-and-affiliates-of-the-intergalactic-computer-network.

Licklider, Joseph Carl Robnet, and Robert W. Taylor. "The Computer as a Communications Device." *Science and Technology for the Technical Men in Management* 1968, 21–31.

Liebowitz, Stan J., and Stephen E. Margolis. *Winners, Losers & Microsoft: Competition and Antitrust in High Technology*. Oakland, CA: The Independent Institute, 2001.

Light, Jennifer S. "When Computers Were Women." *Technology and Culture* 40, no. 3 (July 1999): 455–483.

Lih, Andrew. *The Wikipedia Revolution*. New York: Hyperion, 2009.

Lingel, Jessa. "Socio-technical Transformations in Secondary Markets: A Comparison of Craigslist and Varage-Sale." *Internet Histories* 3, no. 2 (2019).

Linzmayer, Owen W. *Apple Confidential 2.0: The Definitive History of the World's Most Colorful Company*. San Francisco, CA: No Starch Press, 2004.

Lions, John. *Lion's Commentary on UNIX*. San Jose, CA: Peer-to-Peer Communications, 1996.

Lohr, Steve. *Go To: The Story of the Math Majors, Bridge Players, Engineers, Chess Wizards, Maverick Scientists and Iconoclasts—The Programmers Who Created the Software Revolution*. New York: Basic Books, 2001.

Longo, Bernadette. *Edmund Berkeley and the Social Responsibility of Computer Professionals*. San Rafael, CA: Morgan & Claypool (ACM Books), 2015.

Lord, Kenniston W. *Using the Radio Shack TRS-80 in Your Home*. New York: Van Nostrand Reinhold, 1981.

Lowood, Henry. "Game Engine." In *Debugging Game History: A Critical Lexicon*, edited by Henry Lowood and Raiford Guins, 202–209. Cambridge, MA: MIT Press, 2016.

Lowood, Henry. "Videogames in Computer Space: The Complex History of Pong." *IEEE Annals of the History of Computing* 31, no. 3 (July–September 2009): 5–19.

Luckoff, Herman. *From Dits to Bits: A Personal History of the Electronic Computer*. Portland, OR: Robotics Press, 1979.

"Machine of the Year: The Computer Moves In." *Time Magazine*, January 3, 1983, 14–37.

Mackenzie, C. E. *Coded Character Sets: History & Development*. Reading, MA: Addison-Wesley, 1980.

"Mac Portable's Pluses Outweigh the Negatives." *Computerworld*, December 11, 1989, 45.

Maddox, J. L., J. B. O'Toole, and S. Y. Wong. "The Transac S-1000 Computer." In *Proceedings of the 1956 Eastern Joint Computer Conference*, 13–16. New York: Association for Computing Machinery, 1956.

Madrigal, Alexis C. "The Servant Economy." *The Atlantic*, March 6, 2019. https://www.theatlantic.com/technology/archive/2019/03/what-happened-uber-x-companies/584236/.

Maher, Jimmy. *The Future Was Here: The Commodore Amiga*. Cambridge, MA: MIT Press, 2012.

Mahoney, Michael S., and Thomas Haigh, ed. *Histories of Computing*. Cambridge, MA: Harvard University Press, 2011.

Mailland, Julien, and Kevin Driscoll. *Minitel: Welcome to the Internet*. Cambridge, MA: MIT Press, 2017.

Manes, Stephen, and Paul Andrews. *Gates: How Microsoft's Mogul Reinvented an Industry—and Made Himself the Richest Man in America*. New York: Touchstone, 2002.

Manjoo, Farhad. "Psst. WhatsApp Needs Fixing. Pass It On." *New York Times*, October 24, 2018, B1.

Mann, Charles C. "Is the Internet Doomed?" *Inc* 17, no. 9 (June 13, 1995): 47–50, 52, 54.

Mar, Jerry. "Word Processing on the Apple with WordStar and Diablo." *Creative Computing* 9, no. 3 (March 1983): 81.

Marbach, William D, and Karen Springen. "Compaq Chips Away at IBM's Strength." *Newsweek*, September 22, 1986, 64.

Margolis, Jane, and Allan Fisher. *Unlocking the Clubhouse: Women in Computing*. Cambridge, MA: MIT Press, 2001.

Markoff, John. "Five Window Managers for the IBM PC." *Byte* 9, no. 9 (1984): 65–87.

Markoff, John. "Now, PC's That Read a Page and Store It." *New York Times*, August 17, 1988, D6.

Mauchly, John W. "Preparation of Problems for EDVAC-Type Machines." In *Proceedings of a Symposium on Large-Scale Digital Calculating Machinery, 7–10 January 1947*, edited by William Aspray, 203–207. Cambridge, MA: MIT Press, 1985.

Mazor, Stanley. "Intel 8080 CPU Chip Development." *IEEE Annals of the History of Computing* 29, no. 2 (April 2007): 70–73.

McAfee, Andrew, and Erik Brynjolfsson. *The Second Machine Age: Work, Progress, and Prosperity in a Time of Brilliant Technologies*. New York: W. W. Norton, 2014.

McCarthy, John. "John McCarthy's 1959 Memorandum." *IEEE Annals of the History of Computing* 14, no. 1 (January–March 1992): 19–23.

McCarthy, John. "Time-Sharing Computer Systems." In *Computers and the World of Tomorrow*, edited by Martin Greenberger, 221–248. Cambridge, MA: MIT Press, 1962.

McClellan, Stephen T. *The Coming Computer Industry Shakeout: Winners, Losers, and Survivors*. New York: John Wiley & Sons, 1984.

McCullough, Brian. *How the Internet Happened: From Netscape to the iPhone*. New York: Liveright, 2018.

McDonough, Jimmy. *Shakey: Neil Young's Biography*. New York: Random House, 2002.

McDowell, Edwin. "'No Problem' Machine Poses a Presidential Problem." *New York Times*, March 24, 1981, C7.

McElheny, Victor K. "Xerox Fights to Stay Ahead in the Copier Field." *New York Times*, February 21, 1977, 33–34.

McHugh, Josh. "The Firefox Explosion." *Wired* 13, no. 2 (February 2005).

McIlroy, M. D. "Oral History with Michael S. Mahoney," UNIX Oral History Project, 1989. https://www.princeton.edu/~hos/mike/transcripts/mcilroy.htm.

McIlroy, M. D., E. N. Pinson, and B. A. Tague. "Foreword (to special issue on UNIX Time-Sharing System)." *The Bell System Technical Journal* 57, no. 6, part 2 (July–August 1978): 1899–1904.

McKinsey and Company. *Unlocking the Computer's Profit Potential*. New York: McKinsey & Company, 1968.

McPherson, James C. "Census Experience Operating a UNIVAC System." In *Symposium on Managerial Aspects of Digital Computer Installations*, 30–36. Washington, DC: US Office of Naval Research, 1953.

McShane, Clay. *Down the Asphalt Path*. New York: Columbia University Press, 1994.

McWilliams, Peter A. *The Word Processing Book: A Short Course in Computer Literacy*. 5th ed. Los Angeles: Prelude Press, 1983.

Mead, Carver, and Lynn Conway. *Introduction to VLSI Systems*. Reading, MA: Addison-Wesley, 1980.

Mendelsohn, Andrew. "The Oracle Story: 1984–2001." *IEEE Annals of the History of Computing* 35, no. 2 (April–June 2013): 10–23.

Metcalfe, Robert M. "How Ethernet Was Invented." *IEEE Annals of the History of Computing* 16, no. 4 (October–December 1994): 81–88.

Metcalfe, Robert, and David R. Boggs. "Ethernet: Distributed Packet Switching for Local Computer Networks." *Communications of the ACM* 19, no. 7 (July 1976).

Miaskowski, Stan. "Software Review: Microsoft Flight Simulator." *Byte* 9, no. 3 (1984): 224–232.

The Michigan Terminal System: Volume 1, Reference R1001. Ann Arbor, MI: University of Michigan, Information Technology Division, 1991.

Miller, Michael J. "Computers: More Than One Billion Sold." *PC Magazine*, September 3, 2002, 7.

Mills, Mara. "Hearing Aids and the History of Electronics Miniaturization." *IEEE Annals of the History of Computing* 33, no. 2 (April–June 2011): 24–44.

Mims, Forrest, III. "The Tenth Anniversary of the Altair 8800." *Computers and Electronics*, January 1985, 62.

Mindell, David. *Digital Apollo: Human and Machine in Spaceflight*. Cambridge, MA: MIT Press, 2008.

"Minuteman Is Top Semiconductor User." *Aviation Week and Space Technology*, July 26, 1965.

Misa, Thomas J. "Military Needs, Commercial Realities, and the Development of the Transistor, 1948–1958." In *Military Enterprise and Technological Change*, edited by Merritt Roe Smith, 253–287. Cambridge, MA: MIT Press, 1985.

Misa, Thomas J. "Understanding 'How Computing Has Changed the World.'" *IEEE Annals of the History of Computing* 29, no. 4 (Oct–Dec 2007): 52–63.

Misa, Thomas J., ed. *Communities of Computing: Computer Science and Society in the ACM*. San Rafael, CA: Morgan & Claypool (ACM Books), 2017.

Mollick, Ethan R. "Establishing Moore's Law." *IEEE Annals of the History of Computing* 28, no. 3 (July–Sept 2006): 62–75.

Montfort, Nick. *Twisty Little Passages: An Approach to Interactive Fiction*, Cambridge, MA: MIT Press, 2003.

Montfort, Nick, and Ian Bogost. *Racing the Beam: The Atari Video Computer System*. Cambridge, MA: MIT Press, 2009.

Moody, Glyn. *The Rebel Code: The Inside Story of Linux and the Open Source Revolution*. Cambridge, MA: Perseus, 2001.

Moore, Gordon E. "Moore's Law at 40." In *Understanding Moore's Law*, edited by David C Brock, 67–84. Philadelphia: Chemical Heritage Foundation, 2006.

Moritz, Michael. *The Little Kingdom: The Private Story of Apple Computer*. New York: William Morrow, 1984.

Morris, Mitchell E. "Professor RAMAC's Tenure." *Datamation*, April 1981, 195–198.

Mossberg, Walter S. "Apple Brings Its Flair for Smart Designs to Digital Music Player." *Wall Street Journal*, November 1, 2001.

Mossberg, Walter S. "Security, Cool Features of Firefox Web Browser Beat Microsoft's IE." *Wall Street Journal*, December 30, 2004, B1.

Munk, Nina. *Fools Rush In: Steve Case, Jerry Levin, and the Unmaking of AOL Time Warner*. New York: HarperCollins, 2004.

Murray, Charles J. *The Supermen: The Story of Seymour Cray and the Technical Wizards Behind the Supercomputer*. New York: Wiley, 1997.

Naiman, Arthur. *Word Processing Buyer's Guide*. New York: BYTE/McGraw-Hill, 1983.

Naur, Peter, and Brian Randell, eds. *Software Engineering: Report on a Conference Sponsored by the NATO Science Committee, Garmisch, Germany, 7th to 11th October 1968*. Brussels: Scientific Affairs Division, NATO, 1969.

Nebeker, Frederik. *Signal Processing: The Emergence of a Discipline, 1948 to 1998*. New Brunswick, NJ: IEEE History Center, 1998.

Nelson, Theodor H. *Computer Lib/Dream Machines*: Self-published, 1974.

Netcraft. "June 2020 Web Server Survey." June 25, 2020. https://news.netcraft.com/archives/2020/06/25/june-2020-web-server-survey.html.

Netscape Communications. "Netscape Communications Offers New Network Navigator Free on the Internet." October 13, 1994. Accessed May 30, 2006.

Newell, Allen. "Intellectual Issues in the History of Artificial Intelligence (Report CMU-CS-142)." Carnegie-Mellon University, Department of Computer Science, 1982.

Newman, Lily Hay. "George R. R. Martin Writes on a DOS-Based Word Processor From the 1980s." Slate.com, May 14, 2014. http://www.slate.com/blogs/future_tense/2014/05/14/george_r_r_martin_writes_on_dos_based_wordstar_4_0_software_from_the_1980s.html.

Newman, Michael Z. *Atari Age: The Emergence of Video Games in America*. Cambridge, MA: MIT Press, 2017.

"NHS Told to Ditch 'Absurd' Fax Machines." BBC News, December 9, 2018. https://www.bbc.com/news/uk-46497526.

"Nineteen Sixty-Four: The Year Microcircuits Grew Up." *Electronics*, March 13, 1964, 10–11.

Noble, David F. *Forces of Production: A Social History of Industrial Automation*. New York: Alfred A. Knopf, 1984.

Nofre, David, Mark Priestley, and Gerard Alberts. "When Technology Became Language: The Origins of the Linguistic Conception of Computer Programming, 1950–1960." *Technology and Culture* 55, no. 1 (January 2014): 40–75.

Nolan, Richard L., ed. *Managing the Data Resource Function*. New York: West, 1974.

Nooney, Laine. "Let's Begin Again: Sierra On-Line and the 'Origins' of the Graphical Adventure Game." *American Journal of Play* 10, no. 1 (2017): 71–98.

Norberg, Arthur L. *Computers and Commerce: A Study of Technology and Management at Eckert-Mauchly Computer Company, Engineering Research Associates, and Remington Rand, 1946–1957*. Cambridge, MA: MIT Press, 2005.

Norberg, Arthur L., and Judy E. O'Neill. *Transforming Computer Technology: Information Processing for the Pentagon, 1962–1986*. Baltimore: Johns Hopkins University Press, 1996.

Norman, Donald A. "The Trouble with UNIX." *Datamation*, November 1981, 139–150.

Norvig, Peter. "PowerPoint: Shot with Its Own Bullets." *The Lancet* 262, no. 9381 (2003): 343–344.

November, Joseph. *Biomedical Computing: Digitizing Life in the United States*. Baltimore: Johns Hopkins, 2012.

Noyce, Robert. "Integrated Circuits in Military Equipment." *IEEE Spectrum* 1, no. 6 (June 1964): 71–72.

Noyce, Robert, and Marcian Hoff. "A History of Microprocessor Design at Intel." *IEEE Micro* 1, no. 1 (February 1981): 8–21.

Nygaard, Kristen, and Ole-Johan Dahl. "The Development of the Simula Languages." In *History of Programming Languages*, edited by Richard L Wexelblat, 439–480. New York: Academic Press, 1981.

O'Falt, Chris. "Pixelvision: How a Failed '80s Fisher-Price Toy Camera Became One of Auteurs' Favorite '90s Tools." IndieWire, August 2018. https://www.indiewire.com/2018/08/pixelvision-pxl-2000-fisher-price-toy-experimental-film-camera-lincoln-center-series-1201991348/.

O'Neil, Cathy. *Weapons of Math Destruction: How Big Data Increases Inequality and Threatens Democracy*. New York: Crown Books, 2016.

The Onion. "Apple Introduces Revolutionary New Laptop with No Keyboard," January 5, 2009. https://www.theonion.com/apple-introduces-revolutionary-new-laptop-with-no-keybo-1819594761.

O'Reilly, Tim. "What Is Web 2.0," O'Reilly, September 30, 2005. Accessed October 4, 2006. http://www.oreillynet.com/pub/a/oreilly/tim/news/2005/09/30/what-is-web-20.html.

Olle, T William. "Recent CODASYL Reports on Data Base Management." In *Data Base Systems*, edited by Randall Rustin, 175–184. Englewood Cliffs, NJ: Prentice Hall, 1972.

Orloroso, Arsenio Jr. "PLATO Buyer Must Seek Markets." *Crain's Chicago Business*, July 31, 1989, 4.

Osborn, Roddy F. "GE and UNIVAC: Harnessing the High-Speed Computer." *Harvard Business Review* 32, no. 4 (July–August 1954): 99–107.

Osgood, Donna. "The Difference in Higher Education." *Byte* 12, no. 2 (February 1987): 165–178.

Paasonen, Susanna. "Online Pornography." In *The SAGE Handbook of Web History*, edited by Niels Brügger and Ian Milligan, 551–563. Thousand Oaks, CA: Sage, 2019.

Pabst, Thomas. "3D Accelerator Card Reviews: Diamond Monster 3D." Tom's Hardware, November 9, 1997. https://www.tomshardware.com/reviews/3d-accelerator-card-reviews,42-7.html.

Padegs, A. "System/360 and Beyond." *IBM Journal of Research and Development* 25, no. 5 (September 1981): 377–390.

Parks, Marth Smith. *Microelectronics in the 1970s*: Rockwell International, 1974.

Partridge, Craig. "The Technical Development of Internet Email." *IEEE Annals of the History of Computing* 30, no. 2 (April–June 2008): 3–29.

Pasquale, Frank. *The Black Box Society: The Secret Algorithms That Control Money and Information*. Cambridge, MA: Harvard University Press, 2015.

Passell, Peter. "Economic Scene; Michael Milken's Other Accusers." *New York Times*, April 12, 1989, D2.

Patterson, David A. "Reduced Instruction Set Computers." *Communications of the ACM* 28, no. 1 (January 1985): 8–21.

Patterson, David, and John L Hennessy. *Computer Architecture: A Quantitative Approach*. San Mateo, CA: Morgan Kaufmann, 1990.

Pearson, Jamie Parker. *Digital at Work: Snapshots from the First Thirty-Five Years*. Burlington, MA: Digital Press, 1992.

Perez, Sarah. "Apple's Big App Store Purge Is Now Underway." November 15, 2016. https://techcrunch.com/2016/11/15/apples-big-app-store-purge-is-now-underway/.

Perlis, Alan. "Epigrams on Programming." *ACM SIGPLAN Notices* 17, no. 9 (September 1982).

Peters, John Durham. *The Marvelous Clouds: Towards a Philosophy of Elemental Media*. Chicago: University of Chicago Press, 2015.

Peterson, W. E. *Almost Perfect: How a Bunch of Regular Guys Built WordPerfect Corporation*. Rocklin, CA: Prima, 1994.

Petrick, Elizabeth R. *Making Computers Accessible: Disability Rights and Digital Technology*. Baltimore: Johns Hopkins University Press, 2015.

Phipps, Charles. "The Early History of ICs at Texas Instruments: A Personal View." *IEEE Annals of the History of Computing* 34, no. 1 (2012): 37–47.

Pinch, Trevor, and Frank Trocco. *Analog Days: The Invention and Impact of the Moog Synthesizer*. Cambridge, MA: Harvard University Press, 2002.

Pollack, Andrew. "Retreat Set by Texas Instruments." *New York Times*, October 29, 1983, 35.

Ponczek, Sarah, and Vildana Hajric. "Robinhood Market Made Bursting Bubbles Wall Street's Obsession." Bloomberg, June 13, 2020. https://www.bloomberg.com/news/articles/2020-06-13/robinhood-market-made -bursting-bubbles-wall-street-s-obsession.

Poor, Alfred. "25-MHz Computers: Dell System 325." *PC Magazine*, February 24 1989, 109–116.

Poor, Alfred. "Video Technology: Making a Choice in an Era of Change." *PC Magazine*, January 12 1993, 165–218.

Pountain, Dick. "Benchtest: Acorn Archimedes." *Personal Computer World* 9, no. 8 (August 1987): 98–104.

Pournelle, Jerry. "Ulterior Motives, Lobo, Buying Your First Computer, JRT Update." *Byte* 8, no. 5 (May 1983): 298–324.

Preger, Robert. "The Oracle Story, Part 1: 1977–1986." *IEEE Annals of the History of Computing* 34, no. 4 (October– December 2012): 51–57.

Price, Robert M. *The Eye for Innovation*. New Haven, CT: Yale University Press, 2005.

"The Price TI Is Paying for Misreading a Market." *Business Week*, September 19, 1983.

Priestley, Mark. *Routines of Substitution: John von Neumann's Work on Software Development, 1945–1948*. Cham, Switzerland: Springer, 2018.

Priestley, Mark, and Thomas Haigh. "The Media of Programming." In *Exploring the Early Digital*, edited by Thomas Haigh, 135–158. Cham, Switzerland: Springer, 2019.

"Proposed IRS System May Pose Threat to Privacy." *Computerworld*, February 21, 1977, 1, 6.

Pugh, Emerson W. *Memories That Shaped an Industry: Decision Leading to IBM System/360*. Cambridge, MA: MIT Press, 1984.

Pugh, Emerson W., Lyle R. Johnson, and John H. Palmer. *IBM's 360 and Early 370 Systems*. Cambridge, MA: MIT Press, 1991.

Radin, George. "The 801 Minicomputer." *IBM Journal of Research and Development* 27 (May 1983): 237–246.

Radin, George. "The Early History and Characteristics of PL/1." In *History of Programming Languages*, edited by Richard L. Wexelblat, 551–574. New York: Academic Press, 1981.

Randell, Brian. The 1968/69 NATO Software Engineering Reports, from unpublished proceedings of Dagstuhl- Seminar 9635: "History of Software Engineering." August 26–30, 1996. http://www.cs.ncl.ac.uk/people/brian .randell/home.formal/NATO/NATOReports/index.html.

Rankin, Joy Lisi. *A People's History of Computing in the United States*. Cambridge, MA: Harvard University Press, 2018.

Raymond, Eric S. *The Cathedral and the Bazaar*. Sebastopol, CA: O'Reilly, 2001.

Rayward, W. Boyd. "Visions of Xanadu: Paul Otlet (1868–1944) and Hypertext." *Journal of the American Society for Information Science* 45, no. 4 (May 1994): 235–250.

Rayward, W. Boyd, ed. *Information beyond Borders: International Cultural and Intellectual Exchange in the Belle Époque*. Burlington, VT: Ashgate Publishing, 2014.

Redmond, Kent C., and Thomas M. Smith. *Project Whirlwind: The History of a Pioneer Computer*. Bedford, MA: Digital Press, 1980.

Reid, Robert H. *Architects of the Web: 1,000 Days that Built the Future of Business*. New York: John Wiley & Sons, 1997.

Remington Rand. "UNIAC Fac-Tronic System (brochure)," archive.org, c. 1951. https://archive.org/details /UNIVACFacTronicSystemBrochure/mode/2up.

Reynolds, Simon. "Song from the Future: The Story of Donna Summer and Giorgio Moroder's 'I Feel Love,'" Pitchfork, June 29, 2017. https://pitchfork.com/features/article/song-from-the-future-the-story-of-donna-summer -and-giorgio-moroders-i-feel-love/.

Rheingold, Howard. *The Virtual Community: Homesteading on the Electronic Frontier*. Reading, MA: Addison-Wesley, 1993.

Rheingold, Howard. *Virtual Reality*. New York: Summit, 1991.

Rifkin, Glenn, and George Harrar. *The Ultimate Entrepreneur: The Story of Ken Olsen and Digital Equipment Corporation*. Chicago: Contemporary Books, 1988.

Ritchie, Dennis. "An Incomplete History of the QED Text Editor," Bell Labs, n.d. Accessed December 20, 2018. https://archive.org/stream/incomplete-history-qed/Image071517121025_djvu.txt.

Ritchie, Dennis M. "Unix Time-Sharing System: A Retrospective." *Bell System Technical Journal* 57, no. 6 (1978): 1947–1969.

Roberts, H. Edward, and William Yates. "Exclusive! Altair 8800: the Most Powerful Minicomputer Project Ever Presented—Can Be Built for Under $400." *Popular Electronics*, January 1975, 33–38.

Robertson, Adi, and Michael Zelenko. "Voices for a Virtual Past." The Verge, 2014. https://www.theverge.com /a/virtual-reality/oral_history.

Robinson, Phillip, and Jon R. Edwards. "The Atari 1040ST." *Byte* 11, no. 3 (March 1986): 84–93.

Roland, Alex, and Philip Shiman. *Strategic Computing: DARPA and the Quest for Machine Intelligence*. Cambridge, MA: MIT Press, 2002.

Rosch, Winn L. "Playing Hardball Against the XT." *PC Magazine* 3, no. 6 (April 3, 1984): 115–122.

Rose, Frank. *West of Eden: The End of Innocence at Apple Computer*. New York: Penguin, 1989.

Rosen, Saul. "Programming Systems and Languages: A Historical Survey." In *Proceedings of the April 21–23, 1964 Spring Joint Computer Conference*, 1–15. N.p.: AFIPS, 1964.

Rowe, Lawrence. "History of the Ingres Corporation." *IEEE Annals of the History of Computing* 34, no. 4 (October–December 2012): 58–70.

Rubinstein, Seymour. "Recollections: The Rise and Fall of WordStar." *IEEE Annals of the History of Computing* 28, no. 4 (October–December 2006): 64–72.

Russell, Andrew L. "'Rough Consensus and Running Code' and the Internet-OSI Standards War." *IEEE Annals of the History of Computing* 28, no. 3 (July–September 2006): 48–61.

Russell, Andrew L., and Valerie Schafer. "In the Shadow of ARPANET and Internet: Louis Pouzin and the Cyclades Network in the 1970s." *Technology and Culture* 55, no. 4 (October 2014): 880–907.

Ryan, Jeff. *Super Mario: How Nintendo Conquered America*. New York: Portfolio/Penguin, 2011.

Ryckman, G. F. "The Computer Operation Language." In *Proceedings of the Western Joint Computer Conference*, 341–343, 1960.

Salus, Peter. *A Quarter Century of Unix*. New York: Addison-Wesley, 1994.

Sammet, Jean E. "The Early History of Cobol." In *History of Programming Languages*, edited by Richard L. Wexelblat, 199–242. New York: Academic Press, 1981.

Sammet, Jean E. *Programming Languages: History and Fundamentals*. Eaglewood Cliffs, NJ: Prentice Hall, 1969.

Sandberg-Diment, Erik. "The Little IBM Finally Arrives for a Test." December 27, 1983, C3.

Sanger, David E. "IBM Offers a Blitz of New PC's." *New York Times*, April 3, 1987, D1.

Schechner, Sam, and Mark Secada. "Apps Send User Secrets to Facebook." *Wall Street Journal*, February 23, 2019, A1.

Schwarz, Frederic D. "The Casio Effect." *Innovation & Technology* 18, no. 1 (2002).

Scott, Marilyn, and Robert Hoffman. "The Mercury Programming System." In *Proceedings of the Eastern Joint Computer Conference*, 47–53, 1961.

Scull, John, and Hansen Hsu. "The Killer App That Saved the Macintosh." *IEEE Annals of the History of Computing* 41, no. 3 (July–September 2019): 42–52.

Searle, L. "The Bombsight War: Norden vs. Sperry." *IEEE Spectrum* 26, no. 9 (1989): 60–64.

Sharma, Dinesh C. *The Outsourcer: The Story of India's IT Revolution*. Cambridge, MA: MIT Press, 2015.

Shetterly, Margot Lee. *Hidden Figures*. New York: Harper Collins, 2016.

Shorey, Samantha, and Daniela K. Rosner. "A Voice of Process: Re-Presencing the Gendered Labor of Apollo Innovation." *communication +!* 7, no. 2 (2018): article 4.

Siles, Ignacio. "Blogs." In *The SAGE Handbook of Web History*, edited by Niels Brügger and Ian Milligan, 359–371. Thousand Oaks, CA: Sage, 2019.

Silver, Stephen. "The iPod Touch Is a Worthy End to the Iconic Music Line." AppleInsider, September 5, 2018. https://appleinsider.com/articles/18/09/05/the-ipod-touch-is-a-worthy-end-to-the-iconic-music-line.

Sito, Tom. *Moving Innovation: A History of Computer Animation*. Cambridge, MA: MIT Press, 2013.

"Six-Month Shipments Top 4,000 Mark." *Business Automation*, August 1966, 40–42.

"65-Notes." *Newsletter of the HP-65 Users' Club* 2, no. 1 (January 1975): 7.

Slater, Robert. *Portraits in Silicon*. Cambridge, MA: MIT Press, 1987.

Small, James S. *The Analogue Alternative: The Electronic Analogue Computer in Britain and the USA, 1930–1975*. New York: Routledge, 2001.

Smith, David Canfield. "Designing the Star User Interface." *Byte* 7, no. 4 (April 1982): 242–282.

Smith, David Canfield, Charles Irby, Ralph Kimball, and Eric Harslem. "The Star User Interface: An Overview." In *Proceedings of the AFIPS National Computer Conference*, 515–528, N.P.: AFIPS, 1982.

Smith, Douglas K., and Robert C. Alexander. *Fumbling the Future: How Xerox Invented, Then Ignored, the First Personal Computer*. New York: HarperCollins, 1989.

Smith, Gerry. "Who Killed the Great American Cable-TV Bundle?" Bloomberg, 2018. Accessed August 8. https://www.bloomberg.com/news/features/2018-08-08/who-killed-the-great-american-cable-tv-bundle.

Smith, R. W., and J. N. Tatarewicz. "Replacing a Technology: The Large Space Telescope and CCDs." *Proceedings of the IEEE* 73, no. 7 (July 1985): 1221–1235.

Somerson, Paul. "IBM Brings Out the Big Guns." *PC Magazine*, November 13, 1984, 116–133.

Speed, John R. "What Do You Mean I Can't Call Myself a Software Engineer." *IEEE Software* 16, no. 6 (November–December 1999): 45–50.

Stachniak, Zbrigniew. *Inventing the PC: The MCM/70 Story.* Montreal: McGill-Queen's University Press, 2011.

Stachniak, Zbigniew. "This Is Not a Computer: Negotiating the Microprocessor." *IEEE Annals of the History of Computing* 35, no. 4 (October–December 2013): 48–54.

Stallman, Richard M. "What Is a GNU/Linux System?" *GNU's Bulletin* 1, no. 23 (1997): 4–5.

Starnes, Thomas W. "Design Philosophy Behind Motorola's MC68000: Part 1." *Byte* 8, no. 4 (April 1983): 70–92.

Stein, Jesse Adams. "Domesticity, Gender and the 1977 Apple II Personal Computer." *Design and Culture* 3, no. 2 (2011): 193–216.

Stein, Scott. "iPad Pro (2018) Review: A Powerful, Beautiful Tablet That Needs a Software Overhaul." CNet, December 14, 2018. https://www.cnet.com/reviews/apple-ipad-pro-2018-review/.

Steinhilper, Ulrich. *Don't Talk—Do It! From Flying to Word Processing.* Bromley, UK: Independent Books, 2006.

Stern, Nancy Beth. *From ENIAC to UNIVAC: An Appraisal of the Eckert-Mauchly Computers.* Bedford, MA: Digital Press, 1981.

Sterne, Jonathan. *MP3: The Meaning of a Format.* Durham, NC: Duke University Press, 2012.

Stevens, Tim. "Fitbit Review." Engadget, September 15, 2009. https://www.engadget.com/2009/10/15/fitbit-review.

Stevenson, Michael. "Having It Both Ways: Larry Wall, Perl and the Technology and Culture of the Early Web." *Internet Histories* 2, no. 3–4 (2018): 264–280.

Stockham, T. G., T. M. Cannon, and R. B. Ingebretsen. "Blind Deconvolution through Digital Signal Processing." *Proceedings of the IEEE* 63, no. 4 (April 1975): 678–692.

Stoll, Clifford. *The Cuckoo's Egg: Tracking a Spy through the Maze of Computer Espionage.* New York: Doubleday, 1989.

Stone, Brad. *The Everything Store: Jeff Bezos and the Age of Amazon.* New York: Little, Brown, 2013.

Strassmann, Paul. *The Squandered Computer.* New Canaan, CT: Information Economics Press, 1997.

Strassmann, Paul A. *The Computers Nobody Wanted: My Years at Xerox.* New Canaan, CT: Information Economics Press, 2008.

Strauss, Bob. "Wing Commander III: Heart of the Tiger." *Entertainment Weekly*, February 10 1995.

Strecker, William. "VAX-11/780—A Virtual Address Extension to the PDP-11 Family." In *Proceedings of the National Computer Conference*, 967–980. New York: AFIPS Press, 1978.

Strimpel, Oliver. "The Early Model Personal Computer Contest." *The Computer Museum Report*, Fall 1986, 10–11.

Stross, Randall E. *Steve Jobs and the NeXT Big Thing.* New York: Scribner, 1993.

Stroustrup, Bjarne. *The C++ Programming Language.* Reading, MA: Addison-Wesley, 1985.

Sullivan, Jennifer. "Napster: Music Is for Sharing." *Wired*, November 1999.

Swarbrick, Guy. "Cover Benchtest: Atari Stacey." *Personal Computer World* 11, no. 12 (December 1989): 130–136.

Swarbrick, Guy. "Cover Benchtest: Macintosh Portable." *Personal Computer World* 11, no. 130–136 (October 1989): 130–136.

Swisher, Kara. *aol.com: How Steve Case Beat Bill Gates, Nailed the Netheads, and Made Millions in the War for the Web.* New York: Random House, 1998.

Tanenbaum, Andrew S. *Operating Systems: Design and Implementation* Englewood Cliffs, NJ: Prentice Hall, 1987.

Tatarchenko, Ksenia. "The Great Soviet Calculator Hack." *IEEE Spectrum* 55, no. 10 (October 2018): 42–47.

Tatarchenko, Ksenia. "The Man with a Micro-calculator." In *Exploring the Early Digital*, edited by Thomas Haigh, 179–200. Cham, Switzerland: Springer, 2019.

Tedre, Matti. *The Science of Computing: Shaping a Discipline*. New York: CRC Press, 2015.

Templeton, Brad. "Reaction to the DEC Spam of 1978." Templetons.com, n.d. Accessed June 21, 2019. https://www.templetons.com/brad/spamreact.html#msg.

Tesler, Lawrence G. "How Modeless Editing Came to Be." *IEEE Annals of the History of Computing* 40, no. 3 (July–September 2018): 55–67.

Thacker, Charles P. "Personal Distributed Computing: The Alto and Ethernet Hardware." In *A History of Personal Workstations*, edited by Adele Goldberg, 267–289. New York: ACM Press, 1988.

Thackray, Arnold, David Brock, and Rachel Jones. *Moore's Law: The Life of Gordon Moore, Silicon Valley's Quiet Revolutionary*. New York: Basic Books, 2015.

Thesis, John M. "Practical Application of Electronic Equipment." *Journal of Machine Accounting* 6, no. 3 (March 1955): 5, 7–8, 16–17.

Thiruvathukal, George Kuriakose, and Steven E. Jones. *Codename Revolution: The Nintendo Wii Platform*. Cambridge, MA: MIT Press, 2012.

Thomas, David. *Alan Sugar: The Amstrad Story*. London: Century, 1990.

Thompson, Jim. "How Unisys Transitioned from Proprietary to Open Architecture." Enterprise Tech, July 28, 2015. https://www.enterpriseai.news/2015/07/28/how-unisys-transitioned-from-proprietary-to-open-architecture/.

Thompson, Joe. "A Concise History of the Smartwatch." Bloomberg, January 8, 2018. https://www.bloomberg.com/news/articles/2018-01-08/a-concise-history-of-the-smartwatch.

Titus, Jonathan. "Build the Mark-8 Minicomputer." *Radio-Electronics*, July 1974, 29–33.

Toffler, Alvin. *The Third Wave*. New York: William Morrow, 1980.

Tomash, Erwin, and Arnold A. Cohen. "The Birth of an ERA: Engineering Research Associates, Inc., 1946–1955." *Annals of the History of Computing* 1 (October 1979): 83–97.

Torvalds, Linus, and David Diamond. *Just for Fun: The Story of an Accidental Revolutionary*. 1st ed. New York: HarperBusiness, 2001.

Turkle, Sherry. *The Second Self: Computers and the Human Spirit*. New York: Simon & Schuster, 1984.

Turner, Fred. *From Counterculture to Cyberculture: Stewart Brand, the Whole Earth Network, and the Rise of Digital Utopianism*. Chicago: University of Chicago Press, 2006.

Turner, Fred. "Where the Counterculture Met the New Economy: The WELL and the Origins of Virtual Community." *Technology and Culture* 46, no. 3 (July 2005): 485–512.

Tympas, Aristotle. *Calculation and Computation in the Pre-Electronic Era*. Cham, Switzerland: Springer, 2017.

Ullman, Ellen. *Close to the Machine: Technophilia and Its Discontents*. San Francisco: City Lights Books, 1997.

US Census Bureau. "Computer and Internet Use in the United States: 2003," US Department of Commerce, October, 2005. https://www.census.gov/prod/2005pubs/p23-208.pdf.

US Census Bureau. "Home Computer and Internet Use in the United States: August 2000." US Department of Commerce, September, 2001. https://www.census.gov/prod/2001pubs/p23-207.pdf.

Van Deusen, Edmund. "Electronics Goes Modern." *Fortune*, June 1955, 132–136.

Van Gelder, Lindsy. "WordPerfect Reaches for the Star." *PC Magazine* 1, no. 10 (March 1983): 431–437.

Van Vleck, Tom. "Electronic Mail and Text Messaging in CTSS, 1965–1973." *IEEE Annals of the History of Computing* 34, no. 1 2012): 4–6.

Van Vleck, Tom. "Myths," Multicians.org, n.d.

Van Vleck, Tom. "PL/1." Multicians.org, n.d. Accessed June 12, 2019. https://multicians.org/pl1.html#EPL.

Various. "Discussion of the SPREAD Report, June 23, 1982." *Annals of the History of Computing* 5, no. 1 (January 1983): 27–44.

Veit, Stan. *Stan Veit's History of the Personal Computer*. N.p: Worldcomm Press, 1993.

Vleck, David Walden, and Tom Van, eds. *The Compatible Time Sharing System (1961–1973)*. Washington, DC: IEEE Computer Society, 2011.

von Neumann, John. "First Draft of a Report on the EDVAC." *IEEE Annals of the History of Computing* 15, no. 4 (October 1993): 27–75.

von Neumann, John. *Papers of John von Neumann on Computing and Computing Theory (eds. William Aspray and Arthur Burks)*. Cambridge, MA: MIT Press, 1987.

Wall, Larry, and Randal L. Schwartz. *Programming Perl*. Sebastopol, CA: O'Reilly, 1991.

Waller, Larry. "Clean-Room Inventor Whitfield Leaves a Spotless Legacy." *Electronics*, February 4, 1985, 38.

Walters, Donna K. H. "Lotus to Drop Copy Protection for Some: May Be Extended to All Customers Later." *Los Angeles Times*, August 14, 1986.

Wang, An, and Eugene Linden. *Lessons: An Autobiography*. Reading, MA: Addison-Wesley, 1986.

Ward, Trent. "Quake Review." June 22, 1996. https://www.gamespot.com/reviews/quake-review/1900-2532549/.

Warnock, John E. "The Origins of PostScript." *IEEE Annals of the History of Computing* 40, no. 3 (July–September 2018): 68–76.

Warren, Jim C. "First Word on a Floppy-Disc Operating System." *Dr. Dobb's Journal*, April 1976, 5.

Watson, Thomas J., Jr. "Address by Thomas J. Watson, Jr., President, International Business Machines Corp." In *Data Processing (1): 1958 Conference Proceedings*, edited by Charles H. Johnson, 15–19. Chicago: National Machine Accountants Association, 1958.

Watson, Thomas, Jr., and Peter Petre. *Father, Son & Co: My Life at IBM and Beyond*. New York: Bantam, 1990.

Weber, Marc. "Browsers and Browser Wars." In *The SAGE Handbook of Web History*, edited by Niels Brügger and Ian Milligan, 270–296. Thousand Oaks, CA: Sage, 2019.

Weinberger, Matt. "Where Are They Now?" Business Insider, January 26, 2019. https://www.businessinsider.com/microsoft-1978-photo-2016-10.

Weiser, Mark, Rich Gold, and John Seely Brown. "The Origins of Ubiquitous Computing Research at PARC in the Late 1980s." *IBM Systems Journal* 38, no. 4 (1999): 693–696.

Weizenbaum, Joseph. *Computer Power and Human Reason: From Judgment to Calculation*. San Francisco: W. H. Freeman, 1976.

Welch, Mark J. "Expanding on the PC." *Byte* 8, no. 11 (November 1983): 168–184.

Welsh, David, and Theresa Welsh. *Priming the Pump: How TRS-80 Enthusiasts Helped Spark the PC Revolution*. Ferndale, MI: The Seeker, 2011.

"What Is Copyleft?" *GNU's Bulletin* 1, no. 23 (July 1997).

White, Myles. "'Explorer' Closes Gap." *Toronto Star*, September 4, 1997, J3.

Widener, W Robert. "New Concepts of Running a Business." *Business Automation* 13, no. 4 (April 1966): 38–43, 63.

Widdicombe, Lizzie. "The Programmer's Price." *New Yorker*, November 24, 2014.

Wiener, Norbert. *Cybernetics, or Control and Communication in the Animal and the Machine*. Cambridge, MA: Technology Press, 1948.

Wilkes, Maurice. "The Best Way to Design an Automatic Calculating Machine." *Annals of the History of Computing* 8, no. 2 (April–June 1986): 118–121.

Wilkes, Maurice. *Memoirs of a Computer Pioneer*. Cambridge, MA: MIT Press, 1985.

Wilkes, Maurice V., David J. Wheeler, and Stanley Gill. *The Preparation of Programs for an Electronic Digital Computer*. Cambridge, MA: Addison-Wesley, 1951.

Williams, F. C., and T. Kilburn. "A Storage System for Use with Binary-Digital Computing Machines." *Institution of Electrical Engineers, Proc. Part III* 96 (March 1949): 81–100.

Williams, Gregg. "A Closer Look at the IBM Personal Computer." *Byte* 7, no. 1 (January 1982): 36–68.

Williams, Gregg. "The Lisa Computer System." *Byte* 8, no. 2 (February 1983): 33–50.

Williams, Gregg. "Lotus Development Corporation's 1-2-3." *Byte* 7, no. 12 (December 1982): 182–198.

Williams, Michael R. *A History of Computing Technology*. Englewood Cliffs, NJ: Prentice-Hall, 1985.

Williams, Michael R. "The Origins, Uses, and Fate of the EDVAC." *IEEE Annals of the History of Computing* 15, no. 1 (January–March 1993): 22–38.

Williams, Michael R. "A Preview of Things to Come: Some Remarks on the First Generation of Computers." In *The First Computers: History and Architectures*, edited by Raúl Rojas and Ulf Hashagen, 1–16. Cambridge, MA: MIT Press, 2000.

Williams, Sam. *Free as in Freedom: Richard Stallman's Crusade for Free Software*. Sebastopol, CA: O'Reilly, 2002.

Wise, T A. "IBM's $5,000,000,000 Gamble." *Fortune*, September 1966, 118–123; 224, 226, 228.

Witt, Stephen. *How Music Got Free*. New York: Viking Press, 2015.

Wolfe, Gary. "The (Second Phase of the) Revolution Has Begun." *Wired Magazine*, October 1994.

Woodbury, Gregory G. "Net Cultural Assumptions." *Amateur Computerist* 6, no. 2–3 (1994): 7–9.

Worthy, James C. "Control Data Corporation: the Norris Era." *IEEE Annals of the History of Computing* 17, no. 1 (January–March 1995): 47–53.

Wozniak, Stephen. "The Apple II." *Byte* 2, no. 5 (May 1977): 34–43.

Wozniak, Steve, and Gina Smith. *iWoz: Computer Geek to Cult Icon: How I Invented the Personal Computer, Co-Founded Apple, and Had Fun Doing It*. New York: W. W. Norton, 2006.

Yager, Tom, and Ben Smith. "Son of SPARCstation." *Byte* 15, no. 13 (December 1990): 140–146.

Yates, JoAnne. *Control Through Communication: The Rise of System in American Management*. Baltimore: Johns Hopkins University Press, 1989.

Yates, JoAnne. *Structuring the Information Age*. Baltimore: Johns Hopkins University Press, 2005.

Yost, Jeffrey R. *Making IT Work: A History of the Computer Services Industry*. Cambridge, MA: MIT Press, 2017.

Zachary, G. Pascal. *Show Stopper! The Breakneck Race to Create Windows NT and the Next Generation at Microsoft*. New York: Free Press, 1994.

Zakon, Robert H'obbes'. "Hobbes' Internet Timeline 25." Accessed July 20, 2020. https://www.zakon.org/robert/internet/timeline/#Growth.

Zuboff, Shoshana. *The Age of Surveillance Capitalism*. New York: Public Affairs, 2019.

INDEX

Air Force, US. *See also* Military command and control
  systems
  Ballistic Missile Early Warning System, 45
  Cromemco computers used by, 208
  GPS (global positioning system) navigation systems,
    392–393
  Minuteman missile guidance, 91, 96–97, 100–102
  Rome Air Development Center, 120
  SAGE (semi-automatic ground environment)
    network, 87–89, 139
  Sunnyvale control center, 413
  Univac computers used by, 26
Ajax, 372, 379–381
Aldus PageMaker, 259–260, 266
Alexa (Amazon), 400
Algol, 42, 43, 128–129
Alibaba, 405
Allen, Paul, 177, 178, 179
ALOHAnet, 151, 233
Alphabet (Google), 411
Alpha processor, 287, 290
Altair computer (MITS), 173–177, 208, 217, 409
Altavista, 341
Alto computer (Xerox)
  client server applications, 250–251
  development of, 245–246
  object-oriented programming, 246–249
  Smalltalk, 246–249
  what you see is what you get (WYSIWYG),
    249–250
Amateur Computer Society, 175
Amazon, 422
  Alexa, 400
  Amazon Prime Video, 371
  Amazon Web Services, 363, 371
  COVID pandemic and, 420–421
  Echo speakers, 399–400
  Fire TV, 399–400
  founding of, 345
  market valuation of, 411
  user community and reviews, 372
  Web payments and, 342
AMD, 291, 363, 369, 413
Amdahl, Gene, 68
Amelio, Gill, 290
American Airlines, 89, 144
American Institute of Electrical Engineers, 22

American Management Association, 208
American National Standards Institute, 70. *See also*
  ASCII standard
American Totalisator Company, 23
America Online (AOL)
  AOL Instant Messenger (AIM), 336
  browser wars and, 350–351
  cost of, 339
  growth of, 335–336
  Netscape acquired by, 350, 461n54
  origins of, 329–330
Ameritech, 163
Ames Research Center (NASA), 48–49, 413
Amiga computer (Commodore), 260–261, 457n24
Amstrad
  PC 1512 computer, 230–232
  PCW 256 computer, 213
Analog computers, 84–86
Anderson, Harlan, 91
Andreesen, Marc, 334
Android, 359, 403–405, 413
Angry Birds, 396–397
Antivirus software, 273
AP-101 computers (IBM), 105–106
APL (programming language), 155, 177
Apollo Guidance Computers, 100–101
Apollo missions, 90, 100–101, 104
Apollo workstations, 256, 453n25
Apple, 388, 413. *See also* Jobs, Steve; VisiCalc
  Apple II, 181–183, 193, 202–203, 208, 220
  Apple IIe, 220, 222, 243
  Apple III, 220
  Apple Pascal, 376
  Applesoft BASIC, 182
  AppleTalk, 258
  App Store, 395–397
  ARM architecture, 291, 319, 365, 388, 394
  FaceTime, 420
  founding of, 181
  GPS exploited by, 392
  Hypercard, 331–332
  iOS, 394
  iPad, 402
  iPhone, 385, 393–397, 401–402, 405–406, 418
  iPod, 316–319, 394, 402
  iTunes, 318, 327
  LaserWriter, 259

Computer Research and Applications group, 51

Computer science

artificial intelligence, 400

complexity theory in, 66

computer science, 6

database research in, 78

development of, 20, 22, 129

women in, 203

Computer Space, 186

Computer supported cooperative work (CSCW), 141–142

Computer utility model, 114, 122

Computerworld, 80

Comshare, 124–125, 160

Conditional branching, 12

Conference on Software Engineering (NATO), 127–130

Consoles, games. *See* Games consoles

Control Data Corporation (CDC)

1604 computer, 51, 142

6600 computer, 51–52

8600 computer, 52

Cyber 70 supercomputer, 143

Cyber 73 supercomputer, 142

market dominance of, 73

Plato, 142–144

Star-100 computer, 54

Control key, 116, 210

Conway, Lynn, 213

Cookies, 345

Cooley, James, 294

Coonen, Jerome, 221

Cooper, Alan, 273

Coopersmith, Jonathan, 306

Copeland, Douglas, 271

Corbató, Fernando J., 114

Core processors (Intel), 365–366

Corel Corporation, 266, 272, 355–356, 377

Core memory, 30–32

Core ropes, 104

Cornell University, 161

Counters, program, 17

COVID-19 pandemic, dependence on technology in, 420–423

Cow Clicker, 374

CP/CMS (control program/Cambridge monitor system), 368–369

CP/M (Control Program for Micros), 179–180, 208, 212, 219, 228, 376

CPT, 209

Crabtree, Bob, 90

Craigslist, 346, 375

Crandall, Rick, 159–160

Crawford, Perry O. Jr., 86

Cray, Seymour R., 34, 51, 52

Cray Research

architectural advances by, 4

founding of, 52

market dominance of, 73

purchased by Silicon Graphics, 321

Cray Research computers

Cray 1, 29, 51–54, 123

Cray 2, 123

Cray X-MP, 54, 123

processing speeds, 83

Summit, 366

Titan system, 366

Creasy, Robert, 368–369

Creative Cloud (Adobe), 383–384

Creative Labs

Nomad audio players, 316

SoundBlaster cards, 228

Creative Suite (Adobe), 383–384

Creative Technology Limited Nomad Jukebox, 317

Crocker, Dave, 443n24

Cromemco computers, 208

Cross-interleaved Reed-Solomon code (CIRC), 303

Crowther, Will, 127

CRT. *See* Cathode ray tube (CRT)

CSNET, 154, 161

CTSS (compatible time-sharing system), 114–116, 140

Cullinane, 81

Customer relationship management applications, 378–379

Cutler, Dave, 285–286

Cutting, Doug, 363

Cyber 70 supercomputer (CDC), 143

Cyber 73 supercomputer (CDC), 142

Cybercafes, 338

Cybernetics (Wiener), 86

Cyberspace, 320, 458n45

Cybersquatters, 163

CYCLADES, 147, 151

Cyclotron, 2, 28

Czahor, Raymond, 149

Dabney, Ted, 186

Daemon programs, 338

Dahl, Ole-Johan, 247

Daisywheel printer, 209, 214

DARPA (Defense Advanced Research Projects Agency), 400, 415

Dartmouth College, 116–118, 177

Database administrators (DBAs), 77–78

Database management systems (DMBS), 75–78

Database software, dBase, 223, 225, 265, 274

Data centers

  at Google, 361–363

  introduction of, 360

  Itanium chips, 360–361

  multiple core processors, 363–368

  video streaming, 370–371

  virtualization, 368–370

Data definition language, 77

Data dictionaries, 77

Data General, 154

  Adventure played at, 127

  Eclipse MV/8000 computer, 136

  Nova computer, 136, 173, 183

  servers, 360

Data mining algorithms, 422

Datanet computer (GE), 118

Data Processing Management Association, 63

Data processing systems. *See also* System/360 computers (IBM)

  COBOL (common business oriented language), 66–67, 77, 80, 248, 421, 433n29

  database management systems, 75–78

  early computers, 55–57

  IBM big business computers, 57–58

  IBM small computers, 58–59

  at Internal Revenue Service, 78–80

  LEO, 55–57

  Management Information System, 74–75

  planning and installation of, 62

  RAMAC (random access method of accounting and control), 59–62

  software industry and, 80–82

  sorting and report generation applications, 64–66

  work applications, 62–64

David, Edward E., 128, 133

Davidoff, Monte, 177

Davies, Donald, 146

DB/2 (IBM data base 2), 274

dBase, 223, 225, 265, 274

DDP-516 minicomputers (Honeywell), 146

Dealers of Lightning (Hiltzik), 251

DeCastro, Edson, 93

Decision support system (DSS), 274

"Declaration of the Independence of Cyberspace" (Barlow), 337

DECtape, 93, 112–113

Deep learning applications, 366

Defender, 187

Defense Calculator (IBM), 30

Defense Communications Agency, 149

Defense systems. *See* Military command and control systems

Delay line memory, 14, 17–18, 23, 31, 33, 426n9

Dell, Michael, 237

Dell Computer Corporation, 237–238, 240–242, 394

Delphi, 330

Denning, Peter, 121–122

Department of Defense, US, 66, 115, 244. *See also* ARPANET; Military command and control systems

Deskpro 386 computer (Compaq), 236–237

Desktop publishing, 259–260

Desktop-replacement laptops, 280

Deutsch, Peter, 124

Developing world, smartphones in, 404–405

Dialog, 155

Diamond, Neil, 304

Diamond Rio, 316

Dibbell, Julian, 337

Dictating machines, 209

Diebold, John, 28

Diffie, Whitfield, 346

Diffie-Hellman key exchange, 346

Digital cameras, 308–311

Digital certificates, 346

Digital divide, 338

Digital Equipment Corporation (DEC), 49, 91–96

  Altavista, 341

  email system, 157

  Ethernet and, 233

  founding of, 91

  growth of, 134

  purchased by Compaq, 237

Digital Equipment Corporation computers

  Alpha chip, 287, 290

Processors. *See also* AMD; ARM; Intel products;
    MOS (metal-oxide semiconductor) chips; Mostek;
    Motorola
  64-bit computing, 363
  architectures of, 14–17, 18–20, 23–24, 29–30, 32–33,
    44, 46, 49–54, 68–69, 90, 92, 122–123, 170–172,
    282–285
  microprocessors, 106, 171–176, 220–221, 260–261,
    266–286, 290–291, 365–366
  Moore's Law, 107, 170, 213, 363, 417
  multiple, 54, 119–120, 122–123, 363–368
  scalable processor architecture (SPARC), 284–285
Prodigy, 204
Productivity paradox, 359
Professional engineers, 382
Professor RAMAC, 61–62
Program branching, 11
Programmability, 10–11
Programming languages, 381–382. *See also individual*
    languages
  data definition language, 77
  event-driven code, 248
Programming methodology, 131
Programming Perl (Wall and Schwartz), 346
Programming tools, 116–118, 272
  compilers, 37–38, 128
  mathematical software, 41–43
  SHARE user group, 39–41, 65–66, 73
  stacks, 43–44
  structured programming, 131, 247
Programs, stored, 21
Project Genie, 124
Project MAC, 115, 119–120, 233
Project SCOOP (Scientific Computation of Optimum
    Problems), 26–27
Protocols, 148
  FTP (file transfer protocol), 152
  IP (internet protocol), 151–152
  OSI (open systems interconnection), 160
  SMTP (simple mail transfer protocol), 151
  TCP (transmission control protocol), 151–152
  Telnet, 152, 159
  UUCP (Unix to Unix Copy Protocol), 152–153
  Z-Wave, 407
PS/2 computers (IBM), 234–236
Pseudomachines, 376
PSINet, 162

Psion Organizer II, 386
"Psycho Killer" (Talking Heads), 299, 457n10
P-system, 376
Public key encryption, 346–347
Publishing, desktop, 259–260
Publishing, online. *See* Web publishing
Punched card machines, 14, 25–26, 39, 62–66, 78–79,
    196
Purdue University, 154
PXL-2000 PixelVision camera (Fisher Price), 309
Python, 381–383

*Q*Bert*, 187
QED text editor, 124
QST, 173
Quake, 323, 326
Quantum Leap computer (Sinclair), 260
Quantum Link, 329–330
Quicksort, 66

R2 software (SAP), 82
Radar, 426n9
Radin, George, 283
Radio, two-way, 389
Radio-Electronics, 173
Radio Shack. *See* Tandy Radio Shack
Rainbow computer (DEC), 229
Rainbow Six, 325
RAMAC (random access method of accounting and
    control), 59–62
RAND Corporation, 18, 39, 89, 116, 131, 146
RAND Corporation Symposia, 167
Randell, Brian, 128, 130
Random-access memory (RAM), 31, 61, 171–172, 236
Rankin, Joy, 118
Raskin, Jef, 256
Raymond, Eric, 355
RAZR V3 (Motorola), 391
RCA, 67, 73, 100
Read-only memory (ROM), 171
RealAudio Player, 349
RealNetworks, 350
Real-time control systems, 83–84
  analog computers, 84–86
  integrated circuits, 96–102
  legacy of, 106–107
  miniaturization, 90–96

Scalable processor architecture (SPARC) processor, 284–285

Scanners, 307–308

Scelbi-8H computer, 173

Schneider computer, 232

Science Citation Index, 341

Scientific Data Systems (SDS), 49, 123–126, 244

Scientific notation, 32

Scientific Time Sharing Corporation (STSC), 155

SCOOP (Scientific Computation of Optimum Problems), 26–27

Scooters, dockless, 399, 417

Scramble, 187

Sculley, John, 387

SDS-940 computers (Scientific Data Systems), 123, 124

Seagate, 413

Search and directory companies, 339–342. *See also* Google

Sears, Roebuck, 204

Seattle Computer Products, 219

Second generation (2G) networks, 389–391

Second system effect, 122

Secure sockets layer (SSL), 346

Sega, 204–205, 305

Segments, memory, 70, 264

Selectric typewriter, 71

Self-driving cars. *See* Tesla

Servelets, 378

Server farms, 360

Servers, 361–362. *See also individual computers*

  clustered, 360

  farms, 360

  Itanium chips, 360–361

  MAXC, 251

  virtualization and, 369–370

Seybold, Jonathan, 250

SGML. *See* Standard generalized markup language (SGML)

Shamir, Adi, 346

Shannon, Claude, 294

SHARE, 39–41, 65–66, 73, 354

Shareware, 225–227

Shell, Unix, 133

Shima, Masatoshi, 171

Shugart, Alan, 179

Shugart Associates, 183

Sierra On-Line, 223

Sigma 7 computer (SDS), 126

Silicon chip. *See* Integrated circuits (ICs)

Silicon Graphics, 284, 289, 291, 320–321, 334, 349

Silicon transistors, 52, 98

Silicon Valley, California, 410–411, 421

Simonyi, Charles, 250, 269

Simple mail transfer protocol (SMTP), 151

Simpson, Tom, 90

Sims, John, 24

Simula 67 language, 247

Simulators, flight, 86–87

Sinclair

  Quantum Leap, 260

  ZX Spectrum, 196, 260

  ZX81 computer, 196–198

Siri (Apple), 400

SixPak card (AST), 227

Sketchpad, 111–112, 320

Skype, 396

Slate, 343

Smalltalk, 246–249, 386

Smarr, Larry, 334

Smart modems, 185

Smartphones, 418. *See also* Android; iPhone

  in developing world, 404–405, 422–423

  future of, 405–406

  investments through, 422

  negative impact of, 419

  rise of, 385

  Symbian, 391

Smart TV, 313

Smart watches, 406–407

Smith, David Canfield, 252

SMS messaging, 391

Sndmsg, 147

Social graph, 373

Social media

  advertising model for, 374–375

  blogging, 372, 374

  early services, 372–373

  Facebook, 373–374

  Twitter, 374

  Web 2.0 concept, 371–372

Social Security numbers, 79

SoftCard, 183

Software, definition of, 128

Software Arts, 214

Software as a Service (SaaS), 378–379

Turoff, Murray, 140–141
TV-Typewriter, 173, 182
Twitter, 374, 419
TX-0 computer (MIT), 91–92, 110, 113, 299
TX-2 computer (MIT), 111
Tymnet, 159
Tymshare, 124–126

Uber, 392, 398, 413, 417
Ubiquitous computing, 294
Ubiquitous photography, 311
Ullman, Ellen, 272, 274, 275
Ultima IV-Quest of the Avatar, 203
Uncommitted logic array, 196
Underwood Corporation, 36
Unibus, 134
Uniform resource locators (URLs), 333
Uniprinter, 27
Unisys Corporation, 85, 360, 366
Unit record equipment. *See* Punched card machines
Univac computers, 366, 409. *See also* Remington
  Rand
  1101 computer, 34
  1103 computer, 45
  1108 computer, 74
  automation with, 28
  compilers, 37–38
  customers, 26–28, 56
  design and production of, 23–24
  printers, 27
  Solid State 80, 45
  Total Management Information System, 74
  Univac 1, 32–33
  Univac II, 67
  in use, 24–28
Universality of computers, 3–4, 66
Universal machine, 3, 20, 423
Universal resource identifiers, 333
Universal serial bus (USB), 280
Universal solvent, computer as, 3
University of California at Berkeley. *See* Berkeley, University of California at
Unix
  Berkeley Software Distribution (BSD), 160, 395
  development of, 132–136, 209, 414
  electronic mail, 140
  sales of, 265

Unix to Unix Copy Protocol (UUCP), 152–153
Usenet, 152–154, 162, 330
User groups, SHARE, 39–41
Utah, University of, 245, 246
Utopia, brought on through computers, 28, 74, 409–411, 419
UUNET, 162

Validation and Verification, 104
Vanguard satellite, 89
van Rossum, Guido, 381
van Vleck, Tom, 140
VAX (virtual address extension) computers, 136–138, 245, 282
VAX/VMS operating system, 136
VCS console (Atari), 189–191, 216
Vector processing, 53–54
Velvet Underground, The, 176–177, 446n34
Verizon, 313, 405
Vernacular creativity, 373
Veronica, 332
Very large scale integration (VLSI), 213
VHS players, 312
VIC-20 computer (Commodore), 196, 198
Victor 9000 computer, 228–229
Video, digital, 311–312. *See also* Games
  streaming, 370–371
  video conferencing, 420
Video Electronics Standards Association (VESA), 278–279
Video games. *See* Games
Video graphics array (VGA), 235, 278
Videotex, 157–158
Viewtron, 158, 444n39
Viking Mars landers, 309
Virtual computers, 368
Virtualization, 50–51, 368–370
Virtual reality, 320–322
Virtual reality modeling language (VRML), 349
Viruses, 273
VisiCalc, 184, 213–216, 265
VisiOn, 265
Visual Basic (Microsoft), 273
Visual Studio suite (Microsoft), 272–274
VMWare, 369
Voice assistants, 399–400
von Braun, Wernher, 89

HISTORY OF COMPUTING

William Aspray and Thomas J. Misa, editors

Emerson W. Pugh, Lyle R. Johnson, and John H. Palmer
IBM's Early Computers: A Technical History

Kent C. Redmond and Thomas M. Smith
From Whirlwind to MITRE: The R&D Story of the SAGE Air Defense Computer

Alex Roland with Philip Shiman
Strategic Computing: DARPA and the Quest for Machine Intelligence, 1983–1993

Raúl Rojas and Ulf Hashagen, editors
The First Computers—History and Architectures

Corinna Schlombs
Productivity Machines: German Appropriations of American Technology from Mass Production to Computer Automation

Dinesh C. Sharma
The Outsourcer: A Comprehensive History of India's IT Revolution

Dorothy Stein
Ada: A Life and a Legacy

Christopher Tozzi
For Fun and Profit: A History of the Free and Open Source Software Revolution

John Vardalas
The Computer Revolution in Canada: Building National Technological Competence, 1945–1980

Maurice V. Wilkes
Memoirs of a Computer Pioneer

Jeffrey R. Yost
Making IT Work: A History of the Computer Services Industry